4/83

D1206454

CAMBRIDGE MONOGRAPHS
ON MATHEMATICAL PHYSICS

General Editors: D.W. Sciama, S. Weinberg, P.V. Landshoff

THE LARGE SCALE STRUCTURE
OF SPACE-TIME

THE LARGE SCALE STRUCTURE
OF SPACE-TIME

S. W. HAWKING, F.R.S.

Lucasian Professor of Mathematics in the University of Cambridge and
Fellow of Gonville and Caius College

AND

G. F. R. ELLIS

Professor of Applied Mathematics, University of Cape Town

CAMBRIDGE UNIVERSITY PRESS

CAMBRIDGE

LONDON   NEW YORK   NEW ROCHELLE
MELBOURNE   SYDNEY

Published by the Press Syndicate of the University of Cambridge
The Pitt Building, Trumpington Street, Cambridge CB2 1RP
32 East 57th Street, New York, NY 10022, USA
296 Beaconsfield Parade, Middle Park, Melbourne 3206, Australia

© Cambridge University Press 1973

Library of Congress catalogue card number: 72-93671

ISBN 0 521 20016 4 hard covers
ISBN 0 521 09906 4 paperback

First published 1973
First paperback edition 1974
Reprinted 1976 1977 1979 1980

Printed in the United States of America
First printed in Great Britain at the University Press, Cambridge
Reprinted by Vail-Ballou Press, Inc., Binghamton, NY

To
D.W. SCIAMA

Contents

[vii]

CONTENTS

Preface

The subject of this book is the structure of space–time on length-scales from 10^{-13} cm, the radius of an elementary particle, up to 10^{28} cm, the radius of the universe. For reasons explained in chapters 1 and 3, we base our treatment on Einstein's General Theory of Relativity. This theory leads to two remarkable predictions about the universe: first, that the final fate of massive stars is to collapse behind an event horizon to form a 'black hole' which will contain a singularity; and secondly, that there is a singularity in our past which constitutes, in some sense, a beginning to the universe. Our discussion is principally aimed at developing these two results. They depend primarily on two areas of study: first, the theory of the behaviour of families of timelike and null curves in space–time, and secondly, the study of the nature of the various causal relations in any space–time. We consider these subjects in detail. In addition we develop the theory of the time-development of solutions of Einstein's equations from given initial data. The discussion is supplemented by an examination of global properties of a variety of exact solutions of Einstein's field equations, many of which show some rather unexpected behaviour.

This book is based in part on an Adams Prize Essay by one of us (S. W. H.). Many of the ideas presented here are due to R. Penrose and R. P. Geroch, and we thank them for their help. We would refer our readers to their review articles in the *Battelle Rencontres* (Penrose (1968)), Midwest Relativity Conference Report (Geroch (1970c)), Varenna Summer School Proceedings (Geroch (1971)), and Pittsburgh Conference Report (Penrose (1972b)). We have benefited from discussions and suggestions from many of our colleagues, particularly B. Carter and D. W. Sciama. Our thanks are due to them also.

Cambridge   S. W. Hawking
January 1973   G. F. R. Ellis

1
The role of gravity

The view of physics that is most generally accepted at the moment is that one can divide the discussion of the universe into two parts. First, there is the question of the local laws satisfied by the various physical fields. These are usually expressed in the form of differential equations. Secondly, there is the problem of the boundary conditions for these equations, and the global nature of their solutions. This involves thinking about the edge of space–time in some sense. These two parts may not be independent. Indeed it has been held that the local laws are determined by the large scale structure of the universe. This view is generally connected with the name of Mach, and has more recently been developed by Dirac (1938), Sciama (1953), Dicke (1964), Hoyle and Narlikar (1964), and others. We shall adopt a less ambitious approach: we shall take the local physical laws that have been experimentally determined, and shall see what these laws imply about the large scale structure of the universe.

There is of course a large extrapolation in the assumption that the physical laws one determines in the laboratory should apply at other points of space–time where conditions may be very different. If they failed to hold we should take the view that there was some other physical field which entered into the local physical laws but whose existence had not yet been detected in our experiments, because it varies very little over a region such as the solar system. In fact most of our results will be independent of the detailed nature of the physical laws, but will merely involve certain general properties such as the description of space–time by a pseudo-Riemannian geometry and the positive definiteness of energy density.

The fundamental interactions at present known to physics can be divided into four classes: the strong and weak nuclear interactions, electromagnetism, and gravity. Of these, gravity is by far the weakest (the ratio Gm^2/e^2 of the gravitational to electric force between two electrons is about 10^{-40}). Nevertheless it plays the dominant role in shaping the large scale structure of the universe. This is because the strong and weak interactions have a very short range ($\sim 10^{-13}$ cm or less), and although electromagnetism is a long range interaction, the repulsion of like charges is very nearly balanced, for bodies of macroscopic dimensions, by the attraction of opposite charges. Gravity on the other hand appears to be always attractive. Thus the gravitational fields of all the particles in a body add up to produce a field which, for sufficiently large bodies, dominates over all other forces.

Not only is gravity the dominant force on a large scale, but it is a force which affects every particle in the same way. This universality was first recognized by Galileo, who found that any two bodies fell with the same velocity. This has been verified to very high precision in more recent experiments by Eotvos, and by Dicke and his collaborators (Dicke (1964)). It has also been observed that light is deflected by gravitational fields. Since it is thought that no signals can travel faster than light, this means that gravity determines the causal structure of the universe, i.e. it determines which events of space–time can be causally related to each other.

These properties of gravity lead to severe problems, for if a sufficiently large amount of matter were concentrated in some region, it could deflect light going out from the region so much that it was in fact dragged back inwards. This was recognized in 1798 by Laplace, who pointed out that a body of about the same density as the sun but 250 times its radius would exert such a strong gravitational field that no light could escape from its surface. That this should have been predicted so early is so striking that we give a translation of Laplace's essay in an appendix.

One can express the dragging back of light by a massive body more precisely using Penrose's idea of a closed trapped surface. Consider a sphere $\mathscr{T}$ surrounding the body. At some instant let $\mathscr{T}$ emit a flash of light. At some later time t, the ingoing and outgoing wave fronts from $\mathscr{T}$ will form spheres $\mathscr{T}_1$ and $\mathscr{T}_2$ respectively. In a normal situation, the area of $\mathscr{T}_1$ will be less than that of $\mathscr{T}$ (because it represents ingoing light) and the area of $\mathscr{T}_2$ will be greater than that of $\mathscr{T}$ (because it represents outgoing light; see figure 1). However if a sufficiently large amount of matter is enclosed within $\mathscr{T}$, the areas of $\mathscr{T}_1$ and $\mathscr{T}_2$ will *both* be less than that of $\mathscr{T}$. The surface $\mathscr{T}$ is then said to be a closed trapped surface. As t increases, the area of $\mathscr{T}_2$ will get smaller and smaller provided that gravity remains attractive, i.e. provided that the energy density of the matter does not become negative. Since the matter inside $\mathscr{T}$ cannot travel faster than light, it will be trapped within a region whose boundary decreases to zero within a finite time. This suggests that something goes badly wrong. We shall in fact show that in such a situation a space–time singularity must occur, if certain reasonable conditions hold.

One can think of a singularity as a place where our present laws of physics break down. Alternatively, one can think of it as representing part of the edge of space–time, but a part which is at a finite distance instead of at infinity. On this view, singularities are not so bad, but one still has the problem of the boundary conditions. In other words, one does not know what will come out of the singularity.

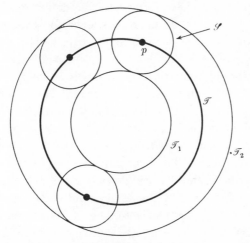

FIGURE 1. At some instant, the sphere $\mathscr{T}$ emits a flash of light. At a later time, the light from a point p forms a sphere $\mathscr{S}$ around p, and the envelopes $\mathscr{T}_1$ and $\mathscr{T}_2$ form the ingoing and outgoing wavefronts respectively. If the areas of *both* $\mathscr{T}_1$ and $\mathscr{T}_2$ are less than the area of $\mathscr{T}$, then $\mathscr{T}$ is a closed trapped surface.

There are two situations in which we expect there to be a sufficient concentration of matter to cause a closed trapped surface. The first is in the gravitational collapse of stars of more than twice the mass of the sun, which is predicted to occur when they have exhausted their nuclear fuel. In this situation, we expect the star to collapse to a singularity which is not visible to outside observers. The second situation is that of the whole universe itself. Recent observations of the microwave background indicate that the universe contains enough matter to cause a time-reversed closed trapped surface. This implies the existence of a singularity in the past, at the beginning of the present epoch of expansion of the universe. This singularity is in principle visible to us. It might be interpreted as the beginning of the universe.

In this book we shall study the large scale structure of space–time on the basis of Einstein's General Theory of Relativity. The predictions of this theory are in agreement with all the experiments so far performed. However our treatment will be sufficiently general to cover modifications of Einstein's theory such as the Brans–Dicke theory.

While we expect that most of our readers will have some acquaintance with General Relativity, we have endeavoured to write this book so that it is self-contained apart from requiring a knowledge of simple calculus, algebra and point set topology. We have therefore devoted chapter 2 to differential geometry. Our treatment is reasonably modern in that we have formulated our definitions in a manifestly coordinate independent manner. However for computational convenience we do use indices at times, and we have for the most part avoided the use of fibre bundles. The reader with some knowledge of differential geometry may wish to skip this chapter.

In chapter 3 a formulation of the General Theory of Relativity is given in terms of three postulates about a mathematical model for space–time. This model is a manifold $\mathscr{M}$ with a metric $\mathbf{g}$ of Lorentz signature. The physical significance of the metric is given by the first two postulates: those of local causality and of local conservation of energy–momentum. These postulates are common to both the General and the Special Theories of Relativity, and so are supported by the experimental evidence for the latter theory. The third postulate, the field equations for the metric $\mathbf{g}$, is less well experimentally established. However most of our results will depend only on the property of the field equations that gravity is attractive for positive matter densities. This property is common to General Relativity and some modifications such as the Brans–Dicke theory.

In chapter 4, we discuss the significance of curvature by considering its effects on families of timelike and null geodesics. These represent the paths of small particles and of light rays respectively. The curvature can be interpreted as a differential or tidal force which induces relative accelerations between neighbouring geodesics. If the energy–momentum tensor satisfies certain positive definite conditions, this differential force always has a net converging effect on non-rotating families of geodesics. One can show by use of Raychaudhuri's equation (4.26) that this then leads to focal or conjugate points where neighbouring geodesics intersect.

To see the significance of these focal points, consider a one-dimensional surface $\mathscr{S}$ in two-dimensional Euclidean space (figure 2). Let p

be a point not on $\mathcal{S}$. Then there will be some curve from $\mathcal{S}$ to p which is shorter than, or as short as, any other curve from $\mathcal{S}$ to p. Clearly this curve will be a geodesic, i.e. a straight line, and will intersect $\mathcal{S}$ orthogonally. In the situation shown in figure 2, there are in fact three geodesics orthogonal to $\mathcal{S}$ which pass through p. The geodesic through the point r is clearly not the shortest curve from $\mathcal{S}$ to p. One way of recognizing this (Milnor (1963)) is to notice that the neighbouring

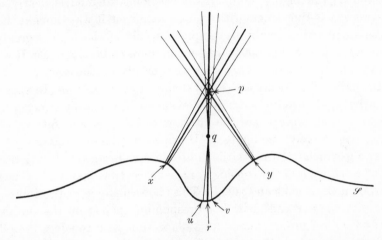

FIGURE 2. The line pr cannot be the shortest line from p to $\mathcal{S}$, because there is a focal point q between p and r. In fact either px or py will be the shortest line from p to $\mathcal{S}$.

geodesics orthogonal to $\mathcal{S}$ through u and v intersect the geodesic through r at a focal point q between $\mathcal{S}$ and p. Then joining the segment uq to the segment qp, one could obtain a curve from $\mathcal{S}$ to p which had the same length as a straight line rp. However as uqp is not a straight line, one could round off the corner at q to obtain a curve from $\mathcal{S}$ to p which was shorter than rp. This shows that rp is not the shortest curve from $\mathcal{S}$ to p. In fact the shortest curve will be either xp or yp.

One can carry these ideas over to the four-dimensional space–time manifold $\mathcal{M}$ with the Lorentz metric $\mathbf{g}$. Instead of straight lines, one considers geodesics, and instead of considering the shortest curve one considers the longest timelike curve between a point p and a spacelike surface $\mathcal{S}$ (because of the Lorentz signature of the metric, there will be no shortest timelike curve but there may be a longest such curve). This longest curve must be a geodesic which intersects $\mathcal{S}$ orthogonally, and there can be no focal point of geodesics orthogonal to $\mathcal{S}$ between

$\mathcal{S}$ and p. Similar results can be proved for null geodesics. These results are used in chapter 8 to establish the existence of singularities under certain conditions.

In chapter 5 we describe a number of exact solutions of Einstein's equations. These solutions are not realistic in that they all possess exact symmetries. However they provide useful examples for the succeeding chapters and illustrate various possible behaviours. In particular, the highly symmetrical cosmological models nearly all possess space–time singularities. For a long time it was thought that these singularities might be simply a result of the high degree of symmetry, and would not be present in more realistic models. It will be one of our main objects to show that this is not the case.

In chapter 6 we study the causal structure of space–time. In Special Relativity, the events that a given event can be causally affected by, or can causally affect, are the interiors of the past and future light cones respectively (see figure 3). However in General Relativity the metric $\mathbf{g}$ which determines the light cones will in general vary from point to point, and the topology of the space–time manifold $\mathcal{M}$ need not be that of Euclidean space R^4. This allows many more possibilities. For instance one can identify corresponding points on the surfaces $\mathcal{S}_1$ and $\mathcal{S}_2$ in figure 3, to produce a space–time with topology $R^3 \times S^1$. This would contain closed timelike curves. The existence of such a curve would lead to causality breakdowns in that one could travel into one's past. We shall mostly consider only space–times which do not permit such causality violations. In such a space–time, given any spacelike surface $\mathcal{S}$, there is a maximal region of space–time (called the Cauchy development of $\mathcal{S}$) which can be predicted from knowledge of data on $\mathcal{S}$. A Cauchy development has a property ('Global hyperbolicity') which implies that if two points in it can be joined by a timelike curve, then there exists a longest such curve between the points. This curve will be a geodesic.

The causal structure of space–time can be used to define a boundary or edge to space–time. This boundary represents both infinity and the part of the edge of space–time which is at a finite distance, i.e. the singular points.

In chapter 7 we discuss the Cauchy problem for General Relativity. We show that initial data on a spacelike surface determines a unique solution on the Cauchy development of the surface, and that in a certain sense this solution depends continuously on the initial data. This chapter is included for completeness and because it uses a number

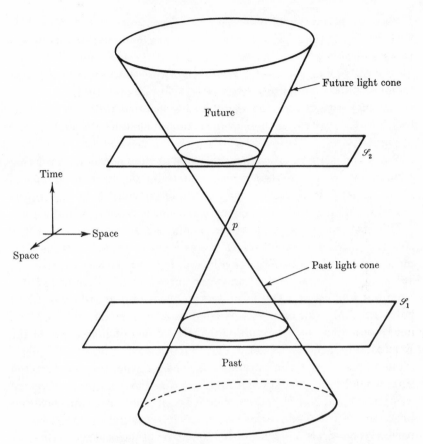

FIGURE 3. In Special Relativity, the light cone of an event p is the set of all light rays through p. The past of p is the interior of the past light cone, and the future of p is the interior of the future light cone.

of results of the previous chapter. However it is not necessary to read it in order to understand the following chapters.

In chapter 8 we discuss the definition of space–time singularities. This presents certain difficulties because one cannot regard the singular points as being part of the space–time manifold $\mathscr{M}$.

We then prove four theorems which establish the occurrence of space–time singularities under certain conditions. These conditions fall into three categories. First, there is the requirement that gravity shall be attractive. This can be expressed as an inequality on the energy–momentum tensor. Secondly, there is the requirement that there is enough matter present in some region to prevent anything escaping from that region. This will occur if there is a closed trapped surface, or if the whole universe is itself spatially closed. The third requirement is that there should be no causality violations. However this requirement is not necessary in one of the theorems. The basic idea of the proofs is to use the results of chapter 6 to prove there must be longest timelike curves between certain pairs of points. One then shows that if there were no singularities, there would be focal points which would imply that there were no longest curves between the pairs of points.

We next describe a procedure suggested by Schmidt for constructing a boundary to space–time which represents the singular points of space–time. This boundary may be different from that part of the causal boundary (defined in chapter 6) which represents singularities.

In chapter 9, we show that the second condition of theorem 2 of chapter 8 should be satisfied near stars of more than $1\frac{1}{2}$ times the solar mass in the final stages of their evolution. The singularities which occur are probably hidden behind an event horizon, and so are not visible from outside. To an external observer, there appears to be a 'black hole' where the star once was. We discuss the properties of such black holes, and show that they probably settle down finally to one of the Kerr family of solutions. Assuming this to be the case, one can place certain upper bounds on the amount of energy which can be extracted from black holes. In chapter 10 we show that the second conditions of theorems 2 and 3 of chapter 8 should be satisfied, in a time-reversed sense, in the whole universe. In this case, the singularities are in our past and constitute a beginning for all or part of the observed universe.

The essential part of the introductory material is that in § 3.1, § 3.2 and § 3.4. A reader wishing to understand the theorems predicting the existence of singularities in the universe need read further only chapter 4, § 6.2–§ 6.7, and § 8.1 and § 8.2. The application of these theorems to collapsing stars follows in § 9.1 (which uses the results of appendix B); the application to the universe as a whole is given in § 10.1, and relies on an understanding of the Robertson–Walker universe models (§ 5.3). Our discussion of the nature of the singularities is contained in § 8.1, § 8.3–§ 8.5, and § 10.2; the example of Taub–NUT space (§ 5.8) plays an important part in this discussion, and the Bianchi I universe model (§ 5.4) is also of some interest.

A reader wishing to follow our discussion of black holes need read only chapter 4, § 6.2–§ 6.6, § 6.9, and § 9.1, § 9.2 and § 9.3. This discussion relies on an understanding of the Schwarzschild solution (§ 5.5) and of the Kerr solution (§ 5.6).

Finally a reader whose main interest is in the time evolution properties of Einstein's equations need read only § 6.2–§ 6.6 and chapter 7. He will find interesting examples given in § 5.1, § 5.2 and § 5.5.

We have endeavoured to make the index a useful guide to all the definitions introduced, and the relations between them.

2

Differential geometry

The space–time structure discussed in the next chapter, and assumed through the rest of this book, is that of a manifold with a Lorentz metric and associated affine connection.

In this chapter, we introduce in § 2.1 the concept of a manifold and in § 2.2 vectors and tensors, which are the natural geometric objects defined on the manifold. A discussion of maps of manifolds in § 2.3 leads to the definitions of the induced maps of tensors, and of submanifolds. The derivative of the induced maps defined by a vector field gives the Lie derivative defined in § 2.4; another differential operation which depends only on the manifold structure is exterior differentiation, also defined in that section. This operation occurs in the generalized form of Stokes' theorem.

An extra structure, the connection, is introduced in § 2.5; this defines the covariant derivative and the curvature tensor. The connection is related to the metric on the manifold in § 2.6; the curvature tensor is decomposed into the Weyl tensor and Ricci tensor, which are related to each other by the Bianchi identities.

In the rest of the chapter, a number of other topics in differential geometry are discussed. The induced metric and connection on a hypersurface are discussed in § 2.7, and the Gauss–Codacci relations are derived. The volume element defined by the metric is introduced in § 2.8, and used to prove Gauss' theorem. Finally, we give a brief discussion in § 2.9 of fibre bundles, with particular emphasis on the tangent bundle and the bundles of linear and orthonormal frames. These enable many of the concepts introduced earlier to be reformulated in an elegant geometrical way. § 2.7 and § 2.9 are used only at one or two points later, and are not essential to the main body of the book.

2.1 Manifolds

A manifold is essentially a space which is locally similar to Euclidean space in that it can be covered by coordinate patches. This structure permits differentiation to be defined, but does not distinguish intrinsically between different coordinate systems. Thus the only concepts defined by the manifold structure are those which are independent of the choice of a coordinate system. We will give a precise formulation of the concept of a manifold, after some preliminary definitions.

Let R^n denote the *Euclidean space of n dimensions*, that is, the set of all n-tuples $(x^1, x^2, ..., x^n)$ $(-\infty < x^i < \infty)$ with the usual topology (open and closed sets are defined in the usual way), and let $\frac{1}{2}R^n$ denote the 'lower half' of R^n, i.e. the region of R^n for which $x^1 \leqslant 0$. A map ϕ of an open set $\mathcal{O} \subset R^n$ (respectively $\frac{1}{2}R^n$) to an open set $\mathcal{O}' \subset R^m$ (respectively $\frac{1}{2}R^m$) is said to be of class C^r if the coordinates $(x'^1, x'^2, ..., x'^m)$ of the image point $\phi(p)$ in $\mathcal{O}'$ are r-times continuously differentiable functions (the rth derivatives exist and are continuous) of the coordinates $(x^1, x^2, ..., x^n)$ of p in $\mathcal{O}$. If a map is C^r for all $r \geqslant 0$, then it is said to be C^∞. By a C^0 map, we mean a continuous map.

A function f on an open set $\mathcal{O}$ of R^n is said to be locally Lipschitz if for each open set $\mathcal{U} \subset \mathcal{O}$ with compact closure, there is some constant K such that for each pair of points $p, q \in \mathcal{U}$, $|f(p) - f(q)| \leqslant K |p - q|$, where by $|p|$ we mean

$$\{(x^1(p))^2 + (x^2(p))^2 + ... + (x^n(p))^2\}^{\frac{1}{2}}.$$

A map ϕ will be said to be locally Lipschitz, denoted by C^{1-}, if the coordinates of $\phi(p)$ are locally Lipschitz functions of the coordinates of p. Similarly, we shall say that a map ϕ is C^{r-} if it is C^{r-1} and if the $(r-1)$th derivatives of the coordinates of $\phi(p)$ are locally Lipschitz functions of the coordinates of p. In the following we shall usually only mention C^r, but similar definitions and results hold for C^{r-}.

If $\mathcal{P}$ is an arbitrary set in R^n (respectively $\frac{1}{2}R^n$), a map ϕ from $\mathcal{P}$ to a set $\mathcal{P}' \subset R^m$ (respectively $\frac{1}{2}R^m$) is said to be a C^r map if ϕ is the restriction to $\mathcal{P}$ and $\mathcal{P}'$ of a C^r map from an open set $\mathcal{O}$ containing $\mathcal{P}$ to an open set $\mathcal{O}'$ containing $\mathcal{P}'$.

A *C^r n-dimensional manifold* $\mathcal{M}$ is a set $\mathcal{M}$ together with a C^r atlas $\{\mathcal{U}_\alpha, \phi_\alpha\}$, that is to say a collection of charts $(\mathcal{U}_\alpha, \phi_\alpha)$ where the $\mathcal{U}_\alpha$ are subsets of $\mathcal{M}$ and the ϕ_α are one–one maps of the corresponding $\mathcal{U}_\alpha$ to open sets in R^n such that

(1) the $\mathcal{U}_\alpha$ cover $\mathcal{M}$, i.e. $\mathcal{M} = \bigcup_\alpha \mathcal{U}_\alpha$,

(2) if $\mathcal{U}_\alpha \cap \mathcal{U}_\beta$ is non-empty, then the map

$$\phi_\alpha \circ \phi_\beta^{-1} \colon \phi_\beta(\mathcal{U}_\alpha \cap \mathcal{U}_\beta) \to \phi_\alpha(\mathcal{U}_\alpha \cap \mathcal{U}_\beta)$$

is a C^r map of an open subset of R^n to an open subset of R^n (see figure 4).

Each $\mathcal{U}_\alpha$ is a *local coordinate neighbourhood* with the local coordinates x^a ($a = 1$ to n) defined by the map ϕ_α (i.e. if $p \in \mathcal{U}_\alpha$, then the coordinates of p are the coordinates of $\phi_\alpha(p)$ in R^n). Condition (2) is the requirement that in the overlap of two local coordinate neighbourhoods, the coordinates in one neighbourhood are C^r functions of the coordinates in the other neighbourhood, and vice versa.

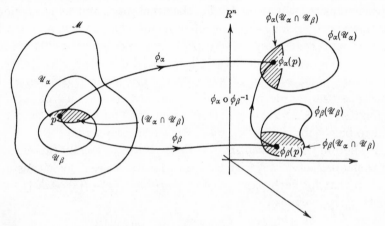

FIGURE 4. In the overlap of coordinate neighbourhoods $\mathcal{U}_\alpha$ and $\mathcal{U}_\beta$, coordinates are related by a C^r map $\phi_\alpha \circ \phi_\beta^{-1}$.

Another atlas is said to be *compatible* with a given C^r atlas if their union is a C^r atlas for all $\mathcal{M}$. The atlas consisting of all atlases compatible with the given atlas is called the *complete atlas* of the manifold; the complete atlas is therefore the set of all possible coordinate systems covering $\mathcal{M}$.

The topology of $\mathcal{M}$ is defined by stating that the open sets of $\mathcal{M}$ consist of unions of sets of the form $\mathcal{U}_\alpha$ belonging to the complete atlas. This topology makes each map ϕ_α into a homeomorphism.

A C^r differentiable manifold with boundary is defined as above, on replacing 'R^n' by '$\frac{1}{2}R^n$'. Then the *boundary of $\mathcal{M}$*, denoted by $\partial\mathcal{M}$, is defined to be the set of all points of $\mathcal{M}$ whose image under a map ϕ_α lies on the boundary of $\frac{1}{2}R^n$ in R^n. $\partial\mathcal{M}$ is an $(n-1)$-dimensional C^r manifold without boundary.

These definitions may seem more complicated than necessary. However simple examples show that one will in general need more than one coordinate neighbourhood to describe a space. The *two-dimensional Euclidean plane* R^2 is clearly a manifold. Rectangular coordinates $(x, y; -\infty < x < \infty, -\infty < y < \infty)$ cover the whole plane in one coordinate neighbourhood, where ϕ is the identity. Polar coordinates (r, θ) cover the coordinate neighbourhood $(r > 0, 0 < \theta < 2\pi)$; one needs at least two such coordinate neighbourhoods to cover R^2. The *two-dimensional cylinder* C^2 is the manifold obtained from R^2 by identifying the points (x, y) and $(x + 2\pi, y)$. Then (x, y) are coordinates in a neighbourhood $(0 < x < 2\pi, -\infty < y < \infty)$ and one needs two such coordinate neighbourhoods to cover C^2. The *Möbius strip* is the manifold obtained in a similar way on identifying the points (x, y) and $(x + 2\pi, -y)$. The *unit two-sphere* S^2 can be characterized as the surface in R^3 defined by the equation $(x^1)^2 + (x^2)^2 + (x^3)^2 = 1$. Then

$$(x^2, x^3; -1 < x^2 < 1, -1 < x^3 < 1)$$

are coordinates in each of the regions $x^1 > 0$, $x^1 < 0$, and one needs six such coordinate neighbourhoods to cover the surface. In fact, it is not possible to cover S^2 by a single coordinate neighbourhood. The *n-sphere* S^n can be similarly defined as the set of points

$$(x^1)^2 + (x^2)^2 + \ldots + (x^{n+1})^2 = 1$$

in R^{n+1}.

A manifold is said to be *orientable* if there is an atlas $\{\mathscr{U}_\alpha, \phi_\alpha\}$ in the complete atlas such that in every non-empty intersection $\mathscr{U}_\alpha \cap \mathscr{U}_\beta$, the Jacobian $|\partial x^i / \partial x'^j|$ is positive, where $(x^1, \ldots, x^n)$ and $(x'^1, \ldots, x'^n)$ are coordinates in $\mathscr{U}_\alpha$ and $\mathscr{U}_\beta$ respectively. The Möbius strip is an example of a non-orientable manifold.

The definition of a manifold given so far is very general. For most purposes one will impose two further conditions, that $\mathscr{M}$ is Hausdorff and that $\mathscr{M}$ is paracompact, which will ensure reasonable local behaviour.

A topological space $\mathscr{M}$ is said to be a *Hausdorff space* if it satisfies the Hausdorff separation axiom: whenever p, q are two distinct points in $\mathscr{M}$, there exist disjoint open sets $\mathscr{U}, \mathscr{V}$ in $\mathscr{M}$ such that $p \in \mathscr{U}, q \in \mathscr{V}$. One might think that a manifold is necessarily Hausdorff, but this is not so. Consider, for example, the situation in figure 5. We identify the points b, b' on the two lines if and only if $x_b = y_{b'} < 0$. Then each point is contained in a (coordinate) neighbourhood homeomorphic to an open subset of R^1. However there are no disjoint open neighbourhoods

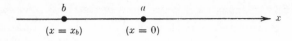

FIGURE 5. An example of a non-Hausdorff manifold. The two lines above are identical for $x = y < 0$. However the two points a ($x = 0$) and a' ($y = 0$) are not identified.

$\mathscr{U}, \mathscr{V}$ satisfying the conditions $a \in \mathscr{U}, a' \in \mathscr{V}$, where a is the point $x = 0$ and a' is the point $y = 0$.

An atlas $\{\mathscr{U}_\alpha, \phi_\alpha\}$ is said to be *locally finite* if every point $p \in \mathscr{M}$ has an open neighbourhood which intersects only a finite number of the sets $\mathscr{U}_\alpha$. $\mathscr{M}$ is said to be *paracompact* if for every atlas $\{\mathscr{U}_\alpha, \phi_\alpha\}$ there exists a locally finite atlas $\{\mathscr{V}_\beta, \psi_\beta\}$ with each $\mathscr{V}_\beta$ contained in some $\mathscr{U}_\alpha$. A connected Hausdorff manifold is paracompact if and only if it has a countable basis, i.e. there is a countable collection of open sets such that any open set can be expressed as the union of members of this collection (Kobayashi and Nomizu (1963), p. 271).

Unless otherwise stated, *all manifolds considered will be paracompact, connected C^∞ Hausdorff manifolds without boundary*. It will turn out later that when we have imposed some additional structure on $\mathscr{M}$ (the existence of an affine connection, see § 2.4) the requirement of paracompactness will be automatically satisfied because of the other restrictions.

A *function* f on a C^k manifold $\mathscr{M}$ is a map from $\mathscr{M}$ to R^1. It is said to be of class C^r ($r \leqslant k$) at a point p of $\mathscr{M}$, if the expression $f \circ \phi_\alpha^{-1}$ of f on any local coordinate neighbourhood $\mathscr{U}_\alpha$ is a C^r function of the local coordinates at p; and f is said to be a C^r *function* on a set $\mathscr{V}$ of $\mathscr{M}$ if f is a C^r function at each point $p \in \mathscr{V}$.

A property of paracompact manifolds we will use later, is the following: given any locally finite atlas $\{\mathscr{U}_\alpha, \phi_\alpha\}$ on a paracompact C^k manifold, one can always (see e.g. Kobayashi and Nomizu (1963), p. 272) find a set of C^k functions g_α such that

(1) $0 \leqslant g_\alpha \leqslant 1$ on $\mathscr{M}$, for each α;

(2) the support of g_α, i.e. the closure of the set $\{p \in \mathscr{M} : g_\alpha(p) \neq 0\}$, is contained in the corresponding $\mathscr{U}_\alpha$;

(3) $\sum_\alpha g_\alpha(p) = 1$, for all $p \in \mathscr{M}$.

Such a set of functions will be called a *partition of unity*. The result is in particular true for C^∞ functions, but is clearly not true for analytic functions (an analytic function can be expressed as a convergent power series in some neighbourhood of each point $p \in \mathcal{M}$, and so is zero everywhere if it is zero on any open neighbourhood).

Finally, the *Cartesian product* $\mathcal{A} \times \mathcal{B}$ of manifolds $\mathcal{A}$, $\mathcal{B}$ is a manifold with a natural structure defined by the manifold structures of $\mathcal{A}$, $\mathcal{B}$: for arbitrary points $p \in \mathcal{A}$, $q \in \mathcal{B}$, there exist coordinate neighbourhoods $\mathcal{U}, \mathcal{V}$ containing p, q respectively, so the point $(p, q) \in \mathcal{A} \times \mathcal{B}$ is contained in the coordinate neighbourhood $\mathcal{U} \times \mathcal{V}$ in $\mathcal{A} \times \mathcal{B}$ which assigns to it the coordinates (x^i, y^j), where x^i are the coordinates of p in $\mathcal{U}$ and y^j are the coordinates of q in $\mathcal{V}$.

2.2 Vectors and tensors

Tensor fields are the set of geometric objects on a manifold defined in a natural way by the manifold structure. A tensor field is equivalent to a tensor defined at each point of the manifold, so we first define tensors at a point of the manifold, starting from the basic concept of a vector at a point.

A C^k *curve* $\lambda(t)$ in $\mathcal{M}$ is a C^k map of an interval of the real line R^1 into $\mathcal{M}$. The *vector* (contravariant vector) $(\partial/\partial t)_\lambda|_{t_0}$ tangent to the C^1 curve $\lambda(t)$ at the point $\lambda(t_0)$ is the operator which maps each C^1 function f at $\lambda(t_0)$ into the number $(\partial f/\partial t)_\lambda|_{t_0}$; that is, $(\partial f/\partial t)_\lambda$ is the derivative of f in the direction of $\lambda(t)$ with respect to the parameter t. Explicitly,

$$\left(\frac{\partial f}{\partial t}\right)_\lambda\bigg|_t = \lim_{s \to 0} \frac{1}{s}\{f(\lambda(t+s)) - f(\lambda(t))\}. \tag{2.1}$$

The curve parameter t clearly obeys the relation $(\partial/\partial t)_\lambda t = 1$.

If $(x^1, \ldots, x^n)$ are local coordinates in a neighbourhood of p,

$$\left(\frac{\partial f}{\partial t}\right)_\lambda\bigg|_{t_0} = \sum_{j=1}^{n} \frac{dx^j(\lambda(t))}{dt}\bigg|_{t=t_0} \cdot \frac{\partial f}{\partial x^j}\bigg|_{\lambda(t_0)} = \frac{dx^j}{dt} \frac{\partial f}{\partial x^j}\bigg|_{\lambda(t_0)}.$$

(Here and throughout this book, we adopt the *summation convention* whereby a repeated index implies summation over all values of that index.) Thus every tangent vector at a point p can be expressed as a linear combination of the coordinate derivatives

$$(\partial/\partial x^1)|_p, \ldots, (\partial/\partial x^n)|_p.$$

Conversely, given a linear combination $V^j(\partial/\partial x^j)|_p$ of these operators, where the V^j are any numbers, consider the curve $\lambda(t)$ defined by

$x^j(\lambda(t)) = x^j(p) + tV^j$, for t in some interval $[-\epsilon, \epsilon]$; the tangent vector to this curve at p is $V^j(\partial/\partial x^j)|_p$. Thus the tangent vectors at p form a vector space over R^1 spanned by the coordinate derivatives $(\partial/\partial x^j)|_p$, where the vector space structure is defined by the relation

$$(\alpha X + \beta Y)f = \alpha(Xf) + \beta(Yf)$$

which is to hold for all vectors X, Y, numbers α, β and functions f. The vectors $(\partial/\partial x^j)_p$ are independent (for if they were not, there would exist numbers V^j such that $V^j(\partial/\partial x^j)|_p = 0$ with at least one V^j non-zero; applying this relation to each coordinate x^k shows

$$V^j \, \partial x^k/\partial x^j = V^k = 0,$$

a contradiction), so the space of all tangent vectors to $\mathcal{M}$ at p, denoted by $T_p(\mathcal{M})$ or simply T_p, is an n-dimensional vector space. This space, representing the set of all directions at p, is called the *tangent vector space* to $\mathcal{M}$ at p. One may think of a vector $\mathbf{V} \in T_p$ as an arrow at p, pointing in the direction of a curve $\lambda(t)$ with tangent vector $\mathbf{V}$ at p, the 'length' of $\mathbf{V}$ being determined by the curve parameter t through the relation $V(t) = 1$. (As $\mathbf{V}$ is an operator, we print it in bold type; its components V^j, and the number $V(f)$ obtained by $\mathbf{V}$ acting on a function f, are numbers, and so are printed in italics.)

If $\{\mathbf{E}_a\}$ ($a = 1$ to n) are any set of n vectors at p which are linearly independent, then any vector $\mathbf{V} \in T_p$ can be written $\mathbf{V} = V^a \mathbf{E}_a$ where the numbers $\{V^a\}$ are the components of $\mathbf{V}$ with respect to the basis $\{\mathbf{E}_a\}$ of vectors at p. In particular one can choose the $\mathbf{E}_a$ as the coordinate basis $(\partial/\partial x^i)|_p$; then the components $V^i = V(x^i) = (dx^i/dt)|_p$ are the derivatives of the coordinate functions x^i in the direction $\mathbf{V}$.

A *one-form* (covariant vector) $\boldsymbol{\omega}$ at p is a real valued linear function on the space T_p of vectors at p. If $\mathbf{X}$ is a vector at p, the number into which $\boldsymbol{\omega}$ maps $\mathbf{X}$ will be written $\langle \boldsymbol{\omega}, \mathbf{X} \rangle$; then the linearity implies that

$$\langle \boldsymbol{\omega}, \alpha \mathbf{X} + \beta \mathbf{Y} \rangle = \alpha \langle \boldsymbol{\omega}, \mathbf{X} \rangle + \beta \langle \boldsymbol{\omega}, \mathbf{Y} \rangle$$

holds for all $\alpha, \beta \in R^1$ and $\mathbf{X}, \mathbf{Y} \in T_p$. The subspace of T_p defined by $\langle \boldsymbol{\omega}, \mathbf{X} \rangle = $ (constant) for a given one-form $\boldsymbol{\omega}$, is linear. One may therefore think of a one-form at p as a pair of planes in T_p such that if $\langle \boldsymbol{\omega}, \mathbf{X} \rangle = 0$ the arrow $\mathbf{X}$ lies in the first plane, and if $\langle \boldsymbol{\omega}, \mathbf{X} \rangle = 1$ it touches the second plane.

Given a basis $\{\mathbf{E}_a\}$ of vectors at p, one can define a unique set of n one-forms $\{\mathbf{E}^a\}$ by the condition: $\mathbf{E}^i$ maps any vector $\mathbf{X}$ to the number X^i (the ith component of $\mathbf{X}$ with respect to the basis $\{\mathbf{E}_a\}$).

Then in particular, $\langle \mathbf{E}^a, \mathbf{E}_b \rangle = \delta^a{}_b$. Defining linear combinations of one-forms by the rules

$$\langle \alpha \boldsymbol{\omega} + \beta \boldsymbol{\eta}, \mathbf{X} \rangle = \alpha \langle \boldsymbol{\omega}, \mathbf{X} \rangle + \beta \langle \boldsymbol{\eta}, \mathbf{X} \rangle$$

for any one-forms $\boldsymbol{\omega}$, $\boldsymbol{\eta}$ and any $\alpha, \beta \in R^1$, $\mathbf{X} \in T_p$, one can regard $\{\mathbf{E}^a\}$ as a basis of one-forms since any one-form $\boldsymbol{\omega}$ at p can be expressed as $\boldsymbol{\omega} = \omega_i \mathbf{E}^i$ where the numbers ω_i are defined by $\omega_i = \langle \boldsymbol{\omega}, \mathbf{E}_i \rangle$. Thus the set of all one forms at p forms an n-dimensional vector space at p, the *dual space* $T^*{}_p$ of the tangent space T_p. The basis $\{\mathbf{E}^a\}$ of one-forms is the *dual basis* to the basis $\{\mathbf{E}_a\}$ of vectors. For any $\boldsymbol{\omega} \in T^*{}_p$, $\mathbf{X} \in T_p$ one can express the number $\langle \boldsymbol{\omega}, \mathbf{X} \rangle$ in terms of the components ω_i, $\mathbf{X}^i$ of $\boldsymbol{\omega}$, $\mathbf{X}$ with respect to dual bases $\{\mathbf{E}^a\}$, $\{\mathbf{E}_a\}$ by the relations

$$\langle \boldsymbol{\omega}, \mathbf{X} \rangle = \langle \omega_i \mathbf{E}^i, X^j \mathbf{E}_j \rangle = \omega_i X^i.$$

Each function f on $\mathcal{M}$ defines a one-form $\mathrm{d}f$ at p by the rule: for each vector $\mathbf{X}$,

$$\langle \mathrm{d}f, \mathbf{X} \rangle = Xf.$$

$\mathrm{d}f$ is called the *differential* of f. If $(x^1, \dots, x^n)$ are local coordinates, the set of differentials $(\mathrm{d}x^1, \mathrm{d}x^2, \dots, \mathrm{d}x^n)$ at p form the basis of one-forms dual to the basis $(\partial/\partial x^1, \partial/\partial x^2, \dots, \partial/\partial x^n)$ of vectors at p, since

$$\langle \mathrm{d}x^i, \partial/\partial x^j \rangle = \partial x^i / \partial x^j = \delta^i{}_j.$$

In terms of this basis, the differential $\mathrm{d}f$ of an arbitrary function f is given by

$$\mathrm{d}f = (\partial f / \partial x^i)\, \mathrm{d}x^i.$$

If $\mathrm{d}f$ is non-zero, the surfaces $\{f = \text{constant}\}$ are $(n-1)$-dimensional manifolds. The subspace of T_p consisting of all vectors $\mathbf{X}$ such that $\langle \mathrm{d}f, \mathbf{X} \rangle = 0$ consists of all vectors tangent to curves lying in the surface $\{f = \text{constant}\}$ through p. Thus one may think of $\mathrm{d}f$ as a normal to the surface $\{f = \text{constant}\}$ at p. If $\alpha \neq 0$, $\alpha \, \mathrm{d}f$ will also be a normal to this surface.

From the space T_p of vectors at p and the space $T^*{}_p$ of one-forms at p, we can form the Cartesian product

$$\Pi_r^s = \underbrace{T^*{}_p \times T^*{}_p \times \dots \times T^*{}_p}_{r \text{ factors}} \times \underbrace{T_p \times T_p \times \dots \times T_p}_{s \text{ factors}},$$

i.e. the ordered set of vectors and one-forms $(\boldsymbol{\eta}^1, \dots, \boldsymbol{\eta}^r, \mathbf{Y}_1, \dots, \mathbf{Y}_s)$ where the $\mathbf{Y}$s and $\boldsymbol{\eta}$s are arbitrary vectors and one-forms respectively.

A *tensor of type* (r, s) *at* p is a function on Π_r^s which is linear in each argument. If $\mathbf{T}$ is a tensor of type (r, s) at p, we write the number into which $\mathbf{T}$ maps the element $(\boldsymbol{\eta}^1, \dots, \boldsymbol{\eta}^r, \mathbf{Y}_1, \dots, \mathbf{Y}_s)$ of Π_r^s as

$$T(\boldsymbol{\eta}^1, \dots, \boldsymbol{\eta}^r, \mathbf{Y}_1, \dots, \mathbf{Y}_s).$$

Then the linearity implies that, for example,

$$T(\eta^1, ..., \eta^r, \alpha X + \beta Y, Y_2, ..., Y_s) = \alpha \cdot T(\eta^1, ..., \eta^r, X, Y_2, ..., Y_s)$$
$$+ \beta \cdot T(\eta^1, ..., \eta^r, Y, Y_2, ..., Y_s)$$

holds for all $\alpha, \beta \in R^1$ and $X, Y \in T_p$.

The space of all such tensors is called the *tensor product*

$$T_s^r(p) = \underbrace{T_p \otimes ... \otimes T_p}_{r \text{ factors}} \otimes \underbrace{T^*{}_p \otimes ... \otimes T^*{}_p}_{s \text{ factors}}.$$

In particular, $T_0^1(p) = T_p$ and $T_1^0(p) = T^*{}_p$.

Addition of tensors of type (r, s) is defined by the rule: $(T + T')$ is the tensor of type (r, s) at p such that for all $Y_i \in T_p$, $\eta^j \in T^*{}_p$,

$$(T + T')(\eta^1, ..., \eta^r, Y_1, ..., Y_s) = T(\eta^1, ..., \eta^r, Y_1, ..., Y_s)$$
$$+ T'(\eta^1, ..., \eta^r, Y_1, ..., Y_s).$$

Similarly, *multiplication of a tensor by a scalar* $\alpha \in R^1$ is defined by the rule: (αT) is the tensor such that for all $Y_i \in T_p$, $\eta^j \in T^*{}_p$,

$$(\alpha T)(\eta^1, ..., \eta^r, Y_1, ..., Y_s) = \alpha \cdot T(\eta^1, ..., \eta^r, Y_1, ..., Y_s).$$

With these rules of addition and scalar multiplication, the tensor product $T_s^r(p)$ is a vector space of dimension n^{r+s} over R^1.

Let $X_i \in T_p$ ($i = 1$ to r) and $\omega^j \in T^*{}_p$ ($j = 1$ to s). Then we shall denote by $X_1 \otimes ... \otimes X_r \otimes \omega^1 \otimes ... \otimes \omega^s$ that element of $T_s^r(p)$ which maps the element $(\eta^1, ..., \eta^r, Y_1, ..., Y_s)$ of Π_r^s into

$$\langle \eta^1, X_1 \rangle \langle \eta^2, X_2 \rangle ... \langle \eta^r, X_r \rangle \langle \omega^1, Y_1 \rangle ... \langle \omega^s, Y_s \rangle.$$

Similarly, if $R \in T_s^r(p)$ and $S \in T_q^p(p)$, we shall denote by $R \otimes S$ that element of $T_{s+q}^{r+p}(p)$ which maps the element $(\eta^1, ..., \eta^{r+p}, Y_1, ..., Y_{s+q})$ of Π_{r+p}^{s+q} into the number

$$R(\eta^1, ..., \eta^s, Y_1, ..., Y_r) S(\eta^{s+1}, ..., \eta^{s+q}, Y_{r+1}, ..., Y_{r+p}).$$

With the product $\otimes$, the tensor spaces at p form an algebra over R.

If $\{E_a\}$, $\{E^a\}$ are dual bases of T_p, $T^*{}_p$ respectively, then

$$\{E_{a_1} \otimes ... \otimes E_{a_r} \otimes E^{b_1} \otimes ... \otimes E^{b_s}\}, \quad (a_i, b_j \text{ run from 1 to } n),$$

will be a basis for $T_s^r(p)$. An arbitrary tensor $T \in T_s^r(p)$ can be expressed in terms of this basis as

$$T = T^{a_1 ... a_r}{}_{b_1 ... b_s} E_{a_1} \otimes ... \otimes E_{a_r} \otimes E^{b_1} \otimes ... \otimes E^{b_s}$$

where $\{T^{a_1 \cdots a_r}{}_{b_1 \ldots b_s}\}$ are the *components* of **T** with respect to the dual bases $\{\mathbf{E}_a\}$, $\{\mathbf{E}^a\}$ and are given by

$$T^{a_1 \cdots a_r}{}_{b_1 \ldots b_s} = T(\mathbf{E}^{a_1}, \ldots, \mathbf{E}^{a_r}, \mathbf{E}_{b_1}, \ldots, \mathbf{E}_{b_s}).$$

Relations in the tensor algebra at p can be expressed in terms of the components of tensors. Thus

$$(T + T')^{a_1 \cdots a_r}{}_{b_1 \ldots b_r} = T^{a_1 \cdots a_r}{}_{b_1 \ldots b_s} + T'^{a_1 \cdots a_r}{}_{b_1 \ldots b_s},$$

$$(\alpha T)^{a_1 \cdots a_r}{}_{b_1 \ldots b_s} = \alpha \cdot T^{a_1 \cdots a_r}{}_{b_1 \ldots b_s},$$

$$(T \otimes T')^{a_1 \cdots a_{r+p}}{}_{b_1 \ldots b_{s+q}} = T^{a_1 \cdots a_r}{}_{b_1 \ldots b_s} T'^{a_{r+1} \cdots a_{r+p}}{}_{b_{s+1} \ldots b_{s+q}}.$$

Because of its convenience, we shall usually represent tensor relations in this way.

If $\{\mathbf{E}_{a'}\}$ and $\{\mathbf{E}^{a'}\}$ are another pair of dual bases for T_p and $T^*{}_p$, they can be represented in terms of $\{\mathbf{E}_a\}$ and $\{\mathbf{E}^a\}$ by

$$\mathbf{E}_{a'} = \Phi_{a'}{}^a \mathbf{E}_a \tag{2.2}$$

where $\Phi_{a'}{}^a$ is an $n \times n$ non-singular matrix. Similarly

$$\mathbf{E}^{a'} = \Phi^{a'}{}_a \mathbf{E}^a \tag{2.3}$$

where $\Phi^{a'}{}_a$ is another $n \times n$ non-singular matrix. Since $\{\mathbf{E}_{a'}\}$, $\{\mathbf{E}^{a'}\}$ are dual bases,

$$\delta^{b'}{}_{a'} = \langle \mathbf{E}^{b'}, \mathbf{E}_{a'} \rangle = \langle \Phi^{b'}{}_b \mathbf{E}^b, \Phi_{a'}{}^a \mathbf{E}_a \rangle = \Phi_{a'}{}^a \Phi^{b'}{}_b \delta_a{}^b = \Phi_{a'}{}^a \Phi^{b'}{}_a,$$

i.e. $\Phi_{a'}{}^a$, $\Phi^{a'}{}_a$ are inverse matrices, and $\delta^a{}_b = \Phi^a{}_{b'} \Phi^{b'}{}_b$.

The components $T^{a'_1 \cdots a'_r}{}_{b'_1 \ldots b'_s}$ of a tensor **T** with respect to the dual bases $\{\mathbf{E}_{a'}\}$, $\{\mathbf{E}^{a'}\}$ are given by

$$T^{a'_1 \cdots a'_r}{}_{b'_1 \ldots b'_s} = T(\mathbf{E}^{a_1'}, \ldots, \mathbf{E}^{a_r'}, \mathbf{E}_{b'_1}, \ldots, \mathbf{E}_{b'_s}).$$

They are related to the components $T^{a_1 \cdots a_r}{}_{b_1 \ldots b_s}$ of **T** with respect to the bases $\{\mathbf{E}_a\}$, $\{\mathbf{E}^a\}$ by

$$T^{a'_1 \cdots a'_r}{}_{b'_1 \ldots b'_s} = T^{a_1 \cdots a_r}{}_{b_1 \ldots b_s} \Phi^{a'_1}{}_{a_1} \cdots \Phi^{a'_r}{}_{a_r} \Phi_{b'_1}{}^{b_1} \cdots \Phi_{b'_s}{}^{b_s}. \tag{2.4}$$

The *contraction* of a tensor **T** of type (r, s), with components $T^{ab \cdots d}{}_{ef \ldots g}$ with respect to bases $\{\mathbf{E}_a\}$, $\{\mathbf{E}^a\}$, on the first contravariant and first covariant indices is defined to be the tensor $C^1_1(\mathbf{T})$ of type $(r-1, s-1)$ whose components with respect to the same basis are $T^{ab \cdots d}{}_{af \ldots g}$, i.e.

$$C^1_1(\mathbf{T}) = T^{ab \cdots d}{}_{af \ldots g} \mathbf{E}_b \otimes \ldots \otimes \mathbf{E}_d \otimes \mathbf{E}^f \otimes \ldots \otimes \mathbf{E}^g.$$

If $\{\mathbf{E}_{a'}\}$, $\{\mathbf{E}^{a'}\}$ are another pair of dual bases, the contraction $C_1^1(\mathbf{T})$ defined by them is

$$C'_1^1(\mathbf{T}) = T^{a'b'\dots d'}{}_{a'f'\dots g'}\, \mathbf{E}_{b'} \otimes \dots \otimes \mathbf{E}_{d'} \otimes \mathbf{E}^{f'} \otimes \dots \otimes \mathbf{E}^{g'}$$

$$= \Phi^{a'}{}_a\, \Phi^a{}_{h'}\, T^{h'b'\dots d'}{}_{a'f'\dots g'}\, \Phi_{b'}{}^b \dots \Phi_{d'}{}^d\, \Phi^{f'}{}_f \dots \Phi^{g'}{}_g$$

$$\qquad\qquad . \, \mathbf{E}_b \otimes \dots \otimes \mathbf{E}_d \otimes \mathbf{E}^f \dots \otimes \mathbf{E}^g$$

$$= T^{ab\dots d}{}_{af\dots g}\, \mathbf{E}_b \otimes \dots \otimes \mathbf{E}_d \otimes \mathbf{E}^f \otimes \dots \otimes \mathbf{E}^g = C_1^1(\mathbf{T}),$$

so the contraction C_1^1 of a tensor is independent of the basis used in its definition. Similarly, one could contract $\mathbf{T}$ over any pair of contravariant and covariant indices. (If we were to contract over two contravariant or covariant indices, the resultant tensor would depend on the basis used.)

The symmetric part of a tensor $\mathbf{T}$ of type $(2, 0)$ is the tensor $S(\mathbf{T})$ defined by

$$S(\mathbf{T})\,(\boldsymbol{\eta}_1, \boldsymbol{\eta}_2) = \frac{1}{2!}\{T(\boldsymbol{\eta}_1, \boldsymbol{\eta}_2) + T(\boldsymbol{\eta}_2, \boldsymbol{\eta}_1)\}$$

for all $\boldsymbol{\eta}_1, \boldsymbol{\eta}_2 \in T^*{}_p$. We shall denote the components $S(\mathbf{T})^{ab}$ of $S(\mathbf{T})$ by $T^{(ab)}$; then

$$T^{(ab)} = \frac{1}{2!}\{T^{ab} + T^{ba}\}.$$

Similarly, the components of the skew-symmetric part of $\mathbf{T}$ will be denoted by

$$T^{[ab]} = \frac{1}{2!}\{T^{ab} - T^{ba}\}.$$

In general, the components of the symmetric or antisymmetric part of a tensor on a given set of covariant or contravariant indices will be denoted by placing round or square brackets around the indices. Thus

$$T_{(a_1 \dots a_r)}{}^{b\dots f}$$

$$= \frac{1}{r!}\{\text{sum over all permutations of the indices } a_1 \text{ to } a_r (T_{a_1 \dots a_r}{}^{b\dots f})\}$$

and

$$T_{[a_1 \dots a_r]}{}^{b\dots f}$$

$$= \frac{1}{r!}\{\text{alternating sum over all permutations of the indices}$$

$$\qquad\qquad\qquad a_1 \text{ to } a_r\, (T_{a_1 \dots a_r}{}^{b\dots f})\}.$$

For example,

$$K^a{}_{[bcd]} = \tfrac{1}{6}\{K^a{}_{bcd} + K^a{}_{dbc} + K^a{}_{cdb} - K^a{}_{bdc} - K^a{}_{chd} - K^a{}_{dcb}\}.$$

A tensor is *symmetric* in a given set of contravariant or covariant indices if it is equal to its symmetrized part on these indices, and is *antisymmetric* if it is equal to its antisymmetrized part. Thus, for example, a tensor **T** of type $(0, 2)$ is symmetric if $T_{ab} = \frac{1}{2}(T_{ab} + T_{ba})$, (which we can also express in the form: $T_{[ab]} = 0$).

A particularly important subset of tensors is the set of tensors of type $(0, q)$ which are antisymmetric on all q positions (so $q \leqslant n$); such a tensor is called a *q-form*. If **A** and **B** are p- and q-forms respectively, one can define a $(p+q)$-form $\mathbf{A} \wedge \mathbf{B}$ from them, where $\wedge$ is the skew-symmetrized tensor product $\otimes$; that is, $\mathbf{A} \wedge \mathbf{B}$ is the tensor of type $(0, p+q)$ with components determined by

$$(A \wedge B)_{a \ldots bc \ldots f} = A_{[a \ldots b} B_{c \ldots f]}.$$

This rule implies $(\mathbf{A} \wedge \mathbf{B}) = (-)^{pq}(\mathbf{B} \wedge \mathbf{A})$. With this product, the space of forms (i.e. the space of all p-forms for all p, including one-forms and defining scalars as zero-forms) constitutes the Grassmann algebra of forms. If $\{\mathbf{E}^a\}$ is a basis of one-forms, then the forms $\mathbf{E}^{a_1} \wedge \ldots \wedge \mathbf{E}^{a_p}$ (a_i run from 1 to n) are a basis of p-forms, as any p-form **A** can be written $\mathbf{A} = A_{a \ldots b} \mathbf{E}^a \wedge \ldots \wedge \mathbf{E}^b$, where $A_{a \ldots b} = A_{[a \ldots b]}$.

So far, we have considered the set of tensors defined at a point on the manifold. A set of local coordinates $\{x^i\}$ on an open set $\mathscr{U}$ in $\mathscr{M}$ defines a basis $\{(\partial/\partial x^i)|_p\}$ of vectors and a basis $\{(\mathrm{d}x^i)|_p\}$ of one-forms at each point p of $\mathscr{U}$, and so defines a basis of tensors of type (r, s) at each point of $\mathscr{U}$. Such a basis of tensors will be called a coordinate basis. A C^k *tensor field* **T** *of type* (r, s) on a set $\mathscr{V} \subset \mathscr{M}$ is an assignment of an element of $T^r_s(p)$ to each point $p \in \mathscr{V}$ such that the components of **T** with respect to any coordinate basis defined on an open subset of $\mathscr{V}$ are C^k functions.

In general one need not use a coordinate basis of tensors, i.e. given any basis of vectors $\{\mathbf{E}_a\}$ and dual basis of forms $\{\mathbf{E}^a\}$ on $\mathscr{V}$, there will not necessarily exist any open set in $\mathscr{V}$ on which there are local coordinates $\{x^a\}$ such that $\mathbf{E}_a = \partial/\partial x^a$ and $\mathbf{E}^a = \mathrm{d}x^a$. However if one does use a coordinate basis, certain specializations will result; in particular for any function f, the relations $E_a(E_b f) = E_b(E_a f)$ are satisfied, being equivalent to the relations $\partial^2 f/\partial x^a \partial x^b = \partial^2 f/\partial x^b \partial x^a$. If one changes from a coordinate basis $\mathbf{E}_a = \partial/\partial x^a$ to a coordinate basis $\mathbf{E}_{a'} = \partial/\partial x^{a'}$, applying (2.2), (2.3) to x^a, $x^{a'}$ shows that

$$\Phi_{a'}{}^a = \frac{\partial x^a}{\partial x^{a'}}, \quad \Phi^{a'}{}_a = \frac{\partial x^{a'}}{\partial x^a}.$$

Clearly a general basis $\{\mathbf{E}_a\}$ can be obtained from a coordinate basis

$\{\partial/\partial x^i\}$ by giving the functions $E_a{}^i$ which are the components of the $\mathbf{E}_a$ with respect to the basis $\{\partial/\partial x^i\}$; then (2.2) takes the form $\mathbf{E}_a = E_a{}^i \partial/\partial x^i$ and (2.3) takes the form $\mathbf{E}^a = E^a{}_i \, dx^i$, where the matrix $E^a{}_i$ is dual to the matrix $E_a{}^i$.

2.3 Maps of manifolds

In this section we define, via the general concept of a C^k manifold map, the concepts of 'imbedding', 'immersion', and of associated tensor maps, the first two being useful later in the study of submanifolds, and the last playing an important role in studying the behaviour of families of curves as well as in studying symmetry properties of manifolds.

A map ϕ from a C^k n-dimensional manifold $\mathcal{M}$ to a $C^{k'}$ n'-dimensional manifold $\mathcal{M}'$ is said to be a C^r map ($r \leqslant k$, $r \leqslant k'$) if, for any local coordinate systems in $\mathcal{M}$ and $\mathcal{M}'$, the coordinates of the image point $\phi(p)$ in $\mathcal{M}'$ are C^r functions of the coordinates of p in $\mathcal{M}$. As the map will in general be many–one rather than one–one (e.g. it cannot be one–one if $n > n'$), it will in general not have an inverse; and if a C^r map does have an inverse, this inverse will in general not be C^r (e.g. if ϕ is the map $R^1 \to R^1$ given by $x \to x^3$, then ϕ^{-1} is not differentiable at the point $x = 0$).

If f is a function on $\mathcal{M}'$, the mapping ϕ defines the function ϕ^*f on $\mathcal{M}$ as the function whose value at the point p of $\mathcal{M}$ is the value of f at $\phi(p)$, i.e.

$$\phi^*f(p) = f(\phi(p)). \tag{2.5}$$

Thus when ϕ maps points from $\mathcal{M}$ to $\mathcal{M}'$, ϕ^* maps functions linearly from $\mathcal{M}'$ to $\mathcal{M}$.

If $\lambda(t)$ is a curve through the point $p \in \mathcal{M}$, then the image curve $\phi(\lambda(t))$ in $\mathcal{M}'$ passes through the point $\phi(p)$. If $r \geqslant 1$, the tangent vector to this curve at $\phi(p)$ will be denoted by $\phi_*(\partial/\partial t)_\lambda|_{\phi(p)}$; one can regard it as the image, under the map ϕ, of the vector $(\partial/\partial t)_\lambda|_p$. Clearly ϕ_* is a linear map of $T_p(\mathcal{M})$ into $T_{\phi(p)}(\mathcal{M}')$. From (2.5) and the definition (2.1) of a vector as a directional derivative, the vector map ϕ_* can be characterized by the relation: for each C^r ($r \geqslant 1$) function f at $\phi(p)$ and vector $\mathbf{X}$ at p,

$$X(\phi^*f)|_p = \phi_* X(f)|_{\phi(p)}. \tag{2.6}$$

Using the vector mapping ϕ_* from $\mathcal{M}$ to $\mathcal{M}'$, we can if $r \geqslant 1$ define a linear one-form mapping ϕ^* from $T^*_{\phi(p)}(\mathcal{M}')$ to $T^*_p(\mathcal{M})$ by the condition: vector–one-form contractions are to be preserved under the maps. Then the one-form $\mathbf{A} \in T^*_{\phi(p)}$ is mapped into the one-form $\phi^*\mathbf{A} \in T^*_p$ where, for arbitrary vectors $\mathbf{X} \in T_p$,

$$\langle \phi^*\mathbf{A}, \mathbf{X} \rangle|_p = \langle \mathbf{A}, \phi_* \mathbf{X} \rangle|_{\phi(p)}.$$

A consequence of this is that

$$\phi^*(\mathrm{d}f) = \mathrm{d}(\phi^*f). \tag{2.7}$$

The maps ϕ_* and ϕ^* can be extended to maps of contravariant tensors from $\mathcal{M}$ to $\mathcal{M}'$ and covariant tensors from $\mathcal{M}'$ to $\mathcal{M}$ respectively, by the rules $\phi_*\colon \mathbf{T} \in T^r_0(p) \to \phi_* \mathbf{T} \in T^r_0(\phi(p))$ where for any $\boldsymbol{\eta}^i \in T^*_{\phi(p)}$,

$$T(\phi^*\boldsymbol{\eta}^1, \ldots, \phi^*\boldsymbol{\eta}^r)|_p = \phi_* T(\boldsymbol{\eta}^1, \ldots, \boldsymbol{\eta}^r)|_{\phi(p)}$$

and

$$\phi^*\colon \mathbf{T} \in T^0_s(\phi(p)) \to \phi^* \mathbf{T} \in T^0_s(p),$$

where for any $\mathbf{X}_i \in T_p$,

$$\phi^*T(\mathbf{X}_1, \ldots, \mathbf{X}_s)|_p = T(\phi_* \mathbf{X}_1, \ldots, \phi_* \mathbf{X}_s)|_{\phi(p)}.$$

When $r \geqslant 1$, the C^r map ϕ from $\mathcal{M}$ to $\mathcal{M}'$ is said to be of *rank s* at p if the dimension of $\phi_*(T_p(\mathcal{M}))$ is s. It is said to be *injective* at p if $s = n$ (and so $n \leqslant n'$) at p; then no vector in T_p is mapped to zero by ϕ_*. It is said to be *surjective* if $s = n'$ (so $n \geqslant n'$).

A C^r map ϕ ($r \geqslant 0$) is said to be an *immersion* if it and its inverse are C^r maps, i.e. if for each point $p \in \mathcal{M}$ there is a neighbourhood $\mathcal{U}$ of p in $\mathcal{M}$ such that the inverse ϕ^{-1} restricted to $\phi(\mathcal{U})$ is also a C^r map. This implies $n \leqslant n'$. By the implicit function theorem (Spivak (1965), p. 41), when $r \geqslant 1$, ϕ will be an immersion if and only if it is injective at every point $p \in \mathcal{M}$; then ϕ_* is an isomorphism of T_p into the image $\phi_*(T_p) \subset T_{\phi(p)}$. The image $\phi(\mathcal{M})$ is then said to be an n-dimensional *immersed submanifold* in $\mathcal{M}'$. This submanifold may intersect itself, i.e. ϕ may not be a one–one map from $\mathcal{M}$ to $\phi(\mathcal{M})$ although it is one–one when restricted to a sufficiently small neighbourhood of $\mathcal{M}$. An immersion is said to be an *imbedding* if it is a homeomorphism onto its image in the induced topology. Thus an imbedding is a one–one immersion; however not all one–one immersions are imbeddings, cf. figure 6. A map ϕ is said to be a proper map if the inverse image $\phi^{-1}(\mathcal{K})$ of any compact set $\mathcal{K} \subset \mathcal{M}'$ is compact. It can be shown that a proper one–one immersion is an imbedding. The image $\phi(\mathcal{M})$ of $\mathcal{M}$ under an imbedding ϕ is said to be an n-dimensional *imbedded submanifold* of $\mathcal{M}'$.

The map ϕ from $\mathcal{M}$ to $\mathcal{M}'$ is said to be a C^r *diffeomorphism* if it is a one–one C^r map and the inverse ϕ^{-1} is a C^r map from $\mathcal{M}'$ to $\mathcal{M}$. In this case, $n = n'$, and ϕ is both injective and surjective if $r \geqslant 1$; conversely, the implicit function theorem shows that if ϕ_* is both injective and surjective at p, then there is an open neighbourhood $\mathcal{U}$ of p such that $\phi \colon \mathcal{U} \to \phi(\mathcal{U})$ is a diffeomorphism. Thus ϕ is a local diffeomorphism near p if ϕ_* is an isomorphism from T_p to $T_{\phi(p)}$.

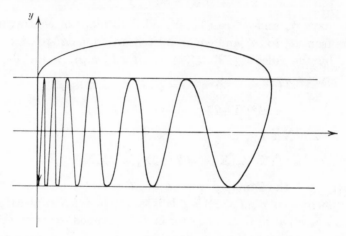

FIGURE 6. A one–one immersion of R^1 in R^2 which is not an imbedding, obtained by joining smoothly part of the curve $y = \sin(1/x)$ to the curve

$$\{(y, 0); \; -\infty < y < 1\}.$$

When the map ϕ is a C^r ($r \geqslant 1$) diffeomorphism, ϕ_* maps $T_p(\mathcal{M})$ to $T_{\phi(p)}(\mathcal{M}')$ and $(\phi^{-1})^*$ maps $T^*_p(\mathcal{M})$ to $T^*_{\phi(p)}(\mathcal{M}')$. Thus we can define a map ϕ_* of $T^r_s(p)$ to $T^r_s(\phi(p))$ for any r, s, by

$$T(\boldsymbol{\eta}^1, \ldots, \boldsymbol{\eta}^s, \mathbf{X}_1, \ldots, \mathbf{X}_r)\big|_p$$

$$= \phi_* T((\phi^{-1})^* \boldsymbol{\eta}^1, \ldots, (\phi^{-1})^* \boldsymbol{\eta}^s, \phi_* \mathbf{X}_1, \ldots, \phi_* \mathbf{X}_r)\big|_{\phi(p)}$$

for any $\mathbf{X}_i \in T_p$, $\boldsymbol{\eta}^i \in T^*_p$. This map of tensors of type (r, s) on $\mathcal{M}$ to tensors of type (r, s) on $\mathcal{M}'$ preserves symmetries and relations in the tensor algebra; e.g. the contraction of $\phi_* \mathbf{T}$ is equal to ϕ_* (the contraction of $\mathbf{T}$).

2.4 Exterior differentiation and the Lie derivative

We shall study three differential operators on manifolds, the first two being defined purely by the manifold structure while the third is defined (see § 2.5) by placing extra structure on the manifold.

The *exterior differentiation* operator d maps r-form fields linearly to $(r+1)$-form fields. Acting on a zero-form field (i.e. a function) f, it gives the one-form field df defined by (cf. §2.2)

$$\langle df, \mathbf{X} \rangle = Xf \text{ for all vector fields } \mathbf{X} \tag{2.8}$$

and acting on the r-form field

$$\mathbf{A} = A_{ab\ldots d}\, dx^a \wedge dx^b \wedge \ldots \wedge dx^d$$

it gives the $(r+1)$-form field $d\mathbf{A}$ defined by

$$d\mathbf{A} = dA_{ab\ldots d} \wedge dx^a \wedge dx^b \wedge \ldots \wedge dx^d. \tag{2.9}$$

To show that this $(r+1)$-form field is independent of the coordinates $\{x^a\}$ used in its definition, consider another set of coordinates $\{x^{a'}\}$. Then

$$\mathbf{A} = A_{a'b'\ldots d'}\, dx^{a'} \wedge dx^{b'} \wedge \ldots \wedge dx^{d'},$$

where the components $A_{a'b'\ldots d'}$ are given by

$$A_{a'b'\ldots d'} = \frac{\partial x^a}{\partial x^{a'}} \frac{\partial x^b}{\partial x^{b'}} \cdots \frac{\partial x^d}{\partial x^{d'}} A_{ab\ldots d}.$$

Thus the $(r+1)$-form $d\mathbf{A}$ defined by these coordinates is

$$d\mathbf{A} = dA_{a'b'\ldots d'}\, dx^{a'} \wedge dx^{b'} \wedge \ldots \wedge dx^{d'}$$

$$= d\left(\frac{\partial x^a}{\partial x^{a'}} \frac{\partial x^b}{\partial x^{b'}} \cdots \frac{\partial x^d}{\partial x^{d'}} A_{ab\ldots d} \right) \wedge dx^{a'} \wedge dx^{b'} \wedge \ldots \wedge dx^{d'}$$

$$= \frac{\partial x^a}{\partial x^{a'}} \frac{\partial x^b}{\partial x^{b'}} \cdots \frac{\partial x^d}{\partial x^{d'}} dA_{ab\ldots d} \wedge dx^{a'} \wedge dx^{b'} \wedge \ldots \wedge dx^{d'}$$

$$+ \frac{\partial^2 x^a}{\partial x^{a'}\,\partial x^{e'}} \frac{\partial x^b}{\partial x^{b'}} \cdots \frac{\partial x^d}{\partial x^{d'}} A_{ab\ldots d}\, dx^{e'} \wedge dx^{a'} \wedge dx^{b'} \wedge \ldots \wedge dx^{d'} + \ldots + \ldots$$

$$= dA_{ab\ldots d} \wedge dx^a \wedge dx^b \wedge \ldots \wedge dx^d$$

as $\partial^2 x^a / \partial x^{a'}\,\partial x^{e'}$ is symmetric in a' and e', but $dx^{e'} \wedge dx^{a'}$ is skew. Note that this definition only works for *forms*; it would not be independent of the coordinates used if the $\wedge$ product were replaced by a tensor product. Using the relation $d(fg) = g\,df + f\,dg$, which holds for arbitrary functions f, g, it follows that for any r-form $\mathbf{A}$ and form $\mathbf{B}$, $d(\mathbf{A} \wedge \mathbf{B}) = d\mathbf{A} \wedge \mathbf{B} + (-)^r \mathbf{A} \wedge d\mathbf{B}$. Since (2.8) implies that the local coordinate expression for df is $df = (\partial f/\partial x^i)\,dx^i$, it follows that $d(df) = (\partial^2 f/\partial x^i\,\partial x^j)\,dx^i \wedge dx^j = 0$, as the first term is symmetric and the second skew-symmetric. Similarly it follows from (2.9) that

$$d(d\mathbf{A}) = 0$$

holds for any r-form field $\mathbf{A}$.

The operator d commutes with manifold maps, in the sense: if $\phi: \mathcal{M} \to \mathcal{M}'$ is a C^r ($r \geqslant 2$) map and A is a C^k ($k \geqslant 2$) form field on $\mathcal{M}'$, then (by (2.7))

$$d(\phi^*A) = \phi^*(dA)$$

(which is equivalent to the chain rule for partial derivatives).

The operator d occurs naturally in the general form of Stokes' theorem on a manifold. We first define integration of n-forms: let $\mathcal{M}$ be a compact, orientable n-dimensional manifold with boundary $\partial\mathcal{M}$ and let $\{f_\alpha\}$ be a partition of unity for a finite oriented atlas $\{\mathcal{U}_\alpha, \phi_\alpha\}$. Then if A is an n-form field on $\mathcal{M}$, the integral of A over $\mathcal{M}$ is defined as

$$\int_{\mathcal{M}} A = (n!)^{-1} \sum_\alpha \int_{\phi_\alpha(\mathcal{U}_\alpha)} f_\alpha A_{12\ldots n} \mathrm{d}x^1 \mathrm{d}x^2 \ldots \mathrm{d}x^n, \qquad (2.10)$$

where $A_{12\ldots n}$ are the components of A with respect to the local co-ordinates in the coordinate neighbourhood $\mathcal{U}_\alpha$, and the integrals on the right-hand side are ordinary multiple integrals over open sets $\phi_\alpha(\mathcal{U}_\alpha)$ of R^n. Thus integration of forms on $\mathcal{M}$ is defined by mapping the form, by local coordinates, into R^n and performing standard multiple integrals there, the existence of the partition of unity ensuring the global validity of this operation.

The integral (2.10) is well-defined, since if one chose another atlas $\{\mathcal{V}_\beta, \psi_\beta\}$ and partition of unity $\{g_\beta\}$ for this atlas, one would obtain the integral

$$(n!)^{-1} \sum_\beta \int_{\psi_\beta(\mathcal{V}_\beta)} g_\beta A_{1'2'\ldots n'} \, \mathrm{d}x^{1'} \, \mathrm{d}x^{2'} \ldots \mathrm{d}x^{n'},$$

where $x^{i'}$ are the corresponding local coordinates. Comparing these two quantities in the overlap $(\mathcal{U}_\alpha \cap \mathcal{V}_\beta)$ of coordinate neighbourhoods belonging to two atlases, the first expression can be written

$$(n!)^{-1} \sum_\alpha \sum_\beta \int_{\phi_\alpha(\mathcal{U}_\alpha \cap \mathcal{V}_\beta)} f_\alpha g_\beta A_{12\ldots n} \, \mathrm{d}x^1 \, \mathrm{d}x^2 \ldots \mathrm{d}x^n,$$

and the second can be written

$$(n!)^{-1} \sum_\alpha \sum_\beta \int_{\psi_\beta(\mathcal{U}_\alpha \cap \mathcal{V}_\beta)} f_\alpha g_\beta A_{1'2'\ldots n'} \, \mathrm{d}x^{1'} \, \mathrm{d}x^{2'} \ldots \mathrm{d}x^{n'}.$$

Comparing the transformation laws for the form A and the multiple integrals in R^n, these expressions are equal at each point, so $\int_{\mathcal{M}} A$ is independent of the atlas and partition of unity chosen.

Similarly, one can show that this integral is invariant under diffeomorphisms:

$$\int_{\mathcal{M}'} \phi_* \mathbf{A} = \int_{\mathcal{M}} \mathbf{A}$$

if ϕ is a C^r diffeomorphism $(r \geqslant 1)$ from $\mathcal{M}$ to $\mathcal{M}'$.

Using the operator d, the *generalized Stokes' theorem* can now be written in the form: if $\mathbf{B}$ is an $(n-1)$-form field on $\mathcal{M}$, then

$$\int_{\partial \mathcal{M}} \mathbf{B} = \int_{\mathcal{M}} d\mathbf{B},$$

which can be verified (see e.g. Spivak (1965)) from the definitions above; it is essentially a general form of the fundamental theorem of calculus. To perform the integral on the left, one has to define an orientation on the boundary $\partial \mathcal{M}$ of $\mathcal{M}$. This is done as follows: if $\mathcal{U}_\alpha$ is a coordinate neighbourhood from the oriented atlas of $\mathcal{M}$ such that $\mathcal{U}_\alpha$ intersects $\partial \mathcal{M}$, then from the definition of $\partial \mathcal{M}$, $\phi_\alpha(\mathcal{U}_\alpha \cap \partial \mathcal{M})$ lies in the plane $x^1 = 0$ in R^n and $\phi_\alpha(\mathcal{U}_\alpha \cap \mathcal{M})$ lies in the lower half $x^1 \leqslant 0$. The coordinates $(x^2, x^3, ..., x^n)$ are then oriented coordinates in the neighbourhood $\mathcal{U}_\alpha \cap \partial \mathcal{M}$ of $\partial \mathcal{M}$. It may be verified that this gives an oriented atlas on $\partial \mathcal{M}$.

The other type of differentiation defined naturally by the manifold structure is *Lie differentiation*. Consider any C^r $(r \geqslant 1)$ vector field $\mathbf{X}$ on $\mathcal{M}$. By the fundamental theorem for systems of ordinary differential equations (Burkill (1956)) there is a unique maximal curve $\lambda(t)$ through each point p of $\mathcal{M}$ such that $\lambda(0) = p$ and whose tangent vector at the point $\lambda(t)$ is the vector $\mathbf{X}|_{\lambda(t)}$. If $\{x^i\}$ are local coordinates, so that the curve $\lambda(t)$ has coordinates $x^i(t)$ and the vector $\mathbf{X}$ has components X^i, then this curve is locally a solution of the set of differential equations

$$dx^i/dt = X^i(x^1(t), ..., x^n(t)).$$

This curve is called the *integral curve* of $\mathbf{X}$ with initial point p. For each point q of $\mathcal{M}$, there is an open neighbourhood $\mathcal{U}$ of q and an $\epsilon > 0$ such that $\mathbf{X}$ defines a family of diffeomorphisms $\phi_t \colon \mathcal{U} \to \mathcal{M}$ whenever $|t| < \epsilon$, obtained by taking each point p in $\mathcal{U}$ a parameter distance t along the integral curves of $\mathbf{X}$ (in fact, the ϕ_t form a one-parameter local group of diffeomorphisms, as $\phi_{t+s} = \phi_t \circ \phi_s = \phi_s \circ \phi_t$ for $|t|, |s|, |t+s| < \epsilon$, so $\phi_{-t} = (\phi_t)^{-1}$ and ϕ_0 is the identity). This diffeomorphism maps each tensor field $\mathbf{T}$ at p of type (r, s) into $\phi_{t*} \mathbf{T}|_{\phi_t(p)}$.

The *Lie derivative* $L_{\mathbf{X}} \mathbf{T}$ of a tensor field $\mathbf{T}$ with respect to $\mathbf{X}$ is defined to be minus the derivative with respect to t of this family of tensor fields, evaluated at $t = 0$, i.e.

$$L_{\mathbf{X}}\mathbf{T}|_p = \lim_{t \to 0} \frac{1}{t}\{\mathbf{T}|_p - \phi_{t*}\mathbf{T}|_p\}.$$

From the properties of ϕ_*, it follows that

(1) $L_{\mathbf{X}}$ preserves tensor type, i.e. if $\mathbf{T}$ is a tensor field of type (r, s), then $L_{\mathbf{X}}\mathbf{T}$ is also a tensor field of type (r, s);

(2) $L_{\mathbf{X}}$ maps tensors linearly and preserves contractions.

As in ordinary calculus, one can prove Leibniz' rule:

(3) For arbitrary tensors $\mathbf{S}, \mathbf{T}, L_{\mathbf{X}}(\mathbf{S} \otimes \mathbf{T}) = L_{\mathbf{X}}\mathbf{S} \otimes \mathbf{T} + \mathbf{S} \otimes L_{\mathbf{X}}\mathbf{T}$.

Direct from the definitions:

(4) $L_{\mathbf{X}}f = Xf$, where f is any function.

Under the map ϕ_t, the point $q = \phi_{-t}(p)$ is mapped into p. Therefore ϕ_{t*} is a map from T_q to T_p. Thus, by (2.6),

$$(\phi_{t*} Y)f|_p = Y(\phi_t{}^*f)|_q.$$

If $\{x^i\}$ are local coordinates in a neighbourhood of p, the coordinate components of $\phi_{t*}Y$ at p are

$$(\phi_{t*} Y)^i|_p = \phi_{t*} Y|_p x^i = Y^j|_q \frac{\partial}{\partial x^j(q)} (x^i(p))$$

$$= \frac{\partial x^i(\phi_t(q))}{\partial x^j(q)} Y^j|_q.$$

Now
$$\frac{\mathrm{d}x^i(\phi_t(q))}{\mathrm{d}t} = X^i|_{\phi_t(q)},$$

therefore
$$\frac{\mathrm{d}}{\mathrm{d}t}\left(\frac{\partial x^i(\phi_t(q))}{\partial x^j(q)}\right)\Bigg|_{t=0} = \frac{\partial X^i}{\partial x^j}\Bigg|_p,$$

so
$$(L_{\mathbf{X}} Y)^i = -\frac{\mathrm{d}}{\mathrm{d}t}(\phi_{t*} Y)^i|_{t=0} = \frac{\partial Y^i}{\partial x^j} X^j - \frac{\partial X^i}{\partial x^j} Y^j. \tag{2.11}$$

One can rewrite this in the form

$$(L_{\mathbf{X}} Y)f = X(Yf) - Y(Xf)$$

for all C^2 functions f. We shall sometimes denote $L_{\mathbf{X}}Y$ by $[\mathbf{X}, \mathbf{Y}]$, i.e.

$$L_{\mathbf{X}}\mathbf{Y} = -L_{\mathbf{Y}}\mathbf{X} = [\mathbf{X}, \mathbf{Y}] = -[\mathbf{Y}, \mathbf{X}].$$

If the Lie derivative of two vector fields $\mathbf{X}, \mathbf{Y}$ vanishes, the vector fields are said to commute. In this case, if one starts at a point p, goes a parameter distance t along the integral curves of $\mathbf{X}$ and then a parameter distance s along the integral curves of $\mathbf{Y}$, one arrives at the

same point as if one first went a distance s along the integral curves
of **Y** and then a parameter distance t along the integral curves of **X**
(see figure 7). Thus the set of all points which can be reached along
integral curves of **X** and **Y** from a given point p will then form an
immersed two-dimensional submanifold through p.

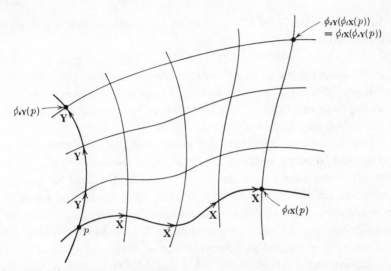

FIGURE 7. The transformations generated by commuting vector fields **X**, **Y**
move a point p to points $\phi_{t\mathbf{X}}(p)$, $\phi_{s\mathbf{Y}}(p)$ respectively. By successive applications
of these transformations, p is moved to the points of a two-surface.

The components of the Lie derivative of a one-form **ω** may be found
by contracting the relation

$$L_{\mathbf{X}}(\boldsymbol{\omega} \otimes \mathbf{Y}) = L_{\mathbf{X}}\boldsymbol{\omega} \otimes \mathbf{Y} + \boldsymbol{\omega} \otimes L_{\mathbf{X}}\mathbf{Y}$$

(Lie derivative property (3)) to obtain

$$L_{\mathbf{X}}\langle\boldsymbol{\omega}, \mathbf{Y}\rangle = \langle L_{\mathbf{X}}\boldsymbol{\omega}, \mathbf{Y}\rangle + \langle\boldsymbol{\omega}, L_{\mathbf{X}}\mathbf{Y}\rangle$$

(by property (2) of Lie derivatives), where **X**, **Y** are arbitrary C^1
vector fields, and then choosing **Y** as a basis vector $\mathbf{E}_i$. One finds the
coordinate components (on choosing $\mathbf{E}_i = \partial/\partial x^i$) to be

$$(L_{\mathbf{X}}\boldsymbol{\omega})_i = (\partial\omega_i/\partial x^j)\,X^j + \omega_j(\partial X^j/\partial x^i)$$

because (2.11) implies

$$(L_{\mathbf{X}}(\partial/\partial x^i))^j = -\partial X^j/\partial x^i.$$

Similarly, one can find the components of the Lie derivative of any
C^r $(r \geqslant 1)$ tensor field **T** of type (r, s) by using Leibniz' rule on

$$L_{\mathbf{X}}(\mathbf{T} \otimes \mathbf{E}^a \otimes \dots \otimes \mathbf{E}^d \otimes \mathbf{E}_e \otimes \dots \otimes \mathbf{E}_g),$$

and then contracting on all positions. One finds the coordinate components to be

$$(L_{\mathbf{X}}T)^{ab\ldots d}{}_{ef\ldots g} = (\partial T^{ab\ldots d}{}_{ef\ldots g}/\partial x^i)\,X^i - T^{ib\ldots d}{}_{ef\ldots g}\,\partial X^a/\partial x^i$$

$$- \text{(all upper indices)} + T^{ab\ldots d}{}_{if\ldots g}\,\partial X^i/\partial x^e + \text{(all lower indices)}.$$

$$(2.12)$$

Because of (2.7), any Lie derivative commutes with d, i.e. for any p-form field $\boldsymbol{\omega}$,

$$\mathrm{d}(L_{\mathbf{X}}\boldsymbol{\omega}) = L_{\mathbf{X}}(\mathrm{d}\boldsymbol{\omega}).$$

From these formulae, as well as from the geometrical interpretation, it follows that the Lie derivative $L_{\mathbf{X}}\mathbf{T}|_p$ of a tensor field $\mathbf{T}$ of type (r, s) depends not only on the direction of the vector field $\mathbf{X}$ at the point p, but also on the direction of $\mathbf{X}$ at neighbouring points. Thus the two differential operators defined by the manifold structure are too limited to serve as the generalization of the concept of a partial derivative one needs in order to set up field equations for physical quantities on the manifold; d operates only on forms, while the ordinary partial derivative is a directional derivative depending only on a direction at the point in question, unlike the Lie derivative. One obtains such a generalized derivative, the covariant derivative, by introducing extra structure on the manifold. We do this in the next section.

2.5 Covariant differentiation and the curvature tensor

The extra structure we introduce is a (affine) connection on $\mathcal{M}$. A *connection* ∇ at a point p of $\mathcal{M}$ is a rule which assigns to each vector field $\mathbf{X}$ at p a differential operator $\nabla_{\mathbf{X}}$ which maps an arbitrary C^r $(r \geqslant 1)$ vector field $\mathbf{Y}$ into a vector field $\nabla_{\mathbf{X}}\mathbf{Y}$, where:

(1) $\nabla_{\mathbf{X}}\mathbf{Y}$ is a tensor in the argument $\mathbf{X}$, i.e. for any functions f, g, and C^1 vector fields $\mathbf{X}$, $\mathbf{Y}$, $\mathbf{Z}$,

$$\nabla_{f\mathbf{X}+g\mathbf{Y}}\mathbf{Z} = f\nabla_{\mathbf{X}}\mathbf{Z} + g\nabla_{\mathbf{Y}}\mathbf{Z};$$

(this is equivalent to the requirement that the derivative $\nabla_{\mathbf{X}}$ at p depends only on the direction of $\mathbf{X}$ at p);

(2) $\nabla_{\mathbf{X}}\mathbf{Y}$ is linear in $\mathbf{Y}$, i.e. for any C^1 vector fields $\mathbf{Y}$, $\mathbf{Z}$ and $\alpha, \beta \in R^1$,

$$\nabla_{\mathbf{X}}(\alpha\mathbf{Y} + \beta\mathbf{Z}) = \alpha\nabla_{\mathbf{X}}\mathbf{Y} + \beta\nabla_{\mathbf{X}}\mathbf{Z};$$

(3) for any C^1 function f and C^1 vector field $\mathbf{Y}$,

$$\nabla_{\mathbf{X}}(f\mathbf{Y}) = \mathbf{X}(f)\mathbf{Y} + f\nabla_{\mathbf{X}}\mathbf{Y}.$$

Then $\nabla_{\mathbf{X}}\mathbf{Y}$ is the *covariant derivative* (with respect to ∇) of $\mathbf{Y}$ *in the direction* $\mathbf{X}$ at p. By (1), we can define $\nabla\mathbf{Y}$, the *covariant derivative* of $\mathbf{Y}$, as that tensor field of type $(1, 1)$ which, when contracted with $\mathbf{X}$, produces the vector $\nabla_{\mathbf{X}}\mathbf{Y}$. Then we have

$$(3) \Leftrightarrow \nabla(f\mathbf{Y}) = \mathrm{d}f \otimes \mathbf{Y} + f\nabla\mathbf{Y}.$$

A C^r *connection* ∇ on a C^k manifold $\mathscr{M}$ $(k \geqslant r+2)$ is a rule which assigns a connection ∇ to each point such that if $\mathbf{Y}$ is a C^{r+1} vector field on $\mathscr{M}$, then $\nabla\mathbf{Y}$ is a C^r tensor field.

Given any C^{r+1} vector basis $\{\mathbf{E}_a\}$ and dual one-form basis $\{\mathbf{E}^a\}$ on a neighbourhood $\mathscr{U}$, we shall write the components of $\nabla\mathbf{Y}$ as $Y^a{}_{;b}$, so

$$\nabla\mathbf{Y} = Y^a{}_{;b}\,\mathbf{E}^b \otimes \mathbf{E}_a.$$

The connection is determined on $\mathscr{U}$ by n^3 C^r functions $\Gamma^a{}_{bc}$ defined by

$$\Gamma^a{}_{bc} = \langle \mathbf{E}^a, \nabla_{\mathbf{E}_b}\mathbf{E}_c \rangle \Leftrightarrow \nabla\mathbf{E}_c = \Gamma^a{}_{bc}\,\mathbf{E}^b \otimes \mathbf{E}_a.$$

For any C^1 vector field $\mathbf{Y}$,

$$\nabla\mathbf{Y} = \nabla(Y^c\mathbf{E}_c) = \mathrm{d}Y^c \otimes \mathbf{E}_c + Y^c\Gamma^a{}_{bc}\,\mathbf{E}^b \otimes \mathbf{E}_a.$$

Thus the components of $\nabla\mathbf{Y}$ with respect to coordinate bases $\{\partial/\partial x^a\}$, $\{\mathrm{d}x^b\}$ are

$$Y^a{}_{;b} = \partial Y^a/\partial x^b + \Gamma^a{}_{bc}\,Y^c.$$

The transformation properties of the functions $\Gamma^a{}_{bc}$ are determined by connection properties (1), (2), (3); for

$$\Gamma^{a'}{}_{b'c'} = \langle \mathbf{E}^{a'}, \nabla_{\mathbf{E}_{b'}}\mathbf{E}_{c'} \rangle = \langle \Phi^{a'}{}_a \mathbf{E}^a, \nabla_{\Phi_{b'}{}^b\mathbf{E}_b}(\Phi_{c'}{}^c\mathbf{E}_c) \rangle$$
$$= \Phi^{a'}{}_a \Phi_{b'}{}^b(E_b(\Phi_{c'}{}^a) + \Phi_{c'}{}^c\,\Gamma^a{}_{bc})$$

if $\mathbf{E}_{a'} = \Phi_{a'}{}^a\,\mathbf{E}_a$, $\mathbf{E}^{a'} = \Phi^{a'}{}_a\,\mathbf{E}^a$. One can rewrite this as

$$\Gamma^{a'}{}_{b'c'} = \Phi^{a'}{}_a(E_{b'}(\Phi_{c'}{}^a) + \Phi_{b'}{}^b\,\Phi_{c'}{}^c\,\Gamma^a{}_{bc}).$$

In particular, if the bases are coordinate bases defined by coordinates $\{x^a\}$, $\{x^{a'}\}$, the transformation law is

$$\Gamma^{a'}{}_{b'c'} = \frac{\partial x^{a'}}{\partial x^a}\left(\frac{\partial^2 x^a}{\partial x^{b'}\,\partial x^{c'}} + \frac{\partial x^b}{\partial x^{b'}}\frac{\partial x^c}{\partial x^{c'}}\,\Gamma^a{}_{bc} \right).$$

Because of the term $E_{b'}(\Phi_{c'}{}^a)$, the $\Gamma^a{}_{bc}$ do not transform as the components of a tensor. However if $\nabla\mathbf{Y}$ and $\hat{\nabla}\mathbf{Y}$ are covariant derivatives obtained from two different connections, then

$$\nabla\mathbf{Y} - \hat{\nabla}\mathbf{Y} = (\Gamma^a{}_{bc} - \hat{\Gamma}^a{}_{bc})\,Y^c\mathbf{E}^b \otimes \mathbf{E}_a$$

will be a tensor. Thus the difference terms $(\Gamma^a{}_{bc} - \hat{\Gamma}^a{}_{bc})$ will be the components of a tensor.

The definition of a covariant derivative can be extended to any C^r tensor field if $r \geqslant 1$ by the rules (cf. the Lie derivative rules):

(1) if $\mathbf{T}$ is a C^r tensor field of type (q, s), then $\nabla \mathbf{T}$ is a C^{r-1} tensor field of type $(q, s+1)$;

(2) ∇ is linear and commutes with contractions;

(3) for arbitrary tensor fields $\mathbf{S}$, $\mathbf{T}$, Liebniz' rule holds, i.e.

$$\nabla(\mathbf{S} \otimes \mathbf{T}) = \nabla \mathbf{S} \otimes \mathbf{T} + \mathbf{S} \otimes \nabla \mathbf{T};$$

(4) $\nabla f = \mathrm{d}f$ for any function f.

We write the components of $\nabla \mathbf{T}$ as $(\nabla_{\mathbf{E}_h} T)^{a...d}{}_{e...g} = T^{a...d}{}_{e...g;h}$. As a consequence of (2) and (3),

$$\nabla_{\mathbf{E}_b} \mathbf{E}^c = - \Gamma^c{}_{ba} \mathbf{E}^a,$$

where $\{\mathbf{E}^a\}$ is the dual basis to $\{\mathbf{E}_a\}$, and methods similar to those used in deriving (2.12) show that the coordinate components of $\nabla \mathbf{T}$ are

$$T^{ab...d}{}_{ef...g;h} = \partial T^{ab...d}{}_{ef...g}/\partial x^h + \Gamma^a{}_{hj} T^{jb...d}{}_{ef...g}$$
$$+ \text{(all upper indices)} - \Gamma^j{}_{he} T^{ab...d}{}_{jf...g} - \text{(all lower indices)}. \quad (2.13)$$

As a particular example, the unit tensor $\mathbf{E}_a \otimes \mathbf{E}^a$, which has components $\delta^a{}_b$, has vanishing covariant derivative, and so the generalized unit tensors with components $\delta^{(a_1}{}_{b_1} \delta^{a_2}{}_{b_2} \ldots \delta^{a_s)}{}_{b_s}$, $\delta^{[a_1}{}_{b_1} \delta^{a_2}{}_{b_2} \ldots \delta^{a_p]}{}_{b_p}$ $(p \leqslant n)$ also have vanishing covariant derivatives.

If $\mathbf{T}$ is a C^r $(r \geqslant 1)$ tensor field defined along a C^r curve $\lambda(t)$, one can define $\mathrm{D}\mathbf{T}/\partial t$, the *covariant derivative of* $\mathbf{T}$ *along* $\lambda(t)$, as $\nabla_{\partial/\partial t} \mathbf{\bar{T}}$ where $\mathbf{\bar{T}}$ is any C^r tensor field extending $\mathbf{T}$ onto an open neighbourhood of λ. $\mathrm{D}\mathbf{T}/\partial t$ is a C^{r-1} tensor field defined along $\lambda(t)$, and is independent of the extension $\mathbf{\bar{T}}$. In terms of components, if $\mathbf{X}$ is the tangent vector to $\lambda(t)$, then $\mathrm{D}T^{a...d}{}_{e...g}/\partial t = T^{a...d}{}_{e...g;h} X^h$. In particular one can choose local coordinates so that $\lambda(t)$ has the coordinates $x^a(t)$, $X^a = \mathrm{d}x^a/\mathrm{d}t$, and then for a vector field $\mathbf{Y}$

$$\mathrm{D}Y^a/\partial t = \partial Y^a/\partial t + \Gamma^a{}_{bc} Y^c \, \mathrm{d}x^b/\mathrm{d}t. \quad (2.14)$$

The tensor $\mathbf{T}$ is said to be *parallelly transported* along λ if $\mathrm{D}\mathbf{T}/\partial t = 0$. Given a curve $\lambda(t)$ with endpoints p, q, the theory of solutions of ordinary differential equations shows that if the connection ∇ is at least C^{1-} one obtains a unique tensor at q by parallelly transferring any given tensor from p along λ. Thus parallel transfer along λ is a linear map from $T^r_s(p)$ to $T^r_s(q)$ which preserves all tensor products and tensor contractions, so in particular if one parallelly transfers a basis of vectors along a given curve from p to q, this determines an isomorphism of T_p to T_q. (If there are self-intersections in the curve, p and q could be the *same* point.)

A particular case is obtained by considering the covariant derivative of the tangent vector itself along λ. The curve $\lambda(t)$ is said to be a *geodesic curve* if

$$\nabla_{\mathbf{X}}\mathbf{X} = \frac{\mathrm{D}}{\partial t}\left(\frac{\partial}{\partial t}\right)_{\lambda}$$

is parallel to $(\partial/\partial t)_{\lambda}$, i.e. if there is a function f (perhaps zero) such that $X^a{}_{;b}X^b = fX^a$. For such a curve, one can find a new parameter $v(t)$ along the curve such that

$$\frac{\mathrm{D}}{\partial v}\left(\frac{\partial}{\partial v}\right)_{\lambda} = 0;$$

such a parameter is called an *affine parameter*. The associated tangent vector $\mathbf{V} = (\partial/\partial v)_{\lambda}$ is parallel to $\mathbf{X}$ but has its scale determined by $V(v) = 1$; it obeys the equations

$$V^a{}_{;b}V^b = 0 \Leftrightarrow \frac{\mathrm{d}^2x^a}{\mathrm{d}v^2} + \Gamma^a{}_{bc}\frac{\mathrm{d}x^b}{\mathrm{d}v}\frac{\mathrm{d}x^c}{\mathrm{d}v} = 0, \qquad (2.15)$$

the second expression being the local coordinate expression obtainable from (2.14) applied to the vector $\mathbf{V}$. The affine parameter of a geodesic curve is determined up to an additive and a multiplicative constant, i.e. up to transformations $v' = av+b$ where a, b are constants; the freedom of choice of b corresponds to the freedom to choose a new initial point $\lambda(0)$, the freedom of choice in a corresponding to the freedom to renormalize the vector $\mathbf{V}$ by a constant scale factor, $\mathbf{V}' = (1/a)\mathbf{V}$. The curve parametrized by any of these affine parameters is said to be a *geodesic*.

Given a C^r ($r \geqslant 0$) connection, the standard existence theorems for ordinary differential equations applied to (2.15) show that for any point p of $\mathcal{M}$ and any vector $\mathbf{X}_p$ at p, there exists a maximal geodesic $\lambda_{\mathbf{X}}(v)$ in $\mathcal{M}$ with starting point p and initial direction $\mathbf{X}_p$, i.e. such that $\lambda_{\mathbf{X}}(0) = p$ and $(\partial/\partial v)_{\lambda}|_{v=0} = \mathbf{X}_p$. If $r \geqslant 1-$, this geodesic is unique and depends continuously on p and $\mathbf{X}_p$. If $r \geqslant 1$, it depends differentiably on p and $\mathbf{X}_p$. This means that if $r \geqslant 1$, one can define a C^r map exp: $T_p \to \mathcal{M}$, where for each $\mathbf{X} \in T_p$, exp $(\mathbf{X})$ is the point in $\mathcal{M}$ a unit parameter distance along the geodesic $\lambda_{\mathbf{X}}$ from p. This map may not be defined for all $\mathbf{X} \in T_p$, since the geodesic $\lambda_{\mathbf{X}}(v)$ may not be defined for all v. If v does take all values, the geodesic $\lambda(v)$ will be said to be a *complete* geodesic. The manifold $\mathcal{M}$ is said to be *geodesically complete* if all geodesics on $\mathcal{M}$ are complete, that is if exp is defined on all T_p for every point p of $\mathcal{M}$.

Whether $\mathcal{M}$ is complete or not, the map $\exp_p$ is of rank n at p. Therefore by the implicit function theorem (Spivak (1965)) there exists an open neighbourhood $\mathcal{N}_0$ of the origin in T_p and an open neighbourhood $\mathcal{N}_p$ of p in $\mathcal{M}$ such that the map exp is a C^r diffeomorphism of $\mathcal{N}_0$ onto $\mathcal{N}_p$. Such a neighbourhood $\mathcal{N}_p$ is called a *normal neighbourhood* of p. Further, one can choose $\mathcal{N}_p$ to be *convex*, i.e. to be such that any point q of $\mathcal{N}_p$ can be joined to any other point r in $\mathcal{N}_p$ by a unique geodesic starting at q and totally contained in $\mathcal{N}_p$. Within a convex normal neighbourhood $\mathcal{N}$ one can define coordinates $(x^1, ..., x^n)$ by choosing any point $q \in \mathcal{N}$, choosing a basis $\{\mathbf{E}_a\}$ of T_q, and defining the coordinates of the point r in $\mathcal{N}$ by the relation $r = \exp(x^a \mathbf{E}_a)$ (i.e. one assigns to r the coordinates, with respect to the basis $\{\mathbf{E}_a\}$, of the point $\exp^{-1}(r)$ in T_q.) Then $(\partial/\partial x^i)|_q = \mathbf{E}_i$ and (by (2.15)) $\Gamma^i{}_{(jk)}|_q = 0$. Such coordinates will be called *normal coordinates* based on q. The existence of normal neighbourhoods has been used by Geroch (1968c) to prove that a connected C^3 Hausdorff manifold $\mathcal{M}$ with a C^1 connection has a countable basis. Thus one may infer the property of paracompactness of a C^3 manifold from the existence of a C^1 connection on the manifold. The 'normal' local behaviour of geodesics in these neighbourhoods is in contrast to the behaviour of geodesics in the large in a general space, where on the one hand two arbitrary points cannot in general be joined by any geodesic, and on the other hand some of the geodesics through one point may converge to 'focus' at some other point. We shall later encounter examples of both types of behaviour.

Given a C^r connection ∇, one can define a C^{r-1} tensor field $\mathbf{T}$ of type $(1, 2)$ by the relation

$$\mathbf{T}(\mathbf{X}, \mathbf{Y}) = \nabla_{\mathbf{X}} \mathbf{Y} - \nabla_{\mathbf{Y}} \mathbf{X} - [\mathbf{X}, \mathbf{Y}],$$

where $\mathbf{X}$, $\mathbf{Y}$ are arbitrary C^r vector fields. This tensor is called the *torsion tensor*. Using a coordinate basis, its components are

$$T^i{}_{jk} = \Gamma^i{}_{jk} - \Gamma^i{}_{kj}.$$

We shall deal only with *torsion-free* connections, i.e. we shall assume $\mathbf{T} = 0$. In this case, the coordinate components of the connection obey $\Gamma^i{}_{jk} = \Gamma^i{}_{kj}$, so such a connection is often called a symmetric connection. A connection is torsion-free if and only if $f_{;ij} = f_{;ji}$ for all functions f. From the geodesic equation (2.15) it follows that a torsion-free connection is completely determined by a knowledge of the geodesics on $\mathcal{M}$.

When the torsion vanishes, the covariant derivatives of arbitrary C^1 vector fields $\mathbf{X}$, $\mathbf{Y}$ are related to their Lie derivative by

$$[\mathbf{X}, \mathbf{Y}] = \nabla_{\mathbf{X}} \mathbf{Y} - \nabla_{\mathbf{Y}} \mathbf{X} \Leftrightarrow (L_{\mathbf{X}} \mathbf{Y})^a = Y^a{}_{;b} X^b - X^a{}_{;b} Y^b, \quad (2.16)$$

and for any C^1 tensor field $\mathbf{T}$ of type (r, s) one finds

$$(L_{\mathbf{X}}T)^{ab...d}{}_{ef...g} = T^{ab...d}{}_{ef...g;h}X^h - T^{jb...d}{}_{ef...g}X^a{}_{;j}$$

$$- \text{(all upper indices)} + T^{ab...d}{}_{jf...g}X^j{}_{;e} + \text{(all lower indices)}. \quad (2.17)$$

One can also easily verify that the exterior derivative is related to the covariant derivative by

$$d\mathbf{A} = A_{a...c;\,d}\,dx^d \wedge dx^a \wedge ... \wedge dx^c \Leftrightarrow (dA)_{a...cd} = (-)^p A_{[a...c;\,d]},$$

where $\mathbf{A}$ is any p-form. Thus equations involving the exterior derivative or Lie derivative can always be expressed in terms of the covariant derivative. However, because of their definitions, the Lie derivative and exterior derivative are independent of the connection.

If one starts from a given point p and parallelly transfers a vector $\mathbf{X}_p$ along a curve γ that ends at p again, one will obtain a vector $\mathbf{X}'_p$ which is in general different from $\mathbf{X}_p$; if one chooses a different curve γ', the new vector one obtains at p will in general be different from $\mathbf{X}_p$ and $\mathbf{X}'_p$. This non-integrability of parallel transfer corresponds to the fact that the covariant derivatives do not generally commute. The *Riemann (curvature) tensor* gives a measure of this non-commutation. Given C^{r+1} vector fields $\mathbf{X}, \mathbf{Y}, \mathbf{Z}$, a C^{r-1} vector field $\mathbf{R}(\mathbf{X}, \mathbf{Y})\mathbf{Z}$ is defined by a C^r connection ∇ by

$$\mathbf{R}(\mathbf{X}, \mathbf{Y})\mathbf{Z} = \nabla_{\mathbf{X}}(\nabla_{\mathbf{Y}}\mathbf{Z}) - \nabla_{\mathbf{Y}}(\nabla_{\mathbf{X}}\mathbf{Z}) - \nabla_{[\mathbf{X},\mathbf{Y}]}\mathbf{Z}. \quad (2.18)$$

Then $\mathbf{R}(\mathbf{X}, \mathbf{Y})\mathbf{Z}$ is linear in $\mathbf{X}, \mathbf{Y}, \mathbf{Z}$ and it may be verified that the value of $\mathbf{R}(\mathbf{X}, \mathbf{Y})\mathbf{Z}$ at p depends only on the values of $\mathbf{X}, \mathbf{Y}, \mathbf{Z}$ at p, i.e. it is a C^{r-1} tensor field of type $(3, 1)$. To write (2.18) in component form, we define the second covariant derivative $\nabla\nabla\mathbf{Z}$ of the vector $\mathbf{Z}$ as the covariant derivative $\nabla(\nabla\mathbf{Z})$ of $\nabla\mathbf{Z}$; it has components

$$Z^a{}_{;\,bc} = (Z^a{}_{;\,b})_{;\,c}.$$

Then (2.18) can be written

$$R^a{}_{bcd}X^cY^dZ^b = (Z^a{}_{;\,d}Y^d)_{;\,c}X^c - (Z^a{}_{;\,d}X^d)_{;\,c}Y^c$$

$$- Z^a{}_{;\,d}(Y^d{}_{;\,c}X^c - X^d{}_{;\,c}Y^c)$$

$$= (Z^a{}_{;\,dc} - Z^a{}_{;\,cd})X^cY^d,$$

where the Riemann tensor components $R^a{}_{bcd}$ with respect to dual bases $\{\mathbf{E}_a\}, \{\mathbf{E}^a\}$ are defined by $R^a{}_{bcd} = \langle \mathbf{E}^a, \mathbf{R}(\mathbf{E}_c, \mathbf{E}_d)\mathbf{E}_b \rangle$. As $\mathbf{X}, \mathbf{Y}$ are arbitrary vectors, $$Z^a{}_{;\,dc} - Z^a{}_{;\,cd} = R^a{}_{bcd}Z^b \quad (2.19)$$

expresses the non-commutation of second covariant derivatives of $\mathbf{Z}$ in terms of the Riemann tensor.

Since

$$\nabla_{\mathbf{X}}(\eta \otimes \nabla_{\mathbf{Y}} Z) = \nabla_{\mathbf{X}}\eta \otimes \nabla_{\mathbf{Y}} Z + \eta \otimes \nabla_{\mathbf{X}}\nabla_{\mathbf{Y}} Z$$

$$\Rightarrow \langle \eta, \nabla_{\mathbf{X}}\nabla_{\mathbf{Y}} Z \rangle = X(\langle \eta, \nabla_{\mathbf{Y}} Z \rangle) - \langle \nabla_{\mathbf{X}}\eta, \nabla_{\mathbf{Y}} Z \rangle$$

holds for any C^2 one-form field η and vector fields $\mathbf{X}, \mathbf{Y}, \mathbf{Z}$, (2.18) implies

$$\langle \mathbf{E}^a, \mathbf{R}(\mathbf{E}_c, \mathbf{E}_d)\,\mathbf{E}_b \rangle = E_c(\langle \mathbf{E}^a, \nabla_{\mathbf{E}_d}\mathbf{E}_b \rangle) - E_d(\langle \mathbf{E}^a, \nabla_{\mathbf{E}_c}\mathbf{E}_b \rangle)$$

$$- \langle \nabla_{\mathbf{E}_c}\mathbf{E}^a, \nabla_{\mathbf{E}_d}\mathbf{E}_b \rangle + \langle \nabla_{\mathbf{E}_d}\mathbf{E}^a, \nabla_{\mathbf{E}_c}\mathbf{E}_b \rangle - \langle \mathbf{E}^a, \nabla_{[\mathbf{E}_c, \mathbf{E}_d]}\mathbf{E}_b \rangle.$$

Choosing the bases as coordinate bases, one finds the expression

$$R^a{}_{bcd} = \partial \Gamma^a{}_{db}/\partial x^c - \partial \Gamma^a{}_{cb}/\partial x^d + \Gamma^a{}_{cf}\Gamma^f{}_{db} - \Gamma^a{}_{df}\Gamma^f{}_{cb} \qquad (2.20)$$

for the coordinate components of the Riemann tensor, in terms of the coordinate components of the connection.

It can be verified from these definitions that in addition to the symmetry

$$R^a{}_{bcd} = -R^a{}_{bdc} \Leftrightarrow R^a{}_{b(cd)} = 0 \qquad (2.21a)$$

the curvature tensor has the symmetry

$$R^a{}_{[bcd]} = 0 \Leftrightarrow R^a{}_{bcd} + R^a{}_{dbc} + R^a{}_{cdb} = 0. \qquad (2.21b)$$

Similarly the first covariant derivatives of the Riemann tensor satisfy *Bianchi's identities*

$$R^a{}_{b[cd;\,e]} = 0 \Leftrightarrow R^a{}_{bcd;\,e} + R^a{}_{bec;\,d} + R^a{}_{bde;\,c} = 0. \qquad (2.22)$$

It now turns out that parallel transfer of an arbitrary vector along an arbitrary closed curve is locally integrable (i.e. $\mathbf{X}'_p$ is necessarily the same as $\mathbf{X}_p$ for each $p \in \mathcal{M}$) only if $R^a{}_{bcd} = 0$ at all points of $\mathcal{M}$; in this case we say that the connection is *flat*.

By contracting the curvature tensor, one can define the *Ricci tensor* as the tensor of type $(0, 2)$ with components

$$R_{bd} = R^a{}_{bad}.$$

2.6   The metric

A *metric tensor* $\mathbf{g}$ at a point $p \in \mathcal{M}$ is a symmetric tensor of type $(0, 2)$ at p, so a C^r metric on $\mathcal{M}$ is a C^r symmetric tensor field $\mathbf{g}$. The metric $\mathbf{g}$ at p assigns a 'magnitude' $(|g(\mathbf{X}, \mathbf{X})|)^{\frac{1}{2}}$ to each vector $\mathbf{X} \in T_p$ and defines the 'cos angle'

$$\frac{g(\mathbf{X}, \mathbf{Y})}{(|g(\mathbf{X}, \mathbf{X}) \cdot g(\mathbf{Y}, \mathbf{Y})|)^{\frac{1}{2}}}$$

between any vectors $\mathbf{X}, \mathbf{Y} \in T_p$ such that $g(\mathbf{X}, \mathbf{X}) \cdot g(\mathbf{Y}, \mathbf{Y}) \neq 0$; vectors $\mathbf{X}, \mathbf{Y}$ will be said to be *orthogonal* if $g(\mathbf{X}, \mathbf{Y}) = 0$.

The components of $\mathbf{g}$ with respect to a basis $\{\mathbf{E}_a\}$ are

$$g_{ab} = g(\mathbf{E}_a, \mathbf{E}_b) = g(\mathbf{E}_b, \mathbf{E}_a),$$

i.e. the components are simply the scalar products of the basis vectors $\mathbf{E}_a$. If a coordinate basis $\{\partial/\partial x^a\}$ is used, then

$$\mathbf{g} = g_{ab}\, \mathrm{d}x^a \otimes \mathrm{d}x^b. \tag{2.23}$$

Tangent space magnitudes defined by the metric are related to magnitudes on the manifold by the definition: the *path length* between points $p = \gamma(a)$ and $q = \gamma(b)$ along a C^0, piecewise C^1 curve $\gamma(t)$ with tangent vector $\partial/\partial t$ such that $g(\partial/\partial t, \partial/\partial t)$ has the same sign at all points along $\gamma(t)$, is the quantity

$$L = \int_a^b (|g(\partial/\partial t, \partial/\partial t)|)^{\frac{1}{2}}\, \mathrm{d}t. \tag{2.24}$$

We may symbolically express the relations (2.23), (2.24) in the form

$$\mathrm{d}s^2 = g_{ij}\, \mathrm{d}x^i\, \mathrm{d}x^j$$

used in classical textbooks to represent the length of the 'infinitesimal' arc determined by the coordinate displacement $x^i \rightarrow x^i + \mathrm{d}x^i$.

The metric is said to be *non-degenerate* at p if there is no non-zero vector $\mathbf{X} \in T_p$ such that $g(\mathbf{X}, \mathbf{Y}) = 0$ for all vectors $\mathbf{Y} \in T_p$. In terms of components, the metric is non-degenerate if the matrix (g_{ab}) of components of $\mathbf{g}$ is non-singular. We shall from now on always assume the metric tensor is non-degenerate. Then we can define a unique symmetric tensor of type $(2, 0)$ with components g^{ab} with respect to the basis $\{\mathbf{E}_a\}$ dual to the basis $\{\mathbf{E}^a\}$, by the relations

$$g^{ab}g_{bc} = \delta^a{}_c,$$

i.e. the matrix (g^{ab}) of components is the inverse of the matrix (g_{ab}). It follows that the matrix (g^{ab}) is also non-singular, so the tensors g^{ab}, g_{ab} can be used to give an isomorphism between any covariant tensor argument and any contravariant argument, or to 'raise and lower indices'. Thus, if X^a are the components of a contravariant vector, then X_a are the components of a uniquely associated covariant vector, where $X_a = g_{ab}X^b$, $X^a = g^{ab}X_b$; similarly, to a tensor T_{ab} of type $(0, 2)$ we can associate unique tensors $T^a{}_b = g^{ac}T_{cb}$, $T_a{}^b = g^{bc}T_{ac}$, $T^{ab} = g^{ac}g^{bd}T_{cd}$. We shall in general regard such associated covariant and contravariant tensors as representations of the same geometric object (so in particular, g_{ab}, $\delta_a{}^b$ and g^{ab} may be thought of as representations (with respect to dual bases) of the same geometric object $\mathbf{g}$), although in some cases where we have more than one metric we shall have to distinguish carefully which metric is used to raise or lower indices.

The *signature* of $\mathbf{g}$ at p is the number of positive eigenvalues of the matrix (g_{ab}) at p, minus the number of negative ones. If $\mathbf{g}$ is non-degenerate and continuous, the signature will be constant on $\mathcal{M}$; by suitable choice of the basis $\{\mathbf{E}_a\}$, the metric components can at any point p be brought to the form

$$g_{ab} = \mathrm{diag}\,(\underbrace{+1, +1, \ldots, +1}_{\frac{1}{2}(n+s)\text{ terms}}, \underbrace{-1, \ldots, -1}_{\frac{1}{2}(n-s)\text{ terms}}),$$

where s is the signature of $\mathbf{g}$ and n is the dimension of $\mathcal{M}$. In this case the basis vectors $\{\mathbf{E}_a\}$ form an orthonormal set at p, i.e. each is a unit vector orthogonal to every other basis vector.

A metric whose signature is n is called a *positive definite metric*; for such a metric, $g(\mathbf{X}, \mathbf{X}) = 0 \Rightarrow \mathbf{X} = 0$, and the canonical form is

$$g_{ab} = \mathrm{diag}\,(\underbrace{+1, \ldots, +1}_{n\text{ terms}}).$$

A positive definite metric is a 'metric' on the space, in the topological sense of the word.

A metric whose signature is $(n-2)$ is called a *Lorentz metric*; the canonical form is

$$g_{ab} = \mathrm{diag}\,(\underbrace{+1, \ldots, +1}_{(n-1)\text{ terms}}, -1).$$

With a Lorentz metric on $\mathcal{M}$, the non-zero vectors at p can be divided into three classes: a vector $\mathbf{X} \in T_p$ being said to be *timelike, null,* or *spacelike* according to whether $g(\mathbf{X}, \mathbf{X})$ is negative, zero, or positive, respectively. The null vectors form a double cone in T_p which separates the timelike from the spacelike vectors (see figure 8). If $\mathbf{X}, \mathbf{Y}$ are any two non-spacelike (i.e. timelike or null) vectors in the same half of the light cone at p, then $g(\mathbf{X}, \mathbf{Y}) \leqslant 0$, and equality can only hold if $\mathbf{X}$ and $\mathbf{Y}$ are parallel null vectors (i.e. if $\mathbf{X} = \alpha\mathbf{Y}$, $g(\mathbf{X}, \mathbf{X}) = 0$).

Any paracompact C^r manifold admits a C^{r-1} positive definite metric (that is, one defined on the whole of $\mathcal{M}$). To see this, let $\{f_\alpha\}$ be a partition of unity for a locally finite atlas $\{\mathcal{U}_\alpha, \phi_\alpha\}$. Then one can define g by

$$g(\mathbf{X}, \mathbf{Y}) = \sum_\alpha f_\alpha \langle (\phi_\alpha)_* \mathbf{X}, (\phi_\alpha)_* \mathbf{Y} \rangle,$$

where $\langle\ ,\ \rangle$ is the natural scalar product in Euclidean space R^n; thus one uses the atlas to determine the metric by mapping the

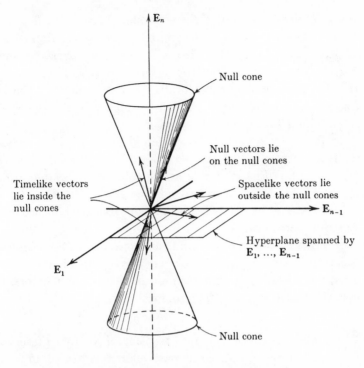

FIGURE 8. The null cones defined by a Lorentz metric.

Euclidean metric into $\mathscr{M}$. This is clearly not invariant under change of atlas, so there are many such positive definite metrics on $\mathscr{M}$.

In contrast to this, a C^r paracompact manifold admits a C^{r-1} Lorentz metric if and only if it admits a non-vanishing C^{r-1} line element field; by a line element field is meant an assignment of a pair of equal and opposite vectors $(\mathbf{X}, -\mathbf{X})$ at each point p of $\mathscr{M}$, i.e. a line element field is like a vector field but with undetermined sign. To see this, let $\hat{\mathbf{g}}$ be a C^{r-1} positive definite metric defined on the manifold. Then one can define a Lorentz metric $\mathbf{g}$ by

$$g(\mathbf{Y}, \mathbf{Z}) = \hat{g}(\mathbf{Y}, \mathbf{Z}) - 2\frac{\hat{g}(\mathbf{X}, \mathbf{Y})\,\hat{g}(\mathbf{X}, \mathbf{Z})}{\hat{g}(\mathbf{X}, \mathbf{X})}$$

at each point p, where $\mathbf{X}$ is one of the pair $(\mathbf{X}, -\mathbf{X})$ at p. (Note that as $\mathbf{X}$ appears an even number of times, it does not matter whether $\mathbf{X}$ or $-\mathbf{X}$ is chosen.) Then $g(\mathbf{X}, \mathbf{X}) = -\hat{g}(\mathbf{X}, \mathbf{X})$, and if $\mathbf{Y}, \mathbf{Z}$ are orthogonal to $\mathbf{X}$ with respect to $\hat{\mathbf{g}}$, they are also orthogonal to $\mathbf{X}$ with respect to $\mathbf{g}$ and $g(\mathbf{Y}, \mathbf{Z}) = \hat{g}(\mathbf{Y}, \mathbf{Z})$. Thus an orthonormal basis for $\hat{\mathbf{g}}$ is also an orthonormal basis for $\mathbf{g}$. As $\hat{\mathbf{g}}$ is not unique, there are in fact many

Lorentz metrics on $\mathcal{M}$ if there is one. Conversely, if $\mathbf{g}$ is a given Lorentz metric, consider the equation $g_{ab}X^b = \lambda\hat{g}_{ab}X^b$ where $\hat{\mathbf{g}}$ is any positive definite metric. This will have one negative and $(n-1)$ positive eigenvalues. Thus the eigenvector field $\mathbf{X}$ corresponding to the negative eigenvalue will locally be a vector field determined up to a sign and a normalizing factor; one can normalize it by $g_{ab}X^aX^b = -1$, so defining a line element field on $\mathcal{M}$.

In fact, any non-compact manifold admits a line element field, while a compact manifold does so if and only if its Euler invariant is zero (e.g. the torus T^2 does, but the sphere S^2 does not, admit a line element field). It will later turn out that a manifold can be a reasonable model of space–time only if it is non-compact, so there will exist many Lorentz metrics on $\mathcal{M}$.

So far, the metric tensor and connection have been introduced as separate structures on $\mathcal{M}$. However given a metric $\mathbf{g}$ on $\mathcal{M}$, there is a unique torsion-free connection on $\mathcal{M}$ defined by the condition: the covariant derivative of $\mathbf{g}$ is zero, i.e.

$$g_{ab;c} = 0. \tag{2.25}$$

With this connection, parallel transfer of vectors preserves scalar products defined by $\mathbf{g}$, so in particular magnitudes of vectors are invariant. For example if $\partial/\partial t$ is the tangent vector to a geodesic, then $g(\partial/\partial t, \partial/\partial t)$ is constant along the geodesic.

From (2.25) it follows that

$$X(g(\mathbf{Y},\mathbf{Z})) = \nabla_{\mathbf{X}}(g(\mathbf{Y},\mathbf{Z})) = \nabla_{\mathbf{X}}g(\mathbf{Y},\mathbf{Z}) + g(\nabla_{\mathbf{X}}\mathbf{Y},\mathbf{Z})$$
$$+ g(\mathbf{Y},\nabla_{\mathbf{X}}\mathbf{Z}) = g(\nabla_{\mathbf{X}}\mathbf{Y},\mathbf{Z}) + g(\mathbf{Y},\nabla_{\mathbf{X}}\mathbf{Z})$$

holds for arbitrary C^1 vector fields $\mathbf{X}, \mathbf{Y}, \mathbf{Z}$. Adding the similar expression for $Y(g(\mathbf{Z},\mathbf{X}))$ and subtracting that for $Z(g(\mathbf{X},\mathbf{Y}))$ shows

$$g(\mathbf{Z},\nabla_{\mathbf{X}}\mathbf{Y}) = \tfrac{1}{2}\{-Z(g(\mathbf{X},\mathbf{Y})) + Y(g(\mathbf{Z},\mathbf{X})) + X(g(\mathbf{Y},\mathbf{Z}))$$
$$+ g(\mathbf{Z},[\mathbf{X},\mathbf{Y}]) + g(\mathbf{Y},[\mathbf{Z},\mathbf{X}]) - g(\mathbf{X},[\mathbf{Y},\mathbf{Z}])\}.$$

Choosing $\mathbf{X}, \mathbf{Y}, \mathbf{Z}$ as basis vectors, one obtains the connection components

$$\Gamma_{abc} = g(\mathbf{E}_a, \nabla_{\mathbf{E}_b}\mathbf{E}_c) = g_{ad}\Gamma^d{}_{bc}$$

in terms of the derivatives of the metric components $g_{ab} = g(\mathbf{E}_a, \mathbf{E}_b)$, and the Lie derivatives of the basis vectors. In particular, on using a coordinate basis these Lie derivatives vanish, so one obtains the usual Christoffel relations

$$\Gamma_{abc} = \tfrac{1}{2}\{\partial g_{ab}/\partial x^c + \partial g_{ac}/\partial x^b - \partial g_{bc}/\partial x^a\} \tag{2.26}$$

for the coordinate components of the connection.

From now on we will assume that the connection on $\mathcal{M}$ is the unique C^{r-1} torsion-free connection determined by the C^r metric $\mathbf{g}$. Using this connection, one can define normal coordinates (§2.5) in a neighbourhood of a point q using an orthonormal basis of vectors at q. In these coordinates the components g_{ab} of $\mathbf{g}$ at q will be $\pm \delta_{ab}$ and the components $\Gamma^a{}_{bc}$ of the connection will vanish at q. By 'normal coordinates', we shall in future mean normal coordinates defined using an orthonormal basis.

The Riemann tensor of the connection defined by the metric is a C^{r-2} tensor with the symmetry

$$R_{(ab)cd} = 0 \Leftrightarrow R_{abcd} = -R_{bacd} \qquad (2.27a)$$

in addition to the symmetries (2.21); as a consequence of (2.21) and (2.27a), the Riemann tensor is also symmetric in the pairs of indices $\{ab\}$, $\{cd\}$, i.e.
$$R_{abcd} = R_{cdab}. \qquad (2.27b)$$

This implies that the Ricci tensor is symmetric:

$$R_{ab} = R_{ba}. \qquad (2.27c)$$

The *curvature scalar* R is the contraction of the Ricci tensor:

$$R = R^a{}_a = R^a{}_{bad} g^{bd}.$$

With these symmetries, there are $\frac{1}{12}n^2(n^2-1)$ algebraically independent components of R_{abcd}, where n is the dimension of M; $\frac{1}{2}n(n+1)$ of them can be represented by the components of the Ricci tensor. If $n = 1$, $R_{abcd} = 0$; if $n = 2$ there is one independent component of R_{abcd}, which is essentially the function R. If $n = 3$, the Ricci tensor completely determines the curvature tensor; if $n > 3$, the remaining components of the curvature tensor can be represented by the *Weyl tensor* C_{abcd}, defined by

$$C_{abcd} = R_{abcd} + \frac{2}{n-2}\{g_{a[d}R_{c]b} + g_{b[c}R_{d]a}\} + \frac{2}{(n-1)(n-2)} Rg_{a[c}g_{d]b}.$$

As the last two terms on the right-hand side have the curvature tensor symmetries (2.21), (2.27), it follows that C_{abcd} also has these symmetries. One can easily verify that in addition,

$$C^a{}_{bad} = 0,$$

i.e. one can think of the Weyl tensor as that part of the curvature tensor such that all contractions vanish.

An alternative characterization of the Weyl tensor is given by the fact that it is a conformal invariant. The metrics $\mathbf{g}$ and $\hat{\mathbf{g}}$ are said to be *conformal* if

$$\hat{\mathbf{g}} = \Omega^2 \mathbf{g} \qquad (2.28)$$

for some non-zero suitably differentiable function Ω. Then for any vectors $\mathbf{X}$, $\mathbf{Y}$, $\mathbf{V}$, $\mathbf{W}$ at a point p,

$$\frac{g(\mathbf{X}, \mathbf{Y})}{g(\mathbf{V}, \mathbf{W})} = \frac{\hat{g}(\mathbf{X}, \mathbf{Y})}{\hat{g}(\mathbf{V}, \mathbf{W})},$$

so angles and ratios of magnitudes are preserved under conformal transformations; in particular, the null cone structure in T_p is preserved by conformal transformations, since

$$g(\mathbf{X}, \mathbf{X}) > 0,\ = 0,\ < 0 \Rightarrow \hat{g}(\mathbf{X}, \mathbf{X}) > 0,\ = 0,\ < 0,$$

respectively. As the metric components are related by

$$\hat{g}_{ab} = \Omega^2 g_{ab}, \quad \hat{g}^{ab} = \Omega^{-2} g^{ab},$$

the coordinate components of the connections defined by the metrics (2.28) are related by

$$\hat{\Gamma}^a{}_{bc} = \Gamma^a{}_{bc} + \Omega^{-1} \left(\delta^a{}_b \frac{\partial \Omega}{\partial x^c} + \delta^a{}_c \frac{\partial \Omega}{\partial x^b} - g_{bc} g^{ad} \frac{\partial \Omega}{\partial x^d} \right). \qquad (2.29)$$

Calculating the Riemann tensor of $\hat{\mathbf{g}}$, one finds

$$\hat{R}^{ab}{}_{cd} = \Omega^{-2} R^{ab}{}_{cd} + \delta^{[a}{}_{[c} \Omega^{b]}{}_{d]},$$

where $\qquad \Omega^a{}_b := 4\Omega^{-1}(\Omega^{-1})_{;be} g^{ae} - 2(\Omega^{-1})_{;c}(\Omega^{-1})_{;d} g^{cd} \delta^a{}_b;$

the covariant derivatives in this equation are those determined by the metric $\mathbf{g}$. Then (assuming $n > 2$)

$$\hat{R}^b{}_d = \Omega^{-2} R^b{}_d + (n-2)\Omega^{-1}(\Omega^{-1})_{;dc} g^{bc} - (n-2)^{-1}\Omega^{-n}(\Omega^{n-2})_{;ac} g^{ac} \delta^b{}_d$$

and $\qquad\qquad\qquad \hat{C}^a{}_{bcd} = C^a{}_{bcd},$

the last equation expressing the fact that the Weyl tensor is conformally invariant. These relations imply

$$\hat{R} = \Omega^{-2} R - 2(n-1)\Omega^{-3}\Omega_{;cd} g^{cd} - (n-1)(n-4)\Omega^{-4}\Omega_{;c}\Omega_{;d} g^{cd}. \quad (2.30)$$

Having split the Riemann tensor into a part represented by the Ricci tensor and a part represented by the Weyl tensor, one can use the Bianchi identities (2.22) to obtain differential relations between the Ricci tensor and the Weyl tensor: contracting (2.22) one obtains

$$R^a{}_{bcd;a} = R_{bd;c} - R_{bc;d} \qquad (2.31)$$

and contracting again one obtains

$$R^a{}_{c;\,a} = \tfrac{1}{2}R_{;\,c}.$$

From the definition of the Weyl tensor, one can (if $n > 3$) rewrite (2.31) in the form

$$C^a{}_{bcd;\,a} = 2\frac{n-3}{n-2}\left(R_{b[d;\,c]} - \frac{1}{2(n-1)}g_{b[d}R_{;\,c]}\right). \qquad (2.32)$$

If $n \leqslant 4$, (2.31) contain all the information in the Bianchi identities (2.22), so if $n = 4$, (2.32) are equivalent to these identities.

A diffeomorphism $\phi\colon \mathcal{M} \to \mathcal{M}$ will be said to be an *isometry* if it carries the metric into itself, that is, if the mapped metric $\phi_*\mathbf{g}$ is equal to $\mathbf{g}$ at every point. Then the map $\phi_*\colon T_p \to T_{\phi(p)}$ preserves scalar products, as

$$g(\mathbf{X},\mathbf{Y})|_p = \phi_*g(\phi_*\mathbf{X},\phi_*\mathbf{Y})|_{\phi(p)} = g(\phi_*\mathbf{X},\phi_*\mathbf{Y})|_{\phi(p)}.$$

If the local one-parameter group of diffeomorphisms ϕ_t generated by a vector field $\mathbf{K}$ is a group of isometries (i.e. for each t, the transformation ϕ_t is an isometry) we call the vector field $\mathbf{K}$ a *Killing vector field*. The Lie derivative of the metric with respect to $\mathbf{K}$ is

$$L_\mathbf{K}\mathbf{g} = \lim_{t\to 0}\frac{1}{t}(\mathbf{g} - \phi_{t\,*}\mathbf{g}) = 0,$$

since $\mathbf{g} = \phi_{t\,*}\mathbf{g}$ for each t. But from (2.17), $L_\mathbf{K}g_{ab} = 2K_{(a;\,b)}$, so a Killing vector field $\mathbf{K}$ satisfies Killing's equation

$$K_{a;\,b} + K_{b;\,a} = 0. \qquad (2.33)$$

Conversely, if $\mathbf{K}$ is a vector field which satisfies Killing's equation, then $L_\mathbf{K}\mathbf{g} = 0$, so

$$\phi_{t\,*}\mathbf{g}|_p = \mathbf{g}|_p + \int_0^t \frac{\mathrm{d}}{\mathrm{d}t'}(\phi_{t'}*\mathbf{g})|_p\,\mathrm{d}t'$$

$$= \mathbf{g}|_p + \int_0^t \frac{\mathrm{d}}{\mathrm{d}s}(\phi_{t'}*\phi_s*\mathbf{g})_{s=0}|_p\,\mathrm{d}t'$$

$$= \mathbf{g}|_p + \int_0^t \left(\phi_{t'}*\frac{\mathrm{d}}{\mathrm{d}s}\phi_s*\mathbf{g}\right)_{s=0}\bigg|_p\,\mathrm{d}t'$$

$$= \mathbf{g}|_p - \int_0^t \phi_{t'}*(L_\mathbf{K}\mathbf{g}|_{\phi_{-t'(p)}})\,\mathrm{d}t' = \mathbf{g}|_p.$$

Thus $\mathbf{K}$ is a Killing vector field if and only if it satisfies Killing's equation. Then one can locally choose coordinates $x^a = (x^\nu, t)$ ($\nu = 1$ to $n-1$)

such that $K^a = \partial x^a/\partial t = \delta^a{}_n$; in these coordinates Killing's equation takes the form
$$\partial g_{ab}/\partial t = 0 \Leftrightarrow g_{ab} = g_{ab}(x^\nu).$$

A general space will not have any symmetries, and so will not admit any Killing vector fields. However a special space may admit r linearly independent Killing vector fields $\mathbf{K}_a$ ($a = 1, ..., r$). It can be shown that the set of all Killing vector fields on such a space forms a Lie algebra of dimension r over R, with the algebra product given by the Lie bracket [,] (see (2.16)), where $0 \leqslant r \leqslant \frac{1}{2}n(n+1)$. (The upper limit may be lessened if the metric is degenerate.) The local group of diffeomorphisms generated by these vector fields is an r-dimensional Lie group of isometries of the manifold $\mathcal{M}$. The full group of isometries of $\mathcal{M}$ may include some discrete isometries (such as reflections in a plane) which are not generated by Killing vector fields; the symmetry properties of the space are completely characterized by this full group of isometries.

2.7 Hypersurfaces

If $\mathcal{S}$ is an $(n-1)$-dimensional manifold and $\theta: \mathcal{S} \to \mathcal{M}$ is an imbedding, the image $\theta(\mathcal{S})$ of $\mathcal{S}$ is said to be a *hypersurface* in $\mathcal{M}$. If $p \in \mathcal{S}$, the image of T_p in $T_{\theta(p)}$ under the map θ_* will be a $(n-1)$-dimensional plane through the origin. Thus there will be some non-zero form $\mathbf{n} \in T^*_{\theta(p)}$ such that for any vector $\mathbf{X} \in T_p$, $\langle \mathbf{n}, \theta_* \mathbf{X} \rangle = 0$. The form $\mathbf{n}$ is unique up to a sign and a normalizing factor, and if $\theta(\mathcal{S})$ is given locally by the equation $f = 0$ where $\mathrm{d}f \neq 0$ then $\mathbf{n}$ may be taken locally as $\mathrm{d}f$. If $\theta(\mathcal{S})$ is two-sided in $\mathcal{M}$, one can choose $\mathbf{n}$ to be a nowhere zero one-form field on $\theta(\mathcal{S})$. This will be the situation if $\mathcal{S}$ and $\mathcal{M}$ are both orientable manifolds. In this case, the choice of a direction of $\mathbf{n}$ will relate the orientations of $\theta(\mathcal{S})$ and of $\mathcal{M}$: if $\{x^i\}$ are local coordinates from the oriented atlas of $\mathcal{M}$ such that locally $\theta(\mathcal{S})$ has the equation $x^1 = 0$ and $\mathbf{n} = \alpha\,\mathrm{d}x^1$ where $\alpha > 0$, then $(x^2, ..., x^n)$ are oriented local coordinates for $\theta(\mathcal{S})$.

If $\mathbf{g}$ is a metric on $\mathcal{M}$, the imbedding will induce a metric $\theta^*\mathbf{g}$ on $\mathcal{S}$, where if $\mathbf{X}, \mathbf{Y} \in T_p$, $\theta^*g(\mathbf{X}, \mathbf{Y})|_p = g(\theta_*\mathbf{X}, \theta_*\mathbf{Y})|_{\theta(p)}$. This metric is sometimes called the first fundamental form of $\mathcal{S}$. If $\mathbf{g}$ is positive definite the metric $\theta^*\mathbf{g}$ will be positive definite, while if $\mathbf{g}$ is Lorentz, $\theta^*\mathbf{g}$ will be

(a) Lorentz if $g^{ab}n_a n_b > 0$ (in this case, $\theta(\mathcal{S})$ will be said to be a *timelike hypersurface*),

(b) degenerate if $g^{ab}n_a n_b = 0$ (in this case, $\theta(\mathcal{S})$ will be said to be a *null hypersurface*),

(c) positive definite if $g^{ab}n_a n_b < 0$ (in this case, $\theta(\mathcal{S})$ will be said to be a *spacelike hypersurface*).

To see this, consider the vector $N^b = n_a g^{ab}$. This will be orthogonal to all the vectors tangent to $\theta(\mathcal{S})$, i.e. to all vectors in the subspace $H = \theta_*(T_p)$ in $T_{\theta(p)}$. Suppose first that **N** does not itself lie in this subspace. Then if $(\mathbf{E}_2, ..., \mathbf{E}_n)$ are a basis for T_p, $(\mathbf{N}, \theta_*(\mathbf{E}_2), ..., \theta_*(\mathbf{E}_n))$ will be linearly independent and so will be a basis for $T_{\theta(p)}$. The components of $\mathbf{g}$ with respect to this basis will be

$$g_{ab} = \begin{pmatrix} g(\mathbf{N}, \mathbf{N}) & 0 \\ 0 & g(\theta_*(\mathbf{E}_i), \theta_*(\mathbf{E}_j)) \end{pmatrix} = \begin{pmatrix} g(\mathbf{N}, \mathbf{N}) & 0 \\ 0 & \theta^* g(\mathbf{E}_i, \mathbf{E}_j) \end{pmatrix}.$$

As the metric $\mathbf{g}$ is assumed to be non-degenerate, this shows that $g(\mathbf{N}, \mathbf{N}) \neq 0$. If $\mathbf{g}$ is positive definite, $g(\mathbf{N}, \mathbf{N})$ must be positive and so the induced metric $\theta^* \mathbf{g}$ must also be positive definite. If $\mathbf{g}$ is Lorentz and $g(\mathbf{N}, \mathbf{N}) = g^{ab}n_a n_b < 0$, then $\theta^* \mathbf{g}$ must be positive definite since the matrix of the components of $\mathbf{g}$ has only one negative eigenvalue. Similarly if $g(\mathbf{N}, \mathbf{N}) = g^{ab}n_a n_b > 0$, then $\theta^* \mathbf{g}$ will be a Lorentz metric. Now suppose that **N** is tangent to $\theta(\mathcal{S})$. Then there is some non-zero vector $\mathbf{X} \in T_p$ such that $\theta_*(\mathbf{X}) = \mathbf{N}$. But $g(\mathbf{N}, \theta_* \mathbf{Y}) = 0$ for all $\mathbf{Y} \in T_p$, which implies $\theta^* g(\mathbf{X}, \mathbf{Y}) = 0$. Thus $\theta^* \mathbf{g}$ is degenerate. Also, taking **Y** to be **X**, $g(\mathbf{N}, \mathbf{N}) = g^{ab}n_a n_b = 0$.

If $g^{ab}n_a n_b \neq 0$, one can normalize the normal form **n** to have unit magnitude, i.e. $g^{ab}n_a n_b = \pm 1$. In this case the map $\theta^*: T^*_{\theta(p)} \to T^*_p$ will be one–one on the $(n-1)$-dimensional subspace $H^*_{\theta(p)}$ of $T^*_{\theta(p)}$ consisting of all forms $\boldsymbol{\omega}$ at $\theta(p)$ such that $g^{ab}n_a \omega_b = 0$, because $\theta^* \mathbf{n} = 0$ and **n** does not lie in H^*. Therefore the inverse $(\theta^*)^{-1}$ will be a map $\tilde{\theta}_*$ of T^*_p onto $H^*_{\theta(p)}$, and so into $T^*_{\theta(p)}$.

This map can be extended in the usual way to a map of covariant tensors on $\mathcal{S}$ to covariant tensors on $\theta(\mathcal{S})$ in $\mathcal{M}$; as there already is a map θ_* of contravariant tensors on $\mathcal{S}$ to $\theta(\mathcal{S})$, one can extend θ_* to a map $\tilde{\theta}_*$ of arbitrary tensors on $\mathcal{S}$ to $\theta(\mathcal{S})$. This map has the property that $\tilde{\theta}_* \mathbf{T}$ has zero contraction with **n** on all indices, i.e.

$$(\tilde{\theta}_* T)^{a...b}{}_{c...d} n_a = 0 \quad \text{and} \quad (\tilde{\theta}_* T)^{a...b}{}_{c...d} g^{ce} n_e = 0$$

for any tensor $\mathbf{T} \in T^r_s(\mathcal{S})$.

The tensor **h** on $\theta(\mathcal{S})$ is defined by $\mathbf{h} = \tilde{\theta}_*(\theta^* \mathbf{g})$. In terms of the normalized form **n** (remember $g^{ab}n_a n_b = \pm 1$),

$$h_{ab} = g_{ab} \mp n_a n_b$$

since this implies $\theta^* \mathbf{h} = \theta^* \mathbf{g}$ and $h_{ab} g^{bc} n_c = 0$.

The tensor $h^a{}_b = g^{ac}h_{cb}$ is a projection operator, i.e. $h^a{}_b h^b{}_c = h^a{}_c$. It projects a vector $\mathbf{X} \in T_{\theta(p)}$ into its part lying in the subspace $H = \theta_*(T_p)$ of $T_{\theta(p)}$ tangent to $\theta(\mathscr{S})$,

$$X^a = h^a{}_b X^b \pm n^a n_b X^b,$$

where the second term represents the part of $\mathbf{X}$ orthogonal to $\theta(\mathscr{S})$. Also $h^a{}_b$ projects a form $\boldsymbol{\omega} \in T^*{}_{\theta(p)}$ into its part lying in the subspace $H^*{}_{\theta(p)}$:

$$\omega_a = h^b{}_a \omega_b \pm n_a n^b \omega_b.$$

Similarly one can project any tensor $\mathbf{T} \in T^r_s(\theta(p))$ into its part in

$$H^r_s(\theta(p)) = \underbrace{H_{\theta(p)} \otimes \ldots \otimes H_{\theta(p)}}_{r \text{ factors}} \otimes \underbrace{H^*_{\theta(p)} \otimes \ldots \otimes H^*_{\theta(p)}}_{s \text{ factors}},$$

i.e. its part which is orthogonal to $\mathbf{n}$ on all indices.

The map θ_* is one–one from T_p to $H_{\theta(p)}$. Therefore one can define a map $\tilde{\theta}^*$ from $T_{\theta(p)}$ to T_p by first projecting with $h^a{}_b$ into $H_{\theta(p)}$ and then using the inverse $(\theta)_*{}^{-1}$. As one already has a map θ^* of forms on $\theta(\mathscr{S})$ to forms on $\mathscr{S}$, one can extend the definition of θ^* to a map $\tilde{\theta}^*$ of tensors of any type on $\theta(\mathscr{S})$ to tensors on $\mathscr{S}$. This map has the property that $\tilde{\theta}^*(\tilde{\theta}_*\mathbf{T}) = \mathbf{T}$ for any tensor $\mathbf{T} \in T^r_s(p)$ and $\tilde{\theta}_*(\tilde{\theta}^*\mathbf{T}) = \mathbf{T}$ for any tensor $\mathbf{T} \in H^r_s(\theta(p))$. We shall identify tensors on $\mathscr{S}$ with tensors in H^r_s on $\theta(\mathscr{S})$ if they correspond under the maps $\tilde{\theta}_*$, $\tilde{\theta}^*$. In particular, $\mathbf{h}$ can then be regarded as the induced metric on $\theta(\mathscr{S})$.

If $\bar{\mathbf{n}}$ is any extension of the unit normal $\mathbf{n}$ onto an open neighbourhood of $\theta(\mathscr{S})$ then the tensor $\boldsymbol{\chi}$ defined on $\theta(\mathscr{S})$ by

$$\chi_{ab} = h^c{}_a h^d{}_b \bar{n}_{c;\,d}$$

is called the *second fundamental form* of $\mathscr{S}$. It is independent of the extension, since the projections by $h^a{}_b$ restrict the covariant derivatives to directions tangent to $\theta(\mathscr{S})$. Locally the field $\bar{\mathbf{n}}$ can be expressed in the form $\bar{\mathbf{n}} = \alpha \, \mathrm{d}f$ where f and α are functions on $\mathscr{M}$ and $f = 0$ on $\theta(\mathscr{S})$. Therefore χ_{ab} must be symmetric, since $f_{;ab} = f_{;ba}$ and $f_{;a} h^a{}_b = 0$.

The induced metric $\mathbf{h} = \theta^*\mathbf{g}$ on $\mathscr{S}$ defines a connection on $\mathscr{S}$. We shall denote covariant differentiation with respect to this connection by a double stroke, $\|$. For any tensor $\mathbf{T} \in H^r_s$,

$$T^{a\ldots b}{}_{c\ldots d\|e} = \bar{T}^{i\ldots j}{}_{k\ldots l;\,m} h^a{}_i \ldots h^b{}_j h^k{}_c \ldots h^l{}_d h^m{}_e,$$

where $\bar{\mathbf{T}}$ is any extension of $\mathbf{T}$ to a neighbourhood of $\theta(\mathscr{S})$. This definition is independent of the extension, as the hs restrict the covariant differentiation to directions tangential to $\theta(\mathscr{S})$. To see this is the correct formula, one has only to show that the covariant derivative of the induced metric is zero and that the torsion vanishes. This follows because

$$h_{ab\|c} = (g_{ef} \mp \bar{n}_e \bar{n}_f)_{;g} h^e{}_a h^f{}_b h^g{}_c = 0,$$

and $\qquad f_{\|ab} = h^e{}_a h^g{}_b f_{;eg} = h^e{}_a h^g{}_b f_{;ge} = f_{\|ba}.$

The curvature tensor $R'^a{}_{bcd}$ of the induced metric $\mathbf{h}$ can be related to the curvature tensor $R^a{}_{bcd}$ on $\theta(\mathscr{S})$ and the second fundamental form $\boldsymbol{\chi}$ as follows. If $\mathbf{Y} \in H$ is a vector field on $\theta(\mathscr{S})$, then

$$R'^a{}_{bcd} Y^b = Y^a{}_{\|dc} - Y^a{}_{\|cd}.$$

Now

$$Y^a{}_{\|dc} = (Y^a{}_{\|d})_{\|c} = (\bar{Y}^e{}_{;f} h^g{}_e h^f{}_i)_{;k} h^a{}_g h^i{}_d h^k{}_c$$

$$= \bar{Y}^e{}_{;fk} h^a{}_e h^f{}_d h^k{}_c \mp \bar{Y}^e{}_{;f} \bar{n}_e \bar{n}^g{}_{;k} h^f{}_d h^a{}_g h^k{}_c \mp \bar{Y}^e{}_{;f} \bar{n}^f \bar{n}_i{}_{;k} h^a{}_e h^i{}_d h^k{}_c$$

and $\qquad \bar{Y}^e{}_{;f} \bar{n}_e h^f{}_d = (\bar{Y}^e \bar{n}_e)_{;f} h^f{}_d - \bar{Y}^e \bar{n}_{e;f} h^f{}_d = - \bar{Y}^e \bar{n}_{e;f} h^f{}_d,$

since $\bar{Y}^e \bar{n}_e = 0$ on $\theta(\mathscr{S})$, therefore

$$R'^a{}_{bcd} Y^b = (R^e{}_{bkf} h^a{}_e h^k{}_c h^f{}_d \pm \chi_{bd} \chi^a{}_c \mp \chi_{bc} \chi^a{}_d) Y^b.$$

Since this holds for all $\mathbf{Y} \in H$,

$$R'^a{}_{bcd} = R^e{}_{fgh} h^a{}_e h^f{}_b h^g{}_c h^h{}_d \pm \chi^a{}_c \chi_{bd} \mp \chi^a{}_d \chi_{bc}. \qquad (2.34)$$

This is known as Gauss' equation.

Contracting this equation on a and c and multiplying by h^{bd}, one obtains the curvature scalar R' of the induced metric:

$$R' = R \mp 2R_{ab} n^a n^b \pm (\chi^a{}_a)^2 \mp \chi^{ab} \chi_{ab}. \qquad (2.35)$$

One can derive another relation between the second fundamental form and the curvature tensor $R^a{}_{bcd}$ on $\theta(\mathscr{S})$ by subtracting the expressions

$$(\chi^a{}_a)_{\|b} = (\bar{n}^a{}_{;d} h^d{}_a)_{;e} h^e{}_b$$

and $\qquad (\chi^a{}_b)_{\|a} = (\bar{n}^c{}_{;d} h^a{}_c h^d{}_e)_{;f} h^f{}_a h^e{}_b,$

finding $\qquad \chi^a{}_{b\|a} - \chi^a{}_{a\|b} = R_{ef} n^f h^e{}_b. \qquad (2.36)$

This is known as Codacci's equation.

2.8 The volume element and Gauss' theorem

If $\{\mathbf{E}^a\}$ is a basis of one-forms, one can form from it the n-form

$$\boldsymbol{\epsilon} = n! \mathbf{E}^1 \wedge \mathbf{E}^2 \wedge \ldots \wedge \mathbf{E}^n.$$

If $\{\mathbf{E}^{a'}\}$, related to $\{\mathbf{E}^a\}$ by $\mathbf{E}^{a'} = \Phi^{a'}{}_a \mathbf{E}^a$, is another basis, the n-form $\boldsymbol{\epsilon}'$ defined by this basis will be related to $\boldsymbol{\epsilon}$ by

$$\boldsymbol{\epsilon}' = \det(\Phi^{a'}{}_a)\,\boldsymbol{\epsilon},$$

so this form is not unique. However, one can use the existence of the metric to define (in a given basis) the form

$$\boldsymbol{\eta} = |g|^{\frac{1}{2}}\,\boldsymbol{\epsilon}$$

where $g \equiv \det(g_{ab})$. This form has components

$$\eta_{ab...d} = n!\,|g|^{\frac{1}{2}}\,\delta^1{}_{[a}\,\delta^2{}_b \ldots \delta^n{}_{d]}.$$

The transformation law for g will just cancel the determinant, $\det(\Phi^{a'}{}_a)$, provided that $\det(\Phi^{a'}{}_a) > 0$. Therefore if $\mathcal{M}$ is orientable the n-forms $\boldsymbol{\eta}$ defined by coordinate bases of an oriented atlas will be identical, i.e. given an orientation of $\mathcal{M}$, one can define a unique n-form field $\boldsymbol{\eta}$, the *canonical n-form*, on $\mathcal{M}$.

The contravariant antisymmetric tensor

$$\eta^{ab...d} = g^{ae}g^{bf} \ldots g^{dh}\eta_{ef...h}$$

has components

$$\eta^{ab...d} = (-)^{\frac{1}{2}(n-s)}n!\,|g|^{\frac{1}{2}}\,\delta^{[a}{}_1\,\delta^b{}_2 \ldots \delta^{d]}{}_n,$$

where s is the signature of $\mathbf{g}$ (so $\frac{1}{2}(n-s)$ is the number of negative eigenvalues of the matrix of metric components (g_{ab})). Therefore these tensors satisfy the relations

$$\eta^{ab...d}\eta_{ef...h} = (-)^{\frac{1}{2}(n-s)}n!\,\delta^a{}_{[e}\delta^b{}_f \ldots \delta^d{}_{h]}. \qquad (2.37)$$

The Christoffel relations imply that the covariant derivatives of $\eta_{ab...d}$ and $\eta^{ab...d}$ with respect to the connection defined by the metric vanish, i.e.

$$\eta^{ab...d}{}_{;e} = 0 = \eta_{ab...d;e}.$$

Using the canonical n-form, one can define the volume (with respect to the metric $\mathbf{g}$) of an n-dimensional submanifold $\mathcal{U}$ as $\dfrac{1}{n!}\displaystyle\int_{\mathcal{U}} \boldsymbol{\eta}$. Thus $\boldsymbol{\eta}$ can be regarded as a positive definite volume measure on $\mathcal{M}$. We shall often use it in this sense, and shall denote it by $\mathrm{d}v$. Note that d is not meant to represent the exterior differential operator here; $\mathrm{d}v$ is simply a measure on $\mathcal{M}$. If f is a function on $\mathcal{M}$, one can define its integral over $\mathcal{U}$ with respect to this volume measure as

$$\int_{\mathcal{U}} f\,\mathrm{d}v = \frac{1}{n!}\int_{\mathcal{U}} f\,\boldsymbol{\eta}.$$

With respect to local oriented coordinates $\{x^a\}$, this can be expressed as the multiple integral

$$\int_{\mathscr{U}} f |g|^{\frac{1}{2}} \, dx^1 \, dx^2 \ldots dx^n,$$

which is invariant under a change of coordinates.

If $\mathbf{X}$ is a vector field on $\mathscr{M}$, its contraction with $\boldsymbol{\eta}$ will be an $(n-1)$-form field $\mathbf{X} \cdot \boldsymbol{\eta}$, where

$$(\mathbf{X} \cdot \boldsymbol{\eta})_{b \ldots d} = X^a \eta_{ab \ldots d}.$$

This $(n-1)$-form may be integrated over any $(n-1)$-dimensional compact orientable submanifold $\mathscr{V}$. We write this integral as

$$\int_{\mathscr{V}} X^a \, d\sigma_a = \frac{1}{(n-1)!} \int_{\mathscr{V}} \mathbf{X} \cdot \boldsymbol{\eta},$$

where the canonical form $\boldsymbol{\eta}$ is regarded as defining a measure-valued form $d\sigma_a$ on the submanifold $\mathscr{V}$. If the orientation of $\mathscr{V}$ is given by the direction of the normal form n_a, then $d\sigma_a$ can be expressed as $n_a \, d\sigma$ where $d\sigma$ is a positive definite volume measure on the submanifold $\mathscr{V}$. The volume measure $d\sigma$ is not unique unless the normal n_a is normalized. If n_a is normalized to unit magnitude in a metric $\mathbf{g}$ on $\mathscr{M}$, i.e. $n_a n_b g^{ab} = \pm 1$, then $d\sigma$ is equal to the volume measure on $\mathscr{V}$ defined by the induced metric on $\mathscr{V}$ (to see this, simply choose an orthonormal basis with $n_a g^{ab}$ as one of the basis vectors).

Using the canonical form, one can derive Gauss' formula from Stokes' theorem: for any compact n-dimensional submanifold $\mathscr{U}$ of $\mathscr{M}$,

$$\int_{\partial \mathscr{U}} X^a \, d\sigma_a = \frac{1}{(n-1)!} \int_{\partial \mathscr{U}} \mathbf{X} \cdot \boldsymbol{\eta} = \frac{1}{(n-1)!} \int_{\mathscr{U}} d(\mathbf{X} \cdot \boldsymbol{\eta}).$$

But

$$(d(\mathbf{X} \cdot \boldsymbol{\eta}))_{a \ldots de} = (-)^{n-1} (X^g \eta_{g[a \ldots d})_{; e]}$$

$$= (-)^{n-1} \delta^s_{[a} \ldots \delta^t_d \delta^u_{e]} \eta_{gs \ldots t} X^g_{; u}$$

$$= (-)^{(n-1) - \frac{1}{2}(n-s)} \frac{1}{n!} \eta^{s \ldots tu} \eta_{a \ldots de} \eta_{gs \ldots t} X^g_{; u}$$

$$= \eta_{a \ldots de} \delta^s_{[s} \ldots \delta^t_t \delta^u_{g]} X^g_{; u}$$

$$= n^{-1} \eta_{a \ldots de} X^g_{; g},$$

on using relation (2.37) twice. Therefore

$$\int_{\partial \mathscr{U}} X^a \, d\sigma_a = \int_{\mathscr{U}} X^g_{; g} \, dv$$

holds for any vector field $\mathbf{X}$; this is Gauss' theorem. Note that the orientation on $\mathcal{U}$ for which this theorem is valid is that given by the normal form $\boldsymbol{\eta}$ such that $\langle \mathbf{n}, \mathbf{X} \rangle$ is positive if $\mathbf{X}$ is a vector which points out of $\mathcal{U}$. If the metric $\mathbf{g}$ is such that $g^{ab}n_a n_b$ is negative, the vector $g^{ab}n_b$ will point into $\mathcal{U}$.

2.9 Fibre bundles

Some of the geometrical properties of a manifold $\mathcal{M}$ can be most easily examined by constructing a manifold called a fibre bundle, which is locally a direct product of $\mathcal{M}$ and a suitable space. In this section we shall give the definition of a fibre bundle and shall consider four examples that will be used later: the tangent bundle $T(\mathcal{M})$, the tensor bundle $T_s^r(\mathcal{M})$, the bundle of linear frames or bases $L(\mathcal{M})$, and the bundle of orthonormal frames $O(\mathcal{M})$.

A C^k *bundle* over a C^s $(s \geqslant k)$ manifold $\mathcal{M}$ is a C^k manifold $\mathcal{E}$ and a C^k surjective map $\pi \colon \mathcal{E} \to \mathcal{M}$. The manifold $\mathcal{E}$ is called the total space, $\mathcal{M}$ is called the base space and π the projection. Where no confusion can arise, we will denote the bundle simply by $\mathcal{E}$. In general, the inverse image $\pi^{-1}(p)$ of a point $p \in \mathcal{M}$ need not be homeomorphic to $\pi^{-1}(q)$ for another point $q \in \mathcal{M}$. The simplest example of a bundle is a *product bundle* $(\mathcal{M} \times \mathcal{A}, \mathcal{M}, \pi)$ where $\mathcal{A}$ is some manifold and the projection π is defined by $\pi(p, v) = p$ for all $p \in \mathcal{M}$, $v \in \mathcal{A}$. For example, if one chooses $\mathcal{M}$ as the circle S^1 and $\mathcal{A}$ as the real line R^1, one constructs the cylinder C^2 as a product bundle over S^1.

A bundle which is locally a product bundle is called a fibre bundle. Thus a bundle is a *fibre bundle* with fibre $\mathcal{F}$ if there exists a neighbourhood $\mathcal{U}$ of each point q of $\mathcal{M}$ such that $\pi^{-1}(\mathcal{U})$ is isomorphic with $\mathcal{U} \times \mathcal{F}$, in the sense that for each point $p \in \mathcal{U}$ there is a diffeomorphism ϕ_p of $\pi^{-1}(p)$ onto $\mathcal{F}$ such that the map ψ defined by $\psi(u) = (\pi(u), \phi_{\pi(u)})$ is a diffeomorphism $\psi \colon \pi^{-1}(\mathcal{U}) \to \mathcal{U} \times \mathcal{F}$. Since $\mathcal{M}$ is paracompact, we can choose a locally finite covering of $\mathcal{M}$ by such open sets $\mathcal{U}_\alpha$. If $\mathcal{U}_\alpha$ and $\mathcal{U}_\beta$ are two members of such a covering, the map

$$(\phi_{\alpha, p}) \circ (\phi_{\beta, p}^{-1})$$

is a diffeomorphism of $\mathcal{F}$ onto itself for each $p \in (\mathcal{U}_\alpha \cap \mathcal{U}_\beta)$. The inverse images $\pi^{-1}(p)$ of points $p \in \mathcal{M}$ are therefore necessarily all diffeomorphic to $\mathcal{F}$ (and so to each other). For example, the Möbius strip is a fibre bundle over S^1 with fibre R^1; we need two open sets $\mathcal{U}_1$, $\mathcal{U}_2$

to give a covering by sets of the form $\mathcal{U}_i \times R^1$. This example shows that if a manifold is locally the direct product of two other manifolds, it is nevertheless not, in general, a product manifold; it is for this reason that the concept of a fibre bundle is so useful.

The *tangent bundle* $T(\mathcal{M})$ is the fibre bundle over a C^k manifold $\mathcal{M}$ obtained by giving the set $\mathscr{E} = \bigcup_{p \in \mathcal{M}} T_p$ its natural manifold structure and its natural projection into $\mathcal{M}$. Thus the projection π maps each point of T_p into p. The manifold structure in $\mathscr{E}$ is defined by local coordinates $\{z^A\}$ in the following way. Let $\{x^i\}$ be local coordinates in an open set $\mathcal{U}$ of $\mathcal{M}$. Then any vector $\mathbf{V} \in T_p$ (for any $p \in \mathcal{U}$) can be expressed as $\mathbf{V} = V^i \, \partial/\partial x^i|_p$. The coordinates $\{z^A\}$ are defined in $\pi^{-1}(\mathcal{U})$ by $\{z^A\} = \{x^i, V^a\}$. On choosing a covering of $\mathcal{M}$ by coordinate neighbourhoods $\mathcal{U}_\alpha$, the corresponding charts define a C^{k-1} atlas on $\mathscr{E}$ which turn it into a C^{k-1} manifold (of dimension n^2); to check this, one needs only note that in any overlap $(\mathcal{U}_\alpha \cap \mathcal{U}_\beta)$ the coordinates $\{x^i{}_\alpha\}$ of a point are C^k functions of the coordinates $\{x^i{}_\beta\}$ of the point, and the components $\{V^a{}_\alpha\}$ of a vector field are C^{k-1} functions of the components $\{V^a{}_\beta\}$ of the vector field. Thus in $\pi^{-1}(\mathcal{U}_\alpha \cap \mathcal{U}_\beta)$, the coordinates $\{z^A{}_\alpha\}$ are C^{k-1} functions of the coordinates $\{z^A{}_\beta\}$.

The fibre $\pi^{-1}(p)$ is T_p, and so is a vector space of dimension n. This vector space structure is preserved by the map $\phi_{\alpha,\,p} \colon T_p \to R^n$, which is given by $\phi_{\alpha,\,p}(u) = V^a(u)$, i.e. $\phi_{\alpha,\,p}$ maps a vector at p into its components with respect to the coordinates $\{x^a{}_\alpha\}$. If $\{x^a{}_\beta\}$ are another set of local coordinates then the map $(\phi_{\alpha,\,p}) \circ (\phi_{\beta,\,p}{}^{-1})$ is a linear map of R^n onto itself. Thus it is an element of the general linear group $GL(n, R)$ (the group of all non-singular $n \times n$ matrices).

The *bundle of tensors of type* (r, s) over $\mathcal{M}$, denoted by $T^r_s(\mathcal{M})$, is defined in a very similar way. One forms the set $\mathscr{E} = \bigcup_{p \in \mathcal{M}} T^r_s(p)$, defines the projection π as mapping each point in $T^r_s(p)$ into p, and, for any coordinate neighbourhood $\mathcal{U}$ in $\mathcal{M}$, assigns local coordinates $\{z^A\}$ to $\pi^{-1}(\mathcal{U})$ by $\{z^A\} = \{x^i, T^{a\ldots b}{}_{c\ldots d}\}$ where $\{x^i\}$ are the coordinates of the point p and $\{T^{a\ldots b}{}_{c\ldots d}\}$ are the coordinate components of $\mathbf{T}$ (that is, $\mathbf{T} = T^{a\ldots b}{}_{c\ldots d} \, \partial/\partial x^a \otimes \ldots \otimes dx^d|_p$). This turns $\mathscr{E}$ into a C^{k-1} manifold of dimension n^{r+s+1}; any point u in $T^r_s(\mathcal{M})$ corresponds to a unique tensor $\mathbf{T}$ of type (r, s) at $\pi(u)$.

The *bundle of linear frames* (or bases) $L(\mathcal{M})$ is a C^{k-1} fibre bundle defined as follows: the total space $\mathscr{E}$ consists of all bases at all points of $\mathcal{M}$, that is all sets of non-zero linearly independent n-tuples of vectors $\{\mathbf{E}_a\}$, $\mathbf{E}_a \in T_p$, for each $p \in \mathcal{M}$ (a runs from 1 to n). The projection

π is the natural one which maps a basis at a point p to the point p. If $\{x^i\}$ are local coordinates in an open set $\mathcal{U} \subset \mathcal{M}$, then

$$\{z^A\} = \{x^a, E_1{}^j, E_2{}^k, ..., E_n{}^m\}$$

are local coordinates in $\pi^{-1}(\mathcal{U})$, where $E_a{}^j$ is the jth components of the vector $\mathbf{E}_a$ with respect to the coordinate bases $\partial/\partial x^i$. The general linear group $GL(n, R)$ acts on $L(\mathcal{M})$ in the following way: if $\{\mathbf{E}_a\}$ is a basis at $p \in \mathcal{M}$, then $\mathbf{A} \in GL(n, R)$ maps $u = \{p, \mathbf{E}_a\}$ to

$$A(u) = \{p, A_{ab}\mathbf{E}_b\}.$$

When there is a metric $\mathbf{g}$ of signature s on $\mathcal{M}$, one can define a sub-bundle of $L(\mathcal{M})$, the *bundle of orthonormal frames* $O(\mathcal{M})$, which consists of orthonormal bases (with respect to $\mathbf{g}$) at all points of $\mathcal{M}$. $O(\mathcal{M})$ is acted on by the subgroup $O(\frac{1}{2}(n+s), \frac{1}{2}(n-s))$ of $GL(n, R)$. This consists of the non-singular real matrices A_{ab} such that

$$A_{ab}G_{bc}A_{dc} = G_{ad},$$

where G_{bc} is the matrix

$$\mathrm{diag}\,(\underbrace{+1, +1, ..., +1}_{\frac{1}{2}(n+s) \text{ terms}}, \underbrace{-1, -1, ..., -1}_{\frac{1}{2}(n-s) \text{ terms}}).$$

It maps $(p, \mathbf{E}_a) \in O(\mathcal{M})$ to $(p, A_{ab}\mathbf{E}_b) \in O(\mathcal{M})$. In the case of a Lorentz metric (i.e. $s = n-2$), the group $O(n-1, 1)$ is called the n-dimensional Lorentz group.

A C^r *cross-section* of a bundle is a C^r map $\Phi: \mathcal{M} \to \mathcal{E}$ such that $\pi \circ \Phi$ is the identity map on $\mathcal{M}$; thus a cross-section is a C^r assignment to each point p of $\mathcal{M}$ of an element $\Phi(p)$ of the fibre $\pi^{-1}(p)$. A cross-section of the tangent bundle $T(\mathcal{M})$ is a vector field on $\mathcal{M}$; a cross-section of $T^r_s(\mathcal{M})$ is a tensor field of type (r, s) on $\mathcal{M}$; a cross-section of $L(\mathcal{M})$ is a set of n non-zero vector fields $\{\mathbf{E}_a\}$ which are linearly independent at each point, and a cross-section of $O(\mathcal{M})$ is a set of orthonormal vector fields on $\mathcal{M}$.

Since the zero vectors and tensors define cross-sections in $T(\mathcal{M})$ and $T^r_s(\mathcal{M})$, these fibre bundles will always admit cross-sections. If $\mathcal{M}$ is orientable and non-compact, or is compact with vanishing Euler number, there will exist nowhere zero vector fields, and hence cross-sections of $T(\mathcal{M})$ which are nowhere zero. The bundles $L(\mathcal{M})$ and $O(\mathcal{M})$ may or may not admit cross-sections; for example $L(S^2)$ does not, but $L(R^n)$ does. If $L(\mathcal{M})$ admits a cross-section, $\mathcal{M}$ is said to be *parallelizable*. R. P. Geroch has shown (1968c) that a non-compact four-dimensional Lorentz manifold $\mathcal{M}$ admits a spinor structure if and only if it is parallelizable.

One can describe a connection on $\mathcal{M}$ in an elegant geometrical way in terms of the fibre bundle $L(\mathcal{M})$. A connection on $\mathcal{M}$ may be regarded as a rule for parallelly transporting vectors along any curve $\gamma(t)$ in $\mathcal{M}$. Thus if $\{\mathbf{E}_a\}$ is a basis at a point $p = \gamma(t_0)$, i.e. $\{p, \mathbf{E}_a\}$ is a point u in $L(\mathcal{M})$, one can obtain a unique basis at any other point $\gamma(t)$, i.e. a unique point $\overline{\gamma}(t)$ in the fibre $\pi^{-1}(\gamma(t))$, by parallelly transporting $\{\mathbf{E}_a\}$ along $\gamma(t)$. Therefore there is a unique curve $\overline{\gamma}(t)$ in $L(\mathcal{M})$, called the *lift* of $\gamma(t)$, such that:

(1) $\overline{\gamma}(t_0) = u$,

(2) $\pi(\overline{\gamma}(t)) = \gamma(t)$,

(3) the basis represented by the point $\overline{\gamma}(t)$ is parallelly transported along the curve $\gamma(t)$ in $\mathcal{M}$.

In terms of the local coordinates $\{z^A\}$, the curve $\overline{\gamma}(t)$ is given by $\{x^a(\gamma(t)), E_m{}^i(t)\}$, where

$$\frac{\mathrm{d}E_m{}^i(t)}{\mathrm{d}t} + E_m{}^j \, \Gamma^i{}_{aj} \frac{\mathrm{d}x^a(\gamma(t))}{\mathrm{d}t} = 0.$$

Consider the tangent space $T_u(L(\mathcal{M}))$ to the fibre bundle $L(\mathcal{M})$ at the point u. This has a coordinate basis $\{\partial/\partial z^A|_u\}$. The n-dimensional subspace spanned by the tangent vectors $\{(\partial/\partial t)_{\overline{\gamma}(t)}|_u\}$ to the lifts of all curves $\gamma(t)$ through p is called the *horizontal subspace* H_u of $T_u(L(\mathcal{M}))$. In terms of local coordinates,

$$\left(\frac{\partial}{\partial t}\right)_{\overline{\gamma}} = \frac{\mathrm{d}x^a(\gamma(t))}{\mathrm{d}t} \frac{\partial}{\partial x^a} + \frac{\mathrm{d}E_m{}^i}{\mathrm{d}t} \frac{\partial}{\partial E_m{}^i}$$

$$= \frac{\mathrm{d}x^a(\gamma(t))}{\mathrm{d}t} \left(\frac{\partial}{\partial x^a} - E_m{}^j \, \Gamma^i{}_{aj} \frac{\partial}{\partial E_m{}^i}\right),$$

so a coordinate basis of H_u is $\{\partial/\partial x^a - E_m{}^j \, \Gamma^i{}_{aj} \, \partial/\partial E_m{}^i\}$. Thus the connection in $\mathcal{M}$ determines the horizontal subspaces in the tangent spaces at each point of $L(\mathcal{M})$. Conversely, a connection in $\mathcal{M}$ may be defined by giving an n-dimensional subspace of $T_u(L(\mathcal{M}))$ for each $u \in L(\mathcal{M})$ with the properties:

(1) If $\mathbf{A} \in GL(n, R^1)$, then the map $A_*: T_u(L(\mathcal{M})) \to T_{A(u)}(L(\mathcal{M}))$ maps the horizontal subspace H_u into $H_{A(u)}$;

(2) H_u contains no non-zero vector belonging to the vertical subspace V_u.

Here, the vertical subspace V_u is defined as the n^2-dimensional subspace of $T_u(L(\mathcal{M}))$ spanned by the vectors tangent to curves in the fibre $\pi^{-1}(\pi(u))$; in terms of local coordinates, V_u is spanned by the vectors $\{\partial/\partial E_m{}^i\}$. Property (2) implies that T_u is the direct sum of H_u and V_u.

The projection map $\pi \colon L(\mathcal{M}) \to \mathcal{M}$ induces a surjective linear map $\pi_* \colon T_u(L(\mathcal{M})) \to T_{\pi(u)}(\mathcal{M})$, such that $\pi_*(V_u) = 0$ and π_* restricted to H_u is 1–1 onto $T_{\pi(u)}$. Thus the inverse $\pi_*{}^{-1}$ is a linear map of $T_{\pi(u)}(\mathcal{M})$ onto H_u. Therefore for any vector $\mathbf{X} \in T_p(\mathcal{M})$ and point $u \in \pi^{-1}(p)$, there is a unique vector $\overline{\mathbf{X}} \in H_u$, called the *horizontal lift* of $\mathbf{X}$, such that $\pi_*(\overline{\mathbf{X}}) = \mathbf{X}$. Given a curve $\gamma(t)$ in $\mathcal{M}$, and an initial point u in $\pi^{-1}(\gamma(t_0))$, one can construct a unique curve $\overline{\gamma}(t)$ in $L(\mathcal{M})$, where $\overline{\gamma}(t)$ is the curve through u whose tangent vector is the horizontal lift of the tangent vector of $\gamma(t)$ in $\mathcal{M}$. Thus knowing the horizontal subspaces at each point in $L(\mathcal{M})$, one can define parallel propagation of bases along any curve $\gamma(t)$ in $\mathcal{M}$. One can then define the covariant derivative along $\gamma(t)$ of any tensor field $\mathbf{T}$ by taking the ordinary derivatives with respect to t, of the components of $\mathbf{T}$ with respect to a parallelly propagated basis.

If there is a metric $\mathbf{g}$ on $\mathcal{M}$ whose covariant derivative is zero, then orthonormal frames are parallelly propagated into orthonormal frames. Thus the horizontal subspaces are tangent to $O(\mathcal{M})$ in $L(\mathcal{M})$, and define a connection in $O(\mathcal{M})$.

Similarly a connection on $\mathcal{M}$ defines n-dimensional horizontal subspaces in the tangent spaces to the bundles $T(\mathcal{M})$ and $T_s^r(\mathcal{M})$, by parallel propagation of vectors and tensors. These horizontal subspaces have coordinate bases

$$\left\{\frac{\partial}{\partial x^a} - V^e \Gamma^f{}_{ae} \frac{\partial}{\partial V^f}\right\}$$

and

$$\left\{\frac{\partial}{\partial x^e} - \left(T^{f\ldots b}{}_{c\ldots d}\,\Gamma^a{}_{ef} + \text{(all upper indices)} \right.\right.$$
$$\left.\left. - T^{a\ldots b}{}_{f\ldots d}\,\Gamma^f{}_{ec} - \text{(all lower indices)}\right)\frac{\partial}{\partial T^{a\ldots b}{}_{c\ldots d}}\right\}$$

respectively. As with $L(\mathcal{M})$, π_* maps these horizontal subspaces one–one onto $T_{\pi(u)}(\mathcal{M})$; thus again π_* can be inverted to give a unique horizontal lift $\overline{\mathbf{X}} \in T_u$ of any vector $\mathbf{X} \in T_{\pi(u)}$. In the particular case of $T(\mathcal{M})$, u itself corresponds to a unique vector $\mathbf{W} \in T_{\pi(u)}(\mathcal{M})$, and so there is an intrinsic horizontal vector field $\overline{\mathbf{W}}$ defined on $T(\mathcal{M})$ by the connection. In terms of local coordinates $\{x^a, V^b\}$,

$$\overline{\mathbf{W}} = V^a\left(\frac{\partial}{\partial x^a} - V^e \Gamma^f{}_{ae} \frac{\partial}{\partial V^f}\right).$$

This vector field may be interpreted as follows: the integral curve of $\bar{W}$ through $u = (p, X) \in T(\mathcal{M})$ is the horizontal lift of the geodesic in $\mathcal{M}$ with tangent vector X at p. Thus the vector field $\bar{W}$ represents all geodesics on $\mathcal{M}$. In particular, the family of all geodesics through $p \in \mathcal{M}$ is the family of integral curves of $\bar{W}$ through the fibre $\pi^{-1}(p) \subset T(\mathcal{M})$; the curves in $\mathcal{M}$ have self intersections at least at p, but the curves in $T(\mathcal{M})$ are non-intersecting everywhere.

3

General Relativity

In order to discuss the occurrence of singularities and the possible breakdown of General Relativity, it is important to have a precise statement of the theory and to indicate to what extent it is unique. We shall therefore present the theory as a number of postulates about a mathematical model for space–time.

In § 3.1 we introduce the mathematical model and in § 3.2 the first two postulates, local causality and local energy conservation. These postulates are common to both Special and General Relativity, and thus may be regarded as tested by the many experiments that have been performed to check the former. In § 3.3 we derive the equations of the matter fields and obtain the energy–momentum tensor from a Lagrangian.

The third postulate, the field equations, is given in § 3.4. This is not so well established experimentally as the first two postulates, but we shall see that any alternative equations would seem to have one or more undesirable properties, or else require the existence of extra fields which have not yet been detected experimentally.

3.1 The space–time manifold

The mathematical model we shall use for space–time, i.e. the collection of all events, is a pair $(\mathcal{M}, \mathbf{g})$ where $\mathcal{M}$ is a connected four-dimensional Hausdorff C^∞ manifold and $\mathbf{g}$ is a Lorentz metric (i.e. a metric of signature $+2$) on $\mathcal{M}$.

Two models $(\mathcal{M}, \mathbf{g})$ and $(\mathcal{M}', \mathbf{g}')$ will be taken to be equivalent if they are *isometric*, that is if there is a diffeomorphism $\theta: \mathcal{M} \to \mathcal{M}'$ which carries the metric $\mathbf{g}$ into the metric $\mathbf{g}'$, i.e. $\theta_* \mathbf{g} = \mathbf{g}'$. Strictly speaking then, the model for space–time is not just one pair $(\mathcal{M}, \mathbf{g})$ but a whole equivalence class of all pairs $(\mathcal{M}', \mathbf{g}')$ which are equivalent to $(\mathcal{M}, \mathbf{g})$. We shall normally work with just one representative member $(\mathcal{M}, \mathbf{g})$ of the equivalence class, but the fact that this pair is defined only up to equivalence is important in some situations, in particular in the discussion of the Cauchy problem in chapter 7.

The manifold $\mathcal{M}$ is taken to be connected since we would have no knowledge of any disconnected component. It is taken to be Hausdorff since this seems to accord with normal experience. However in chapter 5 we shall consider an example in which one might dispense with this condition. Together with the existence of a Lorentz metric, the Hausdorff condition implies that $\mathcal{M}$ is paracompact (Geroch (1968c)).

A manifold corresponds naturally to our intuitive ideas of the continuity of space and time. So far this continuity has been established for distances down to about 10^{-15} cm by experiments on pion scattering (Foley et al. (1967)). It may be difficult to extend this to much smaller lengths as to do so would require a particle of such high energy that several other particles might be created and confuse the experiment. Thus it may be that a manifold model for space–time is inappropriate for distances less than 10^{-15} cm and that we should use theories in which space–time has some other structure on this scale. However such breakdowns of the manifold picture would not be expected to affect General Relativity until the typical gravitational length scale became of that order. This would happen when the density became about 10^{58} gm cm^{-3}, which is a condition so extreme as to be completely beyond our present knowledge. Nevertheless, by adopting a manifold model for space–time, and making certain other reasonable assumptions, we shall show in chapters 8–10 that some breakdowns of General Relativity must occur. It may be the field equations that go wrong, or it may be that quantization of the metric is needed, or it may be a breakdown of the manifold structure itself that occurs.

The metric $\mathbf{g}$ enables the non-zero vectors at a point $p \in \mathcal{M}$ to be divided into three classes: a non-zero vector $\mathbf{X} \in T_p$ being said to be timelike, spacelike or null according to whether $g(\mathbf{X}, \mathbf{X})$ is negative, positive or zero respectively (cf. figure 5).

The order of differentiability, r, of the metric ought to be sufficient for the field equations to be defined. They can be defined in a distributional sense if the metric coordinate components g_{ab} and g^{ab} are continuous and have locally square integrable generalized first derivatives with respect to the local coordinates. (A set of functions $f_{;a}$ on R^n are said to be the generalized derivatives of a function f on R^n if, for any C^∞ function ψ on R^n with compact support,

$$\int f_{;a} \psi \, \mathrm{d}^n x = -\int f(\partial \psi / \partial x^a) \, \mathrm{d}^n x.)$$

However this condition is too weak, since it guarantees neither the existence nor the uniqueness of geodesics, for which a C^{2-} metric is required. (A C^{2-} metric is one for which the first coordinate derivatives of the metric coordinate components satisfy a local Lipschitz condition, see § 2.1.) We shall in fact assume for most of the book that the metric is at least C^2. This allows the field equations (which involve the second derivatives of the metric) to be defined at every point. In § 8.4 we shall weaken the condition on the metric to C^{2-} and show that this does not affect the results on the occurrence of singularities.

In chapter 7, we use a different kind of differentiability condition in order to show that the time development of the field equations is determined by suitable initial conditions. We require there that the metric components and their generalized first derivatives up to order $m(m \geqslant 4)$ are locally square integrable. This would certainly be true if the metric were C^4.

In fact, the order of differentiability of the metric is probably not physically significant. Since one can never measure the metric exactly, but only with some margin of error, one could never determine that there was an actual discontinuity in its derivatives of any order. Thus one can always represent one's measurements by a C^∞ metric.

If the metric is assumed to be C^r, the atlas of the manifold must be C^{r+1}. However, one can always find an analytic subatlas in any C^s atlas ($s \geqslant 1$) (Whitney (1936), cf. Munkres (1954)). Thus it is no restriction to assume from the start that the atlas is analytic, even though one could physically determine only a C^{r+1} atlas if the metric were C^r.

We have to impose some condition on our model $(\mathcal{M}, \mathfrak{g})$ to ensure that it includes all the non-singular points of space–time. We shall say that the C^r pair $(\mathcal{M}', \mathfrak{g}')$ is a C^r-*extension* of $(\mathcal{M}, \mathfrak{g})$ if there is an isometric C^r imbedding $\mu \colon \mathcal{M} \to \mathcal{M}'$. If there were such an extension $(\mathcal{M}', \mathfrak{g}')$ we should have to regard points of $\mathcal{M}'$ as also being points of space–time. We therefore require that the model $(\mathcal{M}, \mathfrak{g})$ is C^r-*inextendible*, that is there is no C^r extension $(\mathcal{M}', \mathfrak{g}')$ of $(\mathcal{M}, \mathfrak{g})$ where $\mu(\mathcal{M})$ does not equal $\mathcal{M}'$.

As an example of a pair $(\mathcal{M}_1, \mathfrak{g}_1)$ which is not inextendible, consider two-dimensional Euclidean space with the x-axis removed between $x_1 = -1$ and $x_1 = +1$. The obvious way to extend this would simply be to replace the missing points, but one could also extend it by taking another copy $(\mathcal{M}_2, \mathfrak{g}_2)$ of the space, and identifying the bottom side of the x_1-axis for $|x_1| < 1$ with the top side of the x_2-axis for $|x_2| < 1$, and also identifying the top side of the x_1-axis for $|x_1| < 1$ with the bottom side of the x_2-axis for $|x_2| < 1$. The resultant space $(\mathcal{M}_3, \mathbf{g}_3)$ is inextendible but not complete as we have left out the points $x_1 = \pm 1$, $y_1 = 0$. We cannot put these points back in because we were perverse enough to extend the top and bottom sides of the x-axis on different sheets. If however one takes the subset $\mathcal{U}$ of $\mathcal{M}_3$ defined by $1 < x_1 < 2$, $-1 < y_1 < 1$, then one could extend the pair $(\mathcal{U}, \mathbf{g}_3|_{\mathcal{U}})$ and put back the point $x_1 = 1$, $y_1 = 0$. This motivates a rather stronger definition of inextendibility: a pair $(\mathcal{M}, \mathbf{g})$ is said to be C^r-*locally inextendible* if there is no open set $\mathcal{U} \subset \mathcal{M}$ with non-compact closure in $\mathcal{M}$, such that the pair $(\mathcal{U}, \mathbf{g}|_{\mathcal{U}})$ has an extension $(\mathcal{U}', \mathbf{g}')$ in which the closure of the image of $\mathcal{U}$ is compact.

3.2 The matter fields

There will be various fields on $\mathcal{M}$, such as the electromagnetic field, the neutrino field, etc., which describe the matter content of space–time. These fields will obey equations which can be expressed as relations between tensors on $\mathcal{M}$ in which all derivatives with respect to position are covariant derivatives with respect to the symmetric connection defined by the metric $\mathbf{g}$. This is so because the only relations defined by a manifold structure are tensor relations, and the only connection defined so far is that given by the metric. If there were another connection on $\mathcal{M}$, the difference between the two connections would be a tensor and could be regarded as another physical field. Similarly another metric on $\mathcal{M}$ could be regarded as a further physical field. (The equations of the matter fields are sometimes expressed as relations between spinors on $\mathcal{M}$. We do not deal with such relations in this book, as they are not needed for the problems we wish to consider. In fact, all spinor equations can be replaced by rather more complicated tensor equations; see e.g. Ruse (1937).)

The theory one obtains depends on what matter fields one incorporates in it. One should of course include all such fields which have been experimentally observed, but one might postulate the existence of as yet undetected fields. Thus for example Brans and Dicke (Dicke (1964), appendix 7) postulate the existence of a long range scalar field which is weakly coupled to the trace of the energy–momentum tensor. In the form given in Dicke (1964) appendix 2, the Brans–Dicke theory can be regarded simply as General Relativity with an extra scalar field. Whether this scalar field has been experimentally detected or not is at present under dispute.

We shall denote the matter fields included in the theory by $\Psi_{(i)}{}^{a...b}{}_{c...d}$, where the subscript (i) numbers the fields considered. The following two postulates on the nature of the equations obeyed by the $\Psi_{(i)}{}^{a...b}{}_{c...d}$ are common to both the Special and the General Theories of Relativity.

Postulate (a): Local causality

The equations governing the matter fields must be such that if $\mathcal{U}$ is a convex normal neighbourhood and p and q are points in $\mathcal{U}$ then a signal can be sent in $\mathcal{U}$ between p and q if and only if p and q can be joined by a C^1 curve lying entirely in $\mathcal{U}$, whose tangent vector is everywhere non-zero and is either timelike or null; we shall call such a curve, *non-spacelike*. (Our formulation of relativity excludes the possibility of particles such as tachyons, which move on spacelike curves.) Whether the signal is sent from p to q or from q to p will depend on the direction of time in $\mathcal{U}$. The problem of whether a consistent direction of time can be assigned at all points of space–time will be considered in § 6.2.

A more precise statement of this postulate can be given in terms of the Cauchy problem of the matter fields. Let $p \in \mathcal{U}$ be such that every non-spacelike curve through p intersects the spacelike surface $x^4 = 0$ within $\mathcal{U}$. Let $\mathscr{F}$ be the set of points in the surface $x^4 = 0$ which can be reached by non-spacelike curves in $\mathcal{U}$ from p. Then we require that the values of the matter fields at p must be uniquely determined by the values of the fields and their derivatives up to some finite order on $\mathscr{F}$, and that they are not uniquely determined by the values on any proper subset of $\mathscr{F}$ to which it can be continuously retracted. (For a fuller discussion of the Cauchy problem, see chapter 7.)

It is this postulate which sets the metric $\mathbf{g}$ apart from the other fields on $\mathcal{M}$ and gives it its distinctive geometrical character. If $\{x^a\}$ are normal coordinates in $\mathcal{U}$ about p, it is intuitively fairly obvious (and is proved in chapter 4) that the points which can be reached from p by non-spacelike curves in $\mathcal{U}$ are those whose coordinates satisfy

$$(x^1)^2 + (x^2)^2 + (x^3)^2 - (x^4)^2 \leqslant 0.$$

The boundary of these points is formed by the image of the null cone of p under the exponential map, that is the set of all null geodesics through p. Thus by observing which points can communicate with p, one can determine the null cone N_p in T_p. Once N_p is known, the metric at p may be determined up to a conformal factor. This may be seen as follows: let $\mathbf{X}, \mathbf{Y} \in T_p$ be respectively timelike and spacelike vectors. The equation

$$g(\mathbf{X} + \lambda\mathbf{Y}, \mathbf{X} + \lambda\mathbf{Y}) = g(\mathbf{X}, \mathbf{X}) + 2\lambda g(\mathbf{X}, \mathbf{Y}) + \lambda^2 g(\mathbf{Y}, \mathbf{Y})$$

$$= 0$$

will have two real roots λ_1 and λ_2 as $g(\mathbf{X}, \mathbf{X}) < 0$ and $g(\mathbf{Y}, \mathbf{Y}) > 0$. If N_p is known, λ_1 and λ_2 may be determined. But

$$\lambda_1 \lambda_2 = g(\mathbf{X}, \mathbf{X})/g(\mathbf{Y}, \mathbf{Y}).$$

Thus the ratio of the magnitudes of a timelike vector and a spacelike vector may be found from the null cone. Then if $\mathbf{W}$ and $\mathbf{Z}$ are any two non-null vectors at p,

$$g(\mathbf{W}, \mathbf{Z}) = \tfrac{1}{2}(g(\mathbf{W}, \mathbf{W}) + g(\mathbf{Z}, \mathbf{Z}) - g(\mathbf{W} + \mathbf{Z}, \mathbf{W} + \mathbf{Z})).$$

Each of the magnitudes on the right-hand side may be compared with the magnitude of either $\mathbf{X}$ or $\mathbf{Y}$, and so $g(\mathbf{W}, \mathbf{Z})/g(\mathbf{X}, \mathbf{X})$ may be found. (If $\mathbf{W} + \mathbf{Z}$ is null, the corresponding expression involving $\mathbf{W} + 2\mathbf{Z}$ could be used.) Thus observation of local causality enables one to measure the metric up to a conformal factor. In practice this measurement is performed most conveniently using the experimental fact that no signal has been observed to travel faster than electromagnetic radiation. This means that light must travel on null geodesics. This however is a consequence of the particular equations the electromagnetic field obeys, not of the theory of relativity itself. Causality will be considered further in chapter 6. Among other results, it will be shown that causal relations may be used to determine the topological structure of $\mathcal{M}$. The conformal factor in the metric may be determined using postulate (b) below; thus all the elements of the theory will be physically observable.

Postulate (b): Local conservation of energy and momentum

The equations governing the matter fields are such that there exists a symmetric tensor T^{ab}, called the energy–momentum tensor, which depends on the fields, their covariant derivatives, and the metric, and which has the properties:

(i) T^{ab} vanishes on an open set $\mathcal{U}$ if and only if all the matter fields vanish on $\mathcal{U}$,

(ii) T^{ab} obeys the equation

$$T^{ab}{}_{;b} = 0. \tag{3.1}$$

Condition (i) expresses the principle that all fields have energy. One might possibly object to the 'only if' on the grounds that there might be two non-zero fields, one of whose energy–momentum tensor exactly cancelled that of the other. This possibility is related to that of the existence of negative energy which will be discussed in § 3.3.

If the metric admits a Killing vector field $\mathbf{K}$, equations (3.1) can be integrated to give a conservation law. To see this, define P^a to be the vector whose components are $P^a = T^{ab}K_b$. Then,

$$P^a{}_{;a} = T^{ab}{}_{;a}K_b + T^{ab}K_{b;a}.$$

The first term is zero by the conservation equations, and the second vanishes as T^{ab} is symmetric and $2K_{(a;b)} = L_{\mathbf{K}}g_{ab} = 0$, since $\mathbf{K}$ is a Killing vector. Thus if $\mathscr{D}$ is a compact orientable region with boundary $\partial\mathscr{D}$, Gauss' theorem (§ 2.7) shows

$$\int_{\partial\mathscr{D}} P^b\,\mathrm{d}\sigma_b = \int_{\mathscr{D}} P^b{}_{;b}\,\mathrm{d}v = 0. \tag{3.2}$$

This may be interpreted as saying that the total flux over a closed surface of the $\mathbf{K}$-component of energy–momentum is zero.

When the metric is flat, as it is in the Special Theory of Relativity, one may choose coordinates $\{x^a\}$ in which the components of the metric are $g_{ab} = e_a\delta_{ab}$ (no summation) where δ_{ab} is the Kronecker delta and e_a is -1 if $a = 4$ and is $+1$ if $a = 1, 2, 3$. Then the following are Killing vectors:
$$\mathop{\mathbf{L}}_{\alpha} = \partial/\partial x^\alpha \quad (\alpha = 1, 2, 3, 4)$$

(these generate four translations) and

$$\mathop{\mathbf{M}}_{\alpha\beta} = e_\alpha x^\alpha \frac{\partial}{\partial x^\beta} - e_\beta x^\beta \frac{\partial}{\partial x^\alpha} \quad \text{(no summation; } \alpha, \beta = 1, 2, 3, 4)$$

(these generate six 'rotations' in space–time). These isometries form the ten-parameter Lie group of isometries of flat space–time known as the inhomogeneous Lorentz group. One may use them to define ten vectors $\mathop{P^a}_{\alpha}$ and $\mathop{P^a}_{\alpha\beta}$ which will obey (3.2). We may think of $\mathbf{P}$ as representing the flow of energy and $\mathop{\mathbf{P}}_{1}, \mathop{\mathbf{P}}_{2}, \mathop{\mathbf{P}}_{3}$ as the flow of the three components of linear momentum. The $\mathop{\mathbf{P}}_{\alpha\beta}$ can be interpreted as the flow of angular momentum.

If the metric is not flat there will not, in general, be any Killing vectors and so the above integral conservation laws will not hold. However, in a suitable neighbourhood of a point q one may introduce normal coordinates $\{x^a\}$. Then at q the components g_{ab} of the metric are $e_a \delta_{ab}$ (no summation), and the components $\Gamma^a{}_{bc}$ of the connection are zero. One may take a neighbourhood $\mathscr{D}$ of q in which the g_{ab} and $\Gamma^a{}_{bc}$ differ from their values at q by an arbitrarily small amount; then the $L_{(a; b)}$ and $M_{(a; b)}$ will not exactly vanish in $\mathscr{D}$, but will in this neigh-_{α} $_{\alpha\beta}$ bourhood differ from zero by an arbitrarily small amount. Thus

$$\int_{\partial\mathscr{D}} \underset{\alpha}{P^b} \, d\sigma_b \quad \text{and} \quad \int_{\partial\mathscr{D}} \underset{\alpha\beta}{P^b} \, d\sigma_b$$

will still be zero in the first approximation; that is to say, one still has approximate conservation of energy, momentum and angular momentum in a small region of space–time. Using this it can be shown that a small isolated body moves approximately on a timelike geodesic curve independent of its internal constitution provided that the energy density of matter in it is non-negative (for an account of the motion of a small body in relativity, see Dixon (1970)). This may be thought of as Galileo's principle that all bodies fall equally fast. In Newtonian terms one would say that the inertial mass (the m in $\mathbf{F} = m\mathbf{a}$) and the passive gravitational mass (the mass acted on by a gravitational field) are equal for all bodies. This has been verified to a high order of accuracy in experiments by Eötvos and by Dicke (1964).

Postulate (a) enables one to measure the metric up to a conformal factor at each point. Using postulate (b) one may relate these factors at different points, for the conservation equations $T^{ab}{}_{;b} = 0$ would not in general hold for a connection derived from a metric $\hat{\mathbf{g}} = \Omega^2 \mathbf{g}$. One way of doing this would be to observe the paths of small 'test' particles and so to determine the timelike geodesic curves. Then if $\gamma(t)$ is such a curve with tangent vector $\mathbf{K} = (\partial/\partial t)_\gamma$, one has from (2.29)

$$\frac{\hat{D}}{\partial t} K^a = \frac{D}{\partial t} K^a + 2\Omega^{-1}\Omega_{;b} K^b K^a - \Omega^{-1}(K^b K^c \hat{g}_{bc}) \hat{g}^{ad}\Omega_{;d}.$$

Since $\gamma(t)$ is a geodesic with respect to the space–time metric $\mathbf{g}$, $K^{[b}(D/\partial t) K^{a]} = 0$. Thus

$$K^{[b}\frac{\hat{D}}{\partial t} K^{a]} = -(K^c K^d \hat{g}_{cd}) K^{[b}\hat{g}^{a]e} (\log \Omega)_{;e}. \tag{3.3}$$

Knowing the conformal structure, one can choose a metric $\hat{\mathbf{g}}$ which represents the conformal equivalence class of metrics and can evaluate the left-hand side of (3.3) for any test particle. Then the right-hand side of (3.3) determines $(\log \Omega)_{;b}$ up to the addition of a multiple of $K^a \hat{g}_{ab}$.

By considering another curve $\gamma'(t)$ whose tangent vector K'^a is not parallel to K^a, one can find $(\log \Omega)_{;b}$ and so can determine Ω everywhere up to a constant multiplying factor. This constant factor specifies one's units of measurement, and so can be chosen arbitrarily.

This is, of course, not the way one measures the conformal factor in practice; one makes use of the fact that there exist a large number of similar systems (such as the electronic states of atoms) whose internal motions define a number of events along the timelike curve which represents their position in space–time. The intervals between these events seem to be independent of their past history in the sense that the intervals measured by two nearby systems correspond. If one can effectively isolate them against external matter fields (so they must move on geodesic curves) and if one assumes their internal motion is independent of the curvature of space–time, then the only thing it can depend on is the metric. Thus the arc-length between two successive events on a curve must be the same for each pair of successive events on any such curve. If one takes this arc-length as one's unit of measurement, one can determine the conformal factor at any point of space–time.

In fact it may not be possible to isolate a system from external matter fields. Thus for example in the Brans–Dicke theory there is a scalar field which is non-zero everywhere. However the conformal factor can still be determined by the requirement that the conservation equation $T^{ab}{}_{;b} = 0$ should hold. Thus knowledge of the energy–momentum tensor T_{ab} determines the conformal factor.

3.3 Lagrangian formulation

The conditions (i) and (ii) of postulate (b) do not tell one how to construct the energy–momentum tensor for a given set of fields, or whether it is unique. In practice one relies heavily on one's intuitive knowledge of what energy and momentum are. However, there is a definite and unique formula for the energy–momentum tensor in the case that the equations of the fields can be derived from a Lagrangian.

Let L be the Lagrangian which is some scalar function of the fields $\Psi_{(i)}{}^{a\ldots b}{}_{c\ldots d}$, their first covariant derivatives, and the metric. One obtains the equations of the fields by requiring that the action

$$I = \int_{\mathscr{D}} L \, \mathrm{d}v$$

be stationary under variations of the fields in the interior of a compact four-dimensional region $\mathcal{D}$. By a *variation of the fields* $\Psi_{(i)}{}^{a\ldots b}{}_{c\ldots d}$ in $\mathcal{D}$ we mean a one-parameter family of fields $\Psi_{(i)}(u, r)$ where $u \in (-\epsilon, \epsilon)$ and $r \in \mathcal{M}$, such that

   (i) $\Psi_{(i)}(0, r) = \Psi_{(i)}(r)$,

   (ii) $\Psi_{(i)}(u, r) = \Psi_{(i)}(r)$ when $r \in \mathcal{M} - \mathcal{D}$.

We denote $\partial\Psi_{(i)}(u, r)/\partial u \big|_{u=0}$ by $\Delta\Psi_{(i)}$.

Then

$$\frac{\partial I}{\partial u}\bigg|_{u=0} = \sum_{(i)} \int_{\mathcal{D}} \left(\frac{\partial L}{\partial \Psi_{(i)}{}^{a\ldots b}{}_{c\ldots d}} \Delta\Psi_{(i)}{}^{a\ldots b}{}_{c\ldots d} \right.$$
$$\left. + \frac{\partial L}{\partial \Psi_{(i)}{}^{a\ldots b}{}_{c\ldots d;\, e}} \Delta(\Psi_{(i)}{}^{a\ldots b}{}_{c\ldots d;\, e}) \right) dv,$$

where $\Psi_{(i)}{}^{a\ldots b}{}_{c\ldots d;\, e}$ are the components of the covariant derivatives of $\Psi_{(i)}$. But $\Delta(\Psi_{(i)}{}^{a\ldots b}{}_{c\ldots d;\, e}) = (\Delta\Psi_{(i)}{}^{a\ldots b}{}_{c\ldots d})_{;\, e}$, thus the second term can be expressed as

$$\sum_{(i)} \int_{\mathcal{D}} \left[\left(\frac{\partial L}{\partial \Psi_{(i)}{}^{a\ldots b}{}_{c\ldots d;\, e}} \Delta\Psi_{(i)}{}^{a\ldots b}{}_{c\ldots d} \right)_{;\, e} \right.$$
$$\left. - \left(\frac{\partial L}{\partial \Psi_{(i)}{}^{a\ldots b}{}_{c\ldots d;\, e}} \right)_{;\, e} \Delta\Psi_{(i)}{}^{a\ldots b}{}_{c\ldots d} \right] dv.$$

The first term in this expression can be written as

$$\int_{\mathcal{D}} Q^a{}_{;\, a}\, dv = \int_{\partial\mathcal{D}} Q^a\, d\sigma_a,$$

where $\mathbf{Q}$ is a vector whose components are

$$Q^e = \sum_{(i)} \frac{\partial L}{\partial \Psi_{(i)}{}^{a\ldots b}{}_{c\ldots d;\, e}} \Delta\Psi_{(i)}{}^{a\ldots b}{}_{c\ldots d}.$$

This integral is zero as condition (ii) is the statement that $\Delta\Psi_{(i)}$ vanish at the boundary $\partial\mathcal{D}$. Thus in order that $\partial I/\partial u\big|_{u=0}$ should vanish for all variations on all volumes $\mathcal{D}$, it is necessary and sufficient that the *Euler–Lagrange equations*,

$$\frac{\partial L}{\partial \Psi_{(i)}{}^{a\ldots b}{}_{c\ldots d}} - \left(\frac{\partial L}{\partial \Psi_{(i)}{}^{a\ldots b}{}_{c\ldots d;\, e}} \right)_{;\, e} = 0, \tag{3.4}$$

hold for all i. These are the equations of the fields.

We obtain the energy–momentum tensor from the Lagrangian by considering the change in the action induced by a change in the metric.

Suppose a variation $g_{ab}(u, r)$ leaves the fields $\Psi_{(i)}{}^{a\cdots b}{}_{c\ldots d}$ unchanged but alters the components g_{ab} of the metric. Then

$$\left.\frac{\partial I}{\partial u}\right|_{u=0} = \int_{\mathscr{D}} \left(\sum_{(i)} \frac{\partial L}{\partial \Psi_{(i)}{}^{a\cdots b}{}_{c\ldots d;\, e}} \Delta(\Psi_{(i)}{}^{a\cdots b}{}_{c\ldots d;\, e}) + \frac{\partial L}{\partial g_{ab}} \Delta g_{ab} \right) dv$$

$$+ \int_{\mathscr{D}} L \frac{\partial(dv)}{\partial g_{ab}} \Delta g_{ab}. \quad (3.5)$$

The last term arises because the volume measure dv depends on the metric, and so will vary when the metric is varied. To evaluate this term, recall that dv is in fact the four-form $(4!)^{-1}\eta$ whose components are $\eta_{abcd} = (-g)^{\frac{1}{2}} 4! \, \delta_{[a}{}^1 \delta_b{}^2 \delta_c{}^3 \delta_{d]}{}^4$, where $g \equiv \det(g_{ab})$. Therefore

$$\frac{\partial \eta_{abcd}}{\partial g_{ef}} = -\tfrac{1}{2}(-g)^{-\frac{1}{2}} \frac{\partial g}{\partial g_{ef}} 4! \, \delta_{[a}{}^1 \delta_b{}^2 \delta_c{}^3 \delta_{d]}{}^4$$

$$= -\tfrac{1}{2}(-g)^{-\frac{1}{2}} g^{ef} g \, 4! \, \delta_{[a}{}^1 \delta_b{}^2 \delta_c{}^3 \delta_{d]}{}^4$$

$$= \tfrac{1}{2} g^{ef} \eta_{abcd}.$$

Thus

$$\frac{\partial(dv)}{\partial g_{ab}} = \tfrac{1}{2} g^{ab} \, dv.$$

The first term in (3.5) arises because $\Delta(\Psi_{(i)}{}^{a\cdots b}{}_{c\ldots d;\, e})$ will not necessarily be zero even though $\Delta\Psi_{(i)}{}^{a\cdots b}{}_{c\ldots d}$ is, since the variation in the metric will induce a variation in the components $\Gamma^a{}_{bc}$ of the connection. As the difference between two connections transforms like a tensor, $\Delta\Gamma^a{}_{bc}$ may be regarded as the components of a tensor. They are related to the variation in the components of the metric by

$$\Delta\Gamma^a{}_{bc} = \tfrac{1}{2} g^{ad}\{(\Delta g_{db})_{;\, c} + (\Delta g_{dc})_{;\, b} - (\Delta g_{bc})_{;\, d}\}.$$

(The easiest way to derive this formula is to note that since it is a tensor relation, it must be valid in any coordinate system. In particular, one could choose normal coordinates about a point p. For these coordinates the components $\Gamma^a{}_{bc}$ and the coordinate derivatives of the components g_{ab} vanish at p. The formula given can then be verified to hold at p.) Using this relation, $\Delta\Psi_{(i)}{}^{a\cdots b}{}_{c\ldots d;\, e}$ may be expressed in terms of $(\Delta g_{bc})_{;\, d}$ and the usual integration by parts employed to give an integrand involving Δg_{ab} only. Thus we may write $\partial I/\partial u$ as

$$\frac{1}{2} \int_{\mathscr{D}} (T^{ab} \Delta g_{ab}) \, dv,$$

where T^{ab} are the components of a symmetric tensor which is taken to be the energy–momentum tensor of the fields. (See Rosenfeld (1940) for the relation between this tensor and the so-called canonical energy–momentum tensor.)

This energy–momentum tensor satisfies the conservation equations as a consequence of the field equations obeyed by the $\Psi_{(i)}{}^{a\cdots b}{}_{c\cdots d}$. For suppose one has a diffeomorphism $\phi\colon \mathcal{M} \to \mathcal{M}$ which is the identity everywhere except in the interior of $\mathcal{D}$. Then, by the invariance of integrals under a differential map,

$$I = \int_{\mathcal{D}} L\,dv = \frac{1}{4!}\int_{\mathcal{D}} L\eta = \frac{1}{4!}\int_{\phi(\mathcal{D})} L\eta = \frac{1}{4!}\int_{\mathcal{D}} \phi^*(L\eta).$$

Thus
$$\frac{1}{4!}\int_{\mathcal{D}} (L\eta - \phi^*(L\eta)) = 0.$$

If the diffeomorphism ϕ is generated by a vector field $\mathbf{X}$ (non-zero only in the interior of $\mathcal{D}$) it follows that

$$\frac{1}{4!}\int_{\mathcal{D}} L_{\mathbf{X}}(L\eta) = 0.$$

But
$$\frac{1}{4!}\int_{\mathcal{D}} L_{\mathbf{X}}(L\eta) = \sum_{(i)}\int_{\mathcal{D}} \left(\frac{\partial L}{\partial \Psi_{(i)}{}^{a\cdots b}{}_{c\cdots d}} - \left(\frac{\partial L}{\partial \Psi_{(i)}{}^{a\cdots b}{}_{c\cdots d;e}}\right)_{;e}\right)$$
$$\times L_{\mathbf{X}}\Psi_{(i)}{}^{a\cdots b}{}_{c\cdots d}\,dv + \frac{1}{2}\int_{\mathcal{D}} T^{ab} L_{\mathbf{X}} g_{ab}\,dv.$$

The first term vanishes as a consequence of the field equations. In the second term, $L_{\mathbf{X}} g_{ab} = 2X_{(a;b)}$. Thus

$$\int_{\mathcal{D}} (T^{ab} L_{\mathbf{X}} g_{ab})\,dv = 2\int_{\mathcal{D}} ((T^{ab} X_a)_{;b} - T^{ab}{}_{;b} X_a)\,dv.$$

The first contribution may be transformed into an integral over the boundary of $\mathcal{D}$ which vanishes as $\mathbf{X}$ is zero there. Since the second term must therefore be zero for arbitrary $\mathbf{X}$, it follows that $T^{ab}{}_{;b} = 0$.

We shall now give as examples Lagrangians for some fields which will be of interest later.

Example 1: *A scalar field* ψ

This can represent, for example, the π^0-meson. The Lagrangian is

$$L = -\tfrac{1}{2}\psi_{;a}\psi_{;b}g^{ab} - \frac{1}{2}\frac{m^2}{\hbar^2}\psi^2$$

where m, $\hbar$ are constants. The Euler–Lagrange equations (3.4) are

$$\psi_{;ab}g^{ab} - \frac{m^2}{\hbar^2}\psi = 0.$$

The energy–momentum tensor is

$$T_{ab} = \psi_{;a}\psi_{;b} - \tfrac{1}{2}g_{ab}\left(\psi_{;c}\psi_{;d}g^{cd} + \frac{m^2}{\hbar^2}\psi^2\right). \qquad (3.6)$$

Example 2: The electromagnetic field

This is described by a one-form **A**, called the potential, which is defined up to the addition of a gradient of a scalar function. The Lagrangian is

$$L = -\frac{1}{16\pi}\,F_{ab}F_{cd}g^{ac}g^{bd},$$

where the electromagnetic field tensor F is defined as $2\,d\mathbf{A}$, i.e. $F_{ab} = 2A_{[b;\,a]}$. Varying A_a, the Euler–Lagrange equations (3.4) are

$$F_{ab;c}g^{bc} = 0.$$

This and $F_{[ab;c]} = 0$ (which is the equation $d\mathbf{F} = d(d\mathbf{A}) = 0$) are the Maxwell equations for the source-free electromagnetic field. The energy–momentum tensor is

$$T_{ab} = \frac{1}{4\pi}(F_{ac}F_{bd}g^{cd} - \tfrac{1}{4}g_{ab}F_{ij}F_{kl}g^{ik}g^{jl}). \qquad (3.7)$$

Example 3: A charged scalar field

This is really a combination of two real scalar fields ψ_1 and ψ_2. These are combined into a complex scalar field $\psi = \psi_1 + i\psi_2$, which could represent, for example, π^+ and π^- mesons. The total Lagrangian of the scalar field and electromagnetic field is

$$L = -\tfrac{1}{2}(\psi_{;a} + ieA_a\psi)g^{ab}(\overline{\psi}_{;b} - ieA_b\overline{\psi}) - \frac{1}{2}\frac{m^2}{\hbar^2}\psi\overline{\psi} - \frac{1}{16\pi}F_{ab}F_{cd}g^{ac}g^{bd},$$

where e is a constant and $\overline{\psi}$ is the complex conjugate of ψ. Varying ψ, $\overline{\psi}$ and A_a independently, one obtains

$$\psi_{;ab}g^{ab} - \frac{m^2}{\hbar^2}\psi + ieA_a g^{ab}(2\psi_{;b} + ieA_b\psi) + ieA_{a;b}g^{ab}\psi = 0,$$

and its complex conjugate, and

$$\frac{1}{4\pi}F_{ab;c}g^{bc} - ie\psi(\overline{\psi}_{;a} - ieA_a\overline{\psi}) + ie\overline{\psi}(\psi_{;a} + ieA_a\psi) = 0.$$

The energy–momentum tensor is

$$T_{ab} = \tfrac{1}{2}(\psi_{;a}\overline{\psi}_{;b} + \overline{\psi}_{;a}\psi_{;b}) + \tfrac{1}{2}(-\psi_{;a}ieA_b\overline{\psi} + \overline{\psi}_{;b}ieA_a\psi$$
$$+ \overline{\psi}_{;a}ieA_b\psi - \psi_{;b}ieA_a\overline{\psi}) + \frac{1}{4\pi}F_{ac}F_{bd}g^{cd} + e^2A_aA_b\psi\overline{\psi} + Lg_{ab}.$$

Example 4: *An isentropic perfect fluid*

The technique here is rather different. The fluid is described by a function ρ, called the density, and a congruence of timelike curves, called the flow lines. By a congruence of curves, is meant a family of curves, one through each point of $\mathcal{M}$. If $\mathscr{D}$ is a sufficiently small compact region, one can represent a congruence by a diffeomorphism $\gamma: [a, b] \times \mathcal{N} \to \mathscr{D}$ where $[a, b]$ is some closed interval of R^1 and $\mathcal{N}$ is some three-dimensional manifold with boundary. The curves are said to be timelike if their tangent vector $\mathbf{W} = (\partial/\partial t)_\gamma$, $t \in [a, b]$, is timelike everywhere. The tangent vector $\mathbf{V}$ is defined by $\mathbf{V} = (-g(\mathbf{W}, \mathbf{W}))^{-\frac{1}{2}}\mathbf{W}$, so $g(\mathbf{V}, \mathbf{V}) = -1$, and the fluid current vector is defined by $\mathbf{j} = \rho\mathbf{V}$. It is required that this is conserved, i.e. $j^a{}_{;a} = 0$. The behaviour of the fluid is determined by prescribing the elastic potential (or internal energy) ϵ as a function of ρ. The Lagrangian is taken to be

$$L = -\rho(1 + \epsilon)$$

and the action I is required to be stationary when the flow lines are varied and ρ is adjusted to keep j^a conserved. A variation of the flow lines is a differentiable map $\gamma: (-\delta, \delta) \times [a, b] \times \mathcal{N} \to \mathscr{D}$ such that

$$\gamma(0, [a, b], \mathcal{N}) = \gamma([a, b], \mathcal{N})$$

and $\quad \gamma(u, [a, b], \mathcal{N}) = \gamma([a, b], \mathcal{N}) \quad$ on $\quad \mathcal{M} - \mathscr{D}, \quad (u \in (-\delta, \delta))$.

Then it follows that $\Delta\mathbf{W} = L_\mathbf{K}\mathbf{W}$ where the vector $\mathbf{K}$ is $\mathbf{K} = (\partial/\partial u)_\gamma$. This vector may be thought of as representing the displacement, under the variation, of a point of the flow line. It follows that

$$\Delta V^a = V^a{}_{;b}K^b - K^a{}_{;b}V^b - V^aV^bK_{b;c}V^c.$$

Using the fact that $\Delta(j^a{}_{;a}) = 0 = (\Delta j^a)_{;a}$, one has

$$(\Delta\rho)_{;a}V^a + \Delta\rho V^a{}_{;a} + \rho_{;a}\Delta V^a + \rho(\Delta V^a)_{;a} = 0.$$

Substituting for ΔV^a and integrating along the flow lines, one finds

$$\Delta\rho = (\rho K^b)_{;b} + \rho K_{b;c}V^bV^c.$$

Therefore the variation of the action integral is

$$\frac{\partial I}{\partial u}\bigg|_{u=0} = -\int_\mathscr{D}\left\{((\rho K^b)_{;b} + \rho K_{b;c}V^bV^c)\left(1 + \frac{\mathrm{d}(\rho\epsilon)}{\mathrm{d}\rho}\right)\right\}\mathrm{d}v.$$

Integrating by parts,

$$\frac{\partial I}{\partial u}\bigg|_{u=0} = \int_\mathscr{D}\left\{\left(\rho\left(1 + \frac{\mathrm{d}(\epsilon\rho)}{\mathrm{d}\rho}\right)\dot{V}^a + \rho\left(\frac{\mathrm{d}(\epsilon\rho)}{\mathrm{d}\rho}\right)_{;c}(g^{ca} + V^cV^a)\right)K_a\right\}\mathrm{d}v,$$

where $\dot{V}^a \equiv V^a_{\ ;b} V^b$. If this is zero for all **K**, it follows that

$$(\mu + p)\, \dot{V}^a = -p_{;b}(g^{ba} + V^b V^a),$$

where $\mu = \rho(1 + \epsilon)$ is the energy density and $p = \rho^2(\mathrm{d}\epsilon/\mathrm{d}\rho)$ is the pressure. Thus $\dot{V}^a$, the acceleration of the flow lines, is given by the pressure gradient orthogonal to the flow lines.

To obtain the energy–momentum tensor one varies the metric. The calculations may be simplified by noting that the conservation of the current may be expressed as

$$(j^a)_{;a} = \frac{1}{(\sqrt{-g})}\frac{\partial}{\partial x^a}((\sqrt{-g})j^a) = 0.$$

Given the flow lines, the conservation equations determine j^a uniquely at each point on a flow line in terms of its initial value at some given point on the same flow line. Therefore $(\sqrt{-g})j^a$ is unchanged when the metric is varied. But

$$\rho^2 = g^{-1}((\sqrt{-g})j^a\,(\sqrt{-g})j^b)\,g_{ab},$$

so
$$2\rho\Delta\rho = (j^a j^b - j^c j_c g^{ab})\,\Delta g_{ab},$$

and thus
$$T^{ab} = \left\{\rho(1+\epsilon) + \rho^2\frac{\mathrm{d}\epsilon}{\mathrm{d}\rho}\right\} V^a V^b + \rho^2\frac{\mathrm{d}\epsilon}{\mathrm{d}\rho}g^{ab}$$

$$= (\mu + p)\, V^a V^b + p g^{ab}. \tag{3.8}$$

We shall call any matter whose energy–momentum tensor is of the above form (whether or not it is derived from a Lagrangian) a *perfect fluid*. From the energy and momentum conservation equations (3.1) applied to (3.8) one finds

$$\mu_{;a} V^a + (\mu + p)\, V^a_{\ ;a} = 0, \tag{3.9}$$

$$(\mu + p)\, \dot{V}^a + (g^{ab} + V^a V^b)p_{;b} = 0. \tag{3.10}$$

These are the same as the equations derived from the Lagrangian. We shall call a perfect fluid *isentropic* if the pressure p is a function of the energy density μ only. In this case one can introduce a conserved density ρ and an internal energy ϵ and derive the equations and the energy–momentum tensor from a Lagrangian.

One may also give the fluid a conserved electric charge e (i.e. $J^a_{\ ;a} = 0$ where **J** = e **V** is the electric current). The Lagrangian for the charged fluid and the electromagnetic field is

$$L = -\frac{1}{16\pi}F_{ab}F_{cd}g^{ac}g^{bd} - \rho(1+\epsilon) - \tfrac{1}{2}J^a A_a.$$

The last term gives the interaction between the fluid and the field.
Then varying $\mathbf{A}$, the flow lines and the metric respectively, one finds

$$F^{ab}{}_{;b} = 4\pi J^a,$$

$$(\mu + p)\, \dot{V}^a = -p_{;b}(g^{ab} + V^a V^b) + F^a{}_b J^b,$$

$$T^{ab} = (\mu + p)\, V^a V^b + p g^{ab} + \frac{1}{4\pi}(F^a{}_c F^{bc} - \tfrac{1}{4} g^{ab} F_{cd} F^{cd}).$$

3.4 The field equations

So far, the metric $\mathbf{g}$ has not been specified. In the Special Theory of
Relativity, which does not include gravitational effects, it is taken to
be flat. One might think that one could include gravitation by keeping
the metric flat and by introducing an extra field on space–time. However, experiments have shown that light rays travelling near the sun
are deflected. Since light rays are null geodesics, this shows that the
space–time metric cannot be flat or even conformal to a flat metric.
One therefore has to give some prescription for the curvature of
space–time. It turns out that this prescription can be chosen so as to
reproduce the results of Newtonian gravitation theory in the limit of
small slowly varying curvature. It is therefore not necessary to introduce an extra field to describe gravitation. This is not to say that there
could not be an additional field that produced part of the gravitational
effects. Such a scalar field has been suggested by Jordan (1955), and
Brans and Dicke (see Dicke (1964)). However, as mentioned before,
such an additional field could be regarded as simply another matter
field and included in the total energy–momentum tensor. We therefore
adopt the view that the gravitational field is represented by the
space–time metric itself. The problem then becomes one of finding
field equations to relate the metric to the distribution of matter.

These equations should be tensor equations involving the matter
only through its energy–momentum tensor, i.e. should not distinguish
between two different matter fields which have the same distribution
of energy and momentum. This can be regarded as a generalization of
the Newtonian principle that the active gravitational mass of a body
(the mass producing a gravitational field) is equal to the passive gravitational mass (the mass acted on by the gravitational field). This has
been verified experimentally by Kreuzer (1968).

To determine what the field equations should be, we shall consider
the Newtonian limit. Since the Newtonian gravitational field equation
does not involve time, the correspondence with Newtonian theory

should be made in a metric which is static. By a static metric is meant a metric which admits a timelike Killing vector field **K** which is orthogonal to a family of spacelike surfaces. These surfaces may be regarded as surfaces of constant time and may be labelled by the parameter t. We define the unit timelike vector **V** as $f^{-1}\mathbf{K}$, where $f^2 = -K^a K_a$. Then $V^a{}_{;b} = -\dot{V}^a V_b$, where $\dot{V}^a = V^a{}_{;b} V^b = f^{-1}f_{;b}g^{ab}$ represents the departure from geodesity of the integral curves of **V** (which are of course also integral curves of **K**). Note that $\dot{V}^a V_a = 0$.

These integral curves define the static frame of reference, that is to say, the space–time metric seems to be independent of time to a particle whose history is one of these curves. A particle released from rest and following a geodesic would appear to have an initial acceleration of $-\dot{\mathbf{V}}$ with respect to the static frame. If f differs only slightly from unity the initial acceleration of a freely moving particle released from rest is approximately minus the gradient of f. This suggests that one should regard $f-1$ as the quantity analogous to the Newtonian gravitational potential.

One can derive an equation for this potential by considering the divergence of $\dot{V}^a$:

$$\dot{V}^a{}_{;a} = (V^a{}_{;b} V^b)_{;a} = V^a{}_{;b;a} V^b + V^a{}_{;b} V^b{}_{;a}$$
$$= R_{ab} V^a V^b + (V^a{}_{;a})_{;b} V^b + (V_b \dot{V}^b)^2 = R_{ab} V^a V^b.$$

But $\quad\quad \dot{V}^a{}_{;a} = (f^{-1}f_{;b}g^{ab})_{;a} = -f^{-2}f_{;a}f_{;b}g^{ab} + f^{-1}f_{;ba}g^{ab}$

and $\quad\quad f_{;ab} V^a V^b = -f_{;a} V^a{}_{;b} V^b = -f^{-1}f_{;a}f_{;b}g^{ab},$

so one finds $\quad\quad f_{;ab}(g^{ab} + V^a V^b) = fR_{ab} V^a V^b.$

The term on the left is the Laplacian of f with respect to the induced metric in the three-surface $\{t = \text{constant}\}$. If the metric is almost flat, this will correspond to the Newtonian Laplacian of the potential. One would therefore obtain agreement with Newtonian theory in the limit of a weak field (i.e. when $f \simeq 1$) if the term on the right is equal to $4\pi G$ times the matter density plus terms which are small in the weak field limit.

This will be the case if there is a relation of the form

$$R_{ab} = K_{ab}, \tag{3.11}$$

where K_{ab} is a tensorial function of the energy–momentum tensor and the metric, which is such that $(4\pi G)^{-1}K_{ab} V^a V^b$ is equal to the matter density plus terms which are small in the Newtonian limit. We shall for the moment assume a relation of this form.

Since R_{ab} satisfies the contracted Bianchi identities $R_a{}^b{}_{;b} = \frac{1}{2}R_{;a}$, (3.11) implies

$$K_a{}^b{}_{;b} = \frac{1}{2}K_{;b}. \tag{3.12}$$

This shows that the apparently natural equation $K_{ab} = 4\pi G T_{ab}$ cannot be correct, since (3.12) and the conservation equations $T_a{}^b{}_{;b} = 0$ would imply $T_{;a} = 0$. For a perfect fluid, for example, this would mean that $\mu - 3p$ was constant throughout space–time, which is clearly not satisfied by a general fluid.

In fact in general, the only first order identities satisfied by the energy–momentum tensor are the conservation equations. From this it follows that the only tensorial function K_{ab} of the energy-momentum tensor and the metric which obeys the identities (3.12) for all energy–momentum tensors, is

$$K_{ab} = \kappa(T_{ab} - \tfrac{1}{2}Tg_{ab}) + \Lambda g_{ab}, \tag{3.13}$$

where κ and Λ are constants. The values of these constants can be determined from the Newtonian limit. Consider a perfect fluid with energy density μ and pressure p whose flow lines are the integral curves of the Killing vector (i.e. the fluid is at rest in the static frame). The energy–momentum tensor is given by (3.8). Putting this in (3.13) and (3.11), one finds

$$f_{;ab}(g^{ab} + V^a V^b) = f(\tfrac{1}{2}\kappa(\mu + 3p) - \Lambda). \tag{3.14}$$

In the Newtonian limit the pressure p is normally very small compared to the energy density μ. (We are using units in which the speed of light is unity. In units in which the speed of light is c, the expression $\mu + 3p$ should be replaced by $\mu + 3p/c^2$.) One would therefore obtain approximate agreement with Newtonian theory if $\kappa = 8\pi G$ and if $|\Lambda|$ is very small. We shall use units of mass in which $G = 1$. In these units, a mass of 10^{28} gm corresponds to a length of 1 cm. Sandage's (1961, 1968) observations of distant galaxies place limits on $|\Lambda|$ of the order of 10^{-56} cm^{-2}; we shall normally take Λ to be zero, but shall bear in mind the possibility of other values.

One may then integrate (3.14) over a compact region $\mathscr{F}$ of the three-surface $\{t = \text{constant}\}$ and transform the left-hand side into an integral of the gradient of f over the bounding two-surface $\partial\mathscr{F}$:

$$\int_{\mathscr{F}} f(4\pi(\mu + 3p)) \, d\sigma = \int_{\mathscr{F}} f_{;ab}(g^{ab} + V^a V^b) \, d\sigma$$

$$= \int_{\partial\mathscr{F}} f_{;a}(g^{ab} + V^a V^b) \, d\tau_b,$$

where $d\sigma$ is the volume element of the three-surface $\{t = \text{constant}\}$ in the induced metric, and $d\tau_b$ is the surface element of the two-surface $\partial\mathcal{F}$ in the three-surface. This gives the analogue of the Newtonian formula for the total mass contained within a two-surface. There are however two important differences from the Newtonian case:

(i) a factor f appears in the integral on the right-hand side. This means that matter placed in a region where f is considerably less than one (a large negative Newtonian potential) makes a smaller contribution to the total mass than does the same matter in a region where f is almost one (small negative Newtonian potential);

(ii) the pressure contributes to the total mass. This means that in some circumstances it can actually assist rather than prevent gravitational collapse.

The equations
$$R_{ab} = 8\pi(T_{ab} - \tfrac{1}{2}Tg_{ab}) + \Lambda g_{ab}$$
are called the *Einstein equations* and are often written in the equivalent form
$$(R_{ab} - \tfrac{1}{2}Rg_{ab}) + \Lambda g_{ab} = 8\pi T_{ab}. \tag{3.15}$$
Since both sides are symmetric, these form a set of ten coupled non-linear partial differential equations in the metric and its first and second derivatives. However the covariant divergence of each side vanishes identically, that is,
$$(R^{ab} - \tfrac{1}{2}Rg^{ab} + \Lambda g^{ab})_{;b} = 0$$
and
$$T^{ab}{}_{;b} = 0$$
hold independent of the field equations. Thus the field equations really provide only six independent differential equations for the metric. This is in fact the correct number of equations to determine the space–time, since four of the ten components of the metric can be given arbitrary values by use of the four degrees of freedom to make coordinate transformations. Another way of looking at this is that two metrics $\mathbf{g}_1$ and $\mathbf{g}_2$ on a manifold $\mathcal{M}$ define the same space–time if there is a diffeomorphism θ which takes $\mathbf{g}_1$ into $\mathbf{g}_2$. Therefore the field equations should define the metric only up to an equivalence class under diffeomorphisms, and there are four degrees of freedom to make diffeomorphisms.

We shall consider the Cauchy problem for the Einstein equations in chapter 7, and shall show that, together with the equations for the matter fields, they are sufficient to determine the evolution of space–time given suitable initial conditions, and that they satisfy the causality postulate (a).

The Einstein equations can be derived by requiring that the action

$$I = \int_{\mathscr{D}} (A(R - 2\Lambda) + L)\,dv \tag{3.16}$$

be stationary under variations of g_{ab}, where L is the matter Lagrangian and A a suitable constant. For

$$\Delta((R - 2\Lambda)\,dv) = ((R - 2\Lambda)\tfrac{1}{2}g^{ab}\Delta g_{ab} + R_{ab}\Delta g^{ab} + g^{ab}\Delta R_{ab})\,dv.$$

The last term can be written

$$g^{ab}\Delta R_{ab}\,dv = g^{ab}((\Delta\Gamma^c_{ab})_{;\,c} - (\Delta\Gamma^c_{ac})_{;\,b})\,dv$$
$$= (\Delta\Gamma^c_{ab}g^{ab} - \Delta\Gamma^d_{ad}g^{ac})_{;\,c}\,dv.$$

Thus it may be transformed into an integral over the boundary $\partial\mathscr{D}$, which vanishes as $\Delta\Gamma^a_{bc}$ vanishes on the boundary. Therefore

$$\left.\frac{\partial I}{\partial u}\right|_{u=0} = \int_{\mathscr{D}} \{A((\tfrac{1}{2}R - \Lambda)g^{ab} - R^{ab}) + \tfrac{1}{2}T^{ab}\}\Delta g_{ab}\,dv, \tag{3.17}$$

and so if $\partial I/\partial u$ vanishes for all Δg_{ab}, one obtains the Einstein equations on setting $A = (16\pi)^{-1}$.

One might ask whether varying an action derived from some other scalar combination of the metric and curvature tensors might not give a reasonable alternative set of equations. However the curvature scalar is the only such scalar linear in second derivatives of the metric tensor; so only in this case can one transform away a surface integral and be left with an equation involving only second derivatives of the metric. If one tried any other scalar such as $R_{ab}R^{ab}$ or $R_{abcd}R^{abcd}$ one would obtain an equation involving fourth derivatives of the metric tensor. This would seem objectionable, as all other equations of physics are first or second order. If the field equations were fourth order, it would be necessary to specify not only the initial values of the metric and its first derivatives, but also the second and third derivatives, in order to determine the evolution of the metric.

We shall assume the field equations do not involve derivatives of the metric higher than the second. If these field equations are derived from a Lagrangian, then the action must have the form (3.16). One could however obtain a system of equations other than the Einstein equations, if one restricted the form of the variations Δg_{ab} for which the action was required to be stationary.

For example, one could restrict the metric to be conformal to a flat metric, i.e. assume

$$g_{ab} = \Omega^2 \eta_{ab},$$

where η_{ab} is a flat metric as in Special Relativity. Then

$$\Delta g_{ab} = 2\Omega^{-1}\Delta\Omega g_{ab}$$

and the action will be stationary if

$$\{(A(\tfrac{1}{2}R - \Lambda)g^{ab} - R^{ab}) + T^{ab}\}\Delta\Omega g_{ab} = 0$$

for all $\Delta\Omega$, that is if $\qquad R + A^{-1}T = 4\Lambda.$

From (2.30),

$$R = -6\Omega^{-3}\Omega_{|bc}\eta^{bc} = -6\Omega^{-1}\Omega_{;bc}g^{bc} + 12\Omega^{-2}\Omega_{;c}\,\Omega_{;d}g^{cd},$$

where $|$ denotes covariant differentiation with respect to the flat metric η_{ab}. If the metric is static, Ω will be constant along the integral curves of the Killing vector $\mathbf{K}$ (it will be independent of the time t). The magnitude of $\mathbf{K}$ will be proportional to Ω. Therefore

$$f_{;ab}(g^{ab} + V^a V^b)f^{-1} = \Omega_{;ab}(g^{ab} + V^a V^b)\,\Omega^{-1}$$

$$= -\tfrac{1}{6}R + 2\Omega^{-2}\Omega_{;a}\,\Omega_{;b}g^{ab} - \Omega^{-1}\Omega_{;a}V^a_{\ ;b}\,V^b$$

$$= -\tfrac{1}{6}R + f^{-2}f_{;a}f_{;b}g^{ab}.$$

Thus the Laplacian of f will be equal to $-\tfrac{1}{6}R$ plus a term proportional to the square of the gradient of f. This last term may be neglected in a weak field. From the field equations, $-\tfrac{1}{6}R$ will be equal to $\tfrac{1}{6}A^{-1}T - \tfrac{2}{3}\Lambda$. For a perfect fluid, $T = -\mu + 3p$. One will therefore get agreement with Newtonian theory if Λ is small or zero and $A^{-1} = -24\pi$.

This theory in which the metric is restricted to be conformally flat is known as the Nordström theory. It can be reformulated as a theory in which the metric is the flat metric $\boldsymbol{\eta}$ and in which the gravitational interaction is represented by an additional scalar field ϕ. As mentioned before, this sort of theory would be inconsistent with the observed deflection of light by massive objects, and it would not account for the measured advance of the perihelion of Mercury.

One could in fact obtain the observed deflection of light and the advance of the perihelion of Mercury if the metric was restricted to be of the form
$$g_{ab} = \Omega^2(\eta_{ab} + W_a W_b),$$

where W_a is an arbitrary one-form field. This would give the Newtonian limit in a static metric in which W_a was parallel to the timelike Killing vector. There could however also be other static metrics where W_a was not parallel to the Killing vector and these would not give the Newtonian limit. Further this restriction on the form of the metric seems rather artificial. It appears more natural not to restrict the metric, apart from requiring that it be Lorentzian.

We therefore adopt as our third postulate,

Postulate (c): Field equations

Einstein's field equations (3.15) hold on $\mathcal{M}$.

The predictions of these field equations agree, within the experimental errors, with the observations that have been made so far on the deflection of light and the advance of the perihelion of Mercury, though the question of whether there exists a long range scalar field which ought to be included in the energy–momentum tensor remains open at the present time.

4

The physical significance of curvature

In this chapter we consider the effect of space–time curvature on families of timelike and null curves. These could represent flow lines of fluids or the histories of photons. In § 4.1 and § 4.2 we derive the formulae for the rate of change of vorticity, shear and expansion of such families of curves; the equation for the rate of change of expansion (Raychaudhuri's equation) plays a central role in the proofs of the singularity theorems of chapter 8. In § 4.3 we discuss the general inequalities on the energy–momentum tensor which imply that the gravitational effect of matter is always to tend to cause convergence of timelike and of null curves. A consequence of these energy conditions is, as is seen in § 4.4, that conjugate or focal points will occur in families of non-rotating timelike or null geodesics in general space–times. In § 4.5 it is shown that the existence of conjugate points implies the existence of variations of curves between two points which take a null geodesic into a timelike curve, or a timelike geodesic into a longer timelike curve.

4.1 Timelike curves

In chapter 3 we saw that if the metric was static there was a relation between the magnitude of the timelike Killing vector and the Newtonian potential. One was able to tell whether a body was in a gravitational field by whether, if released from rest, it would accelerate with respect to the static frame defined by the Killing vector. However, in general, space–time will not have any Killing vectors. Thus one will not have any special frame against which to measure acceleration; the best one can do is to take two bodies close together and measure their relative acceleration. This will enable one to measure the gradient of the gravitational field. If one thinks of the metric as being analogous to the Newtonian potential, the gradient of the Newtonian field would correspond to the second derivatives of the metric. These are described by the Riemann tensor. Thus one would expect that the relative acceleration of two neighbouring bodies would be related to some components of the Riemann tensor.

In order to investigate this relation more precisely we shall examine the behaviour of a congruence of timelike curves with timelike unit tangent vector $\mathbf{V}$ ($g(\mathbf{V}, \mathbf{V}) = -1$). These curves could represent the histories of small test particles, in which case they would be geodesics, or they might represent the flow lines of a fluid. If this were a perfect fluid, then by (3.10)

$$(\mu + p)\, \dot{V}^a = -p_{;b} h^{ab}, \tag{4.1}$$

where $\dot{V}^a = V^a{}_{;b} V^b$ is the acceleration of the flow lines and $h^a{}_b = \delta^a{}_b + V^a V_b$ is the tensor which projects a vector $\mathbf{X} \in T_q$ into its component in the subspace H_q of T_q orthogonal to V. One may also think of h_{ab} as the metric in H_q (cf. §2.7).

Suppose $\lambda(t)$ is a curve with tangent vector $\mathbf{Z} = (\partial/\partial t)_\lambda$. Then one may construct a family $\lambda(t, s)$ of curves by moving each point of the curve $\lambda(t)$ a distance s along the integral curves of $\mathbf{V}$. If one now defines $\mathbf{Z}$ as $(\partial/\partial t)_{\lambda(t, s)}$ it follows from the definition of the Lie derivative (see §2.4) that $L_{\mathbf{V}} \mathbf{Z} = 0$ or in other words that

$$\frac{\mathrm{D}}{\partial s} Z^a = V^a{}_{;b} Z^b. \tag{4.2}$$

One may interpret $\mathbf{Z}$ as representing the separation of points equal distances from some arbitrary initial points along two neighbouring curves. If one adds a multiple of $\mathbf{V}$ to $\mathbf{Z}$ then this vector will represent the separation of points on the same two curves but at different distances along the curves. It is really only the separation of neighbouring curves that one is interested in, not the separation of particular points on these curves. One is thus concerned only with $\mathbf{Z}$ modulo a component parallel to $\mathbf{V}$, i.e. only with the projection of $\mathbf{Z}$ at each point q into the space Q_q consisting of equivalence classes of vectors which differ only by addition of a multiple of $\mathbf{V}$. This space can be represented as the subspace H_q of T_q consisting of vectors orthogonal to $\mathbf{V}$. The projection of $\mathbf{Z}$ into H_q will be denoted by $_\perp Z^a = h^a{}_b Z^b$. In the case of a fluid one can regard $_\perp \mathbf{Z}$ as the distance between two neighbouring particles of the fluid as measured in their rest frame.

From (4.2) it follows that

$$_\perp \frac{\mathrm{D}}{\partial s} (_\perp Z^a) = V^a{}_{;b} {}_\perp Z^b. \tag{4.3}$$

This gives the rate of change of the separation of two infinitesimally neighbouring curves as measured in H_q. Operating again with $D/\partial s$ and projecting into H_q, one finds

$$h^a{}_b \frac{D}{\partial s} \left(h^b{}_c \frac{D}{\partial s} {}_\perp Z^c \right) = h^a{}_b (V^b{}_{;cd} {}_\perp Z^c V^d + V^b{}_{;c} V^c{}_{;d} V_e Z^e V^d$$

$$+ V^b{}_{;c} V^c V^e{}_{;d} Z_e V^d + V^b{}_{;c} h^c{}_e Z^e{}_{;d} V^d).$$

Changing the order of the derivatives in the first term and using (4.2), this reduces to

$$h^a{}_b \frac{D}{\partial s} \left(h^b{}_c \frac{D}{\partial s} {}_\perp Z^c \right) = - R^a{}_{bcd} {}_\perp Z^c V^b V^d + h^a{}_b \, \dot{V}^b{}_{;c} {}_\perp Z^c + \dot{V}^a \dot{V}_b {}_\perp Z^b. \quad (4.4)$$

This equation, known as the deviation or Jacobi equation, gives the relative acceleration, i.e. the second time derivative of the separation, of two infinitesimally neighbouring curves as measured in H_q. We see that this depends only on the Riemann tensor if the curves are geodesics.

In Newtonian theory, the acceleration of each particle is given by the gradient of the potential Φ and therefore the relative acceleration of two particles with separation Z^a is $\Phi_{;ab} Z^b$. Thus the Riemann tensor term $R_{abcd} V^b V^d$ is analogous to the Newtonian $\Phi_{;ac}$. The effect of this 'tidal force' term can be seen, for example, by considering a sphere of particles freely falling towards the earth. Each particle moves on a straight line through the centre of the earth but those nearer the earth fall faster than those further away. This means that the sphere does not remain a sphere but is distorted into an ellipsoid with the same volume.

In order to investigate the deviation equation further we shall introduce dual orthonormal bases $\mathbf{E_1}$, $\mathbf{E_2}$, $\mathbf{E_3}$, $\mathbf{E_4}$ and $\mathbf{E^1}$, $\mathbf{E^2}$, $\mathbf{E^3}$, $\mathbf{E^4}$ of T_q and T^*_q at some point q on an integral curve $\gamma(s)$ of $\mathbf{V}$, with $\mathbf{E^4} = \mathbf{V}$. One would like to propagate them along $\gamma(s)$ to obtain similar such bases at each point of $\gamma(s)$. However, if one parallelly propagates them along $\gamma(s)$ (i.e. so that $D/\partial s$ of each vector is zero) $\mathbf{E_4}$ will not remain equal to $\mathbf{V}$, and $\mathbf{E_1}$, $\mathbf{E_2}$, $\mathbf{E_3}$ will not remain orthogonal to $\mathbf{V}$, unless $\gamma(s)$ is a geodesic. We therefore introduce a new derivative along $\gamma(s)$ called the *Fermi derivative* $D_F/\partial s$. This is defined for a vector field $\mathbf{X}$ along $\gamma(s)$ by:

$$\frac{D_F \mathbf{X}}{\partial s} = \frac{D\mathbf{X}}{\partial s} - g\left(\mathbf{X}, \frac{D\mathbf{V}}{\partial s} \right) \mathbf{V} + g(\mathbf{X}, \mathbf{V}) \frac{D\mathbf{V}}{\partial s}.$$

It has the properties:

(i) $\dfrac{D_F}{\partial s} = \dfrac{D}{\partial s}$ if $\gamma(s)$ is a geodesic;

(ii) $\dfrac{D_F V}{\partial s} = 0;$

(iii) if $\mathbf{X}$ and $\mathbf{Y}$ are vector fields along $\gamma(s)$ such that

$$\frac{D_F \mathbf{X}}{\partial s} = 0 = \frac{D_F \mathbf{Y}}{\partial s},$$

then $g(\mathbf{X}, \mathbf{Y})$ is constant along $\gamma(s)$;

(iv) if $\mathbf{X}$ is a vector field along $\gamma(s)$ orthogonal to $\mathbf{V}$ then

$$\frac{D_F \mathbf{X}}{\partial s} = {}_{\perp}\!\left(\frac{D\mathbf{X}}{\partial s}\right).$$

(This last property shows that the Fermi derivative is a natural generalization of the derivative $D/\partial s$.)

Thus, if one propagates an orthonormal basis of T_q along $\gamma(s)$ so that the Fermi derivative of each basis vector is zero, one obtains an orthonormal basis at each point of $\gamma(s)$, with $\mathbf{E}_4 = \mathbf{V}$. The vectors $\mathbf{E}_1$, $\mathbf{E}_2$, $\mathbf{E}_3$ may be interpreted as giving a non-rotating set of axes along $\gamma(s)$. These could be realized physically by small gyroscopes pointing in the direction of each vector.

The definition of the Fermi derivative along $\gamma(s)$ can be extended from vector fields to arbitrary tensor fields by the usual rules:

(i) $D_F/\partial s$ is a linear mapping of tensor fields of type (r, s) along $\gamma(s)$ to tensor fields of type (r, s), which commutes with contractions;

(ii) $\dfrac{D_F}{\partial s}(\mathbf{K} \otimes \mathbf{L}) = \dfrac{D_F \mathbf{K}}{\partial s} \otimes \mathbf{L} + \mathbf{K} \otimes \dfrac{D_F \mathbf{L}}{\partial s};$

(iii) $\dfrac{D_F f}{\partial s} = \dfrac{df}{ds},$ where f is a function.

From these rules it follows that the dual basis $\mathbf{E}^1$, $\mathbf{E}^2$, $\mathbf{E}^3$, $\mathbf{E}^4$ of T^*_q is also Fermi-propagated along $\gamma(s)$. Using Fermi derivatives, (4.3) and (4.4) may be written as:

$$\frac{D_F}{\partial s} {}_{\perp}Z^a = V^a{}_{;b}{}_{\perp}Z^b, \tag{4.5}$$

$$\frac{D^2{}_F}{\partial s^2} {}_{\perp}Z^a = -R^a{}_{bcd}{}_{\perp}Z^c\, V^b V^d + h^a{}_b\, \dot{V}^b{}_{;c}{}_{\perp}Z^c + \dot{V}^a \dot{V}_b{}_{\perp}Z^b. \tag{4.6}$$

One may express these equations in terms of the Fermi-propagated dual bases. As $\perp \mathbf{Z}$ is orthogonal to $\mathbf{V}$ it will have components with respect to $\mathbf{E_1}$, $\mathbf{E_2}$, $\mathbf{E_3}$ only. Thus it may be expressed as $Z^\alpha \mathbf{E_\alpha}$ where we adopt the convention that Greek indices take the values 1, 2, 3 only. Then (4.5) and (4.6) can be written in terms of ordinary derivatives:

$$\frac{d}{ds} Z^\alpha = V^\alpha{}_{;\beta} Z^\beta, \qquad (4.7)$$

$$\frac{d^2}{ds^2} Z^\alpha = (-R^\alpha{}_{4\beta4} + \dot{V}^\alpha{}_{;\beta} + \dot{V}^\alpha \dot{V}_\beta) Z^\beta \qquad (4.8)$$

where $V^\alpha{}_{;\beta}$ are the components of $V^a{}_{;b}$ for which $a = \alpha$ and $b = \beta$. As the components Z^α obey the first order linear ordinary differential equation (4.7), they can be expressed in terms of their values at some point q by:

$$Z^\alpha(s) = A_{\alpha\beta}(s) Z^\beta|_q, \qquad (4.9)$$

where $A_{\alpha\beta}(s)$ is a 3×3 matrix which is the unit matrix at q and satisfies

$$\frac{d}{ds} A_{\alpha\beta}(s) = V_{\alpha;\gamma} A_{\gamma\beta}(s). \qquad (4.10)$$

In the case of a fluid the matrix $A_{\alpha\beta}$ can be regarded as representing the shape and orientation of a small element of fluid which is spherical at q. This matrix can be written as

$$A_{\alpha\beta} = O_{\alpha\delta} S_{\delta\beta} \qquad (4.11)$$

where $O_{\alpha\beta}$ is an orthogonal matrix with positive determinant and $S_{\alpha\beta}$ is a symmetric matrix. These will both be chosen to be the unit matrix at q. The matrix $O_{\alpha\beta}$ may be thought of as representing the rotation that neighbouring curves have undergone with respect to the Fermi-propagated basis while $S_{\alpha\beta}$ represents the separation of these curves from $\gamma(s)$. The determinant of $S_{\alpha\beta}$, which equals the determinant of $A_{\alpha\beta}$, may be thought of as representing the three-volume of the element of the surface orthogonal to $\gamma(s)$ marked out by the neighbouring curves.

At q where $A_{\alpha\beta}$ is the unit matrix, $dO_{\alpha\beta}/ds$ is antisymmetric and $dS_{\alpha\beta}/ds$ is symmetric. Thus the rate of rotation of neighbouring curves at q is given by the antisymmetric part of $V_{\alpha;\beta}$ while the rate of change of their separation from $\gamma(s)$ is given by the symmetric part of $V_{\alpha;\beta}$ and the rate of change of volume is given by the trace of $V_{\alpha;\beta}$. We therefore define the vorticity tensor as

$$\omega_{ab} = h_a{}^c h_b{}^d V_{[c;d]}, \qquad (4.12)$$

the expansion tensor as

$$\theta_{ab} = h_a{}^c h_b{}^d V_{(c;d)}, \qquad (4.13)$$

and the volume expansion as

$$\theta = \theta_{ab}h^{ab} = V_{a;b}h^{ab} = V^a{}_{;a}. \tag{4.14}$$

We further define the shear tensor as the trace free part of θ_{ab},

$$\sigma_{ab} = \theta_{ab} - \tfrac{1}{3}h_{ab}\theta, \tag{4.15}$$

and the vorticity vector as

$$\omega^a = \tfrac{1}{2}\eta^{abcd}V_b\omega_{cd} = \tfrac{1}{2}\eta^{abcd}V_b V_{c;d}. \tag{4.16}$$

The covariant derivative of the vector $\mathbf{V}$ may be expressed in terms of these quantities;

$$V_{a;b} = \omega_{ab} + \sigma_{ab} + \tfrac{1}{3}\theta h_{ab} - \dot{V}_a V_b. \tag{4.17}$$

This decomposition of the gradient of the fluid velocity vector is directly analogous to that in Newtonian hydrodynamics.

In the Fermi-propagated orthonormal basis the vorticity and expansion can be expressed in terms of the matrix $A_{\alpha\beta}$ and its inverse $A^{-1}{}_{\alpha\beta}$:

$$\omega_{\alpha\beta} = -A^{-1}{}_{\gamma[\alpha}\frac{\mathrm{d}}{\mathrm{d}s}A_{\beta]\gamma}, \tag{4.18}$$

$$\theta_{\alpha\beta} = A^{-1}{}_{\gamma(\alpha}\frac{\mathrm{d}}{\mathrm{d}s}A_{\beta)\gamma}, \tag{4.19}$$

$$\theta = (\det \mathbf{A})^{-1}\frac{\mathrm{d}}{\mathrm{d}s}(\det \mathbf{A}). \tag{4.20}$$

From the deviation equation (4.8) it follows that

$$\frac{\mathrm{d}^2}{\mathrm{d}s^2}A_{\alpha\beta} = (-R_{\alpha4\gamma4} + \dot{V}_{\alpha;\gamma} + \dot{V}_\alpha\dot{V}_\gamma)A_{\gamma\beta}. \tag{4.21}$$

This equation enables one to calculate the propagation of the vorticity, shear and expansion along the integral curves of $\mathbf{V}$ if one knows the Riemann tensor.

Multiplying by $A^{-1}{}_{\beta\gamma}$ and taking the antisymmetric part, one obtains

$$\frac{\mathrm{d}}{\mathrm{d}s}\omega_{\alpha\beta} = 2\omega_{\gamma[\alpha}\theta_{\beta]\gamma} + \dot{V}_{[\alpha;\beta]}. \tag{4.22}$$

Thus the propagation of vorticity depends on the antisymmetric gradient of the acceleration but not the 'tidal force'. Another form of the above equation is

$$\frac{\mathrm{d}}{\mathrm{d}s}(A_{\gamma\alpha}\omega_{\gamma\delta}A_{\delta\beta}) = A_{\gamma\alpha}\dot{V}_{[\gamma;\delta]}A_{\delta\beta}. \tag{4.23}$$

Therefore $A_{\gamma\alpha}\omega_{\gamma\delta}A_{\delta\beta}$ is a constant matrix if the curves are geodesics; in particular, if the curves are geodesics and the vorticity vanishes at one point on a curve, it will vanish at all points on the curve. If the curves are the flow lines of a perfect fluid it follows from (4.1) that

$$\dot{V}_{[\alpha;\,\beta]} = -\frac{1}{\mu+p}\,\omega_{\alpha\beta}\frac{\mathrm{d}p}{\mathrm{d}s}.$$

If the fluid is isentropic, this implies the conservation law:

$$WA_{\gamma\alpha}\omega_{\gamma\delta}A_{\delta\beta} = \text{constant}, \qquad (4.24)$$

where

$$\log W = \int \frac{\mathrm{d}p}{\mu+p}.$$

This conservation law is the relativistic form of the Newtonian vorticity conservation law. In the geodesic or pressure-free case, this takes the usual form that the magnitude of the vorticity vector is inversely proportional to the area of a cross-section orthogonal to the vorticity vector of an element of the fluid. When the pressure is non-zero, there is an extra relativistic effect arising from the fact that compression of the fluid does work on the fluid and therefore increases the mass and so the inertia of an element of the fluid (cf. (3.9)). This means that the vorticity of a fluid increases less under compression than would otherwise be expected.

Multiplying (4.21) by $A^{-1}{}_{\beta\gamma}$ and taking the symmetric part, one finds

$$\frac{\mathrm{d}}{\mathrm{d}s}\theta_{\alpha\beta} = -R_{\alpha4\beta4} - \omega_{\alpha\gamma}\omega_{\gamma\beta} - \theta_{\alpha\gamma}\theta_{\gamma\beta} + \dot{V}_{(\alpha;\,\beta)} + \dot{V}_\alpha\dot{V}_\beta. \qquad (4.25)$$

(This equation and (4.23) can be expressed in terms of a general, non-orthonormal, non-Fermi-propagated basis by replacing the ordinary derivatives with Fermi derivatives and projecting everything into the subspace orthogonal to $\mathbf{V}$.)

The trace of (4.25) is

$$\frac{\mathrm{d}}{\mathrm{d}s}\theta = -R_{ab}V^aV^b + 2\omega^2 - 2\sigma^2 - \tfrac{1}{3}\theta^2 + \dot{V}^a{}_{;\,a}, \qquad (4.26)$$

where

$$2\omega^2 = \omega_{ab}\omega^{ab} \geqslant 0,$$

$$2\sigma^2 = \sigma_{ab}\sigma^{ab} \geqslant 0.$$

This equation, which was discovered by Landau and independently by Raychaudhuri, will be of great importance later. From it one sees that vorticity induces expansion as might be expected by analogy with centrifugal force while shear induces contraction. By the field equations, the term $R_{ab}V^aV^b = 4\pi(\mu+3p)$ for a perfect fluid whose flow lines have tangent vectors V^a. Thus one would expect this term also to induce contraction. We shall give a general discussion of the sign of this term in §4.3.

The trace-free part of (4.25) is

$$\frac{D_F}{\partial s}\sigma_{ab} = -C_{acbd}V^cV^d + \tfrac{1}{2}h_a{}^c h_b{}^d R_{cd} - \omega_{ac}\omega^c{}_b - \sigma_{ac}\sigma^c{}_b$$

$$-\tfrac{2}{3}\theta\sigma_{ab} + h_a{}^c h_b{}^d \dot{V}_{(c;\,d)} - \tfrac{1}{3}h_{ab}(2\omega^2 - 2\sigma^2 + \dot{V}^a{}_{;\,a} + \tfrac{1}{2}R_{cd}h^{cd}), \quad (4.27)$$

where C_{abcd} is the Weyl tensor. Since this tensor is trace-free it does not enter directly in the expansion equation (4.26). However since the term $-2\sigma^2$ occurs on the right of the expansion equation, the Weyl tensor produces convergence indirectly by inducing shear. The Riemann tensor can be expressed in terms of the Weyl tensor and the Ricci tensor:

$$R_{abcd} = C_{abcd} - g_{a[d}R_{c]b} - g_{b[c}R_{d]a} - \tfrac{1}{3}Rg_{a[c}g_{d]b}.$$

The Ricci tensor is given by the Einstein equations:

$$R_{ab} - \tfrac{1}{2}g_{ab}R + \Lambda g_{ab} = 8\pi T_{ab}.$$

Thus the Weyl tensor is that part of the curvature which is not determined locally by the matter distribution. However it cannot be entirely arbitrary as the Riemann tensor must satisfy the Bianchi identities:

$$R_{ab[cd;\,e]} = 0$$

These can be rewritten as

$$C^{abcd}{}_{;\,d} = J^{abc}, \quad (4.28)$$

where

$$J^{abc} = R^{c[a;\,b]} + \tfrac{1}{6}g^{c[b}R^{;\,a]}. \quad (4.29)$$

These equations are rather similar to Maxwell's equations in electrodynamics:

$$F^{ab}{}_{;\,b} = J^a,$$

where F^{ab} is the electromagnetic field tensor and J^a is the source current. Thus in a sense one could regard the Bianchi identities (4.28) as field equations for the Weyl tensor giving that part of the curvature at a point that depends on the matter distribution at other points. (This approach has been used to analyse the behaviour of gravitational radiation in papers by Newman and Penrose (1962), Newman and Unti (1962) and Hawking (1966a).)

4.2 Null curves

The Riemann tensor will affect the rate of change of separation of null curves as well as that of timelike curves. For simplicity, we shall consider only null geodesics. These could represent the histories of photons; the effect of the Riemann tensor will be to distort or focus small bundles of light rays.

To investigate this, we consider the deviation equation for a congruence of null geodesics with tangent vector $\mathbf{K}$ ($g(\mathbf{K}, \mathbf{K}) = 0$). There are two important differences between this case and that of the timelike curves considered in the previous section. First, one could normalize the tangent vector $\mathbf{V}$ to the timelike curves by requiring $g(\mathbf{V}, \mathbf{V}) = -1$. In effect this means that one parametrized the curves by the arc-length s. However this is clearly impossible with null curves as they have zero arc-lengths. The best one can do is to choose an affine parameter v; then the tangent vector $\mathbf{K}$ will obey

$$\frac{\mathrm{D}}{\mathrm{d}v} K^a = K^a{}_{;b} K^b = 0.$$

However one could multiply v by a function f which was constant along each curve. Then fv would be another affine parameter and the corresponding tangent vector would be $f^{-1}\mathbf{K}$. Thus, given the curves as point sets in the manifold, the tangent vector is only really unique up to a constant factor along each curve. The second difference is that Q_q, the quotient of T_q by $\mathbf{K}$, is not now isomorphic to H_q, the subspace of T_q orthogonal to $\mathbf{K}$, since H_q includes the vector $\mathbf{K}$ itself as $g(\mathbf{K}, \mathbf{K}) = 0$. In fact as will be shown below, one is not really interested in the whole of Q_q but only in the subspace S_q consisting of equivalence classes of vectors in H_q which differ only by a multiple of $\mathbf{K}$. In the case of light rays, one can regard an element of S_q as representing the separation between two neighbouring light rays which were emitted at the same time by a source.

As before we introduce dual bases $\mathbf{E}_1$, $\mathbf{E}_2$, $\mathbf{E}_3$, $\mathbf{E}_4$, and $\mathbf{E}^1$, $\mathbf{E}^2$, $\mathbf{E}^3$, $\mathbf{E}^4$ of T_q and T_q^* at some point q on a curve $\gamma(v)$. However we will not choose them to be orthonormal. We take $\mathbf{E}_4$ equal to $\mathbf{K}$, $\mathbf{E}_3$ to be some other null vector $\mathbf{L}$ having unit negative scalar product with $\mathbf{E}_4$ ($g(\mathbf{E}_3, \mathbf{E}_3,) = 0$, $g(\mathbf{E}_3, \mathbf{E}_4) = -1$) and $\mathbf{E}_1$ and $\mathbf{E}_2$ to be unit spacelike vectors, orthogonal to each other and to $\mathbf{E}_3$ and $\mathbf{E}_4$

$$(g(\mathbf{E}_1, \mathbf{E}_1) = g(\mathbf{E}_2, \mathbf{E}_2) = 1, \quad g(\mathbf{E}_1, \mathbf{E}_2) = g(\mathbf{E}_1, \mathbf{E}_3) = g(\mathbf{E}_1, \mathbf{E}_4) = 0, \text{etc.}).$$

Note that because of the non-orthonormal character of the basis, the form $\mathbf{E}^3$ is in fact equal to the form $-K^a g_{ab}$ and $\mathbf{E}^4$ is $-L^a g_{ab}$. It can be seen that $\mathbf{E}_1$, $\mathbf{E}_2$ and $\mathbf{E}_4$ constitute a basis for H_q while the projections into Q_q of $\mathbf{E}_1$, $\mathbf{E}_2$ and $\mathbf{E}_3$ form a basis of Q_q, and the projections of $\mathbf{E}_1$ and $\mathbf{E}_2$ form a basis of S_q. We shall normally not distinguish between a vector $\mathbf{Z}$ and its projection into Q_q or S_q. We shall call a basis having the properties of $\mathbf{E}_1$, $\mathbf{E}_2$, $\mathbf{E}_3$, $\mathbf{E}_4$, above, *pseudo-orthonormal*. By parallelly transporting them along the geodesic $\gamma(v)$ one obtains a pseudo-orthonormal basis at each point of $\gamma(v)$.

We use this basis to analyse the deviation equation for null geodesics. If $\mathbf{Z}$ is the vector representing the separation of corresponding points on neighbouring curves, one has, as before:

$$L_\mathbf{K} \mathbf{Z} = 0,$$

so

$$\frac{D}{dv} Z^a = K^a{}_{;b} Z^b \tag{4.30}$$

and

$$\frac{D^2}{dv^2} Z^a = -R^a{}_{bcd} Z^c K^b K^d. \tag{4.31}$$

In the pseudo-orthonormal basis $K^a{}_{;4}$ will be zero as $\mathbf{K}$ is geodesic. Therefore one can express the 1, 2 and 3 components of (4.30) as a system of ordinary differential equations:

$$\frac{d}{dv} Z^\alpha = K^\alpha{}_{;\beta} Z^\beta,$$

where as before Greek indices take the values 1, 2, 3. This shows that the projection of $\mathbf{Z}$ into the space Q_q obeys a propagation equation which involves only this projection, and not the component of $\mathbf{Z}$ parallel to $\mathbf{K}$. Further $K^3{}_{;c} = 0$ since $(K^a g_{ab} K^b)_{;c} = 0$. This implies that $Z^3 = -Z^a K_a$ is constant along the geodesic $\gamma(v)$. This can be interpreted as saying that light rays emitted from the same source at different times maintain a constant separation in time. As this is the case, one is more interested in the behaviour of neighbouring null geodesics which have purely spatial separations, i.e. one is interested in vectors $\mathbf{Z}$ for which $Z^3 = 0$. The projections of such vectors will then lie in the subspace S_q and will obey the equation

$$\frac{d}{dv} Z^m = K^m{}_{;n} Z^n,$$

where m, n take the values 1, 2 only. This is similar to (4.7) for the timelike case, except that now one is dealing only with a two-dimensional space of connecting vectors $\mathbf{Z}$.

As in the previous section, one can express Z^m in terms of their values at some point q:
$$Z^m(v) = \hat{A}_{mn}(v) Z^n\big|_q,$$
where $\hat{A}_{mn}(v)$ is a 2×2 matrix which satisfies

$$\frac{\mathrm{d}}{\mathrm{d}v} \hat{A}_{mn}(v) = K_{m;\,p} \hat{A}_{pn}(v), \tag{4.32}$$

$$\frac{\mathrm{d}^2}{\mathrm{d}v^2} \hat{A}_{mn}(v) = - R_{m4p4} \hat{A}_{pn}(v). \tag{4.33}$$

As before we call the antisymmetric part of $K_{m;\,n}$ the vorticity $\hat{\omega}_{mn}$, the symmetric part the rate of separation θ_{mn} and the trace the expansion θ. We also define the shear $\hat{\sigma}_{mn}$ as the trace-free part of θ_{mn}. They obey similar equations to the analogous quantities in the previous section:

$$\frac{\mathrm{d}}{\mathrm{d}v} \hat{\omega}_{mn} = - \theta \hat{\omega}_{mn} + 2\hat{\omega}_{p[m} \hat{\sigma}_{n]p}, \tag{4.34}$$

$$\frac{\mathrm{d}}{\mathrm{d}v} \theta = - R_{ab} K^a K^b + 2\hat{\omega}^2 - 2\hat{\sigma}^2 - \tfrac{1}{2}\theta^2, \tag{4.35}$$

$$\frac{\mathrm{d}}{\mathrm{d}v} \hat{\sigma}_{mn} = - C_{m4n4} - \theta \hat{\sigma}_{mn} - \hat{\sigma}_{mp} \hat{\sigma}_{pn} - \hat{\omega}_{mp} \hat{\omega}_{pn} + \delta_{mn}(\hat{\sigma}^2 - \hat{\omega}^2). \tag{4.36}$$

Equation (4.35) is the analogue of the Raychaudhuri equation for timelike geodesics. One sees again that vorticity causes expansion while shear causes contraction. We shall show in the next section that the Ricci tensor term $- R_{ab} K^a K^b$ will normally be negative, and so cause focussing. As before the Weyl tensor does not affect the expansion directly but causes distortion which in turn causes contraction (cf. Penrose (1966)).

4.3   Energy conditions

In the actual universe the energy–momentum tensor will be made up of contributions from a large number of different matter fields. It would therefore be impossibly complicated to describe the exact energy–momentum tensor even if one knew the precise form of the contribution of each field and the equations of motion governing it. In fact, one has little idea of the behaviour of matter under extreme conditions of density and pressure. Thus it might seem that one has little hope of predicting the occurrence of singularities in the universe from the Einstein equations as one does not know the right-hand side of the equations. However there are certain inequalities which it is physically reasonable to assume for the energy–momentum tensor. These will be discussed in this section. It turns out that in many circumstances these are sufficient to prove the occurrence of singularities, independent of the exact form of the energy–momentum tensor.

The first of these inequalities is:

The weak energy condition

The energy–momentum tensor at each $p \in \mathcal{M}$ obeys the inequality $T_{ab} W^a W^b \geqslant 0$ for any timelike vector $\mathbf{W} \in T_p$. By continuity this will then also be true for any null vector $\mathbf{W} \in T_p$.

To an observer whose world-line at p has unit tangent vector $\mathbf{V}$, the local energy density appears to be $T_{ab} V^a V^b$. Thus this assumption is equivalent to saying that the energy density as measured by any observer is non-negative. This would seem very reasonable physically. To investigate further the significance of this assumption we use the fact that one may express the components T^{ab} of the energy–momentum tensor at p with respect to an orthonormal basis $\mathbf{E}_1$, $\mathbf{E}_2$, $\mathbf{E}_3$, $\mathbf{E}_4$, ($\mathbf{E}_4$ timelike) in one of four canonical forms.

Type I.

$$T^{ab} = \begin{pmatrix} p_1 & & & \\ & p_2 & & 0 \\ & & p_3 & \\ 0 & & & \mu \end{pmatrix}.$$

This is the general case in which the energy–momentum tensor has a timelike eigenvector $\mathbf{E}_4$. This eigenvector is unique unless $\mu = -p_\alpha$ ($\alpha = 1, 2, 3$). The eigenvalue μ represents the energy–density as measured by an observer whose world-line at p has unit tangent vector $\mathbf{E}_4$ and the eigenvalues p_α ($\alpha = 1, 2, 3$) represent the principal pressures in the three spacelike directions $\mathbf{E}_\alpha$. This is the form of the energy–momentum for all observed fields with non-zero rest mass and also for all zero rest mass fields except in special cases when it is type II.

Type II.

$$T^{ab} = \begin{pmatrix} p_1 & 0 & & \\ 0 & p_2 & & 0 \\ & & \nu - \kappa & \nu \\ 0 & & \nu & \nu + \kappa \end{pmatrix}, \quad \nu = \pm 1.$$

This is the special case in which the energy–momentum tensor has a double null eigenvector $(\mathbf{E}_3 + \mathbf{E}_4)$. The only observed occurrence of this form is for zero rest-mass fields when they represent radiation all of which is travelling in the direction $\mathbf{E}_3 + \mathbf{E}_4$. In this case p_1, p_2 and κ are zero.

Type III.

$$T^{ab} = \begin{pmatrix} p & 0 & 0 & 0 \\ 0 & -\nu & 1 & 1 \\ 0 & 1 & -\nu & 0 \\ 0 & 1 & 0 & \nu \end{pmatrix}.$$

This is the special case in which the energy–momentum tensor has a triple null eigenvector $(\mathbf{E}_3 + \mathbf{E}_4)$. There are no observed fields which have energy–momentum tensors of this form.

Type IV.

$$T^{ab} = \begin{pmatrix} p_1 & 0 & & \\ & & 0 & \\ 0 & p_2 & & \\ & & -\kappa & \nu \\ & 0 & & \\ & & \nu & 0 \end{pmatrix}, \quad \kappa^2 < 4\nu^2.$$

This is the general case in which the energy–momentum tensor has no timelike or null eigenvector. There are no observed fields which have energy–momentum tensors of this form.

For type I, the weak energy condition will hold if $\mu \geqslant 0$, $\mu + p_\alpha \geqslant 0$ $(\alpha = 1, 2, 3)$. For type II it will hold if $p_1 \geqslant 0$, $p_2 \geqslant 0$, $\kappa \geqslant 0$, $\nu = +1$. These inequalities are very reasonable requirements and are satisfied by all experimentally detected fields. The condition will not hold for the physically unrealized types III and IV.

The condition will also hold for the scalar field ϕ postulated by Brans and Dicke and by Dicke (see Dicke (1964)). This field is required to be positive everywhere. It has an energy–momentum tensor of the form (3.6) where now $m = 0$. The energy-tensor of the other fields is ϕ times what it would have been had the scalar field not existed.

The condition will not hold for the 'C'-field proposed by Hoyle and Narlikar (1963). This again is a scalar field with m zero, only this time the energy–momentum tensor has the opposite sign and so the energy density is negative. This allows the simultaneous creation of quanta of positive energy fields and of the negative energy C-field. This process occurs in the steady-state model of the universe suggested by Hoyle and Narlikar in which, as particles move apart due to the general expansion of the universe, new matter is continually being created to keep the average density constant. There is, however, a quantum mechanical difficulty associated with such a process. For even if the cross-section for the process were very small, the infinite phase space available to the positive and negative energy quanta would seem to result in an infinite number of such pairs being produced in a finite region of space–time.

Such a catastrophe could not occur if the weak energy condition held. If a slightly stronger condition holds then creation is impossible in the sense that space–time must remain empty if it is empty at one time and no matter comes in from infinity. Conversely, matter present at one time cannot disappear and so must be present at another time. The condition is

The dominant energy condition

For every timelike W_a, $T^{ab}W_a W_b \geqslant 0$, and $T^{ab}W_a$ is a non-spacelike vector.

This may be interpreted as saying that to any observer the local energy density appears non-negative and the local energy flow vector is non-spacelike. An equivalent statement is that in any orthonormal basis the energy dominates the other components of T_{ab}, i.e.

$$T^{00} \geqslant |T^{ab}| \quad \text{for each } a, b.$$

This holds for type I if $\mu \geqslant 0$, $-\mu \leqslant p_\alpha \leqslant \mu$ $(\alpha = 1, 2, 3)$ and for type II if $\nu = +1$, $\kappa \geqslant 0$, $0 \leqslant p_i \leqslant \kappa$ $(i = 1, 2)$. In other words, the dominant energy condition is the weak energy condition with the additional requirement that the pressure should not exceed the energy density. This holds for all known forms of matter and there is in fact good reason for believing that this should be the case in all situations. For the speed of sound waves travelling in the $\mathbf{E}_\alpha$ direction is $dp_\alpha/d\mu$ (adiabatic) times the speed of light. Thus $dp_\alpha/d\mu$ must be less than or equal to one, as by postulate (a) in § 3.2 no signal can propagate faster than light. It follows that $p_\alpha \leqslant \mu$, since, for every known form of matter, the pressures are small when the density is small. (Bludman and Ruderman (1968, 1970) have shown that there might be fields for which mass renormalization could lead to pressure being greater than the density. We feel, however, that this probably indicates a failure of renormalization theory rather than that such a situation would occur.) Now consider the situation depicted in figure 9 in which there is a C^2

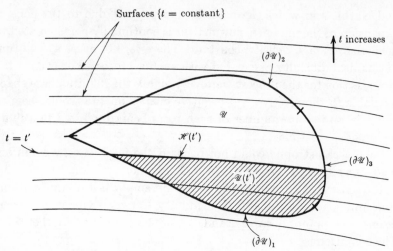

FIGURE 9. A compact region $\mathscr{U}$ of space–time with past and future non-timelike boundaries $(\partial\mathscr{U})_1$, $(\partial\mathscr{U})_2$ and timelike boundary $(\partial\mathscr{U})_3$. The part of $\mathscr{U}$ lying to the past of the surface $\mathscr{H}(t')$ (defined by $t = t'$) is $\mathscr{U}(t')$.

function t whose gradient is everywhere timelike. (It will be shown in § 6.4 that such a function will exist provided space–time is not on the verge of violating causality.) The boundary $\partial\mathscr{U}$ of the compact region $\mathscr{U}$ consists of a part $(\partial\mathscr{U})_1$, whose normal form $\mathbf{n}$ is non-spacelike and such that $n_a t_{;b} g^{ab}$ is positive, a part $(\partial\mathscr{U})_2$ whose normal form $\mathbf{n}$ is non-spacelike and such that $n_a t_{;b} g^{ab}$ is negative, and a remaining part $(\partial\mathscr{U})_3$ (which may be empty). The sign of the normal form $\mathbf{n}$ is given by the requirement that $\langle \mathbf{n}, \mathbf{X} \rangle$ be positive for all vectors $\mathbf{X}$ which point out of $\mathscr{U}$ (cf. § 2.8), $\mathscr{H}(t')$ denotes the surface $t = t'$ and $\mathscr{U}(t')$ denotes the region of $\mathscr{U}$ for which $t < t'$. For later use in § 7.4 we shall establish an inequality which holds not only for the energy–momentum tensor T^{ab} but also for any symmetric tensor S^{ab} which satisfies the dominant energy condition. Applied to the energy–momentum tensor this inequality will show that T^{ab} vanishes everywhere on $\mathscr{U}$ if it vanishes on $(\partial\mathscr{U})_3$ and on the initial surface $(\partial\mathscr{U})_1$.

Lemma 4.3.1

There is some positive constant P such that for any tensor S^{ab} which satisfies the dominant energy condition and vanishes on $(\partial\mathscr{U})_3$,

$$\int_{\mathscr{H}(t)\cap\mathscr{U}} S^{ab} t_{;a}\, \mathrm{d}\sigma_b \leqslant -\int_{(\partial\mathscr{U})_1} S^{ab} t_{;a}\, \mathrm{d}\sigma_b$$
$$+ P\int^t \left(\int_{\mathscr{H}(t')\cap\mathscr{U}} S^{ab} t_{;a}\, \mathrm{d}\sigma_b \right) \mathrm{d}t' + \int^t \left(\int_{\mathscr{H}(t')\cap\mathscr{U}} S^{ab}{}_{;a}\, \mathrm{d}\sigma_b \right) \mathrm{d}t'.$$

Consider the volume integral

$$I(t) = \int_{\mathcal{U}(t)} (S^{ab}t_{;a})_{;b}\,dv = \int_{\mathcal{U}(t)} S^{ab}t_{;ab}\,dv + \int_{\mathcal{U}(t)} S^{ab}_{;b}t_{;a}\,dv.$$

By Gauss' theorem this can be transformed into an integral over the boundary of $\mathcal{U}(t)$:

$$I(t) = \int_{\partial\mathcal{U}(t)} S^{ab}t_{;a}\,d\sigma_b.$$

The boundary of $\mathcal{U}(t)$ will consist of $\mathcal{U}(t) \cap \partial\mathcal{U}$ and $\mathcal{U} \cap \mathcal{H}(t)$. Since S^{ab} is zero on $(\partial\mathcal{U})_3$,

$$I(t) = \int_{\mathcal{U}(t)\,\cap\,(\partial\mathcal{U})_1} + \int_{\mathcal{U}(t)\,\cap\,(\partial\mathcal{U})_2} + \int_{\mathcal{U}\,\cap\,\mathcal{H}(t)}.$$

By the dominant energy condition, $S^{ab}t_{;a}$ is a non-spacelike vector such that $S^{ab}t_{;a}t_{;b} \geqslant 0$. As the normal form to $(\partial\mathcal{U})_2$ is non-spacelike and such that $n_a t_{;b}g^{ab} < 0$, the second term on the right will be non-negative. Thus

$$\int_{\mathcal{U}\,\cap\,\mathcal{H}(t)} S^{ab}t_{;a}\,d\sigma_b \leqslant -\int_{\mathcal{U}(t)\,\cap\,(\partial\mathcal{U})_1} S^{ab}t_{;a}\,d\sigma_b$$
$$+ \int_{\mathcal{U}(t)} (S^{ab}t_{;ab} + S^{ab}_{;b}t_{;a})\,dv.$$

Since $\mathcal{U}$ is compact there will be some upper bound to the components of $t_{;ab}$ in any orthonormal basis whose timelike vector is in the direction of $t_{;a}$. Thus there will be some $P > 0$ such that on $\mathcal{U}$,

$$S^{ab}t_{;ab} \leqslant PS^{ab}t_{;a}t_{;b}$$

for any S^{ab} which obeys the dominant energy condition. The volume integral over $\mathcal{U}(t)$ can be decomposed into a surface integral over $\mathcal{H}(t') \cap \mathcal{U}$ followed by an integral with respect to t':

$$\int_{\mathcal{U}(t)} (PS^{ab}t_{;a}t_{;b} + S^{ab}_{;b}t_{;a})\,dv = \int^t \left\{ \int_{\mathcal{H}(t')\,\cap\,\mathcal{U}} (PS^{ab}t_{;b} + S^{ab}_{;b})\,d\sigma_a \right\} dt',$$

where $d\sigma_a$ is the surface element of $\mathcal{H}(t')$. Thus

$$\int_{\mathcal{H}(t)\,\cap\,\mathcal{U}} S^{ab}t_{;a}\,d\sigma_b \leqslant -\int_{\mathcal{U}(t)\,\cap\,(\partial\mathcal{U})_1} S^{ab}t_{;a}\,d\sigma_b$$
$$+ P\int^t \left(\int_{\mathcal{H}(t')\,\cap\,\mathcal{U}} S^{ab}t_{;a}\,d\sigma_b \right) dt' + \int^t \left(\int_{\mathcal{H}(t')\,\cap\,\mathcal{U}} S^{ab}_{;a}\,d\sigma_b \right) dt'. \quad \square$$

As an immediate consequence of this result one has:

The conservation theorem

If the energy–momentum tensor obeys the dominant energy condition and is zero on $(\partial \mathcal{U})_3$ and on the initial surface $(\partial \mathcal{U})_1$, then it is zero everywhere on $\mathcal{U}$.

Let
$$x(t) = \int_{\mathcal{U}(t)} T^{ab}t_{;a}t_{;b}\,\mathrm{d}v$$

$$= \int^t \left(\int_{\mathscr{H}(t') \cap \mathcal{U}} T^{ab}t_{;a}\,\mathrm{d}\sigma_b \right) \mathrm{d}t' \geqslant 0.$$

Then the above lemma gives $\mathrm{d}x/\mathrm{d}t \leqslant Px$. But for sufficiently early values of t, $\mathscr{H}(t)$ will not intersect $\mathcal{U}$ and so x will vanish. Thus x will vanish for all t which implies that T^{ab} is zero on $\mathcal{U}$.   □

From the conservation theorem it follows that if the energy–momentum tensor vanishes on a set $\mathscr{S}$, then it also vanishes on the

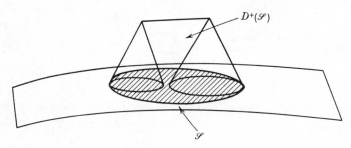

FIGURE 10. The future Cauchy development $D^+(\mathscr{S})$ of a spacelike set $\mathscr{S}$.

future Cauchy development $D^+(\mathscr{S})$, which is defined as the set of all points through which every past-directed non-spacelike curve intersects $\mathscr{S}$ (figure 10) (cf. § 6.5). For if q is any point of $D^+(\mathscr{S})$, the region of $D^+(\mathscr{S})$ to the past of q is compact (proposition 6.6.6) and may be taken as $\mathcal{U}$. This result may be interpreted as saying that the dominant energy condition implies that matter cannot travel faster than light.

For our consideration of singularities, the importance of the weak energy condition is that it implies that matter always has a converging (or more strictly nondiverging) effect on congruences of null geodesics. If the vorticity vanishes, the expansion θ obeys the equation:

$$\frac{\mathrm{d}}{\mathrm{d}v}\theta = -R_{ab}K^a K^b - 2\hat{\sigma}^2 - \tfrac{1}{2}\theta^2.$$

Thus in this case θ will monotonically decrease along the null geodesic if $R_{ab}W^aW^b \geqslant 0$ for any null vector $\mathbf{W}$. We shall call this the *null convergence condition*. From the Einstein equations,

$$R_{ab} - \tfrac{1}{2}g_{ab}R + \Lambda g_{ab} = 8\pi T_{ab},$$

it follows that this condition is implied by the weak energy condition, independent of the value of Λ.

From (4.26) it can be seen that the expansion θ of a timelike geodesic congruence with zero vorticity will monotonically decrease along a geodesic if $R_{ab}W^aW^b \geqslant 0$ for any timelike vector $\mathbf{W}$. We shall call this the *timelike convergence condition*. By the Einstein equation, this condition will be satisfied if the energy–momentum tensor obeys the inequality,

$$T_{ab}W^aW^b \geqslant W^aW_a\left(\tfrac{1}{2}T - \frac{1}{8\pi}\Lambda\right).$$

This will hold for type I if

$$\mu + p_\alpha \geqslant 0, \quad \mu + \Sigma p_\alpha - \frac{1}{4\pi}\Lambda \geqslant 0,$$

and for type II if

$$\nu = +1, \quad \kappa \geqslant 0, \quad p_1 \geqslant 0, \quad p_2 \geqslant 0 \quad \text{and} \quad p_1 + p_2 - \frac{1}{4\pi}\Lambda \geqslant 0.$$

We shall say that the energy–momentum tensor satisfies the *strong energy condition* if it obeys the above inequality for $\Lambda = 0$. This is a stricter requirement than the weak energy condition but it is still physically reasonable for the total energy–momentum tensor. For the general case, type I, it would be violated only by a negative energy density or a large negative pressure (e.g. for a perfect fluid with density 1 gm cm^{-3} it can only be violated if $p < -10^{15}$ atmospheres). It holds for the electromagnetic field and for the scalar field with m zero (in particular, it holds for the scalar field of Brans and Dicke). For m non-zero, the energy–momentum tensor of a scalar field has the form (§3.3):

$$T_{ab} = \phi_{;a}\phi_{;b} - \tfrac{1}{2}g_{ab}(\phi_{;c}\phi_{;d}g^{cd} + m^2\phi^2).$$

Thus if W^a is a unit timelike vector

$$T_{ab}W^aW^b - \tfrac{1}{2}W_aW^aT = (\phi_{;a}W^a)^2 - \frac{1}{2}\frac{m^2}{\hbar^2}\phi^2 \qquad (4.37)$$

which may be negative. However by the equation of the scalar field

$$\frac{1}{2}\frac{m^2}{\hbar^2}\phi^2 = \tfrac{1}{2}\phi\phi_{;ab}g^{ab}.$$

Inserting this in (4.37) and integrating over a region $\mathcal{U}$, one obtains

$$\frac{1}{2}\int_{\mathcal{U}} (g^{ab} + 2W^a W^b)\, \phi_{;a}\phi_{;b}\, \mathrm{d}\sigma - \frac{1}{2}\int_{\partial\mathcal{U}} \phi\phi_{;a}g^{ab}\, \mathrm{d}\sigma_b.$$

The first term will be non-negative since $g^{ab} + 2W^a W^b$ is a positive definite metric and the second term will be small compared to the first if the region $\mathcal{U}$ is large compared to the wavelength h/m. For π mesons, which may be described classically by a scalar field with $m = 6 \times 10^{-25}$ gm, this wavelength is 3×10^{-13} cm. Thus although the energy–momentum tensor of π mesons may not satisfy the strong energy condition at every point, this should not affect the convergence of timelike geodesics over distances greater than 10^{-12} cm. This might possibly lead to a breakdown of the singularity theorems in chapter 8 when the radius of curvature of space–time becomes less than 10^{-12} cm but such a curvature would be so extreme that it might well count as a singularity (§10.2).

4.4 Conjugate points

In §4.1 we saw that the components of the vector which represented the separation between a curve $\gamma(s)$ and a neighbouring curve in a congruence of timelike geodesics, satisfied the Jacobi equation:

$$\frac{\mathrm{d}^2}{\mathrm{d}s^2}Z^\alpha = -R_{\alpha 4 \beta 4}Z^\beta \quad (\alpha, \beta = 1, 2, 3). \tag{4.38}$$

A solution of this equation will be called a *Jacobi field* along $\gamma(s)$. Since a solution may be specified by giving the values of Z^α and $\mathrm{d}Z^\alpha/\mathrm{d}s$ at some point on $\gamma(s)$ there will be six independent Jacobi fields along $\gamma(s)$. There will be three independent Jacobi fields which vanish at some point q of $\gamma(s)$. They may be expressed as:

$$Z^\alpha(s) = A_{\alpha\beta}(s)\frac{\mathrm{d}}{\mathrm{d}s}Z^\beta\big|_q,$$

where

$$\frac{\mathrm{d}^2}{\mathrm{d}s^2}A_{\alpha\beta}(s) = -R_{\alpha 4 \gamma 4}A_{\gamma\beta}(s), \tag{4.39}$$

and $A_{\alpha\beta}(s)$ is a 3×3 matrix which vanishes at q. These Jacobi fields may be thought of as representing the separation of neighbouring geodesics through q. As before one may define the vorticity, shear and expansion of the Jacobi fields along $\gamma(s)$ which vanish at q:

$$\omega_{\alpha\beta} = A^{-1}{}_{\gamma|\beta} \frac{d}{ds} A_{\alpha|\gamma}, \tag{4.40}$$

$$\sigma_{\alpha\beta} = A^{-1}{}_{\gamma(\beta} \frac{d}{ds} A_{\alpha)\gamma} - \tfrac{1}{3}\delta_{\alpha\beta}\theta, \tag{4.41}$$

$$\theta = (\det \mathbf{A})^{-1}\frac{d}{ds}(\det \mathbf{A}). \tag{4.42}$$

These will obey the equations derived in §4.1, with $\dot{V}_\alpha = 0$. In particular

$$A_{\gamma\alpha}\omega_{\gamma\delta}A_{\delta\beta} = \frac{1}{2}\left(A_{\gamma\alpha}\frac{d}{ds}A_{\gamma\beta} - A_{\gamma\beta}\frac{d}{ds}A_{\gamma\alpha}\right)$$

will be constant along $\gamma(s)$. But it vanishes at q where $A_{\alpha\beta}$ is zero. Thus $\omega_{\alpha\beta}$ will be zero wherever $A_{\alpha\beta}$ is non-singular.

We shall say that a point p on $\gamma(s)$ *is conjugate to q along $\gamma(s)$* if there is a Jacobi field along $\gamma(s)$, not identically zero, which vanishes at q and p. One may think of p as a point where infinitesimally neighbouring geodesics through q intersect. (Note, however, that it may be only *infinitesimally* neighbouring geodesics which intersect at p; there need not be two distinct geodesics from q passing through p.) The Jacobi fields along $\gamma(s)$ which vanish at q are described by the matrix $A_{\alpha\beta}$. Thus a point p is conjugate to q along $\gamma(s)$ if and only if $A_{\alpha\beta}$ is singular at p. The expansion θ is defined as $(\det \mathbf{A})^{-1} d(\det \mathbf{A})/ds$. Since $A_{\alpha\beta}$ obeys (4.39) where $R_{\alpha4\gamma4}$ is finite, $d(\det \mathbf{A})/ds$ will be finite. Thus a point p will be conjugate to q along $\gamma(s)$ if θ becomes infinite there. The converse will also be true since $\theta = d\log(\det \mathbf{A})/ds$ and $A_{\alpha\beta}$ can be singular only at isolated points or else it would be singular everywhere.

Proposition 4.4.1

If at some point $\gamma(s_1)$ $(s_1 > 0)$, the expansion θ has a negative value $\theta_1 < 0$ and if $R_{ab}V^aV^b \geqslant 0$ everywhere then there will be a point conjugate to q along $\gamma(s)$ between $\gamma(s_1)$ and $\gamma(s_1 + (3/-\theta_1))$, provided that $\gamma(s)$ can be extended to this parameter value. (This may not be possible if space–time is geodesically incomplete. In chapter 8 we shall interpret such incompleteness as evidence of the existence of a singularity.)

The expansion θ of the matrix $A_{\alpha\beta}$ obeys the Raychaudhuri equation (4.26):

$$\frac{d}{ds}\theta = -R_{ab}V^aV^b - 2\sigma^2 - \tfrac{1}{3}\theta^2$$

where we have used the fact that the vorticity is zero. All the terms on the right-hand side are negative. Thus for $s > s_1$

$$\theta \leqslant \frac{3}{s - (s_1 + (3/-\theta_1))}.$$

So θ will become infinite and there will be a point conjugate to q for some value of s between s_1 and $s_1 + (3/-\theta_1)$. $\square$

In other words, if the timelike convergence condition holds and if the neighbouring geodesics from q start converging on $\gamma(s)$, then some infinitesimally neighbouring geodesic will intersect $\gamma(s)$ providing that $\gamma(s)$ can be extended to large enough values of the parameter s.

Proposition 4.4.2

If $R_{ab} V^a V^b \geqslant 0$ and if at some point $p = \gamma(s_1)$ the tidal force $R_{abcd} V^b V^d$ is non zero, there will be values s_0 and s_2 such that $q = \gamma(s_0)$ and $r = \gamma(s_2)$ will be conjugate along $\gamma(s)$, providing that $\gamma(s)$ can be extended to these values.

A solution of (4.39) along $\gamma(s)$ is uniquely determined by the values of $A_{\alpha\beta}$ and $dA_{\alpha\beta}/ds$ at p. Consider the set P consisting of all such solutions for which $A_{\alpha\beta}|_p = \delta_{\alpha\beta}$, $(dA_{\alpha\beta}/ds)|_p$ is symmetric with trace $\theta|_p \leqslant 0$. For each solution in P there will be some $s_3 > s_1$ for which $A_{\alpha\beta}(s_3)$ is singular, since either $\theta|_p < 0$, in which case this follows from the previous result, or $\theta|_p = 0$, in which case $(d\sigma_{\alpha\beta}/ds)|_p$ is non-zero which will then cause σ^2 to be positive and so cause θ to become negative for $s > s_1$. The members of the set P are in one–one correspondence with the space S of all symmetric 3×3 matrices with non-positive trace (i.e. with the values of $dA_{\alpha\beta}/ds)|_p$). There is thus a map η from S to $\gamma(s)$ which assigns to each initial value $(dA_{\alpha\beta}/ds)|_p$ the point on $\gamma(s)$ where $A_{\alpha\beta}$ first becomes singular. The map η is continuous. Further if any component of $(dA_{\alpha\beta}/ds)|_p$ is very large, the corresponding point on $\gamma(s)$ will lie near p, since in the limit the term $R_{\alpha4\gamma4}$ in (4.39) becomes irrelevant and the solution resembles the flat space case. Thus there is some $C > 0$ and some $s_4 > s_1$ such that if any component of $(dA_{\alpha\beta}/ds)|_p$ is greater than C, the corresponding point on $\gamma(s)$ will be before $\gamma(s_4)$. However the subspace of S consisting of all matrices all of whose components are less than or equal to C, is compact. This shows that there is some $s_5 > s_1$ such that $\eta(S)$ is contained in the segment from $\gamma(s_1)$ to $\gamma(s_5)$. Consider now a point $r = \gamma(s_2)$ where $s_2 > s_5$. If there is no point conjugate to r between r and p, the Jacobi fields which are zero at r

must have an expansion θ which is positive at p (otherwise they would be in the set P which represents all families of Jacobi fields with zero vorticity which have non-positive expansion at p). It follows from the previous result that there is then a point $q = \gamma(s_0)$ $(s_0 < s_1)$ which is conjugate to r along $\gamma(s)$. $\square$

In a physically realistic solution (though not necessarily in an exact one with a high degree of symmetry), one would expect every timelike geodesic to encounter some matter or some gravitational radiation and so to contain some point where $R_{abcd} V^b V^d$ was non-zero. Thus it would be reasonable to assume that in such a solution every timelike geodesic would contain pairs of conjugate points, provided that it could be extended sufficiently far in both directions.

We shall also consider the congruence of timelike geodesics normal to a spacelike three-surface, $\mathscr{H}$. By a *spacelike three-surface*, $\mathscr{H}$, we mean an imbedded three-dimensional submanifold defined locally by $f = 0$ where f is a C^2 function and $g^{ab} f_{;a} f_{;b} < 0$ when $f = 0$. We define $\mathbf{N}$, the unit normal vector to $\mathscr{H}$, by $N^a = (-g^{bc} f_{;b} f_{;c})^{-\frac{1}{2}} g^{ad} f_{;d}$ and the second fundamental tensor $\boldsymbol{\chi}$ of $\mathscr{H}$ by $\chi_{ab} = h_a{}^c h_b{}^d N_{c;d}$, where $h_{ab} = g_{ab} + N_a N_b$ is called the first fundamental tensor (or induced metric tensor) of $\mathscr{H}$ (cf. §2.7). It follows from the definition that $\boldsymbol{\chi}$ is symmetric. The congruence of timelike geodesics orthogonal to $\mathscr{H}$ will consist of the timelike geodesics whose unit tangent vector $\mathbf{V}$ equals the unit normal $\mathbf{N}$ at $\mathscr{H}$. Then one has:

$$V_{a;b} = \chi_{ab} \quad \text{at} \quad \mathscr{H}. \tag{4.43}$$

The vector $\mathbf{Z}$ which represents the separation of a neighbouring geodesic normal to $\mathscr{H}$ from a geodesic $\gamma(s)$ normal to $\mathscr{H}$, will obey the Jacobi equation (4.38). At a point q on $\gamma(s)$ at $\mathscr{H}$ it will satisfy the initial condition:

$$\frac{\mathrm{d}}{\mathrm{d}s} Z^\alpha = \chi_{\alpha\beta} Z^\beta. \tag{4.44}$$

We shall express the Jacobi fields along $\gamma(s)$ which satisfy the above condition as

$$Z^\alpha(s) = A_{\alpha\beta}(s) Z^\beta|_q,$$

where

$$\frac{\mathrm{d}^2}{\mathrm{d}s^2} A_{\alpha\beta} = -R_{\alpha 4 \gamma 4} A_{\gamma\beta} \tag{4.45}$$

and at q, $A_{\alpha\beta}$ is the unit matrix and

$$\frac{\mathrm{d}}{\mathrm{d}s} A_{\alpha\beta} = \chi_{\alpha\gamma} A_{\gamma\beta}. \tag{4.46}$$

We shall say that a point p on $\gamma(s)$ *is conjugate to $\mathscr{H}$ along $\gamma(s)$* if there is a Jacobi field along $\gamma(s)$ not identically zero, which satisfies the initial conditions (4.44) at q and vanishes at p. In other words, p is conjugate to $\mathscr{H}$ along $\gamma(s)$ if and only if $A_{\alpha\beta}$ is singular at p. One may think of p as being a point where neighbouring geodesics normal to $\mathscr{H}$ intersect. As before $A_{\alpha\beta}$ will be singular where and only where the expansion θ becomes infinite. At q, the initial value of $A_{\gamma\alpha}\omega_{\gamma\delta}A_{\delta\beta}$ will be zero, therefore $\omega_{\alpha\beta}$ will be zero on $\gamma(s)$. The initial value of θ will be $\chi_{ab}g^{ab}$.

Proposition 4.4.3

If $R_{ab}V^aV^b \geqslant 0$ and $\chi_{ab}g^{ab} < 0$, there will be a point conjugate to $\mathscr{H}$ along $\gamma(s)$ within a distance $3/(-\chi_{ab}g^{ab})$ from $\mathscr{H}$, provided that $\gamma(s)$ can be extended that far.

This may be proved using the Raychaudhuri equation (4.26) as in proposition 4.4.1. $\square$

We shall call a solution of the equation:

$$\frac{\mathrm{d}^2}{\mathrm{d}v^2}Z^m = -R_{m4n4}Z^n \quad (m, n = 1, 2)$$

along a null geodesic $\gamma(v)$, a *Jacobi field along $\gamma(v)$*. The components Z^m could be thought of as the components, with respect to the basis $\mathbf{E}_1$ and $\mathbf{E}_2$, of a vector in the space S_q at each point q. We shall say that p is conjugate to q along the null geodesic $\gamma(v)$ if there is a Jacobi field along $\gamma(v)$, not identically zero, which vanishes at q and p. If $\mathbf{Z}$ is a vector connecting neighbouring null geodesics which pass through q, the component Z^3 will be zero everywhere. Thus p can be thought of as a point where infinitesimally neighbouring geodesics through q intersect. Representing the Jacobi fields along $\gamma(v)$ which vanish at q by the 2×2 matrix $\hat{A}_{mn}$,

$$Z^m(v) = \hat{A}_{mn}\frac{\mathrm{d}}{\mathrm{d}v}Z^n\big|_q.$$

One has as before: $\hat{A}_{lm}\hat{\omega}_{lk}\hat{A}_{kn} = 0$, so the vorticity of the Jacobi fields which are zero at p vanishes. Also p will be conjugate to q along $\gamma(v)$ if and only if

$$\theta = (\det \hat{A})^{-1}\frac{\mathrm{d}}{\mathrm{d}v}(\det \hat{A})$$

becomes infinite at p. Analogous to proposition 4.4.1, we have:

Proposition 4.4.4

If $R_{ab}K^aK^b \geqslant 0$ everywhere and if at some point $\gamma(v_1)$ the expansion θ has the negative value $\theta_1 < 0$, then there will be a point conjugate to q along $\gamma(v)$ between $\gamma(v_1)$ and $\gamma(v_1 + (2/-\theta_1))$ provided that $\gamma(v)$ can be extended that far.

The expansion θ of the matrix $\hat{A}_{mn}$ obeys (4.35):

$$\frac{\mathrm{d}}{\mathrm{d}v}\theta = -R_{ab}K^aK^b - 2\hat{\sigma}^2 - \tfrac{1}{2}\theta^2,$$

and so the proof proceeds as before. $\square$

Proposition 4.4.5

If $R_{ab}K^aK^b \geqslant 0$ everywhere and if at $p = \gamma(v_1)$, $K^cK^dK_{[a}R_{b]cd[e}K_{f]}$ is non-zero, there will be v_0 and v_2 such that $q = \gamma(v_0)$ and $r = \gamma(v_2)$ will be conjugate along $\gamma(v)$ provided $\gamma(v)$ can be extended to these values.

If $K^cK^dK_{[a}R_{b]cd[e}K_{f]}$ is non zero then so is R_{m4n4}. The proof is then similar to that of proposition 4.4.2. $\square$

As in the timelike case, this condition will be satisfied for a null geodesic which passes through some matter provided that the matter is not pure radiation (energy–momentum tensor type II of §4.3) and moving in the direction of the geodesic tangent vector **K**. It will be satisfied in empty space if the null geodesic contains some point where the Weyl tensor is non-zero and where **K** does not lie in one of the directions (there are at most four such directions) at that point for which $K^cK^dK_{[a}C_{b]cd[e}K_{f]} = 0$. It therefore seems reasonable to assume that in a physically realistic solution every timelike or null geodesic will contain a point at which $K^aK^bK_{[c}R_{d]ab[e}K_{f]}$ is not zero. We shall say that a space–time satisfying this condition satisfies the *generic condition*.

Similarly we may also consider the null geodesics orthogonal to a spacelike two-surface $\mathscr{S}$. By a *spacelike two-surface $\mathscr{S}$*, we mean an imbedded two-dimensional submanifold defined locally by $f_1 = 0$, $f_2 = 0$ where f_1 and f_2 are C^2 functions such that when $f_1 = 0, f_2 = 0$ then $f_{1;a}$ and $f_{2;a}$ are non-vanishing and not parallel and

$$(f_{1;a} + \mu f_{2;a})(f_{1;b} + \mu f_{2;b})g^{ab} = 0$$

for two distinct real values μ_1 and μ_2 of μ. Then any vector lying in the two-surface is necessarily spacelike. We shall define $N_1{}^a$ and $N_2{}^a$, the two null vectors normal to $\mathscr{S}$, as proportional to $g^{ab}(f_{1;b} + \mu_1 f_{2;b})$ and $g^{ab}(f_{1;b} + \mu_2 f_{2;b})$ respectively, and normalize them so that

$$N_1{}^a N_2{}^b g_{ab} = -1.$$

One can complete the pseudo-orthonormal basis by introducing two spacelike unit vectors $Y_1{}^a$ and $Y_2{}^a$ orthogonal to each other and to $N_1{}^a$ and $N_2{}^a$. We define the two null second fundamental tensors of $\mathscr{S}$ as:

$$_n\chi_{ab} = -N_{nc;d}(Y_1{}^c Y_{1a} + Y_2{}^c Y_{2a})(Y_1{}^d Y_{1b} + Y_2{}^d Y_{2b}),$$

where n takes the values 1, 2. The tensors $_1\chi_{ab}$ and $_2\chi_{ab}$ are symmetric.

There will be two families of null geodesics normal to $\mathscr{S}$ corresponding to the two null normals $N_1{}^a$ and $N_2{}^a$. Consider the family whose tangent vector $\mathbf{K}$ equals $\mathbf{N_2}$ at $\mathscr{S}$. We may fix our pseudo-orthogonal basis $\mathbf{E_1, E_2, E_3, E_4}$ by taking $\mathbf{E_1 = Y_1, E_2 = Y_2, E_3 = N_1}$, $\mathbf{E_4 = N_2}$ at $\mathscr{S}$ and parallelly propagating along the null geodesics. The projection into the space S_q of the vector $\mathbf{Z}$ representing the separation of neighbouring null geodesics from the null geodesic $\gamma(v)$ will satisfy (4.30) and the initial conditions

$$\frac{d}{dv} Z^m = {}_2\chi_{mn} Z^n \tag{4.47}$$

at q on $\gamma(v)$ at $\mathscr{S}$. As before the vorticity of these fields will be zero. The initial value of the expansion θ will be $_2\chi_{ab}g^{ab}$. Analogous to proposition 4.4.3 we have:

Proposition 4.4.6

If $R_{ab}K^a K^b \geqslant 0$ everywhere and $_2\chi_{ab}g^{ab}$ is negative there will be a point conjugate to $\mathscr{S}$ along $\gamma(v)$ within an affine distance $2/(-_2\chi_{ab}g^{ab})$ from $\mathscr{S}$.   □

From their definition, the existence of conjugate points implies the existence of self-intersections or caustics in families of geodesics. A further significance of conjugate points will be discussed in the next section.

4.5   Variation of arc-length

In this section we consider timelike and non-spacelike curves which are piecewise C^3 but which may have points at which their tangent vector is discontinuous. We shall require that at such points the two tangent vectors

$$\left.\frac{\partial}{\partial t}\right|_{-} \quad \text{and} \quad \left.\frac{\partial}{\partial t}\right|_{+} \quad \text{satisfy} \quad g\left(\left.\frac{\partial}{\partial t}\right|_{-}, \left.\frac{\partial}{\partial t}\right|_{+}\right) = -1,$$

that is, they point into the same half of the null cone.

Proposition 4.5.1

Let $\mathcal{U}$ be a convex normal coordinate neighbourhood about q. Then the points which can be reached from q by timelike (respectively non-spacelike) curves in $\mathcal{U}$ are those of the form $\exp_q(\mathbf{X})$, $\mathbf{X} \in T_q$ where $g(\mathbf{X}, \mathbf{X}) < 0$ (respectively $\leqslant 0$). (Here, and for the rest of this section, we consider the map exp to be restricted to the neighbourhood of the origin in T_q which is diffeomorphic to $\mathcal{U}$ under $\exp_q$.)

In other words, the null geodesics from q form the boundary of the region in $\mathcal{U}$ which can be reached from q by timelike or non-spacelike curves in $\mathcal{U}$. This is fairly obvious intuitively but because it is fundamental to the concept of causality we shall prove it rigorously. We first establish the following lemma:

Lemma 4.5.2

In $\mathcal{U}$ the timelike geodesics through q are orthogonal to the three-surfaces of constant σ ($\sigma < 0$) where the value of σ at $p \in \mathcal{U}$ is defined to be $g(\exp_q^{-1} p, \exp_q^{-1} p)$.

The pr f is based on the fact that the vector representing the separation of points equal distances along neighbouring geodesics remains orthogonal to the geodesics if it is so initially. More precisely, let $\mathbf{X}(t)$ denote a curve in T_q, where $g(\mathbf{X}(t), \mathbf{X}(t)) = -1$. One must show that the corresponding curves $\lambda(t) = \exp_q(s_0 \mathbf{X}(t))$ (s_0 constant) in $\mathcal{U}$, where defined, are orthogonal to the timelike geodesics $\gamma(s) = \exp_q(s\mathbf{X}(t_0))$ (t_0 constant). Thus in terms of the two-surface α defined by $x(s,t) = \exp_q(s\mathbf{X}(t))$, one must prove that

$$g\left(\left(\frac{\partial}{\partial s}\right)_\alpha, \left(\frac{\partial}{\partial t}\right)_\alpha\right) = 0$$

(see figure 11). Now

$$\frac{\partial}{\partial s} g\left(\frac{\partial}{\partial s}, \frac{\partial}{\partial t}\right) = g\left(\frac{D}{\partial s}\frac{\partial}{\partial s}, \frac{\partial}{\partial t}\right) + g\left(\frac{\partial}{\partial s}, \frac{D}{\partial s}\frac{\partial}{\partial t}\right).$$

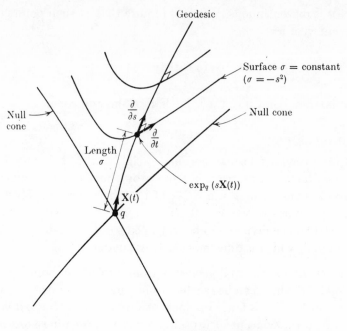

FIGURE 11. In a normal neighbourhood, surfaces at constant distance from q are orthogonal to the geodesics through q.

The first term on the right is zero as $\partial/\partial s$ is the unit tangent vector to the timelike geodesics from q. In the second term one has from the definition of the Lie derivative that

$$\frac{D}{\partial s}\frac{\partial}{\partial t} = \frac{D}{\partial t}\frac{\partial}{\partial s}.$$

Thus $\dfrac{\partial}{\partial s}g\left(\dfrac{\partial}{\partial s}, \dfrac{\partial}{\partial t}\right) = g\left(\dfrac{\partial}{\partial s}, \dfrac{D}{\partial t}\dfrac{\partial}{\partial s}\right) = \dfrac{1}{2}\dfrac{\partial}{\partial t}g\left(\dfrac{\partial}{\partial s}, \dfrac{\partial}{\partial s}\right) = 0.$

Therefore $g(\partial/\partial s, \partial/\partial t)$ is independent of s. But at $s = 0$, $(\partial/\partial t)_\alpha = 0$. Thus $g(\partial/\partial s, \partial/\partial t)$ is identically zero.   □

Proof of proposition 4.5.1. Let C_q denote the set of all timelike vectors at q. These constitute the interior of a solid cone in T_q with vertex at the origin. Let $\gamma(t)$ be a timelike curve in $\mathscr{U}$ from q to p and let $\bar{\gamma}(t)$ be the piecewise C^2 curve in T_q defined by $\bar{\gamma}(t) = \exp_q^{-1}(\gamma(t))$. Then identifying the tangent space to T_q with T_q itself, one has

$$(\partial/\partial t)_\gamma|_q = (\partial/\partial t)_{\bar{\gamma}}|_q.$$

Therefore at q, $(\partial/\partial t)_{\bar{\gamma}}$ will be timelike. This shows that the curve $\bar{\gamma}(t)$ will enter the region C_q. But $\exp_q(C_q)$ is the region of $\mathscr{U}$ on which σ is negative and in which by the previous lemma the surfaces of constant σ are spacelike. Thus σ must monotonically decrease along $\gamma(t)$ since $(\partial/\partial t)_\gamma$ being timelike can never be tangent to the surfaces of constant σ and since at any non-differentiable point of $\gamma(t)$ the two tangent vectors point into the same half of the null cone. Therefore $p \in \exp_q(C_q)$ which completes the proof for timelike curves. To prove that a non-spacelike curve $\gamma(t)$ remains in $\exp_q(\bar{C}_q)$, one performs a small variation of $\gamma(t)$ which makes it into a timelike curve. Let $\mathbf{Y}$ be a vector field on T_q such that in $\mathscr{U}$ the induced vector field $\exp_{q*}(\mathbf{Y})$ is everywhere timelike and such that $g(\mathbf{Y}, (\partial/\partial t)_\gamma|_q) < 0$. For each $\epsilon \geqslant 0$ let $\beta(r, \epsilon)$ be the curve T_q starting at the origin such that the tangent vector $(\partial/\partial r)_\beta$ equals $(\partial/\partial t)_{\bar{\gamma}}|_{t=r} + \epsilon \mathbf{Y}|_{\beta(r, \epsilon)}$. Then $\beta(r, \epsilon)$ depends differentiably on r and ϵ. For each $\epsilon > 0$, $\exp_q(\beta(r, \epsilon))$ is a timelike curve in $\mathscr{U}$ and so is contained in $\exp_q(C_q)$. Thus the non-spacelike curve $\exp_q(\beta(r, 0)) = \gamma(r)$ is contained in $\overline{\exp_q(C_q)} = \exp_q(\bar{C}_q)$. $\qquad\square$

Corollary

If $p \in \mathscr{U}$ can be reached from q by a non-spacelike curve but not by a timelike curve, then p lies on a null geodesic from q. $\qquad\square$

The length of a non-spacelike curve $\gamma(t)$ from q to p is

$$L(\gamma, q, p) = \int_q^p \left[-g\left(\frac{\partial}{\partial t}, \frac{\partial}{\partial t} \right) \right]^{\frac{1}{2}} \mathrm{d}t,$$

where the integral is taken over the differentiable sections of the curve.

In a positive definite metric one may seek the shortest curve between two points but in a Lorentz metric there will not be any shortest curve as any curve can be deformed into a null curve which has zero length. However, in certain cases there will be a longest non-spacelike curve between two points or between a point and a spacelike three-surface. We deal first with the situation when the two points are close together. We shall then derive necessary conditions in the general case when the two points are not close. The sufficient condition in this case will be dealt with in § 6.7.

Proposition 4.5.3

Let q and p lie in a convex normal neighbourhood $\mathscr{U}$. Then, if q and p can be joined by a non-spacelike curve in $\mathscr{U}$, the longest such curve is the unique non-spacelike geodesic curve in $\mathscr{U}$ from q to p. Moreover, defining $\rho(q, p)$ as the length of this curve if it exists, and as zero otherwise, $\rho(q, p)$ is a continuous function on $\mathcal{U} \times \mathcal{U}$.

By the definition of convex normal neighbourhoods (§2.5), there is a unique geodesic $\gamma(t)$ in $\mathcal{U}$ with $\gamma(0) = q$, $\gamma(1) = p$. Since this geodesic depends differentiably on its endpoints, the function

$$\sigma(q, p) = \int_0^1 g\left(\left(\frac{\partial}{\partial t}\right)_\gamma, \left(\frac{\partial}{\partial t}\right)_\gamma\right) \mathrm{d}t$$

will be differentiable on $\mathcal{U} \times \mathcal{U}$. (This function σ is the same as that in lemma 4.5.2.) Thus $\rho(q, p)$ will be continuous on $\mathcal{U} \times \mathcal{U}$ since it equals $[-\sigma(q, p)]^{\frac{1}{2}}$ if $\sigma < 0$ and is zero otherwise. It now remains to show that if q and p can be joined by a timelike curve in $\mathcal{U}$ then the timelike geodesic γ between them is the longest such curve. Let $\alpha(s, t)$ be $\exp_q(s\mathbf{X}(t))$ as before where $g(\mathbf{X}(t), \mathbf{X}(t)) = -1$. If $\lambda(t)$ is a timelike curve in $\mathcal{U}$ from q to p, it can be represented as $\lambda(t) = \alpha(f(t), t)$. Then

$$\left(\frac{\partial}{\partial t}\right)_\lambda = f'(t)\left(\frac{\partial}{\partial s}\right)_\alpha + \left(\frac{\partial}{\partial t}\right)_\alpha.$$

Since the two vectors on the right are mutually orthogonal by lemma 4.5.2. and since $g((\partial/\partial s)_\alpha, (\partial/\partial s)_\alpha) = -1$, this gives

$$g\left(\left(\frac{\partial}{\partial t}\right)_\lambda, \left(\frac{\partial}{\partial t}\right)_\lambda\right) = -(f'(t))^2 + g\left(\left(\frac{\partial}{\partial t}\right)_\alpha, \left(\frac{\partial}{\partial t}\right)_\alpha\right) \geqslant -(f'(t))^2,$$

the equality holding if and only if $(\partial/\partial t)_\alpha = 0$ and hence if and only if λ is a geodesic curve. Thus

$$L(\lambda, q, p) \leqslant \int_q^p f'(t)\,\mathrm{d}t = \rho(q, p),$$

the equality holding if and only if λ is the unique geodesic curve in $\mathcal{U}$ from q to p.   □

We shall now consider the case where q and p are not necessarily contained in a convex normal neighbourhood $\mathcal{U}$. By considering small variations we shall derive necessary conditions for a timelike curve $\gamma(t)$ from q to p to be the longest such curve from q to p. A *variation* α of $\gamma(t)$ is a C^{1-} map $\alpha: (-\epsilon, \epsilon) \times [0, t_p] \to \mathcal{M}$ such that

(1) $\alpha(0, t) = \gamma(t)$;

(2) there is a subdivision $0 = t_1 < t_2 \ldots < t_n = t_p$ of $[0, t_p]$ such that α is C^3 on each $(-\epsilon, \epsilon) \times [t_i, t_{i+1}]$;

(3) $\alpha(u, 0) = q$, $\alpha(u, t_p) = p$;

(4) for each constant u, $\alpha(u, t)$ is a timelike curve.

The vector $(\partial/\partial u)_\alpha|_{u=0}$ will be called the *variation vector* $\mathbf{Z}$. Conversely, given a continuous, piecewise C^2 vector field $\mathbf{Z}$ along $\gamma(t)$ vanishing at q and p, we may define a variation α for which $\mathbf{Z}$ will be the variation vector by:

$$\alpha(u,t) = \exp_r(u\mathbf{Z}|_r),$$

where $u \in (-\epsilon, \epsilon)$ for some $\epsilon > 0$ and $r = \gamma(t)$.

Lemma 4.5.4

The variation of the length from q to p under α is

$$\frac{\partial L}{\partial u}\bigg|_{u=0} = \sum_{i=1}^{n-1} \int_{t_i}^{t_{i+1}} g\left(\frac{\partial}{\partial u}, \left\{f^{-1}\frac{D}{\partial t}\frac{\partial}{\partial t} - f^{-2}\left(\frac{\partial f}{\partial t}\right)\frac{\partial}{\partial t}\right\}\right) dt + \sum_{i=2}^{n-1} g\left(\frac{\partial}{\partial u}, \left[f^{-1}\frac{\partial}{\partial t}\right]\right),$$

where $f^2 = g(\partial/\partial t, \partial/\partial t)$ is the magnitude of the tangent vector and $[f^{-1}\partial/\partial t]$ is the discontinuity at one of the singular points of $\gamma(t)$.

We have:

$$\frac{\partial L}{\partial u}\bigg|_{u=0} = \Sigma \frac{\partial}{\partial u} \int \left(-g\left(\frac{\partial}{\partial t}, \frac{\partial}{\partial t}\right)\right)^{\frac{1}{2}} dt$$

$$= -\Sigma \int g\left(\frac{D}{\partial u}\frac{\partial}{\partial t}, \frac{\partial}{\partial t}\right) f^{-1} dt$$

$$= -\Sigma \int g\left(\frac{D}{\partial t}\frac{\partial}{\partial u}, \frac{\partial}{\partial t}\right) f^{-1} dt$$

$$= -\Sigma \int \left\{\frac{\partial}{\partial t}\left(g\left(\frac{\partial}{\partial u}, \frac{\partial}{\partial t}\right)\right) f^{-1} - g\left(\frac{\partial}{\partial u}, \frac{D}{\partial t}\frac{\partial}{\partial t}\right) f^{-1}\right\} dt.$$

Integrating the first term by parts one has the required formula. $\square$

One may simplify the formula by choosing the parameter t to be the arc-length s. Then $g(\partial/\partial t, \partial/\partial t) = -1$. We shall denote by $\mathbf{V}$ the unit tangent vector $\partial/\partial s$. One has:

$$\frac{\partial L}{\partial u}\bigg|_{u=0} = \sum_{i=1}^{n-1} \int_{t_i}^{t_{i+1}} g(\mathbf{Z}, \dot{\mathbf{V}}) \, ds + \sum_{i=2}^{n-1} g(\mathbf{Z}, [\mathbf{V}])$$

where $\dot{\mathbf{V}} = D\mathbf{V}/\partial s$ is the acceleration. From this one sees again that a necessary condition for $\gamma(t)$ to be the longest curve from q to p is that it *should be an unbroken geodesic curve* as otherwise one could choose a variation which would yield a longer curve.

One may also consider a timelike curve $\gamma(t)$ from a spacelike three-surface $\mathscr{H}$ to a point p. A variation α of this curve is defined as before except that condition (3) is replaced by:

(3) $\alpha(u, 0)$ lies on $\mathscr{H}$, $\alpha(u, t_p) = p$.

Thus at $\mathscr{H}$ the variation vector $\mathbf{Z} = \partial/\partial u$ lies in $\mathscr{H}$.

Lemma 4.5.5

$$\frac{\partial L}{\partial u}\bigg|_{u=0} = \sum_{i=1}^{n-1} \int_{t_i}^{t_{i+1}} g(\dot{\mathbf{V}}, \mathbf{Z})\,ds + \sum_{i=2}^{n-1} g(\mathbf{Z}, [\mathbf{V}]) + g(\mathbf{Z}, \mathbf{V})\big|_{s=0}.$$

The proof is as for lemma 4.5.4. □

From this one sees that a necessary condition for $\gamma(t)$ to be the longest curve from $\mathscr{H}$ to p is that it is an *unbroken geodesic curve orthogonal to $\mathscr{H}$*.

We have seen that, under a variation α, the first derivative of the length of a timelike geodesic curve is zero. To proceed further we shall calculate the second derivative. We define a two-parameter variation α of a geodesic curve $\gamma(t)$ from q to p as a C^1 map:

$$\alpha: (-\epsilon_1, \epsilon_1) \times (-\epsilon_2, \epsilon_2) \times [0, t_p] \to \mathscr{M}$$

such that

(1) $\alpha(0, 0, t) = \gamma(t)$;

(2) there is a subdivision $0 = t_1 < t_2 < \ldots < t_n = t_p$ of $[0, t_p]$ such that α is C^3 on each

$$(-\epsilon_1, \epsilon_1) \times (-\epsilon_2, \epsilon_2) \times [t_i, t_{i+1}];$$

(3) $\alpha(u_1, u_2, 0) = q$, $\alpha(u_1, u_2, t_p) = p$;

(4) for all constant u_1, u_2, $\alpha(u_1, u_2, t)$ is a timelike curve.

We define

$$\mathbf{Z}_1 = \left(\frac{\partial}{\partial u_1}\right)_\alpha \bigg|_{\substack{u_1=0 \\ u_2=0}},$$

$$\mathbf{Z}_2 = \left(\frac{\partial}{\partial u_2}\right)_\alpha \bigg|_{\substack{u_1=0 \\ u_2=0}},$$

as the two variation vectors. Conversely given two continuous, piecewise C^2 vector fields $\mathbf{Z}_1$ and $\mathbf{Z}_2$ along $\gamma(t)$ one may define a variation for which they will be the variation vectors, by:

$$\alpha(u_1, u_2, t) = \exp_r (u_1 \mathbf{Z}_1 + u_2 \mathbf{Z}_2),$$
$$r = \gamma(t).$$

Lemma 4.5.6

Under the two-parameter variation of the geodesic curve $\gamma(t)$, the second derivative of the length will be:

$$\frac{\partial^2 L}{\partial u_2\, \partial u_1}\bigg|_{\substack{u_1=0 \\ u_2=0}} = \sum_{i=1}^{n-1} \int_{t_i}^{t_{i+1}} g\left(\mathbf{Z}_1, \left\{\frac{\mathrm{D}^2}{\partial s^2}(\mathbf{Z}_2 + g(\mathbf{V}, \mathbf{Z}_2)\mathbf{V}) - \mathbf{R}(\mathbf{V}, \mathbf{Z}_2)\mathbf{V}\right\}\right) ds$$

$$+ \sum_{i=2}^{n-1} g\left(\mathbf{Z}_1, \left[\frac{\mathrm{D}}{\partial s}(\mathbf{Z}_2 + g(\mathbf{V}, \mathbf{Z}_2)\mathbf{V})\right]\right).$$

By lemma 4.5.4, one has:

$$\frac{\partial L}{\partial u_1}\Big|_{\substack{u_1=0 \\ u_2=0}} = \Sigma \int g\left(\frac{\partial}{\partial u_1}, \left\{f^{-1}\frac{D}{\partial t}\frac{\partial}{\partial t} - f^{-2}\left(\frac{\partial f}{\partial t}\right)\frac{\partial}{\partial t}\right\}\right)dt + \Sigma g\left(\frac{\partial}{\partial u_1}, \left[f^{-1}\frac{\partial}{\partial t}\right]\right).$$

Therefore

$$\frac{\partial^2 L}{\partial u_2\,\partial u_1}\Big|_{\substack{u_1=0 \\ u_2=0}} = \Sigma \int g\left(\frac{D}{\partial u_2}\frac{\partial}{\partial u_1}, \left\{f^{-1}\frac{D}{\partial t}\frac{\partial}{\partial t} - f^{-2}\left(\frac{\partial f}{\partial t}\right)\frac{\partial}{\partial t}\right\}\right)dt$$

$$-\Sigma \int g\left(\frac{\partial}{\partial u_1}, \left\{f^{-2}\left(\frac{\partial f}{\partial u_2}\right)\frac{D}{\partial t}\frac{\partial}{\partial t} - f^{-1}\frac{D}{\partial u_2}\frac{D}{\partial t}\frac{\partial}{\partial t}\right.\right.$$

$$-2f^{-3}\left(\frac{\partial f}{\partial u_2}\right)\left(\frac{\partial f}{\partial t}\right)\frac{\partial}{\partial t} + f^{-2}\left(\frac{\partial^2 f}{\partial u_2\,\partial t}\right)\frac{\partial}{\partial t} + f^{-2}\left(\frac{\partial f}{\partial t}\right)\frac{D}{\partial u_2}\frac{\partial}{\partial t}\left.\left.\right\}\right)dt$$

$$+\Sigma g\left(\frac{D}{\partial u_2}\frac{\partial}{\partial u_1}, \left[f^{-1}\frac{\partial}{\partial t}\right]\right) + \Sigma g\left(\frac{\partial}{\partial u_1}, \frac{D}{\partial u_2}\left[f^{-1}\frac{\partial}{\partial t}\right]\right).$$

The first and third terms vanish as $\gamma(t)$ is an unbroken geodesic curve. In the second term one can write:

$$\frac{D}{\partial u_2}\frac{D}{\partial t}\frac{\partial}{\partial t} = -R\left(\frac{\partial}{\partial t}, \frac{\partial}{\partial u_2}\right)\frac{\partial}{\partial t} + \frac{D}{\partial t}\frac{D}{\partial u_2}\frac{\partial}{\partial t}$$

$$= -R\left(\frac{\partial}{\partial t}, \frac{\partial}{\partial u_2}\right)\frac{\partial}{\partial t} + \frac{D^2}{\partial t^2}\frac{\partial}{\partial u_2}$$

and

$$\frac{\partial^2 f}{\partial u_2\,\partial t} = -\frac{\partial}{\partial t}\left(f^{-1}g\left(\frac{D}{\partial u_2}\frac{\partial}{\partial t}, \frac{\partial}{\partial t}\right)\right)$$

$$= -\frac{\partial}{\partial t}\left\{f^{-1}\frac{\partial}{\partial t}\left(g\left(\frac{\partial}{\partial u_2}, \frac{\partial}{\partial t}\right)\right) - f^{-1}g\left(\frac{\partial}{\partial u_2}, \frac{D}{\partial t}\frac{\partial}{\partial t}\right)\right\}.$$

In the fourth term:

$$\frac{D}{\partial u_2}\left[f^{-1}\frac{\partial}{\partial t}\right] = \left[f^{-1}\frac{D}{\partial t}\frac{\partial}{\partial u_2} + f^{-3}g\left(\frac{D}{\partial t}\frac{\partial}{\partial u_2}, \frac{\partial}{\partial t}\right)\frac{\partial}{\partial t}\right].$$

Then taking t to be the arc-length s, one obtains the required result. $\square$

Although it is not immediately obvious from the appearance of the expression, one knows from its definition that it is symmetric in the two variation vector fields Z_1 and Z_2. One sees that it only depends on the projections of Z_1 and Z_2 into the space orthogonal to V. Thus we can confine our attention to variations α whose variation vectors are orthogonal to V. We shall define T_γ to be the (infinite-dimensional) vector space consisting of all continuous, piecewise C^2 vector fields along $\gamma(t)$ orthogonal to V and vanishing at q and p. Then $\partial^2 L/\partial u_2\,\partial u_1$

will be a symmetric map of $T_\gamma \times T_\gamma$ to R^1. One may think of it as a symmetric tensor on T_γ and write it as:

$$L(\mathbf{Z}_1, \mathbf{Z}_2) = \frac{\partial^2 L}{\partial u_2\,\partial u_1}\bigg|_{\substack{u_1=0 \\ u_2=0}}, \quad \mathbf{Z}_1, \mathbf{Z}_2 \in T_\gamma.$$

One may also calculate the second derivative of the length from $\mathscr{H}$ to p of a geodesic curve $\gamma(t)$ normal to $\mathscr{H}$. One proceeds as before except that one endpoint of $\gamma(t)$ is allowed to vary over $\mathscr{H}$ instead of being fixed.

Lemma 4.5.7

The second derivative of the length of $\gamma(t)$ from $\mathscr{H}$ to p is:

$$\frac{\partial^2 L}{\partial u_2\,\partial u_1}\bigg|_{\substack{u_1=0 \\ u_2=0}} = \sum_{i=1}^{n-1} \int_{t_i}^{t_{i+1}} g\left(\mathbf{Z}_1, \left\{\frac{\mathbf{D}^2}{\partial s^2}\mathbf{Z}_2 - \mathbf{R}(\mathbf{V}, \mathbf{Z}_2)\,\mathbf{V}\right\}\right) \mathrm{d}s$$

$$+ \sum_{i=2}^{n-1} g\left(\mathbf{Z}_1, \left[\frac{\mathbf{D}}{\partial s}\mathbf{Z}_2\right]\right) + g\left(\mathbf{Z}_1, \frac{\mathbf{D}}{\partial s}\mathbf{Z}_2\right)\bigg|_{\mathscr{H}} - \chi(\mathbf{Z}_1, \mathbf{Z}_2)\bigg|_{\mathscr{H}},$$

where $\mathbf{Z}_1$ and $\mathbf{Z}_2$ have been taken orthogonal to $\mathbf{V}$ and $\chi(\mathbf{Z}_1, \mathbf{Z}_2)$ is the second fundamental tensor of $\mathscr{H}$.

The first two terms are as for lemma 4.5.6. The extra terms are:

$$\frac{\mathbf{D}}{\partial u_2} g\left(\frac{\partial}{\partial u_1}, f^{-1}\frac{\partial}{\partial t}\right)\bigg|_{\mathscr{H}} = f^{-1} g\left(\frac{\mathbf{D}}{\partial u_2}\frac{\partial}{\partial u_1}, \frac{\partial}{\partial t}\right)\bigg|_{\mathscr{H}}$$

$$+ f^{-3} g\left(\frac{\mathbf{D}}{\partial u_2}\frac{\partial}{\partial t}, \frac{\partial}{\partial t}\right) g\left(\frac{\partial}{\partial u_1}, \frac{\partial}{\partial t}\right)\bigg|_{\mathscr{H}} + f^{-1} g\left(\frac{\partial}{\partial u_1}, \frac{\mathbf{D}}{\partial t}\frac{\partial}{\partial u_2}\right)\bigg|_{\mathscr{H}}.$$

The second term vanishes as $\partial/\partial u_1$ is orthogonal to $\partial/\partial t$. If one takes t to be the arc-length s, then $\partial/\partial t$ will be equal to the unit normal $\mathbf{N}$ at $\mathscr{H}$. Since the endpoint of $\gamma(t)$ is restricted to varying over $\mathscr{H}$, $\partial/\partial u_1$ will always be orthogonal to $\mathbf{N}$. Thus

$$g\left(\frac{\mathbf{D}}{\partial u_2}\frac{\partial}{\partial u_1}, \mathbf{N}\right) = \frac{\partial}{\partial u_2} g\left(\frac{\partial}{\partial u_1}, \mathbf{N}\right) - g\left(\frac{\partial}{\partial u_1}, \frac{\mathbf{D}}{\partial u_2}\mathbf{N}\right) = -\chi\left(\frac{\partial}{\partial u_1}, \frac{\partial}{\partial u_2}\right). \quad \square$$

We shall say that a timelike geodesic curve $\gamma(t)$ from q to p is *maximal* if $L(\mathbf{Z}_1, \mathbf{Z}_2)$ is negative semi-definite. In other words, if $\gamma(t)$ is not maximal there is a small variation α which yields a longer curve from p to q. Similarly we shall say that a timelike geodesic curve from $\mathscr{H}$ to p normal to $\mathscr{H}$ is *maximal* if $L(\mathbf{Z}_1, \mathbf{Z}_2)$ is negative semi-definite, so if $\gamma(t)$ is not maximal there is a small variation which yields a longer curve from $\mathscr{H}$ to p.

Proposition 4.5.8

A timelike geodesic curve $\gamma(t)$ from q to p is maximal if and only if there is no point conjugate to q along $\gamma(t)$ in (q, p).

Suppose there is no conjugate point in (q, p). Then introduce a Fermi-propagated orthonormal basis along $\gamma(t)$. The Jacobi fields along $\gamma(t)$ which vanish at q will be represented by a matrix $A_{\alpha\beta}(t)$ which will be non-singular in (q, p), but which will be singular at q and possibly at p. Since conjugate points are isolated, $\mathrm{d}(\log\det \mathbf{A})/\mathrm{d}s$ will be infinite where $A_{\alpha\beta}$ is singular. Thus a C^0, piecewise C^2 vector field $\mathbf{Z} \in T_\gamma$ can be expressed in $[q, p]$ as

$$Z^\alpha = A_{\alpha\beta}\, W^\beta,$$

where W^β is C^0, piecewise C^1 on $[q, p]$. Then,

$$L(\mathbf{Z}, \mathbf{Z}) = \Sigma \int_0^{s_p} A_{\alpha\beta}\, W^\beta \left\{ \frac{\mathrm{d}^2}{\mathrm{d}s^2}(A_{\alpha\delta}\, W^\delta) + R_{\alpha 4\gamma 4} A_{\gamma\delta}\, W^\delta \right\} \mathrm{d}s$$

$$+ \Sigma A_{\alpha\beta}\, W^\beta \left[\frac{\mathrm{d}}{\mathrm{d}s}(A_{\alpha\delta}\, W^\delta) \right]$$

$$= \lim_{\epsilon \to 0+} \Sigma \int_\epsilon^{s_p} A_{\alpha\beta}\, W^\beta \left\{ 2\frac{\mathrm{d}}{\mathrm{d}s} A_{\alpha\delta} \frac{\mathrm{d}}{\mathrm{d}s} W^\delta + A_{\alpha\delta} \frac{\mathrm{d}^2}{\mathrm{d}s^2} W^\delta \right\} \mathrm{d}s$$

$$+ \Sigma A_{\alpha\beta}\, W^\beta A_{\alpha\delta} \left[\frac{\mathrm{d}}{\mathrm{d}s} W^\delta \right]$$

$$= -\Sigma \int_0^{s_p} \left\{ A_{\alpha\beta} \frac{\mathrm{d}}{\mathrm{d}s} W^\beta A_{\alpha\delta} \frac{\mathrm{d}}{\mathrm{d}s} W^\delta + W^\beta \left(\frac{\mathrm{d}}{\mathrm{d}s} A_{\alpha\beta} A_{\alpha\delta} \right. \right.$$

$$\left. \left. - A_{\alpha\beta} \frac{\mathrm{d}}{\mathrm{d}s} A_{\alpha\delta} \right) \frac{\mathrm{d}}{\mathrm{d}s} W^\delta \right\} \mathrm{d}s.$$

(We take the limit because the second derivative of W^δ may not be defined at q.) But

$$\left(\frac{\mathrm{d}}{\mathrm{d}s} A_{\alpha\beta} A_{\alpha\delta} - A_{\alpha\beta} \frac{\mathrm{d}}{\mathrm{d}s} A_{\alpha\delta} \right) = -2A_{\alpha\beta}\omega_{\alpha\gamma}A_{\gamma\delta} = 0.$$

Therefore $L(\mathbf{Z}, \mathbf{Z}) \leqslant 0$.

Conversely, suppose there is a point $r \in (q, p)$ conjugate to q along $\gamma(t)$. Let $\mathbf{W}$ be the Jacobi field along γ which vanishes at q and r. Let $\mathbf{K} \in T_\gamma$ be such that

$$K^a g_{ab} \frac{\mathrm{D}}{\partial s} W^b = -1 \quad \text{at} \quad r.$$

Extend $\mathbf{W}$ to p by putting it zero in $[r, p]$. Let $\mathbf{Z}$ be $\epsilon\mathbf{K} + \epsilon^{-1}\mathbf{W}$, where ϵ is some constant. Then

$$L(\mathbf{Z}, \mathbf{Z}) = \epsilon^2 L(\mathbf{K}, \mathbf{K}) + 2L(\mathbf{K}, \mathbf{W}) + 2\epsilon^{-2}L(\mathbf{W}, \mathbf{W}) = \epsilon^2 L(\mathbf{K}, \mathbf{K}) + 2.$$

Thus by taking ϵ small enough, $L(\mathbf{Z}, \mathbf{Z})$ may be made positive.   □

One may obtain similar results for the case of a timelike geodesic curve $\gamma(t)$ orthogonal to $\mathscr{H}$, from $\mathscr{H}$ to p.

Proposition 4.5.9

A timelike geodesic curve $\gamma(t)$ from $\mathscr{H}$ to p is maximal if and only if there is no point in $(\mathscr{H}, q)$ conjugate to $\mathscr{H}$ along γ.   □

We shall also consider variations of a non-spacelike curve $\gamma(t)$ from q to p. We shall be interested in the circumstances under which it is possible to find a variation α of $\gamma(t)$ which makes $g(\partial/\partial t, \partial/\partial t)$ negative everywhere, or in other words, yields a timelike curve from q to p. Under a variation α:

$$\frac{\partial}{\partial u}\left(g\left(\frac{\partial}{\partial t}, \frac{\partial}{\partial t}\right)\right) = 2g\left(\frac{\mathrm{D}}{\partial u}\frac{\partial}{\partial t}, \frac{\partial}{\partial t}\right) = 2g\left(\frac{\mathrm{D}}{\partial t}\frac{\partial}{\partial u}, \frac{\partial}{\partial t}\right)$$

$$= 2\frac{\partial}{\partial t}\left(g\left(\frac{\partial}{\partial u}, \frac{\partial}{\partial t}\right)\right) - 2g\left(\frac{\partial}{\partial u}, \frac{\mathrm{D}}{\partial t}\frac{\partial}{\partial t}\right). \quad (4.48)$$

In order to obtain a timelike curve from q to p, one requires this to be less than or equal to zero everywhere on $\gamma(t)$.

Proposition 4.5.10

If p and q are joined by a non-spacelike curve $\gamma(t)$ which is not a null geodesic they can also be joined by a timelike curve.

If $\gamma(t)$ is not a null geodesic curve from p to q, there must be some point at which the tangent vector is discontinuous, or there must be some open interval on which the acceleration vector $(\mathrm{D}/\partial t)\,(\partial/\partial t)$ is non-zero and not parallel to $\partial/\partial t$. Consider first the case where there are no discontinuities. One has

$$g\left(\frac{\mathrm{D}}{\partial t}\frac{\partial}{\partial t}, \frac{\partial}{\partial t}\right) = \frac{1}{2}\frac{\partial}{\partial t}\left(g\left(\frac{\partial}{\partial t}, \frac{\partial}{\partial t}\right)\right) = 0.$$

This shows that $(\mathrm{D}/\partial t)\,(\partial/\partial t)$ is a spacelike vector where it is non-zero and not parallel to $\partial/\partial t$. Let $\mathbf{W}$ be a C^2 timelike vector field along $\gamma(t)$

such that $g(\mathbf{W}, \partial/\partial t) < 0$. Then one will obtain a timelike curve from p to q under the variation whose variation vector is

$$\mathbf{Z} = x\mathbf{W} + y\frac{\mathrm{D}}{\partial t}\frac{\partial}{\partial t}$$

with $\qquad\qquad x = c^{-1}e^{b}\int_{t_q}^{t} e^{-b}(1 - \tfrac{1}{2}ya^2)\,\mathrm{d}t,$

where $\qquad a^2 = g\left(\frac{\mathrm{D}}{\partial t}\frac{\partial}{\partial t}, \frac{\mathrm{D}}{\partial t}\frac{\partial}{\partial t}\right),$

$$c = -g\left(\mathbf{W}, \frac{\partial}{\partial t}\right),$$

$$b = -\int_{t_q}^{t} c^{-1}g\left(W, \frac{\mathrm{D}}{\partial t}\frac{\partial}{\partial t}\right)\mathrm{d}t,$$

and y is a C^2 non-negative function on $[p, q]$ such that $y_p = y_q = 0$ and

$$\int_{t_q}^{t_p} e^{-b}(1 - \tfrac{1}{2}ya^2)\,\mathrm{d}t = 0.$$

Suppose now there is some subdivision $t_q < t_1 < t_2 < \ldots < t_p$ such that the tangent vector $\partial/\partial t$ is continuous on each segment $[t_i, t_{i+1}]$. If a segment $[t_i, t_{i+1}]$ is not a null geodesic curve, it can be varied to give a timelike curve between its endpoints. Thus one has only to show that one can obtain a timelike curve from a non-spacelike curve $\gamma(t)$ made up of null geodesic segments whose tangent vectors are not parallel at points of discontinuity $\gamma(t_i)$. The parameter t can be taken to be an affine parameter on each segment $[t_i, t_{i+1}]$. The discontinuity $[\partial/\partial t]|_{t_i}$ will be a spacelike vector, as it is the difference between two non-parallel null vectors in the same half of the null cone. Thus one can find a C^2 vector field $\mathbf{W}$ along $[t_{i-1}, t_{i+1}]$ such that $g(\mathbf{W}, \partial/\partial t) < 0$ on $[t_{i-1}, t_i]$ and $g(\mathbf{W}, \partial/\partial t) > 0$ on $[t_i, t_{i+1}]$. Then a timelike curve between $\gamma(t_{i-1})$ and $\gamma(t_{i+1})$ will be obtained from the variation with variation vector field $\mathbf{Z} = x\mathbf{W}$, where $x = c^{-1}(t_{i+1} - t_i)(t - t_{i-1})$ for $t_{i-1} \leqslant t \leqslant t_i$, and $x = c^{-1}(t_i - t_{i-1})(t_{i+1} - t)$ for $t_i \leqslant t \leqslant t_{i+1}$, where $c = -g(\mathbf{W}, \partial/\partial t)$. $\square$

Thus if $\gamma(t)$ is not a geodesic curve, it can be varied to give a timelike curve. If it is a geodesic curve, the parameter t may be taken to be an affine parameter. One then sees that a necessary, but not sufficient, condition for a variation to yield a timelike curve is that the variation vector $\partial/\partial u$ should be orthogonal to the tangent vector $\partial/\partial t$ everywhere on $\gamma(t)$, since otherwise $(\partial/\partial t)\,g(\partial/\partial u, \partial/\partial t)$ would be positive somewhere on $\gamma(t)$. For such a variation the first derivative $(\partial/\partial u)\,g(\partial/\partial t, \partial/\partial t)$ will be zero and so one will have to examine the second derivative.

We shall therefore consider a two-parameter variation α of a null geodesic $\gamma(t)$ from q to p. The variation α will be defined as before except that, for the reason given above, we shall restrict ourselves to variations whose variation vectors

$$\frac{\partial}{\partial u_1}\bigg|_{\substack{u_1=0 \\ u_2=0}} \quad \text{and} \quad \frac{\partial}{\partial u_2}\bigg|_{\substack{u_1=0 \\ u_2=0}}$$

are orthogonal to the tangent vector $\partial/\partial t$ on $\gamma(t)$.

It is not convenient to study the behaviour of L under such a variation since $(-g(\partial/\partial t, \partial/\partial t))^{\frac{1}{2}}$ is not differentiable when $g(\partial/\partial t, \partial/\partial t) = 0$. Instead we shall consider the variation in:

$$\Lambda \equiv -\sum_{i=1}^{n-1} \int_{t_i}^{t_{i+1}} g\left(\frac{\partial}{\partial t}, \frac{\partial}{\partial t}\right) dt.$$

Clearly a necessary but not sufficient condition that a variation α of $\gamma(t)$ should yield a timelike curve from q to p is that Λ should become positive.

One has

$$\frac{1}{2}\frac{\partial^2}{\partial u_2\,\partial u_1}\left(g\left(\frac{\partial}{\partial t}, \frac{\partial}{\partial t}\right)\right) = \frac{\partial^2}{\partial u_2\,\partial t}\left(g\left(\frac{\partial}{\partial u_1}, \frac{\partial}{\partial t}\right)\right) - \frac{\partial}{\partial u_2}\left(g\left(\frac{\partial}{\partial u_1}, \frac{D}{\partial t}\frac{\partial}{\partial t}\right)\right)$$

$$= \frac{\partial^2}{\partial u_2\,\partial t}\left(g\left(\frac{\partial}{\partial u_1}, \frac{\partial}{\partial t}\right)\right) - g\left(\frac{\partial}{\partial u_1}, \left\{\frac{D^2}{\partial t^2}\frac{\partial}{\partial u_2}\right.\right.$$

$$\left.\left. -\,\mathbf{R}\left(\frac{\partial}{\partial t}, \frac{\partial}{\partial u_2}\right)\frac{\partial}{\partial t}\right\}\right)$$

and so

$$\frac{1}{2}\frac{\partial^2\Lambda}{\partial u_2\,\partial u_1}\bigg|_{\substack{u_1=0 \\ u_2=0}} = \Sigma\int g\left(\frac{\partial}{\partial u_1}, \left\{\frac{D^2}{\partial t^2}\frac{\partial}{\partial u_2} - \mathbf{R}\left(\frac{\partial}{\partial t}, \frac{\partial}{\partial u_2}\right)\frac{\partial}{\partial t}\right\}\right) dt$$

$$+ \Sigma g\left(\frac{\partial}{\partial u_1}, \left[\frac{D}{\partial t}\frac{\partial}{\partial u_2}\right]\right), \quad (4.49)$$

This formula is very similar to that for the variation of the length of a timelike curve. It can be seen that the variation of Λ is zero for a variation vector proportional to the tangent vector $\partial/\partial t$ since $\partial/\partial t$ is null and $\mathbf{R}(\partial/\partial t, \partial/\partial t)(\partial/\partial t) = 0$ as the Riemann tensor is antisymmetric. Such a variation would be equivalent to simply reparametrizing $\gamma(t)$. Thus if one wants a variation which will give a timelike curve one need consider only the projection of the variation vector into the space S_q at each point q of $\gamma(t)$. In other words, introducing a pseudo-orthonormal basis $\mathbf{E}_1$, $\mathbf{E}_2$, $\mathbf{E}_3$, $\mathbf{E}_4$ along $\gamma(t)$ with $\mathbf{E}_4 = \partial/\partial t$, the variation of Λ will depend only on the components Z^m of the variation vector ($m = 1, 2$).

Proposition 4.5.11

If there is no point in $[q, p]$ conjugate to q along $\gamma(t)$ then $\mathrm{d}^2\Lambda/\mathrm{d}u^2|_{u=0}$ will be negative for any variation α of $\gamma(t)$ whose variation vector $\partial/\partial u|_{u=0}$ is orthogonal to the tangent vector $\partial/\partial t$ on $\gamma(t)$ and is not everywhere zero or proportional to $\partial/\partial t$. In other words, if there is no point in $[q, p]$ conjugate to q then there is no small variation of $\gamma(t)$ which gives a timelike curve from q to p.

The proof is similar to that for proposition 4.5.8, using instead the 2×2 matrix $\hat{A}_{mn}$ of §4.2.   □

Proposition 4.5.12

If there is a point r in (q, p) conjugate to q along $\gamma(t)$ then there will be a variation of $\gamma(t)$ which will give a timelike curve from q to p.

The proof is a bit finicky since one has to show that the tangent vector becomes timelike everywhere. Let W^m be the components in the space S (see §4.2) of the Jacobi field which vanishes at q and r. It obeys

$$\frac{\mathrm{d}^2}{\mathrm{d}t^2} W^m = - R_{m4n4}\, W^n,$$

where for convenience t has been taken to be an affine parameter. Since W^m will be at least C^3 and since $\mathrm{d}W^m/\mathrm{d}t$ is not zero at q and r, one can write $W^m = f\hat{W}^m$ where $\hat{W}^m$ is a unit vector and f and $\hat{W}$ are C^2. Then

$$\frac{\mathrm{d}^2}{\mathrm{d}t^2}f + hf = 0,$$

where $$h = \hat{W}^m \frac{\mathrm{d}^2}{\mathrm{d}t^2} \hat{W}^m + R_{m4n4}\, \hat{W}^m \hat{W}^n.$$

Let $x \in [r, p]$ be such that W^m is not zero in $[r, x]$. Let h_1 be the minimum value of h in $[r, x]$. Let $a > 0$ be such that $a^2 + h_1 > 0$ and let $b = \{-f(\mathrm{e}^{at} - 1)^{-1}\}|_x$. Then the field

$$Z^m = \{b(\mathrm{e}^{at} - 1) + f\}\, \hat{W}^m$$

will vanish at q and x and will satisfy

$$Z^m \left(\frac{\mathrm{d}^2}{\mathrm{d}t^2} Z^m + R_{m4n4} Z^n \right) > 0 \quad \text{in} \quad (q, x).$$

We shall choose a variation $\alpha(u, t)$ of $\gamma(t)$ from q to x such that the components in S of its variation vector $\partial/\partial u|_{u=0}$ equals Z^m and such that

$$g\left(\frac{D}{\partial u}\frac{\partial}{\partial u}, \frac{\partial}{\partial t}\right)\bigg|_{u=0}$$

satisfies

$$g\left(\frac{D}{\partial u}\frac{\partial}{\partial u}, \frac{\partial}{\partial t}\right)\bigg|_{u=0} + g\left(\frac{\partial}{\partial u}, \frac{D}{\partial t}\frac{\partial}{\partial u}\right)\bigg|_{u=0} = \begin{cases} -\epsilon t & \text{for} \quad 0 \leqslant t \leqslant \tfrac{1}{4}t_x, \\ \epsilon(t - \tfrac{1}{2}t_x) & \text{for} \quad \tfrac{1}{4}t_x \leqslant t \leqslant \tfrac{3}{4}t_x, \\ \epsilon(t_x - t) & \text{for} \quad \tfrac{3}{4}t_x \leqslant t \leqslant t_x, \end{cases}$$

where t_x is the value of t at x, and $\epsilon > 0$ but less than the least value of Z^m $(\mathrm{d}^2 Z^m/\mathrm{d}t^2 + R_{m4n4}Z^n)$ in the range $\tfrac{1}{4}t_x \leqslant t \leqslant \tfrac{3}{4}t_x$. Then by (4.49) $(\partial^2/\partial u^2)g(\partial/\partial t, \partial/\partial t)$ will be negative everywhere in $[q, x]$ and so for sufficiently small u, α will give a timelike curve from q to x. If one joins this curve to the section of γ from x to p, one will obtain a non-spacelike curve from q to p which is not a null geodesic curve. Thus there will be a variation of this curve which gives a timelike curve from q to p. $\square$

By similar methods one can prove:

Proposition 4.5.13

If $\gamma(t)$ is a null geodesic curve orthogonal to a spacelike two-surface $\mathscr{S}$ from $\mathscr{S}$ to p and if there is no point in $[\mathscr{S}, p]$ conjugate to $\mathscr{S}$ along γ, then no small variation of γ can give a timelike curve from $\mathscr{S}$ to p . $\square$

Proposition 4.5.14

If there is a point in $(\mathscr{S}, p)$ conjugate to $\mathscr{S}$ along p, then there is a variation of γ which gives a timelike curve from $\mathscr{S}$ to p. $\square$

These results on variations of timelike and non-spacelike curves will be used in chapter 8 to show the non-existence of longest geodesics.

5

Exact solutions

Any space–time metric can in a sense be regarded as satisfying Einstein's field equations

$$R_{ab} - \tfrac{1}{2}Rg_{ab} + \Lambda g_{ab} = 8\pi T_{ab}, \qquad (5.1)$$

(where we use the units of chapter 3), because, having determined the left-hand side of (5.1) from the metric tensor of the space–time $(\mathcal{M}, \mathbf{g})$, one can *define* T_{ab} as the right-hand side of (5.1). The matter tensor so defined will in general have unreasonable physical properties; the solution will be reasonable only if the matter content is reasonable.

We shall mean by an *exact solution* of Einstein's equations, a space–time $(\mathcal{M}, \mathbf{g})$ in which the field equations are satisfied with T_{ab} the energy–momentum tensor of some specified form of matter which obeys postulate (*a*) ('local causality') of chapter 3, and one of the energy conditions of § 4.3. In particular, one may look for exact solutions for empty space ($T_{ab} = 0$), for an electromagnetic field (T_{ab} has the form (3.7)), for a perfect fluid (T_{ab} has the form (3.8)), or for a space containing an electromagnetic field and a perfect fluid. Because of the complexity of the field equations, one cannot find exact solutions except in spaces of rather high symmetry. Exact solutions are also idealized in that any region of space–time is likely to contain many forms of matter, while one can obtain exact solutions only for rather simple matter content. Nevertheless, exact solutions give an idea of the qualitative features that can arise in General Relativity, and so of possible properties of realistic solutions of the field equations. The examples we give will show many types of behaviour which will be of interest in later chapters. We shall discuss solutions with particular reference to their global properties. Many of these global properties have only recently been discovered, although the solutions have been known in a local form for some time.

In § 5.1 and § 5.2 we consider the simplest Lorentz metrics: those of constant curvature. The spatially isotropic and homogeneous cosmological models are described in § 5.3, and their simplest anisotropic generalizations are discussed in § 5.4. It is shown that all such simple models will have a singular origin provided that Λ does not take large positive values. The spherically symmetric metrics which describe the field outside a massive charged or neutral body are examined in § 5.5, and the axially symmetric metrics describing the field outside a special class of massive rotating bodies are described in § 5.6. It is shown that some of the apparent singularities are simply due to a bad choice of coordinates. In § 5.7 we describe the Gödel universe and in § 5.8 the Taub–NUT solutions. These probably do not represent the actual universe but they are of interest because of their pathological global properties. Finally some other exact solutions of interest are mentioned in § 5.9.

5.1 Minkowski space–time

Minkowski space–time $(\mathcal{M}, \eta)$ is the simplest empty space–time in General Relativity, and is in fact the space–time of Special Relativity. Mathematically, it is the manifold R^4 with a flat Lorentz metric η. In terms of the natural coordinates (x^1, x^2, x^3, x^4) on R^4, the metric can be expressed in the form

$$ds^2 = -(dx^4)^2 + (dx^1)^2 + (dx^2)^2 + (dx^3)^2. \qquad (5.2)$$

If one uses spherical polar coordinates (t, r, θ, ϕ) where $x^4 = t$, $x^3 = r\cos\theta$, $x^2 = r\sin\theta\cos\phi$, $x^1 = r\sin\theta\sin\phi$, the metric takes the form

$$ds^2 = -dt^2 + dr^2 + r^2(d\theta^2 + \sin^2\theta\, d\phi^2). \qquad (5.3)$$

This metric is apparently singular for $r = 0$ and $\sin\theta = 0$; however this is because the coordinates used are not admissible coordinates at these points. To obtain regular coordinate neighbourhoods one has to restrict the coordinates, e.g. to the ranges $0 < r < \infty$, $0 < \theta < \pi$, $0 < \phi < 2\pi$. One needs two such coordinate neighbourhoods to cover the whole of Minkowski space.

An alternative coordinate system is given by choosing advanced and retarded null coordinates v, w defined by $v = t + r$, $w = t - r$ ($\Rightarrow v \geqslant w$). The metric becomes

$$ds^2 = -dv\, dw + \tfrac{1}{4}(v - w)^2(d\theta^2 + \sin^2\theta\, d\phi^2), \qquad (5.4)$$

where $-\infty < v < \infty$, $-\infty < w < \infty$. The absence in the metric of terms in dv^2, dw^2 corresponds to the fact that the surfaces $\{w = \text{constant}\}$, $\{v = \text{constant}\}$ are null (i.e. $w_{;a}w_{;b}g^{ab} = 0 = v_{;a}v_{;b}g^{ab}$); see figure 12.

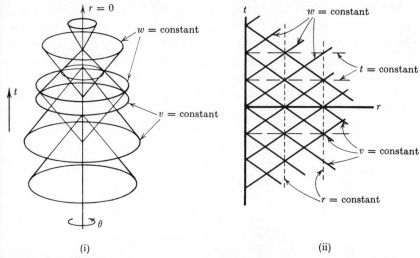

(i)   (ii)

FIGURE 12. Minkowski space. The null coordinate $v(w)$ may be thought of as incoming (outgoing) spherical waves travelling at the speed of light; they are advanced (retarded) time coordinates. The intersection of a surface $\{v = \text{constant}\}$ with a surface $\{w = \text{constant}\}$ is a two-sphere.

(i) The v, w coordinate surfaces (one coordinate is suppressed).

(ii) The (t, r) plane; each point represents a two-sphere of radius r.

In a coordinate system in which the metric takes the form (5.2), the geodesics have the form $x^a(v) = b^a v + c^a$ where b^a and c^a are constants. Thus the exponential map $\exp_p: T_p \to \mathcal{M}$ is given by

$$x^a (\exp_p \mathbf{X}) = X^a + x^a(p),$$

where X^a are the components of $\mathbf{X}$ with respect to the coordinate basis $\{\partial/\partial x^a\}$ of T_p. Since exp is one–one and onto, it is a diffeomorphism between T_p and $\mathcal{M}$. Thus any two points of $\mathcal{M}$ can be joined by a unique geodesic curve. As exp is defined everywhere on T_p for all p, $(\mathcal{M}, \boldsymbol{\eta})$ is geodesically complete.

For a spacelike three-surface $\mathcal{S}$, the future (past) Cauchy development $D^+(\mathcal{S})$ $(D^-(\mathcal{S}))$ is defined as the set of all points $q \in \mathcal{M}$ such that each past-directed (future-directed) inextendible non-spacelike curve through q intersects $\mathcal{S}$, cf. § 6.5. If $D^+(\mathcal{S}) \cup D^-(\mathcal{S}) = \mathcal{M}$, i.e. if every inextendible non-spacelike curve in $\mathcal{M}$ intersects $\mathcal{S}$, then $\mathcal{S}$ is said to be a Cauchy surface. In Minkowski space–time, the surfaces $\{x^4 = \text{constant}\}$ are a family of Cauchy surfaces which cover the whole of $\mathcal{M}$. One can however find inextendible spacelike surfaces which are not Cauchy surfaces; for example the surfaces

$$\mathcal{S}_\sigma: \{-(x^4)^2 + (x^1)^2 + (x^2)^2 + (x^3)^2 = \sigma = \text{constant}\},$$

where $\sigma < 0$, $x^4 < 0$, are spacelike surfaces which lie entirely inside the past null cone of the origin O, and so are not Cauchy surfaces (see figure 13). In fact the future Cauchy development of $\mathscr{S}_\sigma$ is the region bounded by $\mathscr{S}_\sigma$ and the past light cone of the origin. By lemma 4.5.2, the timelike geodesics through the origin O are orthogonal to the surfaces $\mathscr{S}_\sigma$. If $r \in D^+(\mathscr{S}_\sigma) \cup D^-(\mathscr{S}_\sigma)$ then the timelike geodesic through r and O is the longest timelike curve between r and $\mathscr{S}_\sigma$. If

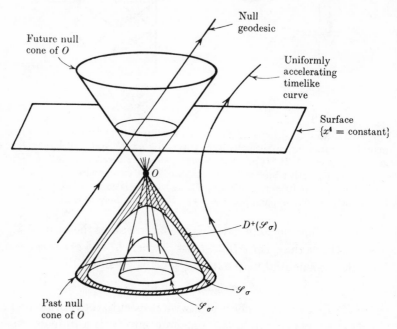

FIGURE 13. A Cauchy surface $\{x^4 = \text{constant}\}$ in Minkowski space–time, and spacelike surfaces $\mathscr{S}_\sigma$, $\mathscr{S}_{\sigma'}$ which are not Cauchy surfaces. The normal geodesics to the surfaces $\mathscr{S}_\sigma$, $\mathscr{S}_{\sigma'}$ all intersect at O.

however r does not lie in $D^+(\mathscr{S}_\sigma) \cup D^-(\mathscr{S}_\sigma)$ there is no longest timelike curve between r and $\mathscr{S}_\sigma$: either r lies in the region $\sigma \geqslant 0$, in which case there is no timelike geodesic through r orthogonal to $\mathscr{S}_\sigma$, or r lies in the region $\sigma < 0$, $x^4 \geqslant 0$, in which case there is a timelike geodesic through r orthogonal to $\mathscr{S}_\sigma$ but this geodesic is not the longest curve between r and $\mathscr{S}_\sigma$ as it contains a conjugate point to $\mathscr{S}_\sigma$ at O (cf. figure 13).

To study the structure of infinity in Minkowski space–time, we shall use the interesting representation of this space–time given by Penrose. From the null coordinates v, w, we define new null coordinates in which the infinities of v, w have been transformed to finite values; thus we define p, q by $\tan p = v$, $\tan q = w$ where $-\frac{1}{2}\pi < p < \frac{1}{2}\pi$, $-\frac{1}{2}\pi < q < \frac{1}{2}\pi$ (and $p \geqslant q$). Then the metric of $(\mathcal{M}, \boldsymbol{\eta})$ takes the form

$$\mathrm{d}s^2 = \sec^2 p \sec^2 q(-\mathrm{d}p\,\mathrm{d}q + \tfrac{1}{4}\sin^2(p-q)\,(\mathrm{d}\theta^2 + \sin^2\theta\,\mathrm{d}\phi^2)).$$

The physical metric $\boldsymbol{\eta}$ is therefore conformal to the metric $\bar{\mathbf{g}}$ given by

$$\mathrm{d}\bar{s}^2 = -4\mathrm{d}p\,\mathrm{d}q + \sin^2(p-q)\,(\mathrm{d}\theta^2 + \sin^2\theta\,\mathrm{d}\phi^2). \tag{5.5}$$

This metric can be reduced to a more usual form by defining

$$t' = p+q, \quad r' = p-q,$$

where $\quad -\pi < t'+r' < \pi, \quad -\pi < t'-r' < \pi, \quad r' \geqslant 0;$ (5.6)

(5.5) is then

$$\mathrm{d}\bar{s}^2 = -(\mathrm{d}t')^2 + (\mathrm{d}r')^2 + \sin^2 r'(\mathrm{d}\theta^2 + \sin^2\theta\,\mathrm{d}\phi^2). \tag{5.7}$$

Thus the whole of Minkowski space-time is given by the region (5.6) of the metric

$$\mathrm{d}s^2 = \tfrac{1}{4}\sec^2(\tfrac{1}{2}(t'+r'))\sec^2(\tfrac{1}{2}(t'-r'))\,\mathrm{d}\bar{s}^2$$

where $\mathrm{d}\bar{s}^2$ is determined by (5.7); the coordinates t, r of (5.3) are related to t', r' by

$$2t = \tan(\tfrac{1}{2}(t'+r')) + \tan(\tfrac{1}{2}(t'-r')),$$

$$2r = \tan(\tfrac{1}{2}(t'+r')) - \tan(\tfrac{1}{2}(t'-r')).$$

Now the metric (5.7) is locally identical to that of the Einstein static universe (see § 5.3), which is a completely homogeneous space-time. One can analytically extend (5.7) to the whole of the Einstein static universe, that is one can extend the coordinates to cover the manifold $R^1 \times S^3$ where $-\infty < t' < \infty$ and r', θ, ϕ are regarded as coordinates on S^3 (with coordinate singularities at $r' = 0$, $r' = \pi$ and $\theta = 0$, $\theta = \pi$ similar to the coordinate singularities in (5.3); these singularities can be removed by transforming to other local coordinates in a neighbourhood of points where (5.7) is singular). On suppressing two dimensions, one can represent the Einstein static universe as the cylinder $x^2 + y^2 = 1$ imbedded in a three-dimensional Minkowski space with metric $\mathrm{d}s^2 = -\mathrm{d}t^2 + \mathrm{d}x^2 + \mathrm{d}y^2$ (the full Einstein static universe can be imbedded as the cylinder $x^2 + y^2 + z^2 + w^2 = 1$ in a five-dimensional Euclidean space with metric $\mathrm{d}s^2 = -\mathrm{d}t^2 + \mathrm{d}x^2 + \mathrm{d}y^2 + \mathrm{d}z^2 + \mathrm{d}w^2$, cf. Robertson (1933)).

One therefore has the situation: the whole of Minkowski space-time is conformal to the region (5.6) of the Einstein static universe, that is,

to the shaded area in figure 14. The boundary of this region may therefore be thought of as representing the conformal structure of infinity
of Minkowski space–time. It consists of the null surfaces $p = \frac{1}{2}\pi$
(labelled $\mathscr{I}^+$) and $q = -\frac{1}{2}\pi$ (labelled $\mathscr{I}^-$) together with points $p = \frac{1}{2}\pi$,
$q = \frac{1}{2}\pi$ (labelled i^+), $p = \frac{1}{2}\pi$, $q = -\frac{1}{2}\pi$ (labelled i^0) and $p = -\frac{1}{2}\pi$,
$q = -\frac{1}{2}\pi$ (labelled i^-). Any future-directed timelike geodesic in

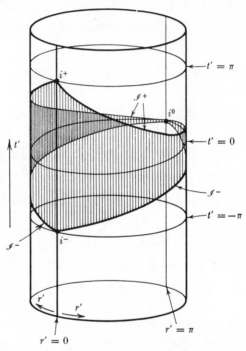

FIGURE 14. The Einstein static universe represented by an imbedded cylinder;
the coordinates θ, ϕ have been suppressed. Each point represents one half
of a two-sphere of area $4\pi \sin^2 r'$. The shaded region is conformal to the whole of
Minkowski space–time; its boundary (part of the null cones of i^+, i^0 and i^-) may
be regarded as the conformal infinity of Minkowski space–time.

Minkowski space approaches i^+ (i^-) for indefinitely large positive
(negative) values of its affine parameter, so one can regard any timelike geodesic as originating at i^- and finishing at i^+ (cf. figure 15(i)).
Similarly one can regard null geodesics as originating at $\mathscr{I}^-$ and ending
at $\mathscr{I}^+$, while spacelike geodesics both originate and end at i^0. Thus one
may regard i^+ and i^- as representing future and past timelike infinity,
$\mathscr{I}^+$ and $\mathscr{I}^-$ as representing future and past null infinity, and i^0 as
representing spacelike infinity. (However non-geodesic curves do not

obey these rules; e.g. non-geodesic timelike curves may start on $\mathscr{I}^-$
and end on $\mathscr{I}^+$.) Since any Cauchy surface intersects all timelike and
null geodesics, it is clear that it will appear as a cross-section of the
space everywhere reaching the boundary at i^0.

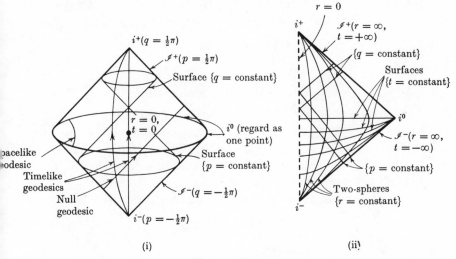

(i)   (ii)

FIGURE 15
(i) The shaded region of figure 14, with only one coordinate suppressed,
representing Minkowski space–time and its conformal infinity.
(ii) The Penrose diagram of Minkowski space–time; each point represents
a two-sphere, except for i^+, i^0 and i^-, each of which is a single point, and points
on the line $r = 0$ (where the polar coordinates are singular).

One can also represent the conformal structure of infinity by
drawing a diagram of the (t', r') plane, see figure 15 (ii). As in figure
12 (ii), each point of this diagram represents a sphere S^2, and radial
null geodesics are represented by straight lines at $\pm 45°$. In fact, the
structure of infinity in any spherically symmetric space–time can be
represented by a diagram of this sort, which we shall call a *Penrose
diagram*. On such diagrams we shall represent infinity by single lines,
the origin of polar coordinates by dotted lines, and irremovable singularities of the metric by double lines.

The conformal structure of Minkowski space we have described is
what one would regard as the 'normal' behaviour of a space–time at
infinity; we shall encounter different types of behaviour in later
sections.

Finally, we mention that one can obtain spaces locally identical to
$(\mathscr{M}, \eta)$ but with different (large scale) topological properties by identi-

fying points in $\mathcal{M}$ which are equivalent under a discrete isometry without a fixed point (e.g. identifying the point (x^1, x^2, x^3, x^4) with the point $(x^1, x^2, x^3, x^4 + c)$, where c is a constant, changes the topological structure from R^4 to $R^3 \times S^1$, and introduces closed timelike lines into the space-time). Clearly, $(\mathcal{M}, \boldsymbol{\eta})$ is the universal covering space for all such derived spaces, which have been studied in detail by Auslander and Markus (1958).

5.2 De Sitter and anti-de Sitter space-times

The space-time metrics of constant curvature are locally characterized by the condition $R_{abcd} = \frac{1}{12}R(g_{ac}g_{bd} - g_{ad}g_{bc})$. This equation is equivalent to $C_{abcd} = 0 = R_{ab} - \frac{1}{4}Rg_{ab}$; thus the Riemann tensor is determined by the Ricci scalar R alone. It follows at once from the contracted Bianchi identities that R is constant throughout space-time; in fact these space-times are homogeneous. The Einstein tensor is

$$R_{ab} - \tfrac{1}{2}Rg_{ab} = -\tfrac{1}{4}Rg_{ab}.$$

One can therefore regard these spaces as solutions of the field equations for an empty space with $\Lambda = \frac{1}{4}R$, or for a perfect fluid with a constant density $R/32\pi$ and a constant pressure $-R/32\pi$. However the latter choice does not seem reasonable, as in this case one cannot have both the density and the pressure positive; in addition, the equation of motion (3.10) is indeterminate for such a fluid.

The space of constant curvature with $R = 0$ is Minkowski space-time. The space for $R > 0$ is *de Sitter space-time*, which has the topology $R^1 \times S^3$ (see Schrödinger (1956) for an interesting account of this space). It is easiest visualized as the hyperboloid

$$-v^2 + w^2 + x^2 + y^2 + z^2 = \alpha^2$$

in flat five-dimensional space R^5 with metric

$$-\mathrm{d}v^2 + \mathrm{d}w^2 + \mathrm{d}x^2 + \mathrm{d}y^2 + \mathrm{d}z^2 = \mathrm{d}s^2$$

(see figure 16). One can introduce coordinates (t, χ, θ, ϕ) on the hyperboloid by the relations

$$\alpha \sinh(\alpha^{-1}t) = v, \quad \alpha \cosh(\alpha^{-1}t)\cos\chi = w,$$

$$\alpha \cosh(\alpha^{-1}t)\sin\chi\cos\theta = x, \quad \alpha \cosh(\alpha^{-1}t)\sin\chi\sin\theta\cos\phi = y,$$

$$\alpha \cosh(\alpha^{-1}t)\sin\chi\sin\theta\sin\phi = z.$$

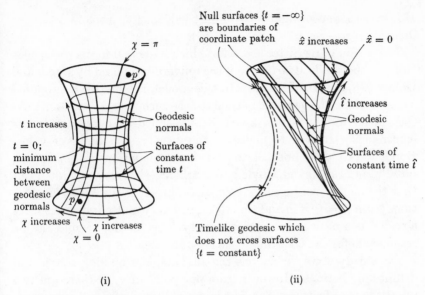

FIGURE 16. De Sitter space–time represented by a hyperboloid imbedded in a five-dimensional flat space (two dimensions are suppressed in the figure).

(i) Coordinates (t, χ, θ, ϕ) cover the whole hyperboloid; the sections $\{t = \text{constant}\}$ are surfaces of curvature $k = +1$.

(ii) Coordinates $(\hat{t}, \hat{x}, \hat{y}, \hat{z})$ cover half the hyperboloid; the surfaces $\{\hat{t} = \text{constant}\}$ are flat three-spaces, their geodesic normals diverging from a point in the infinite past.

In these coordinates, the metric has the form

$$\mathrm{d}s^2 = -\mathrm{d}t^2 + \alpha^2 \cdot \cosh^2(\alpha^{-1}t) \cdot \{\mathrm{d}\chi^2 + \sin^2\chi(\mathrm{d}\theta^2 + \sin^2\theta\,\mathrm{d}\phi^2)\}.$$

The singularities in the metric at $\chi = 0$, $\chi = \pi$ and at $\theta = 0$, $\theta = \pi$, are simply those that occur with polar coordinates. Apart from these trivial singularities, the coordinates cover the whole space for $-\infty < t < \infty$, $0 \leqslant \chi \leqslant \pi$, $0 \leqslant \theta \leqslant \pi$, $0 \leqslant \phi \leqslant 2\pi$. The spatial sections of constant t are spheres S^3 of constant positive curvature and are Cauchy surfaces. Their geodesic normals are lines which contract monotonically to a minimum spatial separation and then re-expand to infinity (see figure 16 (i)).

One can also introduce coordinates

$$\hat{t} = \alpha\log\frac{w+v}{\alpha}, \quad \hat{x} = \frac{\alpha x}{w+v}, \quad \hat{y} = \frac{\alpha y}{w+v}, \quad \hat{z} = \frac{\alpha z}{w+v}$$

on the hyperboloid. In these coordinates, the metric takes the form

$$\mathrm{d}s^2 = -\mathrm{d}\hat{t}^2 + \exp(2\alpha^{-1}\hat{t})\,(\mathrm{d}\hat{x}^2 + \mathrm{d}\hat{y}^2 + \mathrm{d}\hat{z}^2).$$

However these coordinates cover only half the hyperboloid as $\hat{t}$ is not defined for $w + v \leqslant 0$ (see figure 16 (ii)).

The region of de Sitter space for which $v + w > 0$ forms the space-time for the *steady state* model of the universe proposed by Bondi and Gold (1948) and Hoyle (1948). In this model, the matter is supposed to move along the geodesic normals to the surfaces $\{\hat{t} = \text{constant}\}$. As the matter moves further apart, it is assumed that more matter is continuously created to maintain the density at a constant value. Bondi and Gold did not seek to provide field equations for this model, but Pirani (1955), and Hoyle and Narlikar (1964) have pointed out that the metric can be considered as a solution of the Einstein equations (with $\Lambda = 0$) if in addition to the ordinary matter one introduces a scalar field of negative energy density. This 'C'-field would also be responsible for the continual creation of matter.

The steady state theory has the advantage of making simple and definite predictions. However from our point of view there are two unsatisfactory features. The first is the existence of negative energy, which was discussed in § 4.3. The other is the fact that the space–time is extendible, being only half of de Sitter space. Despite these aesthetic objections, the real test of the steady state theory is whether its predictions agree with observations or not. At the moment it seems that they do not, though the observations are not yet quite conclusive.

de Sitter space is geodesically complete; however, there are points in the space which cannot be joined to each other by any geodesic. This is in contrast to spaces with a positive definite metric, when geodesic completeness guarantees that any two points of a space can be joined by at least one geodesic. The half of de Sitter space which represents the steady state universe is not complete in the past (there are geodesics which are complete in the full space, and cross the boundary of the steady state region; they are therefore incomplete in that region).

To study infinity in de Sitter space–time, we define a time coordinate t' by
$$t' = 2 \arctan (\exp \alpha^{-1} t) - \tfrac{1}{2}\pi,$$
where
$$-\tfrac{1}{2}\pi < t' < \tfrac{1}{2}\pi. \tag{5.8}$$
Then
$$ds^2 = \alpha^2 \cosh^2 (\alpha^{-1} t') \cdot d\bar{s}^2,$$

where $d\bar{s}^2$ is given by (5.7) on identifying $r' = \chi$. Thus the de Sitter space is conformal to that part of the Einstein static universe defined by (5.8) (see figure 17 (i)). The Penrose diagram of de Sitter space is accordingly as in figure 17 (ii). One half of this figure gives the Penrose

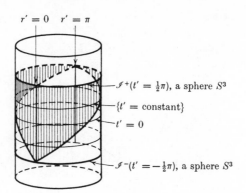

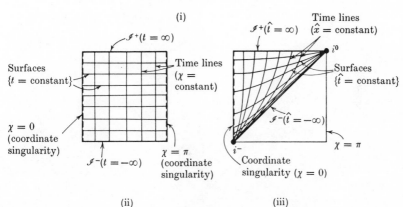

FIGURE 17

(i) De Sitter space–time is conformal to the region $-\tfrac{1}{2}\pi < t' < \tfrac{1}{2}\pi$ of the Einstein static universe. The steady state universe is conformal to the shaded region.

(ii) The Penrose diagram of de Sitter space–time.

(iii) The Penrose diagram of the steady state universe.

In (ii), (iii) each point represents a two-sphere of area $2\pi\sin^2\chi$; null lines are at 45°. $\chi = 0$ and $\chi = \pi$ are identified.

diagram of the half of de Sitter space–time which constitutes the steady state universe (figure 17 (iii)).

One sees that de Sitter space has, in contrast to Minkowski space, a spacelike infinity for timelike and null lines, both in the future and the past. This difference corresponds to the existence in de Sitter space–time of both particle and event horizons for geodesic families of observers.

In de Sitter space, consider a family of particles whose histories are timelike geodesics; these must originate at the spacelike infinity $\mathscr{I}^-$ and end at the spacelike infinity $\mathscr{I}^+$. Let p be some event on the world-

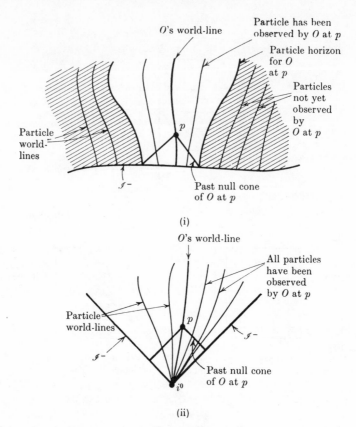

FIGURE 18

(i) The particle horizon defined by a congruence of geodesic curves when past null infinity $\mathscr{I}^-$ is spacelike.

(ii) Lack of such a horizon if $\mathscr{I}^-$ is null.

line of a particle O in this family, i.e. some time in its history (proper time measured along O's world-line). The past null cone of p is the set of events in space–time which can be observed by O at that time. The world-lines of some other particles may intersect this null cone; these particles are visible to O. However, there can exist particles whose world-lines do not intersect this null cone, and so are not yet visible to O. At a later time O can observe more particles, but there still exist particles not visible to O at that time. We say that the division of particles into those seen by O at p and those not seen by O at p, is the *particle horizon* for the observer O at the event p; it represents the history of those particles lying at the limits of O's vision. Note that it is determined only when the world-lines of all the particles in the family are known. If some particle lies on the horizon, then the event p is the event at which the particle's creation light cone intersects O's world-line. In Minkowski space, on the other hand, all the other particles are visible at any event p on O's world-line if they move on timelike geodesics. As long as one considers only families of geodesic observers, one may think of the existence of the particle horizon as a consequence of past null infinity being spacelike (see figure 18).

All events outside the past null cone of p are events which are not, and never have been, observable by O up to the time represented by the event p. There is a limit to O's world-line on $\mathscr{I}^+$. In de Sitter space–time, the past null cone of this point (obtained by a limiting process in the actual space–time, or directly from the conformal space–time) is a boundary between events which will at some time be observable by O, and those that will never be observable by O. We call this surface the *future event horizon* of the world-line. It is the boundary of the past of the world-line. In Minkowski space–time, on the other hand, the limiting null cone of any geodesic observer includes the whole of space–time, so there are no events which a geodesic observer will never be able to see. However if an observer moves with uniform acceleration his world-line may have a future event horizon. One may think of the existence of a future event horizon for a geodesic observer as being a consequence of $\mathscr{I}^+$ being spacelike (see figure 19).

Consider the event horizon for the observer O in de Sitter space–time and suppose that at some proper time (event p) on his world-line, his light cone intersects the world-line of the particle Q. Then Q is always visible to O at times after p. However there is on Q's world-line an event r which lies on O's future event horizon; O can never see later events on Q's world-line than r. Moreover an infinite proper time elapses on O's world-line from any given point till he observes r, but a finite proper time elapses along Q's world-line from any given event to r, which is a perfectly ordinary event on his world-line. Thus O sees a finite part of Q's history in an infinite time; expressed more physically, as O observes Q he sees a redshift which approaches infinity as O observes points on Q's world-line which approach r. Correspondingly, Q never sees beyond some point on O's world-line, and sees nearby points on O's world-line only with a very large redshift.

At any point on O's world-line, the future null cone is the boundary of the set of events in space–time which O can influence at and after that time. To obtain the maximal set of events in space–time that O could at any time influence, we take the future light cone of the limit

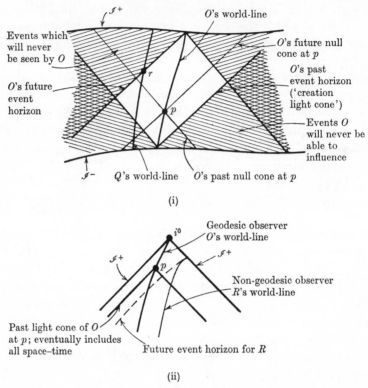

FIGURE 19

(i) The future event horizon for a particle O which exists when future infinity $\mathscr{I}^+$ is spacelike; also the past event horizon which exists when past infinity $\mathscr{I}^-$ is spacelike.

(ii) If future infinity consists of a null $\mathscr{I}^+$ and i^0, there is no future event horizon for a geodesic observer O. However an accelerating observer R may have a future event horizon.

point of O's world-line on past infinity $\mathscr{I}^-$; that is, we take the boundary of the future of the world-line (which can be regarded as O's creation light cone). This has a non-trivial existence for a geodesic observer only if the past infinity $\mathscr{I}^-$ is spacelike (and is in fact then O's past event horizon). It is clear from the above discussion that in the steady state universe, which has a null past infinity for timelike and null geodesics and a spacelike future infinity, any fundamental observer has a future event horizon but no past particle horizon.

One can obtain other spaces which are locally equivalent to the de Sitter space, by identifying points in de Sitter space. The simplest such identification is to identify antipodal points p, p' (see figure 16) on the hyperboloid. The resulting space is not time orientable; if time increases in the direction of the arrow at p, the antipodal identification implies it must increase in the direction of the arrow at p', but one cannot continuously extend this identification of future and past half null cones over the whole hyperboloid. Calabi and Markus (1962) have studied in detail the spaces resulting from such identifications; they show in particular that an arbitrary point in the resulting space can be joined to any other point by a geodesic if and only if it is not time orientable.

The space of constant curvature with $R < 0$ is called *anti-de Sitter space*. It has the topology $S^1 \times R^3$, and can be represented as the hyperboloid

$$-u^2 - v^2 + x^2 + y^2 + z^2 = 1$$

in the flat five-dimensional space R^5 with metric

$$\mathrm{d}s^2 = -(\mathrm{d}u)^2 - (\mathrm{d}v)^2 + (\mathrm{d}x)^2 + (\mathrm{d}y)^2 + (\mathrm{d}z)^2.$$

There are closed timelike lines in this space; however it is not simply connected, and if one unwraps the circle S^1 (to obtain its covering space R^1) one obtains the universal covering space of anti-de Sitter space which does not contain any closed timelike lines. This has the topology of R^4. We shall in future mean by 'anti-de Sitter space', this universal covering space.

It can be represented by the metric

$$\mathrm{d}s^2 = -\mathrm{d}t^2 + \cos^2 t \{\mathrm{d}\chi^2 + \sinh^2 \chi (\mathrm{d}\theta^2 + \sin^2 \theta \, \mathrm{d}\phi^2)\}. \quad (5.9)$$

This coordinate system covers only part of the space, and has apparent singularities at $t = \pm \frac{1}{2}\pi$. The whole space can be covered by coordinates $\{t', r, \theta, \phi\}$ for which the metric has the static form

$$\mathrm{d}s^2 = -\cosh^2 r \, \mathrm{d}t'^2 + \mathrm{d}r^2 + \sinh^2 r (\mathrm{d}\theta^2 + \sin^2 \theta \, \mathrm{d}\phi^2).$$

In this form, the space is covered by the surfaces $\{t' = \text{constant}\}$ which have non-geodesic normals.

To study the structure at infinity, define the coordinate r' by

$$r' = 2 \arctan (\exp r) - \tfrac{1}{2}\pi, \quad 0 \leqslant r' < \tfrac{1}{2}\pi.$$

Then one finds $\mathrm{d}s^2 = \cosh^2 r \, \mathrm{d}\bar{s}^2$, where $\mathrm{d}\bar{s}^2$ is given by (5.7); that is, the whole of anti-de Sitter space is conformal to the region $0 \leqslant r' < \frac{1}{2}\pi$ of the Einstein static cylinder. The Penrose diagram is shown in figure 20; null and spacelike infinity can be thought of as a timelike surface in this case. This surface has the topology $R^1 \times S^2$.

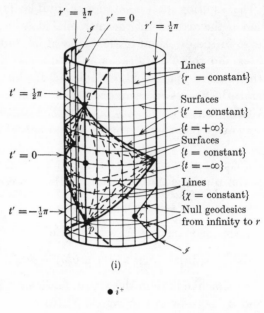

(i)

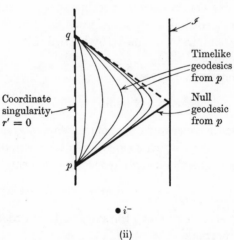

(ii)

FIGURE 20

(i) Universal anti-de Sitter space is conformal to one half of the Einstein static universe. While coordinates (t', r, θ, ϕ) cover the whole space, coordinates (t, χ, θ, ϕ) cover only one diamond-shaped region as shown. The geodesics orthogonal to the surfaces $\{t = \text{constant}\}$ all converge at p and q, and then diverge out into similar diamond-shaped regions.

(ii) The Penrose diagram of universal anti-de Sitter space. Infinity consists of the timelike surface $\mathscr{I}$ and the disjoint points i^+, i^-. The projection of some timelike and null geodesics is shown.

One cannot find a conformal transformation which makes timelike infinity finite without pinching off the Einstein static universe to a point (if a conformal transformation makes the time coordinate finite it also scales the space sections by an infinite factor), so we represent timelike infinity by the disjoint points i^+, i^-.

The lines $\{\chi, \theta, \phi$ constant$\}$ are the geodesics orthogonal to the surfaces $\{t = $ constant$\}$; they all converge to points q (respectively, p) in the future (respectively, past) of the surface, and this convergence is the reason for the apparent (coordinate) singularities in the original metric form. The region covered by these coordinates is the region between the surface $t = 0$ and the null surfaces on which these normals become degenerate.

The space has two further interesting properties. First, as a consequence of the timelike infinity, there exists no Cauchy surface whatever in the space. While one can find families of spacelike surfaces (such as the surfaces $\{t' = $ constant$\}$) which cover the space completely, each surface being a complete cross-section of the space–time, one can find null geodesics which never intersect any given surface in the family. Given initial data on any such surface, one cannot predict beyond the Cauchy development of the surface; thus from the surface $\{t = 0\}$, one can predict only in the region covered by the coordinates t, χ, θ, ϕ. Any attempt to predict beyond this region is prevented by fresh information coming in from the timelike infinity.

Secondly, corresponding to the fact that the geodesic normals from $t = 0$ all converge at p and q, all the past timelike geodesics from p expand out (normal to the surfaces $\{t = $ constant$\}$) and reconverge at q. In fact, all the timelike geodesics from any point in this space (to either the past or future) reconverge to an image point, diverging again from this image point to refocus at a second image point, and so on. The future timelike geodesics from p therefore never reach $\mathscr{I}$, in contrast to the future null geodesics which go to $\mathscr{I}$ from p and form the boundary of the future of p. This separation of timelike and null geodesics results in the existence of regions in the future of p (i.e. which can be reached from p by a future-directed timelike line) which cannot be reached from p by any geodesic. The set of points which can be reached by future-directed timelike lines from p is the set of points lying beyond the future null cone of p; the set of points which can be reached from p by future-directed timelike geodesics is the interior of the infinite chain of diamond-shaped regions similar to that covered by coordinates (t, χ, θ, ϕ). One notes that all points in the Cauchy development of the surface $t = 0$ can be reached from this surface by a unique geodesic normal to this surface, but that a general point outside this Cauchy development cannot be reached by any geodesic normal to the surface.

5.3 Robertson–Walker spaces

So far, we have not considered the relation of exact solutions to the physical universe. Following Einstein, we can ask: can one find space–times which are exact solutions for some suitable form of matter and which give a good representation of the large scale properties of the observable universe? If so, we can claim to have a reasonable 'cosmological model' or model of the physical universe.

However we are not able to make cosmological models without some admixture of ideology. In the earliest cosmologies, man placed himself in a commanding position at the centre of the universe. Since the time of Copernicus we have been steadily demoted to a medium sized planet going round a medium sized star on the outer edge of a fairly average galaxy, which is itself simply one of a local group of galaxies. Indeed we are now so democratic that we would not claim that our position in space is specially distinguished in any way. We shall, following Bondi (1960), call this assumption the *Copernican principle*.

A reasonable interpretation of this somewhat vague principle is to understand it as implying that, when viewed on a suitable scale, the universe is approximately spatially homogeneous.

By spatially homogeneous, we mean there is a group of isometries which acts freely on $\mathcal{M}$, and whose surfaces of transitivity are space-like three-surfaces; in other words, any point on one of these surfaces is equivalent to any other point on the same surface. Of course, the universe is not exactly spatially homogeneous; there are local irregularities, such as stars and galaxies. Nevertheless it might seem reasonable to suppose that the universe is spatially homogeneous on a large enough scale.

While one can build mathematical models fulfilling this requirement of homogeneity (see next section), it is difficult to test homogeneity directly by observation, as there is no simple way of measuring the separation between us and distant objects. This difficulty is eased by the fact that we can, in principle, fairly easily observe *isotropies* in extragalactic observations (i.e. we can see if these observations are the same in different directions, or not), and isotropies are closely con-

nected with homogeneity. Those observational investigations of isotropy which have been carried out so far support the conclusion that
the universe is approximately spherically symmetric about us.

In particular, it has been shown that extragalactic radio sources are
distributed approximately isotropically, and that the recently observed microwave background radiation, where it has been examined,
is very highly isotropic (see chapter 10 for further discussion).

It is possible to write down and examine the metrics of all space–
times which are spherically symmetric; particular examples are the
Schwarzschild and Reissner–Nordström solutions (see § 5.5); however
these are asymptotically flat spaces. In general, there can exist at most
two points in a spherically symmetric space from which the space looks
spherically symmetric. While these may serve as models of space–time
near a massive body, they can only be models of the universe consistent
with the isotropy of our observations if we are located near a very
special position. The exceptional cases are those in which the universe
is isotropic about *every* point in space time; so we shall interpret the
Copernican principle as stating that the universe is approximately
spherically symmetric about every point (since it is approximately
spherically symmetric about us).

As has been shown by Walker (1944), exact spherical symmetry
about every point would imply that the universe is spatially homogeneous and admits a six-parameter group of isometries whose surfaces
of transitivity are spacelike three-surfaces of constant curvature. Such
a space is called a *Robertson–Walker* (or *Friedmann*) space (Minkowski
space, de Sitter space and anti-de Sitter space are all special cases of
the general Robertson–Walker spaces). Our conclusion, then, is that
these spaces are a good approximation to the large scale geometry of
space–time in the region that we can observe.

In the Robertson–Walker spaces, one can choose coordinates so that
the metric has the form

$$\mathrm{d}s^2 = -\,\mathrm{d}t^2 + S^2(t)\,\mathrm{d}\sigma^2,$$

where $\mathrm{d}\sigma^2$ is the metric of a three-space of constant curvature and is
independent of time. The geometry of these three-spaces is qualitatively different according to whether they are three-spaces of constant
positive, negative or zero curvature; by rescaling the function S, one
can normalize this curvature K to be $+1$ or -1 in the first two cases.
Then the metric $\mathrm{d}\sigma^2$ can be written

$$\mathrm{d}\sigma^2 = \mathrm{d}\chi^2 + f^2(\chi)\,(\mathrm{d}\theta^2 + \sin^2\theta\,\mathrm{d}\phi^2),$$

where
$$f(\chi) = \begin{cases} \sin\chi & \text{if} \quad K = +1, \\ \chi & \text{if} \quad K = 0, \\ \sinh\chi & \text{if} \quad K = -1. \end{cases}$$

The coordinate χ runs from 0 to ∞ if $K = 0$ or -1, but runs from 0 to 2π if $K = +1$. When $K = 0$ or -1, the three-spaces are diffeomorphic to R^3 and so are 'infinite', but when $K = +1$ they are diffeomorphic to a three-sphere S^3 and so are compact ('closed' or 'finite'). One could identify suitable points in these three-spaces to obtain other global topologies; it is even possible to do this, in the case of negative or zero curvature, in such a way that the resulting three-space is compact (Löbell (1931)). However such a compact surface of constant negative curvature would have no continuous groups of isometries (Yano and Bochner (1953)) – although Killing vectors exist at each point, they would not determine any global Killing vector fields and the local groups of isometries they generate would not link up to form global groups. In the case of zero curvature, a compact space could only have a three-parameter group of isometries. In neither case would the resulting space–time be isotropic. We shall not make such identifications, as our original reason for considering these spaces was that they were isotropic (and so had a six-parameter group of isometries). In fact the only identifications which would not result in an anisotropic space would be to identify antipodal points on S^3 in the case of constant positive curvature.

The symmetry of the Robertson–Walker solutions requires that the energy–momentum tensor has the form of a perfect fluid whose density μ and pressure p are functions of the time coordinate t only, and whose flow lines are the curves (χ, θ, ϕ) constant (so the coordinates are comoving coordinates). This fluid can be thought of as a smoothed out approximation to the matter in the universe; then the function $S(t)$ represents the separation of neighbouring flow lines, that is, of 'nearby' galaxies.

The equation of conservation of energy (3.9) in these spaces takes the form
$$\dot{\mu} = -3(\mu+p)S^{\cdot}/S. \tag{5.10}$$

The Raychaudhuri equation (4.26) takes the form
$$4\pi(\mu+3p) - \Lambda = -3S^{\cdot\cdot}/S. \tag{5.11}$$

The remaining field equation (which is essentially (2.35)) can be written
$$3S^{\cdot 2} = 8\pi(\mu S^3)/S + \Lambda S^2 - 3K. \tag{5.12}$$

Whenever $S^{\cdot} \neq 0$, (5.12) can in fact be derived, with an arbitrary value of the constant K, as a first integral of (5.10), (5.11); so the real effect of this field equation is to identify the integration constant as the curvature of the metric $d\sigma^2$ of the three-spaces $\{t = \text{constant}\}$.

It is reasonable to assume (cf. the energy conditions, § 4.3) that μ is positive and p is non-negative. (In fact, present estimates are $10^{-29} \text{gm cm}^{-3} \geqslant \mu_0 \geqslant 10^{-31} \text{gm cm}^{-3}$, $\mu_0 \gg p_0 \geqslant 0$). Then, if Λ is zero, (5.11) shows that S cannot be constant; in other words the field equations then imply the universe is either expanding or contracting. Observations of other galaxies show, as first found by Slipher and Hubble, that they are moving away from us, and so indicate that the matter in the universe is expanding at the present time. Current observations give the value of $S^{\cdot}/S$ at the present time as

$$H \equiv (S^{\cdot}/S)|_0 \approx 10^{-10} \text{year}^{-1},$$

believed correct to within a factor 2. From this, (5.11) shows that if Λ is zero, S must have been zero a finite time t_0 ago (that is, a time t_0 measured along the world-line of our galaxy) where

$$t_0 < H^{-1} \approx 10^{10} \text{ years}.$$

From (5.10) it follows that the density decreases as the universe expands, and conversely that the density was higher in the past, increasing without bound as $S \to 0$. This is therefore not merely a coordinate singularity (as for example, in anti-de Sitter universe expressed in coordinates (5.9)); the fact that the density is infinite there shows that some scalar defined by the curvature tensor is also infinite. It is this that makes the singularity so much worse than in the corresponding Newtonian situation; in both cases the world-lines of all the particles intersect in a point and the density becomes infinite, but here space–time itself becomes singular at the point $S = 0$. We must therefore exclude this point from the space–time manifold, as no known physical laws could be valid there.

This singularity is the most striking feature of the Robertson–Walker solutions. It occurs in all models in which $\mu + 3p$ is positive and Λ is negative, zero, or with not too large a positive value. It would imply that the universe (or at least that part of which we can have any physical knowledge) had a beginning a finite time ago. However this result has here been deduced from the assumptions of exact spatial homogeneity and spherical symmetry. While these may be reasonable approximations on a large enough scale at the present time, they certainly do not hold locally. One might think that, as one traced the evolution of the universe back in time, the local irregularities would grow and could prevent the occurrence of a singularity, causing the universe to 'bounce' instead. Whether this could happen, and whether physically realistic solutions with inhomogeneities would contain singularities, is a central question of cosmology and constitutes the principal problem dealt with in this book; it will turn out that there is good evidence to believe that the physical universe does in fact become singular in the past.

If some suitable relation between p and μ is specified, (5.10) can be integrated to give μ as a function of S. In fact the pressure is very small at the present epoch. If one takes it and Λ to be zero, one finds from (5.10)

$$\frac{4\pi}{3}\mu = \frac{M}{S^3},$$

where M is a constant, and (5.12) becomes

$$3S'^2 - 6M/S = -3K \equiv E/M. \tag{5.13}$$

The first equation expresses the conservation of mass when the pressure is zero, while the second (the *Friedmann equation*) is an energy conservation equation for a comoving volume of matter; the constant E represents the sum of the kinetic and potential energies. If E is negative (i.e. K is positive), S will increase to some maximum value and then decrease to zero; if E is positive or zero (i.e. K is negative or zero), S will increase indefinitely.

The explicit solutions of (5.13) have a simple form if given in terms of a rescaled time parameter $\tau(t)$, defined by

$$d\tau/dt = S^{-1}(t); \tag{5.14}$$

they take the form

$$S = (E/3)(\cosh\tau - 1), \quad t = (E/3)(\sinh\tau - \tau), \quad \text{if} \quad K = -1;$$

$$S = \tau^2, \quad t = \tfrac{1}{3}\tau^3, \quad \text{if} \quad K = 0;$$

$$S = (-E/3)(1 - \cos\tau), \quad t = (-E/3)(\tau - \sin\tau), \quad \text{if} \quad K = 1.$$

(The case $K = 0$ is the Einstein–de Sitter universe; clearly $S \propto t^{\frac{2}{3}}$.)

If p is non-zero but positive, the qualitative behaviour is the same. In particular if $p = (\gamma - 1)\mu$ where γ is a constant, $1 \leqslant \gamma \leqslant 2$, one finds

$\frac{4}{3}\pi\mu = M/S^{3\gamma}$, and the solution of (5.12) near the singularity takes the form

$$S \propto t^{2/3\gamma}.$$

If Λ is negative, the solution expands from an initial singularity, reaches a maximum and then recollapses to a second singularity. If Λ is positive, then for $K = 0$ or -1 the solution expands forever and asymptotically approaches the steady state model. For $K = +1$ there are several possibilities. If Λ is greater than some value Λ_{crit} ($\Lambda_{\text{crit}} = (-E/3M)^3/(3M)^2$ if $p = 0$) the solution will start from an initial singularity and will expand forever asymptotically approaching the steady state model. If $\Lambda = \Lambda_{\text{crit}}$ there is a static solution, the *Einstein static universe*. (The metric form (5.7) is that of the particular Einstein static solution for which $\mu + p = (4\pi)^{-1}$, $\Lambda = 1 + 8\pi p$.) There is also a solution which starts from an initial singularity and asymptotically approaches the Einstein universe, and one which starts from the Einstein universe in the infinite past and expands forever. If $\Lambda < \Lambda_{\text{crit}}$ there are two solutions – one expands from an initial singularity and then recollapses to a second singularity; the other contracts from an infinite radius in the infinite past, reaches a minimum radius, and then re-expands. This and the universe asymptotic to the static universe in the infinite past are the only solutions which could represent the observed universe and which do not have a singularity. In these models, $S^{..}$ is always positive, and this seems to be in conflict with observations of redshifts of distant galaxies (Sandage (1961, 1968)). Also, the maximum density in these models would not have been very much larger than the present density. This would make it difficult to understand phenomena such as the microwave background radiation and the cosmic abundance of helium, which seem to point to a very hot dense phase in the history of the universe.

Just as in the previous cases we have studied, one can find conformal mappings of the Robertson–Walker spaces into the Einstein static space. We use the coordinate τ defined by (5.14) as a time coordinate; then the metric takes the form

$$ds^2 = S^2(\tau)\{-d\tau^2 + d\chi^2 + f^2(\chi)(d\theta^2 + \sin^2\theta\, d\phi^2)\}. \qquad (5.15)$$

In the case $K = +1$, this is already conformal to the Einstein static space (put $\tau = t'$, $\chi = r'$ to agree with the notation of (5.7)). Thus these spaces are mapped into precisely that part of the Einstein static space determined by the values taken by τ. When $p = \Lambda = 0$, τ lies in the range $0 < \tau < \pi$, so the whole space is mapped into this region in the Einstein static universe while its boundary is mapped into the three-spheres $\tau = 0$, $\tau = \pi$. (If $p > 0$, it is mapped into a region for which τ takes values $0 < \tau < a < \pi$, for some number a.) In the case $K = 0$, the same coordinates represent the space as conformal to flat space (see (5.15)), so on using the conformal transformations of § 5.1, one obtains these spaces mapped into some part of the diamond representing Minkowski space–time in the Einstein static universe (see figure 14); the actual region is again determined by the values taken by τ. When $\Lambda = 0$, $0 < \tau < \infty$, so this space (which is the Einstein–de Sitter space when $p = 0$) is conformal to the half $t' > 0$ of the diamond which represents Minkowski space–time. In the case $K = -1$, one obtains the metric conformal to part of the region of the Einstein static space for which $\frac{1}{2}\pi \geqslant t' + r' \geqslant -\frac{1}{2}\pi$, $\frac{1}{2}\pi \geqslant t' - r' \geqslant -\frac{1}{2}\pi$, on defining

$$t' = \arctan(\tanh \tfrac{1}{2}(\tau + \chi)) + \arctan(\tanh \tfrac{1}{2}(\tau - \chi)),$$

$$r' = \arctan(\tanh \tfrac{1}{2}(\tau + \chi)) - \arctan(\tanh \tfrac{1}{2}(\tau - \chi)).$$

The part of this diamond-shaped region covered depends on the range of τ; when $\Lambda = 0$, the space is mapped into the upper half.

One thus obtains these spaces and their boundaries conformal to some (generally finite) region of the Einstein static space, see figure 21 (i). However there is an important difference from the previous cases: part of the boundary is not 'infinity' in the sense it was previously, but represents the singularity when $S = 0$. (The conformal factor can be thought of as making infinity finite by giving an infinite compression, but making the singular point $S = 0$ finite by an infinite expansion.) In fact this makes little difference to the conformal diagrams; one can give the Penrose diagrams as before (see figures 21 (ii) and 21 (iii)). In each case when $p \geqslant 0$ the singularity at $t = 0$ is represented by a spacelike surface; this corresponds to the existence of particle horizons (defined precisely as in § 5.2) in these spaces. Also when $K = +1$ the future boundary is spacelike, implying the existence of event horizons for the fundamental observers; when $K = 0$ or -1 and $\Lambda = 0$, future infinity is null and there are no future event horizons for the fundamental observers in these spaces.

At this stage, one should examine the following question: anti-de Sitter space could be expressed in the Robertson–Walker form (5.9) and then expressed conformally as part of the Einstein static universe. When one did so, one found that the Robertson–Walker coordinates

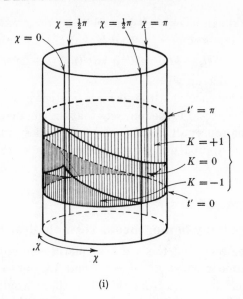

(i)

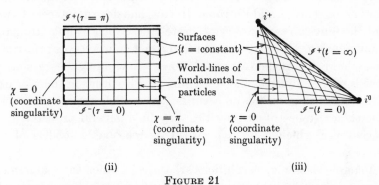

(ii)   (iii)

FIGURE 21

(i) The Robertson–Walker spaces ($p = \Lambda = 0$) are conformal to the regions of the Einstein static universe shown, in the three cases $K = +1$, 0 and -1.

(ii) Penrose diagram of a Robertson–Walker space with $K = +1$ and $p = \Lambda = 0$.

(iii) Penrose diagram of a Robertson–Walker space with $K = 0$ or -1 and $p = \Lambda = 0$.

covered only a small part of the full space–time. That is to say, the space–time described by the Robertson–Walker coordinates could be extended. One should therefore show that the Robertson–Walker universes in which there is matter are in fact inextendible. This follows because one can show that if $\mu > 0$, $p \geqslant 0$ and $\mathbf{X}$ is any vector at any point q, the geodesic $\gamma(v)$ through $q = \gamma(0)$ in the direction of $\mathbf{X}$ is such that either

(i) $\gamma(v)$ can be extended to arbitrary positive values of v, or

(ii) there is some $v_0 > 0$ such that the scalar invariant

$$(R_{ij} - \tfrac{1}{2}Rg_{ij})\,(R^{ij} - \tfrac{1}{2}Rg^{ij}) = (\mu + \Lambda)^2 + 3(p - \Lambda)^2$$

is unbounded on $\gamma([0, v_0))$.

It is now clear that the surfaces $\{t = \text{constant}\}$ are Cauchy surfaces in these spaces. Further one sees that the singularity is universal in the following sense: all timelike and null geodesics through any point in the space approach it for some finite value of their affine parameter.

5.4 Spatially homogeneous cosmological models

We have seen that there are singularities in any Robertson–Walker space–time in which $\mu > 0$, $p \geqslant 0$ and Λ is not too large. However one could not conclude from this that there would be singularities in more realistic world models which allow for the fact that the universe is *not* homogeneous and isotropic. In fact, one does not expect to find that the universe can be very accurately described by *any* attainable exact solution. However one can find exact solutions, less restricted than the Robertson–Walker solutions, which may be reasonable models of the universe, and see if singularities occur in them or not; the fact that singularities do occur in such models gives an indication that the existence of singularities may be a general property of *all* space–times which can be regarded as reasonable models of the universe.

A simple class of such solutions are those in which the requirement of isotropy is dropped but the requirement of *spatial homogeneity* (the strict Copernican principle) is retained (although the universe seems approximately isotropic at the present time, there might have been large anisotropies at an earlier epoch). Thus in these models one assumes there exists a group of isometries G_r whose orbits in some part of the model are spacelike hypersurfaces. (The orbit of a point p under the group G_r is the set of points into which p is moved by the action of all elements of the group.) These models may be constructed locally by well-known methods; see Heckmann and Schücking (1962) for the case $r = 3$, and Kantowski and Sachs (1967) for the case $r = 4$ (if $r > 4$, the space–time is necessarily a Robertson–Walker space).

The simplest spatially homogeneous space–times are those in which the group of isometries is Abelian; the group is then of type I in the classification given by Bianchi (1918), so we call these *Bianchi I* spaces. We discuss Bianchi I spaces in some detail, and then give a theorem showing singularities will occur in all non-empty spatially homogeneous models in which the timelike convergence condition (§ 4.3) is satisfied.

Suppose the spatially homogeneous space–time has an Abelian isometry group; for simplicity we assume $\Lambda = 0$ and that the matter content is a pressure-free perfect fluid ('dust'). Then there exist comoving coordinates (t, x, y, z) such that the metric takes the form

$$ds^2 = -dt^2 + X^2(t)\,dx^2 + Y^2(t)\,dy^2 + Z^2(t)\,dz^2. \qquad (5.16)$$

Defining the function $S(t)$ by $S^3 = XYZ$, the conservation equations show that the density of matter is given by $\frac{4}{3}\pi\mu = M/S^3$, where M is a suitably chosen constant. The general solution of the field equations can be written

$$X = S(t^{\frac{2}{3}}/S)^{2\sin\alpha}, \quad Y = S(t^{\frac{2}{3}}/S)^{2\sin(\alpha+\frac{2}{3}\pi)},$$

$$Z = S(t^{\frac{2}{3}}/S)^{2\sin(\alpha+\frac{4}{3}\pi)},$$

where S is given by $S^3 = \frac{9}{2}Mt(t+\Sigma);$

$\Sigma\ (> 0)$ is a constant determining the magnitude of the anisotropy (we exclude the isotropic case $(\Sigma = 0)$, which is the Einstein–de Sitter universe (§ 5.3)), and $\alpha(-\frac{1}{6}\pi < \alpha \leqslant \frac{1}{2}\pi)$ is a constant determining the direction in which the most rapid expansion takes place. The average rate of expansion is given by

$$\frac{S^{\cdot}}{S} = \frac{2}{3t}\frac{t+\Sigma/2}{t+\Sigma};$$

the expansion in the x-direction is

$$\frac{X^{\cdot}}{X} = \frac{2}{3t}\frac{t+\Sigma(1+2\sin\alpha)/2}{t+\Sigma},$$

and the expansions $Y^{\cdot}/Y$, $Z^{\cdot}/Z$ in the y, z directions are given by similar expressions in which α is replaced by $\alpha+\frac{2}{3}\pi$, $\alpha+\frac{4}{3}\pi$ respectively.

The solution expands from a highly anisotropic singular state at $t = 0$, reaching a nearly isotropic phase for large t when it is nearly the same as the Einstein–de Sitter universe. The average length S increases monotonically as t increases, its initial high rate of change ($S \propto t^{\frac{1}{3}}$ for small t) decreasing steadily ($S \propto t^{\frac{2}{3}}$ for large t). Thus the universe evolves more rapidly, at early times, than its isotropic equivalent.

Suppose one considers the time-reverse of the model, and follows this forward in time towards the singularity. The initially almost isotropic contraction will become very anisotropic at late times. For general values of α, i.e. $\alpha \neq \frac{1}{2}\pi$, the term $1 + 2\sin(\alpha + \frac{4}{3}\pi)$ will be negative. Thus the collapse in the z-direction would halt, and, for sufficiently early times, be replaced by an expansion, the rate of expansion becoming indefinitely large for early enough times. In the x- and y-directions, on the other hand, the collapse would continue monotonically towards the singularity. Thus if one considers the forward direction of time in the original model, one has a 'cigar' singularity: matter collapses in along the z-axis from infinity, halts, and then starts re-expanding, while in the x- and y-directions the matter expands monotonically at all times. If one could receive signals from early enough times in such a model, one would see a maximum redshift in the z-direction, at earlier times matter in this direction being observed with progressively smaller redshifts and then with indefinitely increasing *blue*-shifts.

The behaviour in the exceptional case $\alpha = \frac{1}{2}\pi$ is rather different. In this case, the terms $1 + 2\sin(\alpha + \frac{2}{3}\pi)$ and $1 + 2\sin(\alpha + \frac{4}{3}\pi)$ both vanish. Thus the expansions in the axis directions are

$$\frac{X^{\cdot}}{X} = \frac{2}{3t}\frac{t + 3\Sigma/2}{t + \Sigma}, \quad \frac{Y^{\cdot}}{Y} = \frac{Z^{\cdot}}{Z} = \frac{2}{3}\frac{1}{t + \Sigma}.$$

If one follows the time-reversed model, the rate of collapse in the y- and z-directions slows asymptotically down to zero, while the rate of collapse in the x-direction increases indefinitely. In the original model, one has a 'pancake' singularity: matter expands monotonically in all directions, starting from an indefinitely high expansion rate in the x-direction but from zero expansion rates in the y- and z-directions. Indefinitely high redshifts would be seen in the x-direction, but there would be limiting redshifts in the y- and z-directions.

Further examination shows that in the general ('cigar') case, there is a particle horizon in every direction despite the anisotropic expansion. However in the exceptional ('pancake') case, no horizon occurs in the x-direction; in fact the particles that can be seen by an observer at the origin at time t_0 are characterized by coordinate values (x, y, z) lying within the infinite cylinder

$$x^2 + y^2 < \rho^2$$

where $\rho = \dfrac{2}{3M}\left\{\left(\dfrac{9M}{2}(t_0 + \Sigma)\right)^{\frac{1}{3}} - \left(\dfrac{9M}{2}\Sigma\right)^{\frac{1}{3}}\right\}.$

While we have here considered these models for vanishing pressure and Λ term only, properties of these spaces with more realistic matter contents can easily be obtained; for example if one has either a perfect fluid with $p = (\gamma - 1)\mu$, γ a constant $(1 < \gamma < 2)$, or a mixture of a photon gas and matter with pressure $p \leqslant \frac{1}{3}\mu$, the behaviour near the singularity is the same as in the dust case.

An interesting consequence of the non-existence of a particle horizon in the x-direction in the exceptional ('pancake') case, is that one can extend the solution continuously across the singularity. We shall show this explicitly in the case of the dust solution.

The metric takes the form (5.16) where now

$$X(t) = t(\tfrac{9}{2}M(t+\Sigma))^{-\frac{1}{3}}, \quad Y(t) = Z(t) = (\tfrac{9}{2}M(t+\Sigma))^{\frac{2}{3}}. \tag{5.17}$$

We now choose new coordinates τ, η which satisfy the equations

$$\tanh(2x/9M\Sigma) = \eta/\tau, \quad \exp\left(\frac{4}{9M}\int_0^t \frac{\mathrm{d}t}{X(t)}\right) = \tau^2 - \eta^2.$$

One then finds that the space with metric (5.16), (5.17) is given in the new coordinates by

$$\mathrm{d}s^2 = A^2(t)(-\mathrm{d}\tau^2 + \mathrm{d}\eta^2) + B^2(t)(\mathrm{d}y^2 + \mathrm{d}z^2) \tag{5.18}$$

where

$$A(t) = \exp\left(-\frac{t+\Sigma}{\Sigma}\right) \cdot (\tfrac{9}{2}M(t+\Sigma))^{-\frac{1}{3}}, \quad B(t) = (\tfrac{9}{2}M(t+\Sigma))^{\frac{2}{3}}, \tag{5.19}$$

the whole space (for $t > 0$) being mapped into the region $\mathcal{V}$ defined by $\tau > 0$, $\tau^2 - \eta^2 > 0$. The function $t(\tau, \eta)$ is now defined implicitly as the solution of the equation

$$\tau^2 - \eta^2 = \tfrac{9}{2}Mt^2 \exp\frac{2(t+\Sigma)}{\Sigma} \tag{5.20}$$

for which $t > 0$. The (τ, η) plane is given in conformally flat coordinates. The region $\mathcal{V}$ in this plane, bounded by the surface $t = 0$, is shown in figure 22. In this diagram, the world-lines of the particles are straight lines diverging from the origin.

The functions $A(t)$, $B(t)$ are continuous as $t \to 0$ from above. One can therefore extend the solution continuously to the whole (τ, η) plane by specifying that (5.19) holds everywhere, (5.20) holds inside $\mathcal{V}$, and that

$$t(\tau, \eta) = 0$$

holds outside $\mathcal{V}$. Then (5.18) is a C^0 metric which is a solution of the

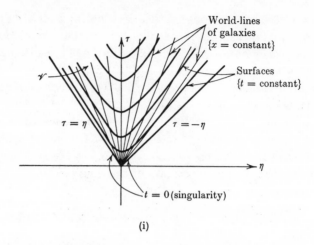

(i)

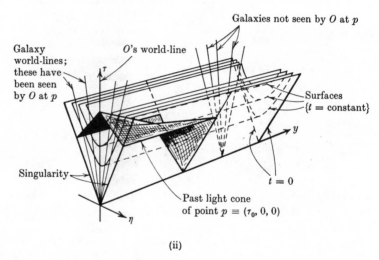

(ii)

FIGURE 22. Dust-filled Bianchi I space with a pancake singularity.
  (i) The (τ, η) plane; null lines are at $\pm\, 45°$.
  (ii) A half-section of the space in $(\tau,\ \eta,\ y)$ coordinates (the z-coordinate is suppressed), showing the past light cone of the point $p = (\tau_0,\ 0,\ 0)$. There is a particle horizon in the y-direction but not in the x- (i.e. η) direction.

field equations equivalent to (5.16), (5.17) inside $\mathscr{V}$, and is a flat space–time outside $\mathscr{V}$. However the solution is not C^1 across the boundary of $\mathscr{V}$, and in fact the density of matter becomes infinite on this boundary (as $S \to 0$ there). Since the first derivatives are not square integrable, the Einstein field equations cannot be interpreted on the boundary even in a distributional sense (see § 8.4). While the extension onto the boundary is unique, it is in no way unique beyond the boundary. We have carried out the extension in the case of dust; a similar extension could be carried out if one had a mixture of matter and radiation.

Let us now return to considering general non-empty spatially homogeneous models. The existence of a singularity in these models will follow directly from Raychaudhuri's equation if the motion of the matter is geodesic and without rotation (as must be the case, for example, if the world-lines are orthogonal to the surfaces of homogeneity) and the timelike convergence condition is satisfied; however there exist such spaces in which the matter accelerates and rotates, and either of these factors could possibly prevent the existence of a singularity. The following result, which is an improved version of a theorem of Hawking and Ellis (1965), shows that in fact neither acceleration nor rotation can prevent the existence of singularities in these models.

Theorem

$(\mathcal{M}, g)$ cannot be timelike geodesically complete if:

(1) $R_{ab} K^a K^b > 0$ for all timelike and null vectors $\mathbf{K}$ (this is true if the energy–momentum tensor is type I (§ 4.3) and $\mu + p_i > 0$, $\mu + \sum_i p_i - 4\pi\Lambda > 0$);

(2) there exist equations of motion for the matter fields such that the Cauchy problem has a unique solution (see chapter 7);

(3) the Cauchy data on some spacelike three-surface $\mathcal{H}$ is invariant under a group of diffeomorphisms of $\mathcal{H}$ which is transitive on $\mathcal{H}$.

Since the intrinsic geometry of $\mathcal{H}$ is invariant under a transitive group of diffeomorphisms, these are isometries and $\mathcal{H}$ is complete, i.e. cannot have any boundary. It can be shown (see § 6.5) that if there is a non-spacelike curve which intersects $\mathcal{H}$ more than once, then there exists a covering manifold $\hat{\mathcal{M}}$ of $\mathcal{M}$ in which each connected component of the image of $\mathcal{H}$ will not intersect any non-spacelike curve more than once. We shall assume that $\hat{\mathcal{M}}$ is timelike geodesically complete, and show that this is inconsistent with conditions (1), (2) and (3).

Let $\hat{\mathcal{H}}$ be a connected component of the image of $\mathcal{H}$ in $\hat{\mathcal{M}}$. By (3), the Cauchy data on $\hat{\mathcal{H}}$ is homogeneous. Therefore by condition (2), the Cauchy development of any region of $\hat{\mathcal{H}}$ is isometric to the Cauchy development of any other similar region of $\hat{\mathcal{H}}$. This implies that the surfaces $\{s = \text{constant}\}$ are homogeneous if they lie within the Cauchy development of $\hat{\mathscr{H}}$, where s is the distance from $\hat{\mathscr{H}}$ measured along the geodesic normals to $\hat{\mathscr{H}}$. These surfaces must lie either entirely within or entirely outside the Cauchy development of $\hat{\mathscr{H}}$, as otherwise there would be equivalent regions in $\hat{\mathscr{H}}$ which had inequivalent Cauchy evolutions. The surfaces $\{s = \text{constant}\}$ will lie in the Cauchy development of $\hat{\mathscr{H}}$ as long as they remain spacelike, because the boundary of the Cauchy development of $\hat{\mathscr{H}}$ (if it exists) must be null (§ 6.5).

The geodesics orthogonal to $\hat{\mathscr{H}}$ will be orthogonal to the surfaces $\{s = \text{constant}\}$, as a vector representing the separation of points equal distances along neighbouring geodesics will remain orthogonal to the geodesics if it is so initially. As in § 4.1, one can represent the spatial separation of neighbouring geodesics orthogonal to $\hat{\mathscr{H}}$ by a matrix $\mathbf{A}$ which is the unit matrix on $\hat{\mathscr{H}}$. By homogeneity, it will be constant on the surfaces $\{s = \text{constant}\}$ while these lie in the Cauchy development of $\hat{\mathscr{H}}$. While $\mathbf{A}$ is non-degenerate, the map from $\hat{\mathscr{H}}$ to a surface $\{s = \text{constant}\}$ defined by the normal geodesics will be of rank three and so the surfaces will be spacelike three-surfaces contained within the Cauchy development of $\hat{\mathscr{H}}$. The expansion

$$\theta = (\det \mathbf{A})^{-1} \, \mathrm{d} \, (\det \mathbf{A})/\mathrm{d}s$$

of these geodesics obeys Raychaudhuri's equation (4.26) with the vorticity and acceleration zero. By condition (1), $R_{ab} V^a V^b$ is positive for all timelike vectors V^a. Thus θ will become infinite and $\mathbf{A}$ will be degenerate for some finite positive or negative value s_0 of s. The map from $\hat{\mathscr{H}}$ to the surface $s = s_0$ can have at most rank two; there will therefore be at least one vector field $\mathbf{Z}$ on $\hat{\mathscr{H}}$ such that $\mathbf{AZ} = 0$. The integral curves of this vector field are curves in $\hat{\mathscr{H}}$ which are mapped by the geodesic normals to one point in the surface $s = s_0$. Thus this surface will be at most two-dimensional. As the geodesics lie in the Cauchy development of $\hat{\mathscr{H}}$ for $|s| < |s_0|$, the surface $s = s_0$ will lie in the Cauchy development or on the boundary of the Cauchy development of $\hat{\mathscr{H}}$. By condition (1), the energy–momentum tensor has a unique timelike eigenvector at each point. These eigenvectors will form a C^1 timelike vector field whose integral curves may be thought of as representing the flow lines of the matter. As the surface $s = s_0$ lies in the Cauchy development of $\hat{\mathscr{H}}$ or on its boundary, all the flow lines that pass through it must intersect $\hat{\mathscr{H}}$. But then as $\hat{\mathscr{H}}$ is homogeneous, all the flow lines that pass through $\hat{\mathscr{H}}$ must pass through $s = s_0$. Thus the flow lines define a diffeomorphism between $\hat{\mathscr{H}}$ and the surface

$s = s_0$. This is impossible, as $\hat{\mathscr{H}}$ is three-dimensional and $s = s_0$ is two-dimensional.   □

In fact, if all the flow lines were to pass through a two-dimensional surface, one would expect the matter density to become infinite. We have now seen that a large scale rotation or acceleration cannot, by itself, prevent the occurrence of singularities in a universe model obeying the strict Copernican principle. In later theorems we shall see that irregularities are in general also unable to prevent the occurrence of singularities in world models.

5.5 The Schwarzschild and Reissner–Nordström solutions

While the spatially homogeneous solutions may be good models for the large scale distribution of matter in the universe, they are inadequate for describing, for example, the local geometry of space–time in the solar system. One can describe this geometry to a good approximation by the Schwarzschild solution, which represents the spherically symmetric empty space–time outside a spherically symmetric massive body. In fact, all the experiments which have so far been carried out to test the difference between the General Theory of Relativity and Newtonian theory are based on predictions by this solution.

The metric can be given in the form

$$ds^2 = -\left(1 - \frac{2m}{r}\right)dt^2 + \left(1 - \frac{2m}{r}\right)^{-1}dr^2 + r^2(d\theta^2 + \sin^2\theta\, d\phi^2), \quad (5.21)$$

where $r > 2m$. It can be seen that this space–time is static, i.e. $\partial/\partial t$ is a timelike Killing vector which is a gradient, and is spherically symmetric, i.e. is invariant under the group of isometries $SO(3)$ operating on the spacelike two-spheres $\{t, r$ constant$\}$ (cf. appendix B). The coordinate r in this metric form is intrinsically defined by the requirement that $4\pi r^2$ is the area of these surfaces of transitivity. The solution is asymptotically flat as the metric has the form $g_{ab} = \eta_{ab} + O(1/r)$ for large r. Comparison with Newtonian theory (cf. §3.4) shows that m should be regarded as the gravitational mass, as measured from infinity, of the body producing the field. It should be emphasized that this solution is unique: if any solution of the vacuum field equations is spherically symmetric, it is locally isometric to the Schwarzschild solution (although it may of course look totally different if it is given in some other coordinate system; see appendix B and Bergmann, Cahen and Komar (1965)).

Normally one would regard the Schwarzschild metric for r greater than some value $r_0 > 2m$ as being the solution outside some spherical body, the metric inside the body $(r < r_0)$ having a different form determined by the energy–momentum tensor of the matter in the body. However it is interesting to see what happens when the metric is regarded as an empty space solution for all values of r.

The metric is then singular when $r = 0$ and when $r = 2m$ (there are also the trivial singularities of polar coordinates when $\theta = 0$ and $\theta = \pi$). One must therefore cut $r = 0$ and $r = 2m$ out of the manifold defined by the coordinates (t, r, θ, ϕ), since in § 3.1 we took space–time to be represented by a manifold with a Lorentz metric. Cutting out the surface $r = 2m$ divides the manifold into two disconnected components for which $0 < r < 2m$ and $2m < r < \infty$. Since we took the space–time manifold to be connected, we must consider only one of these components and the obvious one to choose is the one for $r > 2m$, which represents the external field. One must then ask whether this manifold $\mathcal{M}$ with the Schwarzschild metric $\mathbf{g}$ is extendible, i.e. whether there is a larger manifold $\mathcal{M}'$ into which $\mathcal{M}$ can be imbedded and a suitably differentiable Lorentz metric $\mathbf{g}'$ on $\mathcal{M}'$ which coincides with $\mathbf{g}$ on the image of $\mathcal{M}$. The obvious place where $\mathcal{M}$ might be extended is where r tends to $2m$. A calculation shows that although the metric is singular at $r = 2m$ in the Schwarzschild coordinates (t, r, θ, ϕ), no scalar polynomials of the curvature tensor and the metric diverge as $r \to 2m$. This suggests that the singularity at $r = 2m$ is not a real physical singularity, but rather one which is a result of a bad choice of coordinates.

To confirm this, and to show that $(\mathcal{M}, \mathbf{g})$ can be extended, define

$$r^* \equiv \int \frac{\mathrm{d}r}{1 - 2m/r} = r + 2m \log{(r - 2m)}.$$

Then
$$v \equiv t + r^*$$

is an advanced null coordinate, and

$$w \equiv t - r^*$$

is a retarded null coordinate. Using coordinates (v, r, θ, ϕ) the metric takes the Eddington–Finkelstein form $\mathbf{g}'$ given by

$$\mathrm{d}s^2 = -\left(1 - \frac{2m}{r}\right)\mathrm{d}v^2 + 2\mathrm{d}v\,\mathrm{d}r + r^2(\mathrm{d}\theta^2 + \sin^2\theta\,\mathrm{d}\phi^2). \tag{5.22}$$

The manifold $\mathcal{M}$ is the region $2m < r < \infty$, but the metric (5.22) is non-singular and indeed analytic on the larger manifold $\mathcal{M}'$ for which

$0 < r < \infty$. The region of $(\mathcal{M}', \mathbf{g}')$ for which $0 < r < 2m$ is in fact isometric to the region of the Schwarzschild metric for which $0 < r < 2m$. Thus by using different coordinates, i.e. by taking a different manifold, we have extended the Schwarzschild metric so that it is no longer singular at $r = 2m$. In the manifold $\mathcal{M}'$ the surface $r = 2m$ is a null surface, as can be seen from the Finkelstein diagram (figure 23). This is a section $(\theta, \phi$ constant) of the space–time; each point represents a two-sphere of area $4\pi r^2$. Some null cones and radial null geodesics are indicated on this diagram. Surfaces $\{t = \text{constant}\}$ are indicated; one sees that t becomes infinite on the surface $r = 2m$.

This representation of the Schwarzschild solution has the odd feature that it is not time symmetric. One might expect this from the cross term $(\mathrm{d}v \, \mathrm{d}r)$ in (5.22); it is qualitatively clear from the Finkelstein diagram. The most obvious asymmetry is that the surface $r = 2m$ acts as a one-way membrane, letting future-directed timelike and null curves cross only from the outside $(r > 2m)$ to the inside $(r < 2m)$. Any past-directed timelike or null curve in the outside region cannot cross into the inside region. No past-directed timelike or null curve within $r = 2m$ can approach $r = 0$. However any future-directed timelike or null curve which crosses the surface $r = 2m$ approaches $r = 0$ within a finite affine distance. As $r \to 0$, the scalar $R^{abcd}R_{abcd}$ diverges as m^2/r^6. Therefore $r = 0$ is a real singularity; the pair $(\mathcal{M}', \mathbf{g}')$ cannot be extended in a C^2 manner or in fact even in a C^0 manner across $r = 0$.

If one uses the coordinate w instead of v, the metric takes the form $\mathbf{g}''$ given by

$$\mathrm{d}s^2 = -\left(1 - \frac{2m}{r}\right)\mathrm{d}w^2 - 2\,\mathrm{d}w\,\mathrm{d}r + r^2(\mathrm{d}\theta^2 + \sin^2\theta\,\mathrm{d}\phi^2).$$

This is analytic on the manifold $\mathcal{M}''$ defined by the coordinates (w, r, θ, ϕ) for $0 < r < \infty$. Again the manifold $\mathcal{M}$ is the region $2m < r < \infty$ and the new region $0 < r < 2m$ is isometric to the region $0 < r < 2m$ of the Schwarzschild metric, but the isometry reverses the direction of time. In the manifold $\mathcal{M}''$, the surface $r = 2m$ is again a null surface which acts as a one-way membrane. However this time it acts in the other direction of time, letting only past-directed time-like or null curves cross from the outside $(r > 2m)$ to the inside $(r < 2m)$.

One can in fact make both extensions $(\mathcal{M}', \mathbf{g}')$ and $(\mathcal{M}'', \mathbf{g}'')$ simultaneously; that is to say, there is a still larger manifold $\mathcal{M}*$ with metric $\mathbf{g}*$ into which both $(\mathcal{M}', \mathbf{g}')$ and $(\mathcal{M}'', \mathbf{g}'')$ can be isometrically imbedded, so that they coincide on the region $r > 2m$ which is

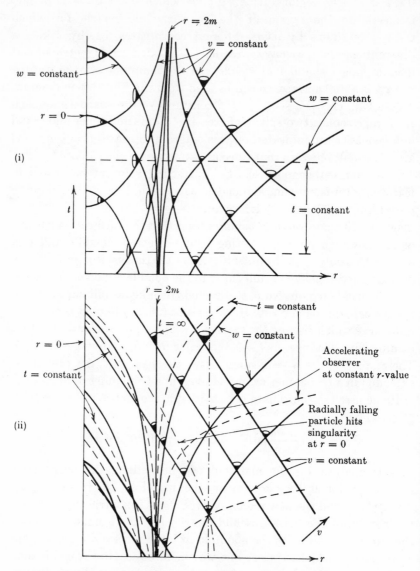

FIGURE 23. Section (θ, ϕ) constant of the Schwarzschild solution.
  (i) Apparent singularity at $r = 2m$ when coordinates (t, r) are used.
  (ii) Finkelstein diagram obtained by using coordinates (v, r) (lines at $45°$ are lines of constant v). Surface $r = 2m$ is a null surface on which $t = \infty$.

isometric to $(\mathcal{M}, \mathbf{g})$. A construction of this larger manifold has been given by Kruskal (1960). To obtain it, consider $(\mathcal{M}, \mathbf{g})$ in the coordinates (v, w, θ, ϕ); then the metric takes the form

$$\mathrm{d}s^2 = -\left(1 - \frac{2m}{r}\right)\mathrm{d}v\,\mathrm{d}w + r^2(\mathrm{d}\theta^2 + \sin^2\theta\,\mathrm{d}\phi^2),$$

where r is determined by

$$\tfrac{1}{2}(v - w) = r + 2m\log{(r - 2m)}.$$

This presents the two-space (θ, ϕ constant) in null conformally flat coordinates, as the space with metric $\mathrm{d}s^2 = -\mathrm{d}v\,\mathrm{d}w$ is flat. The most general coordinate transformation which leaves this two-space expressed in such conformally flat double null coordinates is $v' = v'(v)$, $w' = w'(w)$ where v' and w' are arbitrary C^1 functions. The resulting metric is

$$\mathrm{d}s^2 = -\left(1 - \frac{2m}{r}\right)\frac{\mathrm{d}v}{\mathrm{d}v'}\frac{\mathrm{d}w}{\mathrm{d}w'}\mathrm{d}v'\,\mathrm{d}w' + r^2(\mathrm{d}\theta^2 + \sin^2\theta\,\mathrm{d}\phi^2).$$

To reduce this to a form corresponding to that obtained earlier for Minkowski space–time, define

$$x' = \tfrac{1}{2}(v' - w'), \quad t' = \tfrac{1}{2}(v' + w').$$

The metric takes the final form

$$\mathrm{d}s^2 = F^2(t', x')\,(-\mathrm{d}t'^2 + \mathrm{d}x'^2) + r^2(t', x')\,(\mathrm{d}\theta^2 + \sin^2\theta\,\mathrm{d}\phi^2). \tag{5.23}$$

The choice of the functions v', w' determines the precise form of the metric. Kruskal's choice was $v' = \exp{(v/4m)}$, $w' = -\exp{(-w/4m)}$. Then r is determined implicitly by the equation

$$(t')^2 - (x')^2 = -(r - 2m)\exp{(r/2m)} \tag{5.24}$$

and F is given by

$$F^2 = \exp{(-r/2m)} \,.\, 16m^2/r. \tag{5.25}$$

On the manifold $\mathcal{M}^*$ defined by the coordinates (t', x', θ, ϕ) for $(t')^2 - (x')^2 < 2m$, the functions r and F (defined by (5.24), (5.25)) are positive and analytic. Defining the metric $\mathbf{g}^*$ by (5.23), the region I of $(\mathcal{M}^*, \mathbf{g}^*)$ defined by $x' > |t'|$ is isometric to $(\mathcal{M}, \mathbf{g})$, the region of the Schwarzschild solution for which $r > 2m$. The region defined by $x' > -t'$ (regions I and II in figure 24) is isometric to the advanced Finkelstein extension $(\mathcal{M}', \mathbf{g}')$. Similarly the region defined by $x' > t'$ (regions I and II' in figure 24) is isometric to the retarded Finkelstein extension $(\mathcal{M}'', \mathbf{g}'')$. There is also a region I', defined by $x' < -|t'|$,

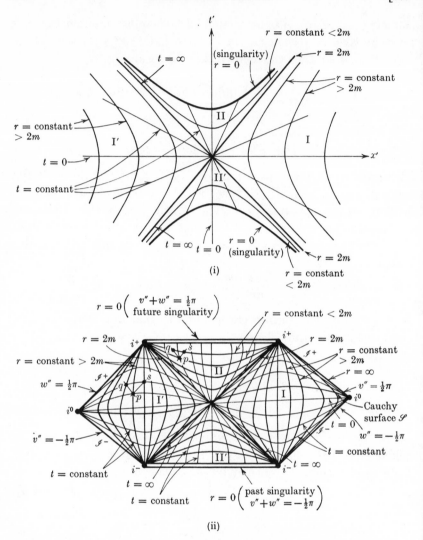

FIGURE 24. The maximal analytic Schwarzschild extension. The θ, ϕ coordinates are suppressed; null lines are at $\pm 45°$. Surfaces $\{r = \text{constant}\}$ are homogeneous.

(i) The Kruskal diagram, showing asymptotically flat regions I and I' and regions II, II' for which $r < 2m$.

(ii) Penrose diagram, showing conformal infinity as well as the two singularities.

which turns out to be again isometric with the exterior Schwarzschild solution $(\mathcal{M}, \mathbf{g})$. This can be regarded as another asymptotically flat universe on the other side of the Schwarzschild 'throat'. (Consider the section $t = 0$. The two-spheres $\{r = \text{constant}\}$ behave as in Euclidean space, for large r; however for small r, they have an area which decreases to the minimum value $16\pi m^2$ and then increases again, as the two spheres expand into the other asymptotically flat three-space.) The regions I' and II are isometric with the advanced Finkelstein extension of region I', and similarly I' and II' are isometric with the retarded Finkelstein extension of I', as can be seen from figure 24. There are no timelike or null curves which go from region I to region I'. All future-directed timelike or null curves which cross the part of the surface $r = 2m$ represented here by $t' = |x'|$ approach the singularity at $t' = (2m + (x')^2)^{\frac{1}{2}}$, where $r = 0$. Similarly past-directed timelike or null curves which cross $t' = -|x'|$ approach another singularity at $t' = -(2m + (x')^2)^{\frac{1}{2}}$, where again $r = 0$.

The Kruskal extension $(\mathcal{M}^*, \mathfrak{g}^*)$ is the unique analytic and locally inextendible extension of the Schwarzschild solution. One can construct the Penrose diagram of the Kruskal extension by defining new advanced and retarded null coordinates

$$v'' = \arctan{(v'(2m)^{-\frac{1}{2}})}, \quad w'' = \arctan{(w'(2m)^{-\frac{1}{2}})}$$

for $-\pi < v'' + w'' < \pi$ and $-\tfrac{1}{2}\pi < v'' < \tfrac{1}{2}\pi, \; -\tfrac{1}{2}\pi < w'' < \tfrac{1}{2}\pi$

(see figure 24 (ii)). This may be compared with the Penrose diagram for Minkowski space (figure 15 (ii)). One now has future, past and null infinities for each of the asymptotically flat regions I and I'. Unlike Minkowski space, the conformal metric is continuous but not differentiable at the points i^0.

If we consider the future light cone of any point outside $r = 2m$, the radial outwards geodesic reaches infinity but the inwards one reaches the future singularity; if the point lies inside $r = 2m$, both these geodesics hit the singularity, and the entire future of the point is ended by the singularity. Thus the singularity may be avoided by any particle outside $r = 2m$ (so it is not 'universal' as it is in the Robertson–Walker spaces), but once a particle has fallen inside $r = 2m$ (in region II) it cannot evade the singularity. This fact will turn out to be closely related to the following property: each point inside region II represents a two-sphere that is a closed trapped surface. This means the following: consider any two-sphere p (represented by a point in figure 24) and two two-spheres q, s formed by photons emitted radially outwards, inwards at one instant from p. The area of q (which is given by $4\pi r^2$) will be greater than the area of p, but the area of s will be less than the area of p, if all three lie in a region $r > 2m$. However if they all lie in the region II where $r < 2m$, then the areas of *both* q and s will be less than the area of p (in the figure, r decreases as one moves from the bottom to the top of region II). In that case, we say that p is a closed trapped surface. Each point inside region II′ represents a time-reversed closed trapped surface (the existence of trapped surfaces is a necessary consequence of the fact that the surfaces $r = $ constant are spacelike), and correspondingly all particles in region II′ must have come from the singularity in the past. We shall see in chapter 8 that the existence of the singularities is closely related to the existence of the closed trapped surfaces.

The Reissner–Nordström solution represents the space–time outside a spherically symmetric charged body carrying an electric charge (but with no spin or magnetic dipole, so this is not a good representation of the field outside an electron). The energy–momentum tensor is therefore that of the electromagnetic field in the space–time which results from the charge on the body. It is the unique spherically symmetric asymptotically flat solution of the Einstein–Maxwell equations and is locally rather similar to the Schwarzschild solution; there exist coordinates in which the metric has the form

$$ds^2 = -\left(1 - \frac{2m}{r} + \frac{e^2}{r^2}\right) dt^2 + \left(1 - \frac{2m}{r} + \frac{e^2}{r^2}\right)^{-1} dr^2 + r^2(d\theta^2 + \sin^2\theta\, d\phi^2),$$

$$(5.26)$$

where m represents the gravitational mass and e the electric charge of the body. This asymptotically flat solution would normally be regarded as the solution outside the body only, the interior being filled in with some other suitable metric; but it is again interesting to see what happens if we regard it as a solution for all r.

If $e^2 > m^2$ the metric is non-singular everywhere except for the irremovable singularity at $r = 0$; this may be thought of as the point charge which produces the field. If $e^2 \leqslant m^2$, the metric also has singularities at r_+ and r_-, where $r_\pm = m \pm (m^2 - e^2)^{\frac{1}{2}}$; it is regular in the regions defined by $\infty > r > r_+, r_+ > r > r_-$ and $r_- > r > 0$ (if $e^2 = m^2$, only the first and third regions exist). As in the Schwarzschild case, these singularities may be removed by introducing suitable coordinates and extending the manifold to obtain a maximal analytic extension (Graves and Brill (1960), Carter (1966)). The major differences that arise are due to the existence of two zeros in the factor in front of dt^2, rather than one as in the Schwarzschild case. In particular this implies that the first and third regions are both static, whereas the second region (when it exists) is spatially homogeneous but is not static.

To obtain the maximally extended manifold, we proceed in steps analogous to those in the Schwarzschild case. Defining the coordinate r^* by

$$r^* = \int dr \bigg/ \left(1 - \frac{2m}{r} + \frac{e^2}{r^2}\right),$$

then for $r > r_+$,

$$r^* = r + \frac{r_+^2}{(r_+ - r_-)} \log (r - r_+) - \frac{r_-^2}{(r_+ - r_-)} \log (r - r_-) \qquad \text{if} \quad e^2 < m^2,$$

$$r^* = r + m \log ((r - m)^2) - \frac{2}{r - m} \qquad \text{if} \quad e^2 = m^2,$$

$$r^* = r + m \log (r^2 - 2mr + e^2) + \frac{2}{e^2 - m^2} \arctan \left(\frac{r - m}{e^2 - m^2}\right) \qquad \text{if} \quad e^2 > m^2.$$

Defining advanced and retarded coordinates v, w by

$$v = t + r^*, \quad w = t - r^*$$

the metric (5.26) takes the double null form

$$ds^2 = -\left(1 - \frac{2m}{r} + \frac{e^2}{r^2}\right) dv\, dw + r^2 (d\theta^2 + \sin^2 \theta\, d\phi^2). \qquad (5.27)$$

In the case $e^2 < m^2$, define new coordinates v'', w'' by

$$v'' = \arctan \left(\exp \left(\frac{r_+ - r_-}{4r_+^2} v\right)\right), \quad w'' = \arctan \left(-\exp \left(\frac{-r_+ + r_-}{4r_+^2} w\right)\right).$$

Then the metric (5.27) takes the form

$$ds^2 = \left(1 - \frac{2m}{r} + \frac{e^2}{r^2}\right) 64 \frac{r_+^4}{(r_+ - r_-)^2} \operatorname{cosec} 2v'' \operatorname{cosec} 2w''\, dv''\, dw''$$
$$+ r^2 (d\theta^2 + \sin^2 \theta\, d\phi^2), \qquad (5.28)$$

where r is defined implicitly by

$$\tan v'' \tan w'' = -\exp \left(\left(\frac{r_+ - r_-}{2r_+^2}\right) r\right) (r - r_+)^{\frac{1}{2}} (r - r_-)^{-\alpha/2}$$

and $\alpha = (r_+)^{-2} (r_-)^2$. The maximal extension is obtained by taking (5.28) as the metric $\mathbf{g}^*$, and $\mathcal{M}^*$ as the maximal manifold on which this metric is C^2.

The Penrose diagram of the maximal extension is shown in figure 25. There are an infinite number of asymptotically flat regions, where $r > r_+$; these are denoted by I. These are connected by intermediate regions II and III where $r_+ > r > r_-$ and $r_- > r > 0$ respectively. There is still an irremovable singularity at $r = 0$ in each region III,

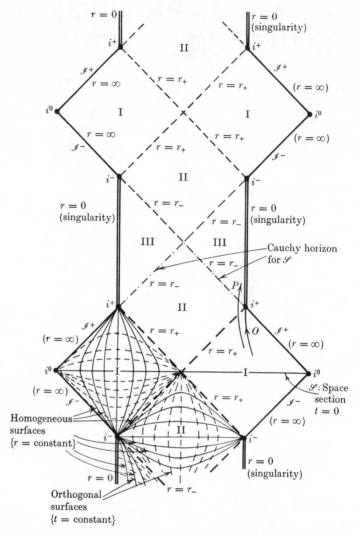

FIGURE 25. Penrose diagram for the maximally extended Reissner–Nordström solution ($e^2 < m^2$). An infinite chain of asymptotically flat regions I ($\infty > r > r_+$) are connected by regions II ($r_+ > r > r_-$) and III ($r_- > r > 0$); each region III is bounded by a timelike singularity at $r = 0$.

but unlike in the Schwarzschild solution, it is timelike and so can be avoided by a future-directed timelike curve from a region I which crosses $r = r_+$. Such a curve can pass through regions II, III and II and re-emerge into another asymptotically flat region I. This raises the intriguing possibility that one might be able to travel to other universes by passing through the 'wormholes' made by charges. Unfortunately it seems that one would not be able to get back again to our universe to report what one had seen on the other side.

The metric (5.28) is analytic everywhere except at $r = r_-$ where it is degenerate but one can define different coordinates v''' and w''' by

$$v''' = \text{arc tan}\left(\exp\left(\frac{r_+ - r_-}{2\,nr_-{}^2}v\right)\right),$$

$$w''' = \text{arc tan}\left(-\exp\left(\frac{-r_+ + r_-}{2nr_-{}^2}w\right)\right),$$

where n is an integer $\geqslant 2(r_+)^2(r_-)^{-2}$. In these coordinates, the metric is analytic everywhere except at $r = r_+$ where it is degenerate. The coordinates v''' and w''' are analytic functions of v'' and w'' for $r \neq r_+$ or r_-. Thus the manifold $\mathcal{M}^*$ can be covered by an analytic atlas, consisting of local coordinate neighbourhoods defined by coordinates v'' and w'' for $r \neq r_-$ and by local coordinate neighbourhoods defined by v''' and w''' for $r \neq r_+$. The metric is analytic in this atlas.

The case $e^2 = m^2$ can be extended similarly; the case $e^2 > m^2$ is already inextendible in the original coordinates. The Penrose diagrams of these two cases are given in figure 26.

In all these cases, the singularity is timelike. This means that, unlike in the Schwarzschild solution, timelike and null curves can always avoid hitting the singularities. In fact the singularities appear to be repulsive: no timelike geodesic hits them, though non-geodesic timelike curves and radial null geodesics can. The spaces are thus timelike (though not null) geodesically complete. The timelike character of the singularity also means that there are no Cauchy surfaces in these spaces: given any spacelike surface, one can find timelike or null curves which run into the singularity and do not cross the surface. For example in the case $e^2 < m^2$, one can find a spacelike surface $\mathcal{S}$ which crosses two asymptotically flat regions I (figure 25). This is a Cauchy surface for the two regions I and the two neighbouring regions II. However in the neighbouring regions III to the future there are past-directed inextendible timelike and null curves which approach the singularity and do not cross the surface $r = r_-$. This surface is therefore said to be the future Cauchy horizon for $\mathcal{S}$. The continuation of the solution beyond $r = r_-$ is not determined by the Cauchy data on $\mathcal{S}$. The continuation we have given is the only locally inextendible analytic one, but there will be other non-analytic C^∞ continuations which satisfy the Einstein–Maxwell equations.

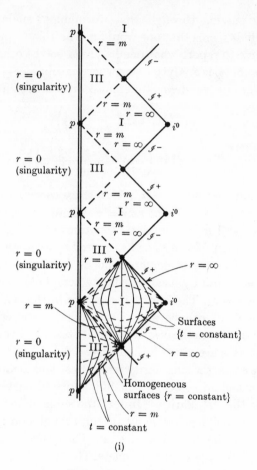

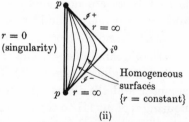

FIGURE 26. Penrose diagrams for the maximally extended Reissner–Nordström solutions:

$$\text{(i)} \ \ e^2 = m^2, \quad \text{(ii)} \ \ e^2 > m^2.$$

In the first case there is an infinite chain or regions I ($\infty > r > m$) connected by regions III ($m > r > 0$). The points p are not part of the singularity at $r = 0$, but are really exceptional points at infinity.

A particle P crossing the surface $r = r_+$ would appear to have infinite redshift to an observer O whose world-line remains outside $r = r_+$ and approaches the future infinity i^+ (figure 25). In the region II between $r = r_+$ and $r = r_-$, the surfaces of constant r are spacelike and so each point of the figure represents a two-sphere which is a closed trapped surface. An observer P crossing the surface $r = r_-$ would see the whole of the history of one of the asymptotically flat regions I in a finite time. Objects in this region would therefore appear to be infinitely blue-shifted as they approached i^+. This suggests that the surface $r = r_-$ would be unstable against small perturbations in the initial data on the spacelike surface $\mathscr{S}$, and that such perturbations would in general lead to singularities on $r = r_-$.

5.6   The Kerr solution

In general, astronomical bodies are rotating and so one would not expect the solution outside them to be exactly spherically symmetric. The Kerr solutions are the only known family of exact solutions which could represent the stationary axisymmetric asymptotically flat field outside a rotating massive object. They will be the exterior solutions only for massive rotating bodies with a particular combination of multipole moments; bodies with different combinations of moments will have other exterior solutions. The Kerr solutions do however appear to be the only possible exterior solutions for black holes (see § 9.2 and § 9.3).

The solutions can be given in Boyer and Lindquist coordinates (r, θ, ϕ, t) in which the metric takes the form

$$\mathrm{d}s^2 = \rho^2 \left(\frac{\mathrm{d}r^2}{\Delta} + \mathrm{d}\theta^2 \right) + (r^2 + a^2)\sin^2\theta\,\mathrm{d}\phi^2 - \mathrm{d}t^2 + \frac{2mr}{\rho^2}(a\sin^2\theta\,\mathrm{d}\phi - \mathrm{d}t)^2,$$

$$(5.29)$$

where $\rho^2(r, \theta) \equiv r^2 + a^2\cos^2\theta$ and $\Delta(r) \equiv r^2 - 2mr + a^2.$

m and a are constants, m representing the mass and ma the angular momentum as measured from infinity (Boyer and Price (1965)); when $a = 0$ the solution reduces to the Schwarzschild solution. This metric form is clearly invariant under simultaneous inversion of t and ϕ, i.e. under the transformation $t \to -t$, $\phi \to -\phi$, although it is not invariant under inversion of t alone (except when $a = 0$). This is what one would expect, since time inversion of a rotating object produces an object rotating in the opposite direction.

When $a^2 > m^2$, $\Delta > 0$ and the above metric is singular only when $r = 0$. The singularity at $r = 0$ is not in fact a point but a ring, as can be seen by transforming to Kerr–Schild coordinates $(x, y, z, \bar{t})$, where

$$x + \mathrm{i}y = (r + \mathrm{i}a)\sin\theta \exp \mathrm{i}\int (\mathrm{d}\phi + a\Delta^{-1}\,\mathrm{d}r),$$

$$z = r\cos\theta, \quad \bar{t} = \int (\mathrm{d}t + (r^2 + a^2)\,\Delta^{-1}\,\mathrm{d}r) - r.$$

In these coordinates, the metric takes the form

$$\mathrm{d}s^2 = \mathrm{d}x^2 + \mathrm{d}y^2 + \mathrm{d}z^2 - \mathrm{d}\bar{t}^2$$

$$+ \frac{2mr^3}{r^4 + a^2 z^2}\left(\frac{r(x\,\mathrm{d}x + y\,\mathrm{d}y) - a(x\,\mathrm{d}y - y\,\mathrm{d}x)}{r^2 + a^2} + \frac{z\,\mathrm{d}z}{r} + \mathrm{d}\bar{t}\right)^2, \quad (5.30)$$

where r is determined implicitly, up to a sign, in terms of x, y, z by

$$r^4 - (x^2 + y^2 + z^2 - a^2)r^2 - a^2 z^2 = 0.$$

For $r \neq 0$, the surfaces $\{r = \text{constant}\}$ are confocal ellipsoids in the (x, y, z) plane, which degenerate for $r = 0$ to the disc $z^2 + y^2 \leqslant a^2$, $z = 0$. The ring $x^2 + y^2 = a^2$, $z = 0$ which is the boundary of this disc, is a real curvature singularity as the scalar polynomial $R_{abcd}R^{abcd}$ diverges there. However no scalar polynomial diverges on the disc except at the boundary ring. The function r can in fact be analytically continued from positive to negative values through the interior of the disc $x^2 + y^2 < a^2$, $z = 0$, to obtain a maximal analytic extension of the solution.

To do this, one attaches another plane defined by coordinates (x', y', z') where a point on the top side of the disc $x^2 + y^2 < a^2$, $z = 0$ in the (x, y, z) plane is identified with a point with the same x and y coordinates on the bottom side of the corresponding disc in the (x', y', z') plane. Similarly a point on the bottom side of the disc in the (x, y, z) plane is identified with a point on the top side of the disc in the (x', y', z') plane (see figure 27). The metric (5.30) extends in the obvious way to this larger manifold. The metric on the (x', y', z') region is again of the form (5.29), but with negative rather than positive values of r. At large negative values of r, the space is again asymptotically flat but this time with negative mass. For small negative values of r near the ring singularity, the vector $\partial/\partial\phi$ is timelike, so the circles $(t = \text{constant}, r = \text{constant}, \theta = \text{constant})$ are closed timelike curves. These closed timelike curves can be deformed to pass through any point of the extended space (Carter (1968a)). This solution is geodesically incomplete at the ring singularity. However the only timelike and null geodesics which reach this singularity are those in the equatorial plane on the positive r side (Carter (1968a)).

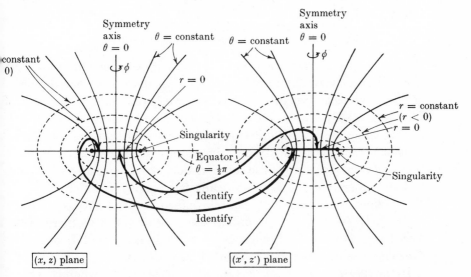

FIGURE 27. The maximal extension of the Kerr solution for $a^2 > m^2$ is obtained by identifying the top of the disc $x^2 + y^2 < a^2$, $z = 0$ in the (x, y, z) plane with the bottom of the corresponding disc in the (x', y', z') plane, and vice versa. The figure shows the sections $y = 0$, $y' = 0$ of these planes. On circling twice round the singularity at $x^2 + y^2 = a^2$, $z = 0$ one passes from the (x, y, z) plane to the (x', y', z') plane (where r is negative) and back to the (x, y, z) plane (where r is positive).

The extension in the case $a^2 < m^2$ is rather more complicated, because of the existence of the two values $r_+ = m + (m^2 - a^2)^{\frac{1}{2}}$ and $r_- = m - (m^2 - a^2)^{\frac{1}{2}}$ of r at which $\Delta(r)$ vanishes. These surfaces are similar to the surfaces $r = r_+$, $r = r_-$ in the Reissner–Nordström solution. To extend the metric across these surfaces, one transforms to the Kerr coordinates (r, θ, ϕ_+, u_+), where

$$\mathrm{d}u_+ = \mathrm{d}t + (r^2 + a^2)\,\Delta^{-1}\,\mathrm{d}r, \quad \mathrm{d}\phi_+ = \mathrm{d}\phi + a\Delta^{-1}\,\mathrm{d}r.$$

The metric then takes the form

$$\mathrm{d}s^2 = \rho^2\,\mathrm{d}\theta^2 - 2a\sin^2\theta\,\mathrm{d}r\,\mathrm{d}\phi_+ + 2\,\mathrm{d}r\,\mathrm{d}u_+$$
$$+ \rho^{-2}[(r^2 + a^2)^2 - \Delta a^2 \sin^2\theta]\sin^2\theta\,\mathrm{d}\phi_+{}^2$$
$$- 4a\rho^{-2}mr\sin^2\theta\,\mathrm{d}\phi_+\,\mathrm{d}u_+ - (1 - 2mr\rho^{-2})\,\mathrm{d}u_+{}^2 \quad (5.31)$$

on the manifold defined by these coordinates, and is analytic at
$r = r_+$ and $r = r_-$. One again has a singularity at $r = 0$, which has the
same ring form and geodesic structure as that described above. The
metric can also be extended on the manifold defined by the coordinates
(r, θ, ϕ_-, u_-) where

$$\mathrm{d}u_- = \mathrm{d}t - (r^2 + a^2)\,\Delta^{-1}\mathrm{d}r, \quad \mathrm{d}\phi_- = \mathrm{d}\phi - a\Delta^{-1}\mathrm{d}r;$$

the metric again takes the form (5.31), with ϕ_+, u_+ replaced by $-\phi_-$,
$-u_-$. The maximal analytic extension can be built up by a combination
of these extensions, as in the Reissner–Nordström case (Boyer and
Lindquist (1967), Carter (1968a)). The global structure is very similar
to that of the Reissner–Nordström solution except that one can now
continue through the ring to negative values of r. Figure 28 (i) shows
the conformal structure of the solution along the symmetry axis. The
regions I represent the asymptotically flat regions in which $r > r_+$.
The regions II $(r_- < r < r_+)$ contain closed trapped surfaces. The
regions III $(-\infty < r < r_-)$ contain the ring singularity; there are
closed timelike curves through every point in a region III, but no
causality violation occurs in the other two regions.

   In the case $a^2 = m^2$, r_+ and r_- coincide and there is no region II. The
maximal extension is similar to that of the Reissner–Nordström solution when $e^2 = m^2$. The conformal structure along the symmetry axis
in this case is shown in figure 28 (ii).

   The Kerr solutions, being stationary and axisymmetric, have a
two-parameter group of isometries. This group is necessarily Abelian
(Carter (1970)). There are thus two independent Killing vector fields
which commute. There is a unique linear combination K^a of these
Killing vector fields which is timelike at arbitrarily large positive and
negative values of r. There is another unique linear combination $\tilde{K}^a$
of the Killing vector fields which is zero on the axis of symmetry. The
orbits of the Killing vector K^a define the stationary frame, that is, an
object moving along one of these orbits appears to be stationary with
respect to infinity. The orbits of the Killing vector $\tilde{K}^a$ are closed curves,
and correspond to the rotational symmetry of the solution.

   In the Schwarzschild and Reissner–Nordström solutions, the
Killing vector K^a which is timelike at large values of r is timelike
everywhere in the region I, becoming null on the surfaces $r = 2m$ and
$r = r_+$ respectively. These surfaces are null. This means that a particle
which crosses one of these surfaces in the future direction cannot
return again to the same region. They are the boundary of the region

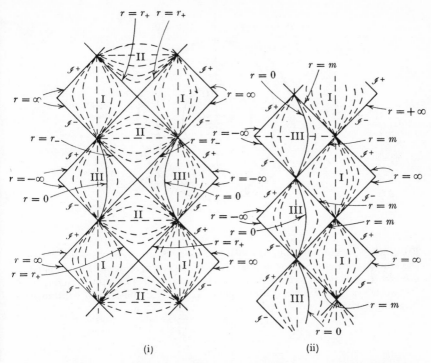

(i)   (ii)

FIGURE 28. The conformal structure of the Kerr solutions along the axis of symmetry, (i) in the case $0 < a^2 < m^2$, (ii) in the case $a^2 = m^2$. The dotted lines are lines of constant r; the regions I, II and III in case (i) are divided by $r = r_+$ and $r = r_-$, and the regions I and III in case (ii) by $r = m$. In both cases, the structure of the space near the ring singularity is as in figure 27.

of the solution from which particles can escape to the infinity $\mathscr{I}^+$ of a particular region I, and are called the *event horizons* of that $\mathscr{I}^+$. (They are in fact the event horizon in the sense of § 5.2 for an observer moving on any of the orbits of the Killing vector K^a in the region I.)

In the Kerr solution on the other hand, the Killing vector K^a is spacelike in a region outside $r = r_+$, called the *ergosphere* (figure 29). The outer boundary of this region is the surface $r = m + (m^2 - a^2 \cos^2 \theta)^{\frac{1}{2}}$ on which K^a is null. This is called the *stationary limit surface* since it is the boundary of the region in which particles travelling on a timelike curve can travel on an orbit of the Killing vector K^a, and so remain at rest with respect to infinity. The stationary limit surface is a timelike surface except at the two points on the axis, where it is null (at these points it coincides with the surface $r = r_+$). Where it is timelike it can be crossed by particles in either the ingoing or the outgoing direction.

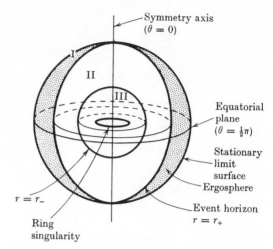

FIGURE 29. In the Kerr solution with $0 < a^2 < m^2$, the ergosphere lies between the stationary limit surface and the horizon at $r = r_+$. Particles can escape to infinity from region I (outside the event horizon $r = r_+$) but not from region II (between $r = r_+$ and $r = r_-$) and region III ($r < r_-$; this region contains the ring singularity).

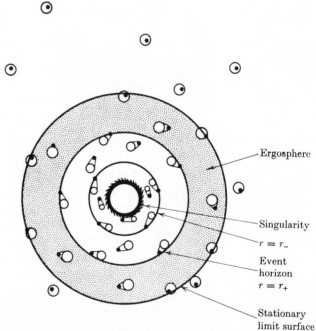

FIGURE 30. The equatorial plane of a Kerr solution with $m^2 > a^2$. The circles represent the position a short time later of flashes of light emitted by the points represented by heavy dots.

It is therefore not the event horizon for $\mathscr{I}^+$. In fact the event horizon is the surface $r = r_+ = m + (m^2 - a^2)^{\frac{1}{2}}$. Figure 30 shows why this is. It shows the equatorial plane $\theta = \frac{1}{2}\pi$; each point in this figure represents an orbit of the Killing vector K^a, i.e. it is stationary with respect to $\mathscr{I}^+$. The small circles represent the position a short time later of flashes of light emitted from the points represented by the heavy black dots. Outside the stationary limit the Killing vector K^a is time-like and so lies within the light cone. This means that the point in figure 30 representing the orbit of emission lies within the wavefront of the light.

On the stationary limit surface, K^a is null and so the point representing the orbit of emission lies on the wavefront. However the wavefront lies partly within and partly outside the stationary limit surface; it is therefore possible for a particle travelling along a timelike curve to escape to infinity from this surface. In the ergosphere between the stationary limit surface and $r = r_+$, the Killing vector K^a is spacelike and so the point representing the orbit of emission lies outside the wavefront. In this region it is impossible for a particle moving on a timelike or null curve to travel along an orbit of the Killing vector and so to remain at rest with respect to infinity. However the positions of the wavefronts are such that the particles can still escape across the stationary limit surface and so out to infinity. On the surface $r = r_+$, the Killing vector K^a is still spacelike. However the wavefront corresponding to a point on this surface lies entirely within the surface. This means that a particle travelling on a timelike curve from a point on or inside the surface cannot get outside the surface and so cannot get out to infinity. The surface $r = r_+$ is therefore the event horizon for $\mathscr{I}^+$ and is a null surface.

Although the Killing vector K^a is spacelike in the ergosphere, the magnitude $K^a \tilde{K}^b K_{[a} \tilde{K}_{b]}$ of the *Killing bivector* $K_{[a} \tilde{K}_{b]}$ is negative everywhere outside $r = r_+$, except on the axis $\tilde{K}^a = 0$ where it vanishes. Therefore K^a and $\tilde{K}^a$ span a timelike two-surface and so at each point outside $r = r_+$ off the axis there is a linear combination of K^a and $\tilde{K}^a$ which is timelike. In a sense, therefore, the solution in the ergosphere is locally stationary, although it is not stationary with respect to infinity. In fact there is no one linear combination of K^a and $\tilde{K}^a$ which is timelike everywhere outside $r = r_+$. The magnitude of the Killing bivector vanishes on $r = r_+$, and is positive just inside this surface. On $r = r_+$, both K^a and $\tilde{K}^a$ are spacelike but there is a linear combination which is null everywhere on $r = r_+$ (Carter (1969)).

The behaviour of the ergosphere and the horizon we have discussed will play an important part in our discussion of black holes in § 9.2 and § 9.3.

Just as the Reissner–Nordström solution can be thought of as a charged version of the Schwarzschild solution, so there is a family of charged Kerr solutions (Carter (1968a)). Their global properties are very similar to those of the uncharged Kerr solutions.

5.7 Gödel's universe

In 1949, Kurt Gödel published a paper (Gödel (1949)) which provided a considerable stimulus to investigation of exact solutions more complex than those examined so far. He gave an exact solution of Einstein's field equations in which the matter takes the form of a pressure-free perfect fluid ($T_{ab} = \rho u_a u_b$ where ρ is the matter density and u_a the normalized four-velocity vector). The manifold is R^4 and the metric can be given in the form

$$ds^2 = -dt^2 + dx^2 - \tfrac{1}{2}\exp\left(2(\sqrt{2})\,\omega x\right)dy^2 + dz^2 - 2\exp\left((\sqrt{2})\,\omega x\right)dt\,dy,$$

where $\omega > 0$ is a constant; the field equations are satisfied if $\mathbf{u} = \partial/\partial x^0$ (i.e. $u^a = \delta^a{}_0$) and

$$4\pi\rho = \omega^2 = -\Lambda.$$

The constant ω is in fact the magnitude of the vorticity of the flow vector u^a.

This space–time has a five-dimensional group of isometries which is transitive, i.e. it is a completely homogeneous space–time. (An action of a group is transitive on $\mathscr{M}$ if it can map any point of $\mathscr{M}$ into any other point of $\mathscr{M}$.) The metric is the direct sum of the metric $\mathbf{g}_1$ given by

$$ds_1{}^2 = -dt^2 + dx^2 - \tfrac{1}{2}\exp\left(2(\sqrt{2})\,\omega x\right)dy^2 - 2\exp\left((\sqrt{2})\,\omega x\right)dt\,dy$$

on the manifold $\mathscr{M}_1 = R^3$ defined by the coordinates (t, x, y), and the metric $\mathbf{g}_2$ given by

$$ds_2{}^2 = dz^2$$

on the manifold $\mathscr{M}_2 = R^1$ defined by the coordinate z. In order to describe the properties of the solution it is sufficient to consider only $(\mathscr{M}_1, \mathbf{g}_1)$.

Defining new coordinates (t', r, ϕ) on $\mathscr{M}_1$ by

$$\exp\left((\sqrt{2})\,\omega x\right) = \cosh 2r + \cos\phi \sinh 2r,$$

$$\omega y \exp\left((\sqrt{2})\,\omega x\right) = \sin\phi \sinh 2r,$$

$$\tan\tfrac{1}{2}(\phi + \omega t - (\sqrt{2})\,t') = \exp\left(-2r\right)\tan\tfrac{1}{2}\phi,$$

the metric $\mathbf{g}_1$ takes the form

$$ds_1{}^2 = 2\omega^{-2}(-dt'^2 + dr^2 - (\sinh^4 r - \sinh^2 r)\,d\phi^2 + 2(\sqrt{2})\sinh^2 r\,d\phi\,dt),$$

where $-\infty < t < \infty$, $0 \leqslant r < \infty$, and $0 \leqslant \phi \leqslant 2\pi$, $\phi = 0$ being identified with $\phi = 2\pi$; the flow vector in these coordinates is $\mathbf{u} = (\omega/(\sqrt{2}))\,\partial/\partial t'$. This form exhibits the rotational symmetry of the solution about the axis $r = 0$. By a different choice of coordinates the axis could be chosen to lie on any flow line of the matter.

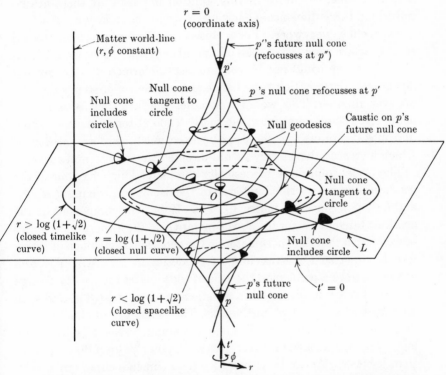

FIGURE 31. Gödel's universe with the irrelevant coordinate z suppressed. The space is rotationally symmetric about any point; the diagram represents correctly the rotational symmetry about the axis $r = 0$, and the time invariance. The light cone opens out and tips over as r increases (see line L) resulting in closed timelike curves. The diagram does not correctly represent the fact that *all* points are in fact equivalent.

The behaviour of $(\mathscr{M}_1, \mathbf{g}_1)$ is illustrated in figure 31. The light cones on the axis $r = 0$ contain the direction $\partial/\partial t'$ (the vertical direction on the diagram) but not the horizontal directions $\partial/\partial r$ and $\partial/\partial\phi$. As one moves away from the axis, the light cones open out and tilt in the

ϕ-direction so that at a radius $r = \log(1 + \sqrt{2})$, $\partial/\partial\phi$ is a null vector and the circle of this radius about the origin is a closed null curve. At greater values of r, $\partial/\partial\phi$ is a timelike vector and circles of constant r, t' are closed timelike curves. As $(\mathcal{M}_1, \mathbf{g}_1)$ has a four-dimensional group of isometries which is transitive, there are closed timelike curves through every point of $(\mathcal{M}_1, \mathbf{g}_1)$, and hence through every point of the Gödel solution $(\mathcal{M}, \mathbf{g})$.

This suggests that the solution is not very physical. The existence of closed timelike curves in this solution implies that there are no imbedded three-dimensional surfaces without boundary in $\mathcal{M}$ which are spacelike everywhere. For a closed timelike curve which crossed such a surface would cross it an odd number of times. This would mean that the curve could not be continuously deformed to zero, since a continuous deformation can change the number of crossings only by an even number. This would contradict the fact that $\mathcal{M}$ is simply connected, being homeomorphic to R^4. The existence of closed time-like lines also shows that there can be no cosmic time coordinate t in $\mathcal{M}$ which increases along every future-directed timelike or null curve.

The Gödel solution is geodesically complete. The behaviour of the geodesics can be described in terms of the decomposition into $(\mathcal{M}_1, \mathbf{g}_1)$ and $(\mathcal{M}_2, \mathbf{g}_2)$. Since the metric $\mathbf{g}_2$ of $\mathcal{M}_2$ is flat, the component of the geodesic tangent vector in $\mathcal{M}_2$ is constant, i.e. the z-coordinate varies linearly with the affine parameter on the geodesic. It is sufficient there-fore to describe the behaviour of geodesics in $(\mathcal{M}_1, \mathbf{g}_1)$. The null geodesics from a point p on the axis of coordinates (figure 31) diverge from the axis initially, reach a caustic at $r = \log(1 + (\sqrt{2}))$, and then reconverge to a point p' on the axis. The behaviour of timelike geo-desics is similar: they reach some maximum value of r less than $\log(1 + (\sqrt{2}))$ and then reconverge to p'. A point q at a radius r greater than $\log(1 + (\sqrt{2}))$ can be joined to p by a timelike curve but not by a timelike or null geodesic.

Further details of Gödel's solution can be found in Gödel (1949), Kundt (1956).

5·8   Taub–NUT space

In 1951, Taub discovered a spatially homogeneous empty space solu-tion of Einstein's equations with topology $R \times S^3$ and metric given by

$$\mathrm{d}s^2 = -U^{-1}\mathrm{d}t^2 + (2l)^2 U(\mathrm{d}\psi + \cos\theta\,\mathrm{d}\phi)^2$$
$$+ (t^2 + l^2)(\mathrm{d}\theta^2 + \sin^2\theta\,\mathrm{d}\phi^2), \quad (5.32)$$

where

$$U(t) \equiv -1 + \frac{2(mt+l^2)}{t^2+l^2}, \quad m \text{ and } l \text{ are positive constants.}$$

Here θ, ϕ, ψ are Euler coordinates on S^3, so $0 \leqslant \psi \leqslant 4\pi$, $0 \leqslant \theta \leqslant \pi$, $0 \leqslant \phi \leqslant 2\pi$. This metric is singular at $t = t_\pm = m \pm (m^2+l^2)^{\frac{1}{2}}$, where $U = 0$. It can in fact be extended across these surfaces to give a space found by Newman, Tamburino and Unti (1963), but before discussing the extension we shall consider a simple two-dimensional example given by Misner (1967) which has many similar properties.

This space has the topology $S^1 \times R^1$ and the metric $\mathbf{g}$ given by

$$ds^2 = -t^{-1}\,dt^2 + t\,d\psi^2$$

where $0 \leqslant \psi \leqslant 2\pi$. This metric is singular when $t = 0$. However if one takes the manifold $\mathcal{M}$ defined by ψ and by $0 < t < \infty$, $(\mathcal{M}, \mathbf{g})$ can be extended by defining $\psi' = \psi - \log t$. The metric then takes the form $\mathbf{g}'$ given by

$$ds^2 = +2\,d\psi'\,dt + t(d\psi')^2.$$

This is analytic on the manifold $\mathcal{M}'$ with topology $S^1 \times R^1$ defined by ψ' and by $-\infty < t < \infty$. The region $t > 0$ of $(\mathcal{M}', \mathbf{g}')$ is isometric with $(\mathcal{M}, \mathbf{g})$. The behaviour of $(\mathcal{M}', \mathbf{g}')$ is shown in figure 32. There are closed timelike lines in the region $t < 0$, but there are none when $t > 0$. One family of null geodesics is represented by the vertical lines in figure 32; these cross the surface $t = 0$. The other family spiral round and round as they approach $t = 0$, but never actually cross this surface, and these geodesics have only finite affine length. Thus the extension $(\mathcal{M}', \mathbf{g}')$ is not symmetric between the two families of null geodesics, although the original space $(\mathcal{M}, \mathbf{g})$ was. However one can define another extension $(\mathcal{M}'', \mathbf{g}'')$ in which the behaviour of the two families of null geodesics is interchanged. To do so define ψ'' by $\psi'' = \psi + \log t$. The metric takes the form $\mathbf{g}''$ given by

$$ds^2 = -2\,d\psi''\,dt + t(d\psi'')^2.$$

This is analytic on the manifold $\mathcal{M}''$ with topology $S^1 \times R^1$ defined by ψ'' and $-\infty < t < \infty$. The region $t > 0$ of $(\mathcal{M}'', \mathbf{g}'')$ is isometric with $(\mathcal{M}, \mathbf{g})$. In a sense, what we have done by defining ψ'' is to untwist the second family of null geodesics so that they become vertical lines, and can be continued beyond $t = 0$. However this twisting winds up the first family of null geodesics so that they spiral around and cannot be continued beyond $t = 0$. One has therefore two inequivalent locally inextendible analytic extensions of $(\mathcal{M}, \mathbf{g})$, both of which are geodesic-

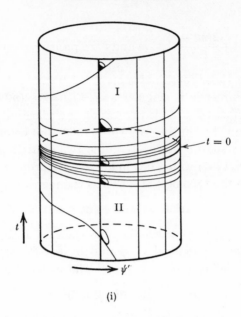

(i)

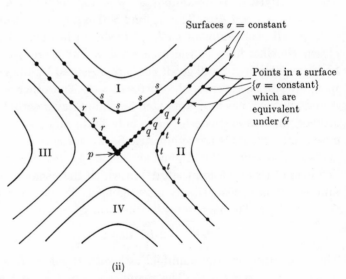

(ii)

FIGURE 32. Misner's two-dimensional example.

(i)   Extension of region I across the boundary $t = 0$ into II. The vertical null geodesics are complete, but the twisted null geodesics are incomplete.

(ii) The universal covering space is two-dimensional Minkowski space. Under the discrete subgroup G of the Lorentz group, points s are equivalent; similarly points r, q and t are equivalent. (i) is obtained by identifying equivalent points in regions I and II.

ally incomplete. The relation between these two extensions can be seen clearly by going to the covering space of $(\mathcal{M}, \tilde{\mathsf{g}})$.

This is in fact the region I of two-dimensional Minkowski space $(\tilde{\mathcal{M}}, \tilde{\eta})$ contained within the future null cone of a point p (figure 32 (ii)). The isometries of $(\tilde{\mathcal{M}}, \tilde{\eta})$ which leave p fixed form a one-dimensional group (the Lorentz group of $\tilde{\eta}$) whose orbits are the hyperbolae $\{\sigma = \text{constant}\}$ where $\sigma \equiv \tilde{t}^2 - \tilde{x}^2$ and $\tilde{t}$, $\tilde{x}$ are the usual Minkowski coordinates. The space $(\mathcal{M}, \tilde{\mathsf{g}})$ is the quotient of $(I, \tilde{\eta})$ by the discrete subgroup G of the Lorentz group consisting of A^n (n integer) where A maps $(\tilde{t}, \tilde{x})$ to

$$(\tilde{t} \cosh \pi + \tilde{x} \sinh \pi, \tilde{x} \cosh \pi + \tilde{t} \sinh \pi),$$

i.e. one identifies the points

$$(\tilde{t} \cosh n\pi + \tilde{x} \sinh n\pi, \tilde{x} \cosh n\pi + \tilde{t} \sinh n\pi)$$

for all integer values of n, and these correspond to the point

$$t = \tfrac{1}{4}(\tilde{t}^2 - \tilde{x}^2), \quad \psi = 2 \operatorname{arc} \tanh (\tilde{x}/\tilde{t}) \quad \text{in } \mathcal{M}.$$

The action of the isometry group G in the region I is properly discontinuous. The action of a group H on a manifold $\mathcal{N}$ is said to be properly discontinuous if:

(1) each point $q \in \mathcal{N}$ has a neighbourhood $\mathcal{U}$ such that $A(\mathcal{U}) \cap \mathcal{U} = \varnothing$ for each $A \in H$ which is not the identity element, and

(2) if $q, r \in \mathcal{N}$ are such that there is no $A \in H$ with $Aq = r$, then there are neighbourhoods $\mathcal{U}$ and $\mathcal{U}'$ of q and r respectively such that there is no $B \in H$ with $B(\mathcal{U}) \cap \mathcal{U}' \neq \varnothing$.

Condition (1) implies that the quotient $\mathcal{N}/H$ is a manifold, and condition (2) implies that it is Hausdorff. Thus the quotient $(I, \tilde{\eta})/G$ is the Hausdorff space $(\mathcal{M}, \tilde{\mathsf{g}})$. The action of G is also properly discontinuous in the regions $I + II$ ($\tilde{t} > -\tilde{x}$). Thus $(I + II, \tilde{\eta})/G$ is also a Hausdorff space; in fact it is $(\mathcal{M}', \tilde{\mathsf{g}}')$. Similarly $(I + III, \tilde{\eta})/G$ is the Hausdorff space $(\mathcal{M}'', \tilde{\mathsf{g}}'')$ where $I + III$ is the region $\tilde{t} > \tilde{x}$. From this it can be seen how it is that one family of null geodesics can be completed in the extension $(\mathcal{M}', \tilde{\mathsf{g}}')$ while the other family can be completed in the extension $(\mathcal{M}'', \tilde{\mathsf{g}}'')$. This suggests that one might perform both extensions at the same time. However the action of the group on the region $(I + II + III)$ (i.e. $\tilde{t} > -|\tilde{x}|$) satisfies condition (1) but condition (2) is not satisfied for points q on the boundary between I and II and points r on the boundary between I and III. Therefore the quotient $(I + II + III, \tilde{\eta})/G$ is not Hausdorff although it is still a manifold.

This kind of non-Hausdorff behaviour is different from that in the example given in § 2.1. In that example, one could have continuous curves which bifurcate, one branch going into one region and another branch going into another region. Such behaviour of an observer's world-line would be very uncomfortable. However the manifold $(I + II + III)/G$ does not have any such bifurcating curves; curves in I can be extended into II or III but not into both simultaneously. Thus one might be prepared to relax the Hausdorff requirement on a space–time model to allow this sort of situation but not the sort in which one gets bifurcating curves. Further work on non-Hausdorff space–times can be found in the papers of Hajicek (1971).

Condition (1) is in fact satisfied by the action of G on $\tilde{\mathcal{M}} - \{p\}$. Thus the space $(\tilde{\mathcal{M}} - \{p\}, \tilde{\eta})/G$ is in some sense the maximal non-Hausdorff extension of $(\mathcal{M}, \mathbf{g})$. However it is still not geodesically complete because there are geodesics which pass through the point p which has been left out. If p is included the action of the group does not satisfy condition (1), and so the quotient $\tilde{\mathcal{M}}/G$ is not even a non-Hausdorff manifold. However consider the bundle of linear frames $L(\tilde{\mathcal{M}})$, i.e. the collection of all pairs $(\mathbf{X}, \mathbf{Y})$, $\mathbf{X}, \mathbf{Y} \in T_q$, of linearly independent vectors at all points $q \in \tilde{\mathcal{M}}$. The action of an element A of the isometry group G on $\tilde{\mathcal{M}}$ induces an action A_* on $L(\tilde{\mathcal{M}})$ which takes the frame $(\mathbf{X}, \mathbf{Y})$ at q to the frame $(A_* \mathbf{X}, A_* \mathbf{Y})$ at $A(q)$. This action satisfies condition (1) because even for $(\mathbf{X}, \mathbf{Y}) \in T_p$, $A_* \mathbf{X} \neq \mathbf{X}$ and $A_* \mathbf{Y} \neq \mathbf{Y}$ unless $A =$ identity, and satisfies condition (2) even if $\mathbf{X}$ and $\mathbf{Y}$ lie on the null cone of p. Thus the quotient $L(\tilde{\mathcal{M}})/G$ is a Hausdorff manifold. It is a fibre bundle over the non-Hausdorff non-manifold $\tilde{\mathcal{M}}/G$. One could in a sense regard it as the bundle of linear frames for this space. The fact that the bundle of frames can be well behaved even though the space is not, suggests that it is useful to look at singularities by using the bundle of linear frames. A general procedure for doing this will be given in § 8.3.

We shall now return to the four-dimensional Taub space $(\mathcal{M}, \mathbf{g})$ where $\mathcal{M}$ is $R^1 \times S^3$ and $\mathbf{g}$ is given by (5.32). As $\mathcal{M}$ is simply connected, one cannot take a covering space as we did in the two-dimensional example. However one can achieve a similar result by considering $\mathcal{M}$ as a fibre bundle over S^2 with fibre $R^1 \times S^1$; the bundle projection $\pi\colon \mathcal{M} \to S^2$ is defined by $(t, \psi, \theta, \phi) \to (\theta, \phi)$. This is in fact the product with the t-axis of the Hopf fibering $S^3 \to S^2$ (Steenrod (1951)) which has fibre S^1. The space $(\mathcal{M}, \mathbf{g})$ admits a four-dimensional group of isometries whose surfaces of transitivity are the three-spheres

{t = constant}. This group of isometries maps fibres of the bundle $\pi \colon \mathcal{M} \to S^2$ into fibres, and so the pairs $(\mathcal{F}, \tilde{\mathbf{g}})$ are all isometric, where $\mathcal{F}$ is a fibre ($\mathcal{F} \approx R^1 \times S^1$) and $\tilde{\mathbf{g}}$ is the metric induced on the fibre by the four-dimensional metric $\mathbf{g}$ on $\mathcal{M}$. The fibre $\mathcal{F}$ can be regarded as the (t, ψ) plane, and the metric $\tilde{\mathbf{g}}$ on $\mathcal{F}$ is obtained from (5.32) by dropping the terms in $\mathrm{d}\theta$ and $\mathrm{d}\phi$; thus $\tilde{\mathbf{g}}$ is given by

$$\mathrm{d}s^2 = -U^{-1}\mathrm{d}t^2 + 4l^2U(\mathrm{d}\psi)^2. \qquad (5.33)$$

The tangent space T_q at the point $q \in \mathcal{M}$ can be decomposed into a vertical subspace V_q which is tangent to the fibre and is spanned by the vectors $\partial/\partial t$ and $\partial/\partial\psi$, and a horizontal subspace H_q which is spanned by the vectors $\partial/\partial\theta$ and $\partial/\partial\phi - \cos\theta\, \partial/\partial\psi$. Any vector $\mathbf{X} \in T_q$ can be split into a part $\mathbf{X}_V$ lying in V_q and a part $\mathbf{X}_H$ lying in H_q. The metric $\mathbf{g}$ on T_q can then be expressed as

$$g(\mathbf{X}, \mathbf{Y}) = g_V(\mathbf{X}_V, \mathbf{Y}_V) + (t^2 + l^2)\,g_H(\pi_*\mathbf{X}_H, \pi_*\mathbf{Y}_H), \qquad (5.34)$$

where $g_V \equiv \tilde{\mathbf{g}}$ and $\mathbf{g}_H$ is the standard metric on the two-sphere given by $\mathrm{d}s^2 = \mathrm{d}\theta^2 + \sin^2\theta\,\mathrm{d}\phi^2$. Thus although the metric $\mathbf{g}$ is not the direct sum of $\mathbf{g}_V$ and $(t^2 + l^2)\mathbf{g}_H$ (because $R^1 \times S^3$ is not the direct product of $R^1 \times S^1$ with S^2) it can nevertheless be regarded as such a sum locally.

The interesting part of the metric $\mathbf{g}$ is contained in $\mathbf{g}_V$ and we shall therefore consider analytic extensions of the pair $(\mathcal{F}, \mathbf{g}_V)$. When combined with the metric $\mathbf{g}_H$ of the two-sphere as in (5.34), these give analytic extensions of $(\mathcal{M}, \mathbf{g})$.

The metric $\mathbf{g}_V$, given by (5.33), has singularities at $t = t_\pm$ where $U = 0$. However if one takes the manifold $\mathcal{F}_0$ defined by ψ and by $t_- < t < t_+$, $(\mathcal{F}_0, \mathbf{g}_V)$ can be extended by defining

$$\psi' = \psi + \frac{1}{2l}\int \frac{\mathrm{d}t}{U(t)}.$$

The metric then takes the form $\mathbf{g}_V{}'$ given by

$$\mathrm{d}s^2 = 4l\,\mathrm{d}\psi'(lU(t)\,\mathrm{d}\psi' - \mathrm{d}t).$$

This is analytic on the manifold $\mathcal{F}'$ with topology $S^1 \times R$ defined by ψ' and by $-\infty < t < \infty$. The region $t_- < t < t_+$ of $(\mathcal{F}', \mathbf{g}_V{}')$ is isometric with $(\mathcal{F}_0, \mathbf{g}_V)$. There are no closed timelike curves in the region $t_- < t < t_+$ but there are for $t < t_-$ and for $t > t_+$. The behaviour is very much as for the space $(\mathcal{M}', \mathbf{g}')$ we considered before, except that there are now two horizons (at $t = t_-$ and $t = t_+$) instead of the one horizon (at $t = 0$). One family of null geodesics crosses both horizons $t = t_-$ and

$t = t_+$ but the other family spirals round near these surfaces and is incomplete.

As before, one can make another extension by defining the coordinate

$$\psi'' = \psi - \frac{1}{2l} \int \frac{\mathrm{d}t}{U(t)}.$$

The metric then takes the form $\mathbf{g}_V''$ given by

$$\mathrm{d}s^2 = 4l\,\mathrm{d}\psi''(lU(t)\,\mathrm{d}\psi'' + \mathrm{d}t)$$

which is analytic on the manifold $\mathscr{F}''$ defined by ψ'' and by $-\infty < t < \infty$, and is again isometric to $(\mathscr{F}_0, \mathbf{g}_V)$ on $t_- < t < t_+$.

Once again one can show the relation between the different extensions by going to the covering space. The covering space of $\mathscr{F}_0$ is the manifold $\tilde{\mathscr{F}}_0$ defined by the coordinates $-\infty < \psi < \infty$ and by $t_- < t < t_+$. On $\tilde{\mathscr{F}}_0$ the metric $\mathbf{g}_V$ can be written in the double null form

$$\mathrm{d}s^2 = 4l^2 U(t)\,\mathrm{d}\psi'\,\mathrm{d}\psi'', \tag{5.35}$$

where $-\infty < \psi' < \infty$, $-\infty < \psi'' < \infty$. One can extend this in a manner similar to that used in the Reissner–Nordström solution. Define new coordinates (u_+, v_+) and (u_-, v_-) on $\tilde{\mathscr{F}}_0$ by

$$u_\pm = \arctan(\exp \psi'/\alpha_\pm), \quad v_\pm = \arctan(-(\exp -\psi''/\alpha_\pm)),$$

where $\qquad \alpha_+ = \dfrac{t_+ - t_-}{4l(mt + l^2)} \quad$ and $\quad \alpha_- = \dfrac{t_+ - t_-}{4nl(mt + l^2)};$

n is some integer greater than $(mt_+ + l^2)/(mt_- + l^2)$. Then the metric $\tilde{\mathbf{g}}_V$ obtained by applying this transformation to (5.35) is analytic on the manifold $\tilde{\mathscr{F}}$ shown in figure 33, where the coordinates (u_+, v_+) are analytic coordinates except at $t = t_-$ where they are at least C^3, and the coordinates (u_-, v_-) are analytic coordinates except at $t = t_+$ where they are at least C^3. This is rather similar to the extension of the (t, r) plane of the Reissner–Nordström solution.

The space $(\tilde{\mathscr{F}}, \tilde{\mathbf{g}}_V)$ has a one-dimensional group of isometries, the orbits of which are shown in figure 33. Near the points p_+, p_- the action of this group is similar to that of the Lorentz group in two-dimensional Minkowski space (figure 32 (ii)). Let G be the discrete subgroup of the isometry group generated by a non-trivial element A of the isometry group. The space $(\mathscr{F}_0, \mathbf{g}_V)$ is the quotient of one of the regions $(\mathrm{II}_+, \tilde{\mathbf{g}}_V)$ by G. The space $(\mathscr{F}', \mathbf{g}_V')$ is the quotient $(\mathrm{I}_- + \mathrm{II}_+ + \mathrm{III}_-, \tilde{\mathbf{g}}_V)/G$, and $(\mathscr{F}'', \mathbf{g}_V'')$ is the quotient

$$(\mathrm{I}_+ + \mathrm{II}_+ + \mathrm{III}_+, \tilde{\mathbf{g}}_V)/G.$$

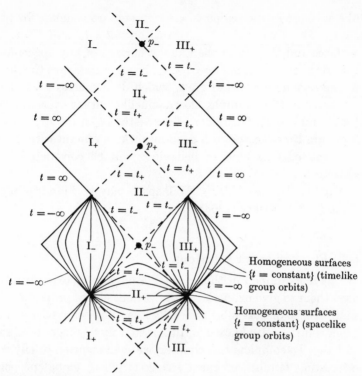

FIGURE 33. Penrose diagram of the maximally extended covering space of a two-dimensional section of Taub–NUT space, showing orbits of the isometry group. Taub–NUT space and its extensions are obtained from part of this space by identification of points under a discrete subgroup of the isometry group.

One would also obtain a Hausdorff manifold by taking the quotient of $(I_+ + II_+ + I_-)$: this corresponds to extending like $(\mathscr{F}', \mathbf{g}_V')$ at the surface $t = t_+$ but extending like $(\mathscr{F}'', \mathbf{g}_V'')$ at the surface $t = t_-$. By taking the quotient of the whole space $\mathscr{F}$ minus the points p_+ and p_- one obtains a non-Hausdorff manifold; and taking the quotient of $\mathscr{F}$ one obtains a non-Hausdorff non-manifold in a way analogous to that in the example above. As in that example, one can take the quotient of the bundle of linear frames over $\mathscr{F}$ and obtain a Hausdorff manifold.

By combining these extensions of the (t, ψ) plane with the coordinates (θ, ϕ) one can obtain corresponding extensions of the four-dimensional space $(\mathscr{M}, \mathbf{g})$. In particular, the two extensions $(\mathscr{F}', \mathbf{g}_V')$ and $(\mathscr{F}'', \mathbf{g}_V'')$ give rise to two different locally inextendible analytic extensions of $(\mathscr{M}, \mathbf{g})$, and both are geodesically incomplete.

Consider one of these extensions, say $(\mathscr{M}', \mathbf{g}')$. The three-spheres which are the surfaces of transitivity of the isometry group are spacelike surfaces in the region $t_- < t < t_+$ and are timelike for $t > t_+$ and $t < t_-$. The two surfaces of transitivity $t = t_-$ and $t = t_+$ are null surfaces and they form the Cauchy horizon of any spacelike surface contained in the region $t_- < t < t_+$, because there are timelike curves in the regions $t < t_-$ and $t > t_+$ which do not cross $t = t_-$ and $t = t_+$ respectively (for example, closed timelike curves exist in the regions $t < t_-$ and $t > t_+$). The region of space–time $t_- \leqslant t \leqslant t_+$ is compact yet there are timelike and null geodesics which remain within it and are incomplete. This kind of behaviour will be considered further in chapter 8.

Further details of Taub–NUT space may be found in Misner and Taub (1969), Misner (1963).

5.9   Further exact solutions

We have examined in this chapter a number of exact solutions and used them to give examples of the various global properties which we shall wish to discuss more generally later. Although a large number of exact solutions are known locally, relatively few have been examined globally. To complete this chapter, we shall mention briefly two other interesting families of exact solutions whose global properties are known.

The first of these are the *plane wave* solutions of the empty space field equations. These are homeomorphic to R^4, and global coordinates (y, z, u, v), which range from $-\infty$ to $+\infty$, can be chosen so that the metric takes the form

$$\mathrm{d}s^2 = 2\,\mathrm{d}u\,\mathrm{d}v + \mathrm{d}y^2 + \mathrm{d}z^2 + H(y, z, u)\,\mathrm{d}u^2,$$

where $$H = (y^2 - z^2)f(u) - 2yzg(u);$$

$f(u)$ and $g(u)$ are arbitrary C^2 functions determining the amplitude and polarization of the wave. These spaces are invariant under a five-parameter group of isometries multiply transitive on the null surfaces $\{u = \text{constant}\}$; a special subclass, in which $f(u) = \cos 2u, g(u) = \sin 2u$, admit an extra Killing vector field, and are homogeneous space–times invariant under a six-parameter group of isometries. These spaces do not contain any closed timelike or null curves; however they admit no Cauchy surfaces (Penrose (1965a)). Local properties of these spaces have been studied in detail by Bondi, Pirani and Robinson (1959), and global properties by Penrose (1965a); Oszváth and Schücking (1962) have studied global properties of the higher symmetry space. The way in which two impulsive plane waves scatter each other and give rise to a singularity has been studied by Khan and Penrose (1971).

The other is the five-parameter family of exact solutions of the source-free Einstein–Maxwell equations found by Carter (1968b) (see also Demianski and Newman (1966)). These include the Schwarzschild, Reissner–Nordström, Kerr, charged Kerr, Taub–NUT, de Sitter and anti-de Sitter solutions as special cases. A description of some of their global properties is given in Carter (1967). Some cases closely related to this family have been examined by Ehlers and Kundt (1962) and Kinnersley and Walker (1970).

6

Causal structure

By postulate (a) of § 3.2, a signal can be sent between two points of $\mathcal{M}$ only if they can be joined by a non-spacelike curve. In this chapter we shall investigate further the properties of such causal relationships, establishing a number of results which will be used in chapter 8 to prove the existence of singularities.

By § 3.2, the study of causal relationships is equivalent to that of the conformal geometry of $\mathcal{M}$, i.e. of the set of all metrics $\tilde{\mathbf{g}}$ conformal to the physical metric $\mathbf{g}$ ($\tilde{\mathbf{g}} = \Omega^2 \mathbf{g}$, where Ω is a non-zero, C^r function). Under such a conformal transformation of the metric a geodesic curve will not, in general, remain a geodesic curve unless it is null, and even in this case an affine parameter along the curve will not remain an affine parameter. Thus in most cases geodesic completeness (i.e. whether all geodesics can be extended to arbitrary values of their affine parameters) will depend on the particular conformal factor and so will not (except in certain special cases described in § 6.4) be a property of the conformal geometry. In fact Clarke (1971) and Siefert (1968) have shown that, provided a physically reasonable causality condition holds, any Lorentz metric is conformal to one in which all null geodesics and all future-directed timelike geodesics are complete. Geodesic completeness will be discussed further in chapter 8 where it forms the basis of a definition of a singularity.

§ 6.1 deals with the question of the orientability of timelike and spacelike bases. In § 6.2 basic causal relations are defined and the definition of a non-spacelike curve is extended from piecewise differentiable to continuous. The properties of the boundary of the future of a set are derived in § 6.3. In § 6.4 a number of conditions which rule out violations or near violations of causality are discussed. The closely related concepts of Cauchy developments and global hyperbolicity are introduced in § 6.5 and § 6.6, and are used in § 6.7 to prove the existence of non-spacelike geodesics of maximum length between certain pairs of points.

In § 6.8 we describe the construction of Geroch, Kronheimer and

[180]

Penrose for attaching a causal boundary to space–time. A particular example of such a boundary is provided by a class of asymptotically flat space–times which are studied in § 6.9.

6.1 Orientability

In our neighbourhood of space–time there is a well-defined arrow of time given by the direction of increase of entropy in quasi-isolated thermodynamic systems. It is not quite clear what the relationship is between this arrow and the other arrows defined by the expansion of the universe and by the direction of electrodynamic radiation; the reader who is interested will find further discussion in Gold (1967), Hogarth (1962), Hoyle and Narlikar (1963) and Ellis and Sciama (1972). Physically it would seem reasonable to suppose that there is a local thermodynamic arrow of time defined continuously at every point of space–time, but we shall only require that it should be possible to define continuously a division of non-spacelike vectors into two classes, which we arbitrarily label future- and past-directed. If this is the case, we shall say that space–time is *time-orientable*. In some space–times it is not possible to define such a time-orientation. An example is the space–time obtained from de Sitter space (§ 5.2) in which points are identified by reflection through the origin of the five-dimensional imbedding space. In this space there are closed curves, non-homotopic to zero, on going round which the orientation of time is reversed. However this difficulty could clearly be resolved by simply unidentifying the points again, and in fact this is always the case: if a space–time $(\mathcal{M}, \mathbf{g})$ is not time-orientable, then it has a double covering space $(\tilde{\mathcal{M}}, \tilde{\mathbf{g}})$ which is. $\tilde{\mathcal{M}}$ may be defined as the set of all pairs (p, α) where $p \in \mathcal{M}$ and α is one of the two orientations of time at p. Then with the natural structure and the projection $\pi: (p, \alpha) \to p$, $\tilde{\mathcal{M}}$ is a double covering of $\mathcal{M}$. If $\tilde{\mathcal{M}}$ consists of two disconnected components then $(\mathcal{M}, \mathbf{g})$ is time-orientable. If $\tilde{\mathcal{M}}$ is connected, then $(\mathcal{M}, \mathbf{g})$ is not time-orientable but $(\tilde{\mathcal{M}}, \tilde{\mathbf{g}})$ is. In the following sections we shall assume that either $(\mathcal{M}, \mathbf{g})$ is time-orientable or we are dealing with the time-orientable covering space. If one can prove the existence of singularities in this space–time then there must also be singularities in $(\mathcal{M}, \mathbf{g})$.

One may also ask whether space–time is *space-orientable*, that is whether it is possible to divide bases of three spacelike axes into right handed and left handed bases in a continuous manner. Geroch (1967a) has pointed out that there is an interesting connection between this and time-orientability which follows because some experiments on elementary particles are not invariant under charge or parity reversals, either singly or together. On the other hand there are theoretical reasons for believing that all interactions are invariant under the combination of charge, parity and time reversals (CPT theorem; see Streater and Wightman (1964)). If one believes that the non-invariance of weak interactions under charge and parity reversals is not merely a local effect but exists at all points of space–time, then it follows that going round any closed curve either the sign of a charge, the orientation of a basis of spacelike axes, and the orientation of time must all reverse, or none of them does. (The ordinary Maxwell theory, in which the electromagnetic field has a definite sign at every point, does not allow the sign of a charge to change on going around a closed curve non-homotopic to zero unless the orientation of time changes. However one could have a theory in which the field was double-valued and changed sign on going round such a curve. This theory would agree with all existing experimental evidence.) In particular if one assumes that space–time is time-orientable then it must also be space-orientable. (This in fact follows on using the experimental evidence alone without appealing to the CPT theorem.)

Geroch (1968c) has also shown that if it is possible to define two-component spinor fields at every point then space–time must be parallelizable, that is it must be possible to introduce a continuous system of bases of the tangent space at every point. (Further consequences of the existence of spinor structures are obtained in Geroch (1970a).)

6.2 Causal curves

Taking space–time to be time-orientable as explained in the previous section, one can divide the non-spacelike vectors at each point into future- and past-directed. For sets $\mathscr{S}$ and $\mathscr{U}$, the *chronological future* $I^+(\mathscr{S}, \mathscr{U})$ *of $\mathscr{S}$ relative to $\mathscr{U}$* can then be defined as the set of all points in $\mathscr{U}$ which can be reached from $\mathscr{S}$ by a future-directed timelike curve in $\mathscr{U}$. (By a curve we mean always one of non-zero extent, not just a single point. Thus $I^+(\mathscr{S}, \mathscr{U})$ may not contain $\mathscr{S}$.) $I^+(\mathscr{S}, \mathscr{M})$ will be denoted by $I^+(\mathscr{S})$, and is an open set, since if $p \in \mathscr{M}$ can be reached by a future-directed timelike curve from $\mathscr{S}$ then there is a small neighbourhood of p which can be so reached.

This definition has a dual in which 'future' is replaced by 'past', and the + by a − ; to avoid repetition, we shall regard dual definitions and results as self-evident.

The *causal future of* $\mathscr{S}$ *relative to* $\mathscr{U}$ is denoted by $J^+(\mathscr{S}, \mathscr{U})$; it is defined as the union of $\mathscr{S} \cap \mathscr{U}$ with the set of all points in $\mathscr{U}$ which can be reached from $\mathscr{S}$ by a future-directed non-spacelike curve in $\mathscr{U}$. We saw in §4.5 that a non-spacelike curve between two points which was not a null geodesic curve could be deformed into a timelike curve between the two points. Thus if $\mathscr{U}$ is an open set and $p, q, r \in \mathscr{U}$, then

$$\left.\begin{array}{ll}\text{either} & q \in J^+(p, \mathscr{U}),\ r \in I^+(q, \mathscr{U}) \\ \text{or} & q \in I^+(p, \mathscr{U}),\ r \in J^+(q, \mathscr{U}) \end{array}\right\} \quad \text{imply} \quad r \in I^+(p, \mathscr{U}).$$

From this it follows that $\overline{I^+}(p, \mathscr{U}) = \overline{J^+}(p, \mathscr{U})$ and $\dot{I}^+(p, \mathscr{U}) = \dot{J}^+(p, \mathscr{U})$ where for any set $\mathscr{K}$, $\overline{\mathscr{K}}$ denotes the closure of $\mathscr{K}$ and

$$\dot{\mathscr{K}} \equiv \overline{\mathscr{K}} \cap \overline{(\mathscr{M} - \mathscr{K})}$$

denotes the boundary of $\mathscr{K}$.

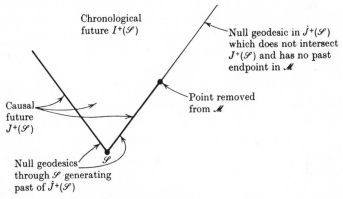

FIGURE 34. When a point has been removed from Minkowski space, the causal future $J^+(\mathscr{S})$ of a closed set $\mathscr{S}$ is not necessarily closed. Further parts of the boundary of the future of $\mathscr{S}$ may be generated by null geodesic segments which have no past endpoints in $\mathscr{M}$.

As before, $J^+(\mathscr{S}, \mathscr{M})$ will be written simply as $J^+(\mathscr{S})$. It is the region of space–time which can be causally affected by events in $\mathscr{S}$. It is not necessarily a closed set even when $\mathscr{S}$ is a single point, as figure 34 shows. This example, incidentally, illustrates a useful technique for constructing space–times with given causal properties: one starts with some simple space–time (unless otherwise indicated this will be Minkowski space), cuts out any closed set and, if desired, pastes it together in an appropriate way (i.e. one makes identifications of points

of $\mathcal{M}$). The result is still a manifold with a Lorentz metric and therefore still a space–time even though it may look rather incomplete
where points have been cut out. As mentioned above, however, this
incompleteness can be cured by an appropriate conformal transformation which sends the cut out points to infinity.

The *future horismos of $\mathscr{S}$ relative to* $\mathscr{U}$, denoted by $E^+(\mathscr{S}, \mathscr{U})$, is
defined as $J^+(\mathscr{S}, \mathscr{U}) - I^+(\mathscr{S}, \mathscr{U})$; we write $E^+(\mathscr{S})$ for $E^+(\mathscr{S}, \mathscr{M})$. (In
some papers the relations $p \in I^+(q)$, $p \in J^+(q)$ and $p \in E^+(q)$ are denoted
by $q \ll p$, $q < p$ and $q \to p$ respectively.) If $\mathscr{U}$ is an open set, points of
$E^+(\mathscr{S}, \mathscr{U})$ must lie on future-directed null geodesics from $\mathscr{S}$ by
proposition 4.5.10, and if $\mathscr{U}$ is a convex normal neighbourhood about p
then it follows from proposition 4.5.1 that $E^+(p, \mathscr{U})$ consists of the
future-directed null geodesics in $\mathscr{U}$ from p, and forms the boundary in
$\mathscr{U}$ of both $I^+(p, \mathscr{U})$ and $J^+(p, \mathscr{U})$. Thus in Minkowski space, the null
cone of p forms the boundary of the causal and chronological futures
of p. However in more complicated space–times this is not necessarily
the case (e.g. see figure 34).

For the purposes of what follows it will be convenient to extend the
definition of timelike and non-spacelike curves from piecewise differentiable to continuous curves. Although such a curve may not have
a tangent vector we can still say that it is non-spacelike if locally
every two points of the curve can be joined by a piecewise differentiable non-spacelike curve. More precisely, we shall say that a continuous curve $\gamma: F \to \mathcal{M}$, where F is a connected interval of R^1, is
future-directed and non-spacelike if for every $t \in F$ there is a neighbourhood G of t in F and a convex normal neighbourhood $\mathscr{U}$ of $\gamma(t)$ in $\mathcal{M}$
such that for any $t_1 \in G$, $\gamma(t_1) \in J^-(\gamma(t), \mathscr{U}) - \gamma(t)$ if $t_1 < t$, and
$\gamma(t_1) \in J^+(\gamma(t), \mathscr{U}) - \gamma(t)$ if $t < t_1$. We shall say that γ is *future-directed
and timelike* if the same conditions hold with J replaced by I. Unless
otherwise specified, we will in future mean by a timelike or non-
spacelike curve such a continuous curve, and shall regard two curves
as equivalent if one is a reparametrization of the other. With this
generalization we can establish a result that will be used repeatedly
in the rest of this chapter. We first give a few more definitions.

A point p will be said to be a *future endpoint* of a future-directed
non-spacelike curve $\gamma: F \to \mathcal{M}$ if for every neighbourhood $\mathscr{V}$ of p there
is a $t \in F$ such that $\gamma(t_1) \in \mathscr{V}$ for every $t_1 \in F$ with $t_1 \geq t$. A non-spacelike
curve is *future-inextendible* (respectively, *future-inextendible in a set $\mathscr{S}$*)
if it has no future endpoint (respectively, no future endpoint in $\mathscr{S}$).
A point p will be said to be a *limit point* of an infinite sequence of non-

spacelike curves λ_n if every neighbourhood of p intersects an infinite number of the λ_n. A non-spacelike curve λ will be said to be a *limit curve* of the sequence λ_n if there is a subsequence λ'_n of the λ_n such that for every $p \in \lambda$, λ'_n converges to p.

Lemma 6.2.1

Let $\mathcal{S}$ be an open set and let λ_n be an infinite sequence of non-spacelike curves in $\mathcal{S}$ which are future-inextendible in $\mathcal{S}$. If $p \in \mathcal{S}$ is a limit point of λ_n, then through p there is a non-spacelike curve λ which is future-inextendible in $\mathcal{S}$ and which is a limit curve of the λ_n.

It is sufficient to consider the case $\mathcal{S} = \mathcal{M}$ since $\mathcal{S}$ can be regarded as a manifold with a Lorentz metric. Let $\mathcal{U}_1$ be a convex normal co-ordinate neighbourhood about p and let $\mathcal{B}(q, a)$ be the open ball of coordinate radius a about q. Let $b > 0$ be such that $\mathcal{B}(p, b)$ is defined and let $\lambda(1, 0)_n$ be a subsequence of $\lambda_n \cap \mathcal{U}_1$ which converges to p. Since $\dot{\mathcal{B}}(p, b)$ is compact it will contain limit points of the $\lambda(1, 0)_n$. Any such limit point y must lie either in $J^-(p, \mathcal{U}_1)$ or $J^+(p, \mathcal{U}_1)$ since otherwise there would be neighbourhoods $\mathcal{V}_1$ of y and $\mathcal{V}_2$ of p between which there would be no non-spacelike curve in $\mathcal{U}_1$. Choose

$$x_{11} \in J^+(p, \mathcal{U}_1) \cap \dot{\mathcal{B}}(p, b)$$

to be one of these limit points (figure 35), and choose $\lambda(1, 1)_n$ to be a subsequence of $\lambda(1, 0)_n$ which converges to x_{11}. The point x_{11} will be a point of our limit curve λ. Continue inductively, defining

$$x_{ij} \in J^+(p, \mathcal{U}_1) \cap \dot{\mathcal{B}}(p, i^{-1}jb)$$

as a limit point of the subsequence $\lambda(i-1, i-1)_n$ for $j = 0$, $\lambda(i, j-1)_n$ for $i \geqslant j \geqslant 1$, and defining $\lambda(i, j)_n$ as a subsequence of the above subsequence which converges to x_{ij}. In other words we are dividing the interval $[0, b]$ into smaller and smaller sections and getting points on our limit curve on the corresponding spheres about p. As any two of the x_{ij} will have non-spacelike separation, the closure of the union of all the x_{ij} $(j \geqslant i)$ will give a non-spacelike curve λ from $p = x_{i0}$ to $x_{11} = x_{ii}$. It now remains to construct a subsequence λ'_n of the λ_n such that for each $q \in \lambda$, λ'_n converges to q. We do this by choosing λ'_m to be a member of the subsequence $\lambda(m, m)_n$ which intersects each of the balls $\mathcal{B}(x_{mj}, m^{-1}b)$ for $0 \leqslant j \leqslant m$. Thus λ will be a limit curve of the λ_n from p to x_{11}. Now let $\mathcal{U}_2$ be a convex normal neighbourhood about x_{11} and repeat the construction using this time the sequence λ'_n. Continuing in this fashion, one can extend λ indefinitely. $\quad\square$

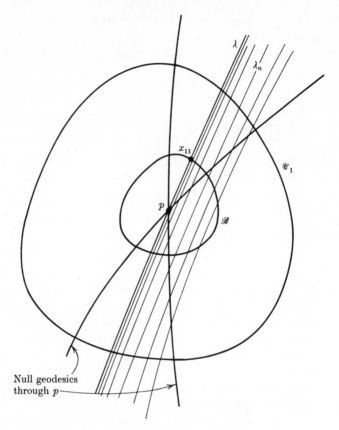

Null geodesics
through p

FIGURE 35. The non-spacelike limit curve λ through p of a family of non-spacelike curves λ_n for which p is a limit point.

6.3 Achronal boundaries

From proposition 4.5.1 it follows that in a convex normal neighbourhood $\mathscr{U}$, the boundary of $I^+(p, \mathscr{U})$ or $J^+(p, \mathscr{U})$ is formed by the future-directed null geodesics from p. To derive the properties of more general boundaries we introduce the concepts of achronal and future sets.

A set $\mathscr{S}$ is said to be *achronal* (sometimes referred to as 'semi-spacelike' in the literature) if $I^+(\mathscr{S}) \cap \mathscr{S}$ is empty, in other words if there are no two points of $\mathscr{S}$ with timelike separation. $\mathscr{S}$ is said to be a *future set* if $\mathscr{S} \supset I^+(\mathscr{S})$. Note that if $\mathscr{S}$ is a future set, $\mathscr{M} - \mathscr{S}$ is a past set. Examples of future sets include $I^+(\mathscr{N})$ and $J^+(\mathscr{N})$, where $\mathscr{N}$ is any set. Examples of achronal sets are given by the following fundamental result.

Proposition 6.3.1

If $\mathscr{S}$ is a future set then $\dot{\mathscr{S}}$, the boundary of $\mathscr{S}$, is a closed, imbedded, achronal three-dimensional C^{1-} submanifold.

If $q \in \dot{\mathscr{S}}$, any neighbourhood of q intersects $\mathscr{S}$ and $\mathcal{M} - \mathscr{S}$. If $p \in I^+(q)$, then there is a neighbourhood of q in $I^-(p)$. Thus $I^+(q) \subset \mathscr{S}$. Similarly $I^-(q) \subset (\mathcal{M} - \mathscr{S})$. If $r \in I^+(q)$, there is a neighbourhood $\mathscr{V}$ of r such that $\mathscr{V} \subset I^+(q) \subset \mathscr{S}$. Thus r cannot belong to $\dot{\mathscr{S}}$. One can introduce normal coordinates (x^1, x^2, x^3, x^4) in a neighbourhood $\mathcal{U}_\alpha$ about q with $\partial/\partial x^4$ timelike and such that the curves $\{x^i = \text{constant} \ (i = 1, 2, 3)\}$ intersect both $I^+(q, \mathcal{U}_\alpha)$ and $I^-(q, \mathcal{U}_\alpha)$. Then each of these curves must contain precisely one point of $\dot{\mathscr{S}}$. The x^4-coordinate of these points must be a Lipschitz function of the x^i $(i = 1, 2, 3)$ since no two points of $\dot{\mathscr{S}}$ have timelike separation. Therefore the one–one map $\phi_\alpha : \dot{\mathscr{S}} \cap \mathcal{U}_\alpha \to R^3$ defined by $\phi_\alpha(p) = x^i(p)$ $(i = 1, 2, 3)$ for $p \in \dot{\mathscr{S}} \cap \mathcal{U}_\alpha$ is a homeomorphism. Thus $(\dot{\mathscr{S}} \cap \mathcal{U}_\alpha, \phi_\alpha)$ is a C^{1-} atlas for $\dot{\mathscr{S}}$. $\square$

We shall call a set with the properties of $\dot{\mathscr{S}}$ listed in proposition 6.3.1, an *achronal boundary*. Such a set can be divided into four disjoint subsets $\dot{\mathscr{S}}_N$, $\dot{\mathscr{S}}_+$, $\dot{\mathscr{S}}_-$, $\dot{\mathscr{S}}_0$ as follows: for a point $q \in \dot{\mathscr{S}}$ there may or may not exist points $p, r \in \dot{\mathscr{S}}$ with $p \in E^-(q) - q$, $r \in E^+(q) - q$. The different possibilities define the subsets of $\dot{\mathscr{S}}$ according to the scheme:

	$\exists p$	$\nexists p$	
$q \in$	$\dot{\mathscr{S}}_N$	$\dot{\mathscr{S}}_-$	$\exists r$
	$\dot{\mathscr{S}}_+$	$\dot{\mathscr{S}}_0$	$\nexists r$

If $q \in \dot{\mathscr{S}}_N$, then $r \in E^+(p)$ since $r \in J^+(p)$ and by proposition 6.3.1, $r \notin I^+(p)$. This means that there is a null geodesic segment in $\dot{\mathscr{S}}$ through q. If $q \in \dot{\mathscr{S}}_+$ (respectively $\dot{\mathscr{S}}_-$) then q is the future (respectively, past) endpoint of a null geodesic in $\dot{\mathscr{S}}$. The subset $\dot{\mathscr{S}}_0$ is spacelike (more strictly, acausal). These divisions are illustrated in figure 36.

A useful condition for a point to lie in $\dot{\mathscr{S}}_N$, $\dot{\mathscr{S}}_+$ or $\dot{\mathscr{S}}_-$ is given in the following lemma due to Penrose (Penrose (1968)):

Lemma 6.3.2

Let $\mathscr{W}$ be a neighbourhood of $q \in \dot{\mathscr{S}}$ where $\mathscr{S}$ is a future set. Then

(i) $I^+(q) \subset I^+(\mathscr{S} - \mathscr{W})$ implies $q \in \dot{\mathscr{S}}_N \cup \dot{\mathscr{S}}_+$,

(ii) $I^-(q) \subset I^-(\mathcal{M} - \mathscr{S} - \mathscr{W})$ implies $q \in \dot{\mathscr{S}}_N \cup \dot{\mathscr{S}}_-$.

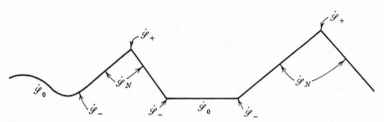

FIGURE 36. An achronal boundary $\dot{\mathscr{S}}$ can be divided into four sets: $\dot{\mathscr{S}}_0$ is space-like, $\dot{\mathscr{S}}_N$ is null, and $\dot{\mathscr{S}}_+$ (respectively, $\dot{\mathscr{S}}_-$) is the future (respectively, past) endpoint of a null geodesic in $\dot{\mathscr{S}}$.

It is sufficient to prove (i) since $\dot{\mathscr{S}}$ can also be regarded as the boundary of the past set $(\mathscr{M} - \mathscr{S})$. Let $\{x_n\}$ be an infinite sequence of points in $I^+(q) \cap \mathscr{W}$ which converge on q. If $I^+(q) \subset I^+(\mathscr{S} - \mathscr{W})$, there will be a past-directed timelike curve λ_n to $\mathscr{S} - \mathscr{W}$ from each x_n. By lemma 6.2.1 there will be a past-directed limit curve λ from q to $(\overline{\mathscr{S} - \mathscr{W}})$. As $I^-(q)$ is open and contained in $\mathscr{M} - \mathscr{S}$, $I^-(q) \cap \mathscr{S}$ is empty. Thus λ must be a null geodesic and must lie in $\dot{\mathscr{S}}$.   □

As an example of the above results, consider $\dot{J}^+(\mathscr{K}) = \dot{I}^+(\mathscr{K})$, the boundary of the future of a closed set $\mathscr{K}$. By proposition 6.3.1 it is an achronal manifold and by the above lemma, every point of $\dot{J}(\mathscr{K}) - \mathscr{K}$ belongs to $[\dot{J}^+(\mathscr{K})]_N$ or $[\dot{J}^+(\mathscr{K})]_+$. This means that $\dot{J}(\mathscr{K}) - \mathscr{K}$ is generated by null geodesic segments which may have future end-points in $\dot{J}^+(\mathscr{K}) - \mathscr{K}$ but which, if they do have past endpoints, can have them only on $\mathscr{K}$ itself. As figure 34 shows, there may be null geodesic generating segments which do not have past endpoints at all but which go out to infinity. This example is admittedly rather artificial but Penrose (1965a) has shown that similar behaviour occurs in something as simple as the plane wave solutions; the anti-de Sitter (§ 5.2) and Reissner–Nordström (§ 5.5) solutions provide other examples. We shall see in § 6.6 that this behaviour is connected with the absence of a Cauchy surface for these solutions.

We shall say that an open set $\mathscr{U}$ is *causally simple* if for every compact set $\mathscr{K} \subset \mathscr{U}$,

$$\dot{J}^+(\mathscr{K}) \cap \mathscr{U} = E^+(\mathscr{K}) \cap \mathscr{U} \quad \text{and} \quad \dot{J}^-(\mathscr{K}) \cap \mathscr{U} = E^-(\mathscr{K}) \cap \mathscr{U}.$$

This is equivalent to saying that $J^+(\mathscr{K})$ and $J^-(\mathscr{K})$ are closed in $\mathscr{U}$.

6.4 Causality conditions

Postulate (a) of § 3.2 required only that causality should hold locally; the global question was left open. Thus we did not rule out the possibility that on a large scale there might be closed timelike curves (i.e. timelike S^1's). However the existence of such curves would seem to lead to the possibility of logical paradoxes: for, one could imagine that with a suitable rocketship one could travel round such a curve and, arriving back before one's departure, one could prevent oneself from setting out in the first place. Of course there is a contradiction only if one assumes a simple notion of free will; but this is not something which can be dropped lightly since the whole of our philosophy of science is based on the assumption that one is free to perform any experiment. It might be possible to form a theory in which there were closed timelike curves and in which the concept of free will was modified (see, for example, Schmidt (1966)) but one would be much more ready to believe that space–time satisfies what we shall call the *chronology condition*: namely, that there are no closed timelike curves. One must however bear in mind the possibility that there might be points (maybe where the density or curvature was very high) of space–time at which this condition does not hold. The set of all such points will be called the *chronology violating* set of $\mathcal{M}$ and has the following character:

Proposition 6.4.1 (*Carter*)

The chronology violating set of $\mathcal{M}$ is the disjoint union of sets of the form $I^+(q) \cap I^-(q)$, $q \in \mathcal{M}$.

If q is in the chronology violating set of $\mathcal{M}$, there must be a future-directed timelike curve λ with past and future endpoints at q. If $r \in I^-(q) \cap I^+(q)$, there will be past- and future-directed timelike curves μ_1 and μ_2 from q to r. Then $(\mu_1)^{-1} \circ \lambda \circ \mu_2$ will be a future-directed timelike curve with past and future endpoints at r. Moreover if

$$r \in [I^-(q) \cap I^+(q)] \cap [I^-(p) \cap I^+(p)]$$

then $p \in I^-(q) \cap I^+(q) = I^-(p) \cap I^+(p)$.

To complete the proof, note that every point r at which chronology is violated is in the set $I^-(r) \cap I^+(r)$.   □

Proposition 6.4.2

If $\mathcal{M}$ is compact, the chronology violating set of $\mathcal{M}$ is non-empty.

$\mathcal{M}$ can be covered by open sets of the form $I^+(q)$, $q \in \mathcal{M}$. If the chronology condition holds at q, then $q \notin I^+(q)$. Thus if the chronology condition held at every point, $\mathcal{M}$ could not be covered by a finite number of sets of the form $I^+(q)$.   □

From this result it would seem reasonable to assume that space–time is non-compact. Another argument against compactness is that any compact, four-dimensional manifold on which there is a Lorentz metric cannot be simply connected. (The existence of a Lorentz metric implies that the Euler number $\chi(\mathcal{M})$ is zero (Steenrod (1951), p. 207). Now $\chi = \sum_{n=0}^{4} (-1)^n B_n$ where $B_n \geqslant 0$ is the nth Betti number of $\mathcal{M}$. By duality (Spanier (1966), p. 297) $B_n = B_{4-n}$. Since $B_0 = B_4 = 1$, this implies that $B_1 \neq 0$ which in turn implies $\pi_1(\mathcal{M}) \neq 0$ (Spanier (1966), p. 398).) Thus a compact space–time is really a non-compact manifold in which points have been identified. It would seem physically reasonable not to identify points but to regard the covering manifold as representing space–time.

We shall say that the *causality condition* holds if there are no closed non-spacelike curves. Similar to proposition 6.4.1, one has:

Proposition 6.4.3

The set of points at which the causality condition does not hold is the disjoint union of sets of the form $J^-(q) \cap J^+(q)$, $q \in \mathcal{M}$.   □

In particular, if the causality condition is violated at $q \in \mathcal{M}$ but the chronology condition holds, there must be a closed null geodesic curve γ through q. Let v be an affine parameter on γ (regarded as a map of an open interval of R^1 to $\mathcal{M}$) and let $\dots, v_{-1}, v_0, v_1, v_2, \dots$ be successive values of v at q. Then we may compare at q the tangent vector $\partial/\partial v|_{v=v_0}$ and the tangent vector $\partial/\partial v|_{v=v_1}$, obtained by parallelly transporting $\partial/\partial v|_{v=v_0}$ round γ. Since they both point in the same direction, they must be proportional: $\partial/\partial v|_{v=v_1} = a \, \partial/\partial v|_{v=v_0}$. The factor a has the following significance: the affine distance covered in the nth circuit of γ, $(v_{n+1} - v_n)$, is equal to $a^{-n}(v_1 - v_0)$. Thus if $a > 1$, v never attains the value $(v_1 - v_0)(1 - a^{-1})^{-1}$ and so γ is geodesically incomplete in the future direction even though one can go round an infinite number of times. Similarly if $a < 1$, γ is incomplete in the past direction, while if $a = 1$, it is complete in both directions. In the two-dimensional model of Taub–NUT space described in § 5.7, there is a closed null geodesic which is an example with $a > 1$. Since the factor a is a conformal invariant, this incompleteness is independent of the conformal factor. This kind of behaviour, however, can happen only if there is a violation of causality in some sense; if the strong causality condition (see below) holds, a suitable conformal transformation of the metric will make all null geodesics complete (Clarke (1971)).

The factor a has a further significance from the following result.

Proposition 6.4.4

If γ is a closed null geodesic curve which is incomplete in the future direction then there is a variation of γ which moves each point of γ towards the future and which yields a closed timelike curve.

By § 2.6, one can find on $\mathcal{M}$ a timelike line-element field $(\mathbf{V}, -\mathbf{V})$ normalized so that $g(\mathbf{V}, \mathbf{V}) = -1$. As we are assuming that $\mathcal{M}$ is time-orientable, one can consistently choose one direction of $(\mathbf{V}, -\mathbf{V})$ and so obtain a future-directed timelike unit vector field $\mathbf{V}$. One can then define a positive definite metric $\mathbf{g}'$ by

$$g'(\mathbf{X}, \mathbf{Y}) = g(\mathbf{X}, \mathbf{Y}) + 2g(\mathbf{X}, \mathbf{V})\, g(\mathbf{Y}, \mathbf{V}).$$

Let t be a (non-affine) parameter on γ which is zero at some point $q \in \gamma$ and which is such that $g(\mathbf{V}, \partial/\partial t) = -2^{-\frac{1}{2}}$. Then t measures proper distance along γ in the metric $\mathbf{g}'$ and has the range $-\infty < t < \infty$. Consider a variation of γ with variation vector $\partial/\partial u$ equal to $x\mathbf{V}$, where x is a function $x(t)$. By § 4.5,

$$\frac{1}{2}\frac{\partial}{\partial u} g\left(\frac{\partial}{\partial t}, \frac{\partial}{\partial t}\right) = \frac{\mathrm{d}}{\mathrm{d}t} g\left(\frac{\partial}{\partial u}, \frac{\partial}{\partial t}\right) - g\left(\frac{\partial}{\partial u}, \frac{\mathrm{D}}{\partial t}\frac{\partial}{\partial t}\right)$$

$$= -2^{-\frac{1}{2}}\left(\frac{\mathrm{d}x}{\mathrm{d}t} - xf\right),$$

where $f\,\partial/\partial t = (\mathrm{D}/\partial t)\,(\partial/\partial t)$. Now suppose v were an affine parameter on γ. Then $\partial/\partial v$ would be proportional to $\partial/\partial t$: $\partial/\partial v = h\,\partial/\partial t$, where $h^{-1}\,\mathrm{d}h/\mathrm{d}t = -f$. On going round one circuit of γ, $\partial/\partial v$ increases by a factor $a > 1$. Thus

$$\oint f\,\mathrm{d}t = -\log a \leqslant 0.$$

Therefore if we take $x(t)$ to be

$$\exp\left(\int_0^t f(t')\,\mathrm{d}t' + b^{-1}t\log a\right),$$

where $b = \oint \mathrm{d}t$, this will give a variation of γ to the future and gives a closed timelike curve. $\qquad\square$

Proposition 6.4.5

If (a) $R_{ab}K^aK^b \geqslant 0$ for every null vector **K**;

(b) the generic condition holds, i.e. every null geodesic contains a point at which $K_{[a}R_{b]cd[e}K_{f]}K^cK^d$ is non-zero, where **K** is the tangent vector;

(c) the chronology condition holds on $\mathscr{M}$,

then the causality condition holds on $\mathscr{M}$.

If there were closed null geodesic curves which were incomplete, then by the previous result they could be varied to give closed timelike curves. If they were complete, then by proposition 4.4.5 they would contain conjugate points and so by proposition 4.5.12 they could again be varied to give closed timelike curves. □

This shows that in physically realistic solutions, the causality and chronology conditions are equivalent.

As well as ruling out closed non-spacelike curves, it would seem reasonable to exclude situations in which there were non-spacelike curves which returned arbitrarily close to their point of origin or which passed arbitrarily close to other non-spacelike curves which then passed arbitrarily close to the origin of the first curve – and so on. In fact Carter (1971a) has pointed out that there is a more than countably infinite hierarchy of such higher degree causality conditions depending on the number and order of the limiting processes involved. We shall describe the first three of these conditions and shall then give the ultimate in causality conditions.

The *future* (respectively, *past*) *distinguishing condition* (Kronheimer and Penrose (1967)) is said to hold at $p \in \mathscr{M}$ if every neighbourhood of p contains a neighbourhood of p which no future (respectively, past) directed non-spacelike curve from p intersects more than once. An equivalent statement is that $I^+(q) = I^+(p)$ (respectively, $I^-(q) = I^-(p)$) implies that $q = p$. Figure 37 shows an example in which the causality and past distinguishing conditions hold everywhere but the future distinguishing condition does not hold at p.

The *strong causality condition* is said to hold at p if every neighbourhood of p contains a neighbourhood of p which no non-spacelike curve intersects more than once. Figure 38 shows an example of violation of this condition.

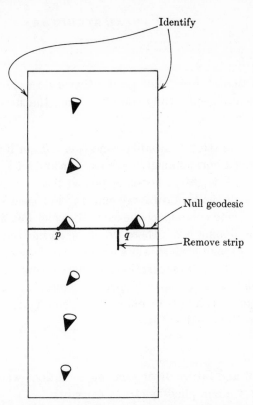

FIGURE 37. A space in which the causality and past distinguishing conditions hold everywhere, but the future distinguishing condition does not hold at p or q (in fact, $I^+(p) = I^+(q)$). The light cones on the cylinder tip over until one null direction is horizontal, and then tip back up; a strip has been removed, thus breaking the closed null geodesic that would otherwise occur.

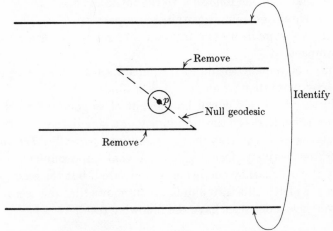

FIGURE 38. A space–time satisfying the causality, future and past distinguishing conditions, but not satisfying the strong causality condition at p. Two strips have been removed from a cylinder; light cones are at $\pm 45°$.

Proposition 6.4.6

If conditions (*a*) to (*c*) of proposition 6.4.5 hold and if in addition, (*d*) $\mathcal{M}$ is null geodesically complete, then the strong causality condition holds on $\mathcal{M}$.

Suppose the strong causality condition did not hold at $p \in \mathcal{M}$. Let $\mathcal{U}$ be a convex normal neighbourhood of p and let $V_n \subset \mathcal{U}$ be an infinite sequence of neighbourhoods of p such that any neighbourhood of p contains all the V_n for n large enough. For each V_n there would be a future-directed non-spacelike curve λ_n which left $\mathcal{U}$ and then returned to V_n. By lemma 6.2.1, there would be an inextendible non-spacelike curve λ through p which was a limit curve of the λ_n. No two points of λ could have timelike separation as otherwise one could join up some λ_n to give a closed non-spacelike curve. Thus λ must be a null geodesic. But by (*a*), (*b*) and (*d*) λ would contain conjugate points and therefore points with timelike separation. $\qquad\qquad\square$

Corollary

The past and future distinguishing conditions would also hold on $\mathcal{M}$ since they are implied by strong causality.

Closely related to these three higher degree causality conditions is the phenomenon of *imprisonment*.

A non-spacelike curve γ that is future-inextendible can do one of three things as one follows it to the future: it can

(i) enter and remain within a compact set $\mathcal{S}$,

(ii) not remain within any compact set but continually re-enter a compact set $\mathcal{S}$,

(iii) not remain within any compact set $\mathcal{S}$ and not re-enter any such set more than a finite number of times.

In the third case γ can be thought of as going off to the edge of space–time, that is either to infinity or a singularity. In the first and second cases we shall say that γ is *totally* and *partially future imprisoned* in $\mathcal{S}$, respectively. One might think that imprisonment could occur only if the causality condition was violated, but the example due to Carter which is illustrated in figure 39 shows that this is not the case. Nevertheless one does have the following result:

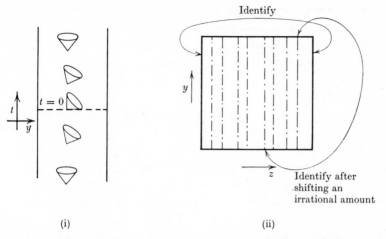

(i)   (ii)

FIGURE 39. A space with imprisoned non-spacelike lines but no closed non-spacelike curves. The manifold is $R^1 \times S^1 \times S^1$ described by coordinates (t, y, z) where (t, y, z) and $(t, y, z+1)$ are identified, and (t, y, z) and $(y, y+1, z+a)$ are identified, where a is an irrational number. The Lorentz metric is given by

$$ds^2 = (\cosh t - 1)^2 (dt^2 - dy^2) + dt\,dy - dz^2.$$

(i) A section $\{z = \text{constant}\}$ showing the orientation of the null cones.
(ii) The section $t = 0$ showing part of a null geodesic.

Proposition 6.4.7

If the strong causality condition holds on a compact set $\mathscr{S}$, there can be no future-inextendible non-spacelike curve totally or partially future imprisoned in $\mathscr{S}$.

$\mathscr{S}$ can be covered by a finite number of convex normal coordinate neighbourhoods $\mathscr{U}_i$ with compact closure, such that no non-spacelike curve intersects any $\mathscr{U}_i$ more than once. (We shall call such neighbourhoods, *local causality neighbourhoods*.) Any future-inextendible non-spacelike curve which intersects one of these neighbourhoods must leave it again and not re-enter it.   □

Proposition 6.4.8

If the future or past distinguishing condition holds on a compact set $\mathscr{S}$, there can be no future-inextendible non-spacelike curve totally future imprisoned in $\mathscr{S}$. (This result is included for its interest but is not needed for what follows.)

Let $\{\mathscr{V}_\alpha\}$, $(\alpha = 1, 2, 3, \dots)$, be a countable basis of open sets for $\mathscr{M}$ (i.e. any open set in $\mathscr{M}$ can be represented as a union of the $\mathscr{V}_\alpha$). As the future or past distinguishing condition holds on $\mathscr{S}$, any point $p \in \mathscr{S}$ will have a convex normal coordinate neighbourhood $\mathscr{U}$ such that no future (respectively, past) directed non-spacelike curve from p intersects $\mathscr{U}$ more than once. We define $f(p)$ to be equal to the least value of α such that $\mathscr{V}_\alpha$ contains p and is contained in some such neighbourhood $\mathscr{U}$.

Suppose there were a future-inextendible non-spacelike curve λ which was totally future imprisoned in $\mathscr{S}$. Let $q \in \lambda$ be such that $\lambda' = \lambda \cap J^+(q)$ is contained in $\mathscr{S}$. Define $\mathscr{A}_0$ to be the closed, non-empty set consisting of all points of $\mathscr{S}$ which are limit points of λ. Let $p_0 \in \mathscr{A}_0$ be such that $f(p_0)$ is equal to the smallest value of $f(p)$ on $\mathscr{A}_0$. Through p_0 there would be an inextendible non-spacelike curve γ_0 every point of which was a limit point of λ'. No two points of γ_0 could have timelike separation since otherwise some segment of λ' could be deformed to give a closed non-spacelike curve. Thus γ_0 would be an inextendible null geodesic which was totally imprisoned in $\mathscr{S}$ in both the past and future directions. Let $\mathscr{A}_1$ be the closed set consisting of all limit points of $\gamma_0 \cap J^+(p_0)$ (or, in the case that the past distinguishing condition holds on $\mathscr{S}$, $\gamma_0 \cap J^-(p_0)$). As every such point would also be a limit point of λ', $\mathscr{A}_1 \subset \mathscr{A}_0$. Since $\mathscr{V}_{f(p_0)}$ could contain no limit point of $\gamma_0 \cap J^+(p_0)$ (respectively, $\gamma_0 \cap J^-(p_0)$), $\mathscr{A}_1$ would be strictly smaller than $\mathscr{A}_0$. We would thus obtain an infinite sequence of closed sets $\mathscr{A}_0 \supset \mathscr{A}_1 \supset \mathscr{A}_2 \supset \dots \supset \mathscr{A}_\beta \supset \dots$. Each $\mathscr{A}_\beta$ would be non-empty, being the set of all limit points of the totally future (respectively, past) imprisoned null geodesic $\gamma_{\beta-1} \cap J^+(p_{\beta-1})$ (respectively, $\gamma_{\beta-1} \cap J^-(p_{\beta-1})$). Let $\mathscr{K} = \bigcap_\beta \mathscr{A}_\beta$. As $\mathscr{S}$ is compact, $\mathscr{K}$ would be non-empty since the intersection of any finite number of the $\mathscr{A}_\beta$ would be non-empty (Hocking and Young (1961), p. 19). Suppose $r \in \mathscr{K}$. Then $f(r) = f(p_\beta)$ for some β. But $\mathscr{V}_{f(p_\beta)} \cap \mathscr{A}_{\beta+1}$ would be empty so r could not be in $\mathscr{A}_{\beta+1}$ and so could not be in $\mathscr{K}$. This shows that there can be no future-inextendible non-spacelike curve totally future imprisoned in $\mathscr{S}$. $\square$

The causal relations on $(\mathscr{M}, \mathbf{g})$ may be used to put a topology on $\mathscr{M}$ called the *Alexandrov topology* This is the topology in which a set is defined to be open if and only if it is the union of one or more sets of the form $I^+(p) \cap I^-(q)$, $p, q \in \mathscr{M}$. As $I^+(p) \cap I^-(q)$ is open in the manifold topology, any set which is open in the Alexandrov topology will be open in the manifold topology, though the converse is not necessarily true.

Suppose however that the strong causality condition holds on $\mathscr{M}$.

Then about any point $r \in \mathcal{M}$ one can find a local causality neighbour-hood $\mathcal{U}$. The Alexandrov topology of $(\mathcal{U}, \mathbf{g}|_{\mathcal{U}})$ regarded as a space–time in its own right, is clearly the same as the manifold topology of $\mathcal{U}$. Thus the Alexandrov topology of $\mathcal{M}$ is the same as the manifold topology since $\mathcal{M}$ can be covered by local causality neighbourhoods. This means that if the strong causality condition holds, one can determine the topological structure of space–time by observation of causal relationships.

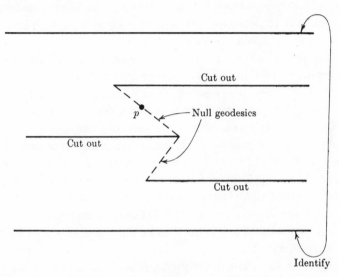

FIGURE 40. A space satisfying the strong causality condition, but in which the slightest variation of the metric would permit there to be closed timelike lines through p. Three strips have been removed from a cylinder; light cones are at $\pm 45°$.

Even imposition of the strong casuality condition does not rule out all causal pathologies, as figure 40 shows one can still have a space–time which is on the verge of violating the chronology condition in that the slightest variation of the metric can lead to closed timelike curves. Such a situation would not seem to be physically realistic since General Relativity is presumably the classical limit of some, as yet unknown, quantum theory of space–time and in such a theory the Uncertainty Principle would prevent the metric from having an exact value at every point. Thus in order to be physically significant, a property of space–time ought to have some form of stability, that is to say, it should also be a property of 'nearby' space–times. In order to give a precise meaning to 'nearby' one has to define a topology on the set of all space–times, that is, all non-compact four-dimensional manifolds and all Lorentz metrics on them. We shall leave the problem of uniting in one connected topological space manifolds of different topologies (this can be done); and shall just consider putting a topology on the set of all C^r Lorentz metrics ($r \geqslant 1$) on a given manifold. There are various ways in which this can be done, depending on whether one requires a 'nearby' metric to be nearby in just its values (C^0 topology) or also in its derivatives up to the kth order (C^k topology) and whether one requires it to be nearby everywhere (open topology) or only on compact sets (compact open topology).

For our purposes here, we shall be interested in the C^0 *open topology*. This may be defined as follows: the symmetric tensor spaces $T_{S2}^0(p)$ of type $(0, 2)$ at every point $p \in \mathcal{M}$ form a manifold (with the natural structure) $T_{S2}^0(\mathcal{M})$, the bundle of symmetric tensors of type $(0, 2)$ over $\mathcal{M}$. A Lorentz metric $\mathbf{g}$ on $\mathcal{M}$ is an assignment of an element of $T_{S2}^0(\mathcal{M})$ at each point $p \in \mathcal{M}$ and so can be regarded as a map or cross-section $\hat{g}: \mathcal{M} \to T_{S2}^0(\mathcal{M})$ such that $\pi \circ \hat{g} = 1$ where π is the projection $T_{S2}^0(\mathcal{M}) \to \mathcal{M}$ which sends $x \in T_{S2}^0(p)$ to p. Let $\mathcal{U}$ be an open set in $T_{S2}^0(\mathcal{M})$ and let $O(\mathcal{U})$ be the set of all C^0 Lorentz metrics $\mathbf{g}$ such that $\hat{g}(\mathcal{M})$ is contained in $\mathcal{U}$ (figure 41). Then the open sets in the C^0 open topology of the C^r Lorentz metrics on $\mathcal{M}$ are defined to be the union of one or more sets of the form $O(\mathcal{U})$.

We say that the *stable causality condition* holds on $\mathcal{M}$ if the space–time metric $\mathbf{g}$ has an open neighbourhood in the C^0 open topology such that there are no closed timelike curves in any metric belonging to the neighbourhood. (It would not make any difference if one used the C^k topology here, but one could not use a compact open topology since in that topology each neighbourhood of any metric contains closed timelike curves.) In other words, what this condition means is that one can expand the light cones slightly at every point without introducing closed timelike curves.

Proposition 6.4.9

The stable causality condition holds everywhere on $\mathcal{M}$ if and only if there is a function f on $\mathcal{M}$ whose gradient is everywhere timelike.

Remark. The function f can be thought of as a sort of cosmic time in the sense that it increases along every future-directed non-spacelike curve.

Proof. The existence of a function f with an everywhere timelike gradient implies the stable causality condition since there can be no closed timelike curves in any metric $\mathbf{h}$ which is sufficiently close to $\mathbf{g}$ that for every point $p \in \mathcal{M}$, the null cone of p in the metric $\mathbf{h}$ intersects the surface $\{f = \text{constant}\}$ through p only at p. To show that the converse is true we introduce a volume measure μ (unrelated to the volume measure defined by the metric $\mathbf{g}$) on $\mathcal{M}$ such that the total volume of

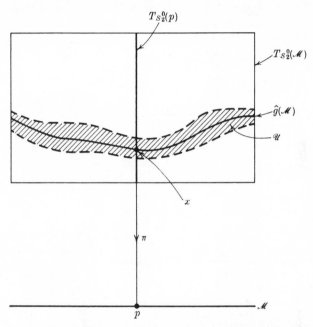

$T_{S_2^0}(p)$

$T_{S_2^0}(\mathcal{M})$

$\hat{g}(\mathcal{M})$

$\mathcal{U}$

x

π

p

$\mathcal{M}$

FIGURE 41. An open set $\mathcal{U}$ in the C^0 open topology on the space $T_{S_2^0}(\mathcal{M})$ of symmetric tensors of type $(0, 2)$ on $\mathcal{M}$.

$\mathcal{M}$ is one. One way of doing this is as follows: choose a countable atlas $(\mathcal{U}_\alpha, \phi_\alpha)$ for $\mathcal{M}$ such that $\overline{\phi_\alpha(\mathcal{U}_\alpha)}$ is compact in R^4. Let μ_0 be the natural Euclidean measure on R^4 and let f_α be a partition of unity for the atlas $(\mathcal{U}_\alpha, \phi_\alpha)$. Then μ may be defined as $\sum_\alpha f_\alpha \, 2^{-\alpha} [\mu_0(\mathcal{U}_\alpha)]^{-1} \phi_\alpha{}^* \mu_0$.

Now if the stable causality condition holds one can find a family of C^r Lorentz metrics $\mathbf{h}(a)$, $a \in [0, 3]$, such that:

(1) $\mathbf{h}(0)$ is the space–time metric $\mathbf{g}$;

(2) there are no closed timelike curves in the metric $\mathbf{h}(a)$ for each $a \in [0, 3]$;

(3) if $a_1, a_2 \in [0, 3]$ with $a_1 < a_2$, then every non-spacelike vector in the metric $\mathbf{h}(a_1)$ is timelike in the metric $\mathbf{h}(a_2)$.

For $p \in \mathscr{M}$, let $\theta(p, a)$ be the volume of $I^-(p, \mathscr{M}, \mathbf{h}(a))$ in the measure μ where we use $I^-(\mathscr{S}, \mathscr{U}, \mathbf{h})$ to denote the past of $\mathscr{S}$ relative to $\mathscr{U}$ in the metric $\mathbf{h}$. For a given value of $a \in (0, 3)$, $\theta(p, a)$ will be a bounded function which increases along every non-spacelike curve. It may not, however, be continuous: as figure 42 shows, it may be possible that a slight alteration of position may allow one to see past an obstruction and so greatly increase the volume of the past. One thus needs some way of smearing out $\theta(p, a)$ so as to obtain a continuous function which

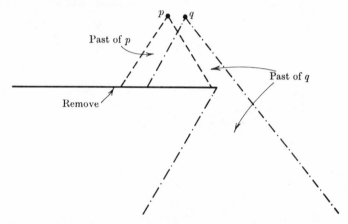

FIGURE 42. A small displacement of a point from p to q results in a large change in the volume of the past of the point. Light cones are at $\pm 45°$ and a strip has been removed as shown.

increases along every curve which is future-directed and non-spacelike in the metric $\mathbf{h}(0)$. One can do this by averaging over a range of a: let

$$\bar{\theta}(p) = \int_1^2 \theta(p, a) \, \mathrm{d}a.$$

We shall show that $\bar{\theta}(p)$ is continuous on $\mathscr{M}$.

First to show that it is upper semi-continuous: given $\epsilon > 0$, let $\mathscr{B}$ be a ball about p such that the volume of $\mathscr{B}$ in the measure μ is less than $\frac{1}{2}\epsilon$. By property (3), for a_1, $a_2 \in [0, 3]$ with $a_1 < a_2$ one can find a neighbourhood $\mathscr{F}(a_1, a_2)$ of p in $\mathscr{B}$ such that

$$[I^-(\mathscr{F}(a_1, a_2), \bar{\mathscr{B}}, \mathbf{h}(a_1)) \cap \dot{\mathscr{B}}] \subset [I^-(p, \bar{\mathscr{B}}, \mathbf{h}(a_2)) \cap \dot{\mathscr{B}}].$$

Let n be a positive integer greater than $2\epsilon^{-1}$. Then we define the set $\mathscr{G}$ to be $\mathscr{G} = \bigcap_i \mathscr{F}(1 + \frac{1}{2}in^{-1}, 1 + \frac{1}{2}(i+1)n^{-1})$, $i = 0, 1, \ldots, 2n$. $\mathscr{G}$ will be a neighbourhood of p and will be contained in $\mathscr{F}(a, a+n^{-1})$ for any $a \in [1, 2]$. Therefore $I^-(q, \mathscr{M}, \mathbf{h}(a)) - \overline{\mathscr{B}}$ will be contained in

$$I^-(p, \mathscr{M}, \mathbf{h}(a+n^{-1})) - \overline{\mathscr{B}} \quad \text{for} \quad q \in \mathscr{G} \quad \text{and} \quad a \in [1, 2].$$

Thus $\qquad\qquad\qquad \theta(q, a) \leqslant \theta(p, a + \tfrac{1}{2}\epsilon) + \tfrac{1}{2}\epsilon$

and so $\overline{\theta}(q) \leqslant \overline{\theta}(p) + \epsilon$, showing that $\overline{\theta}$ is upper semi-continuous. The proof that it is lower semi-continuous is similar. To obtain a differentiable function one can average $\overline{\theta}$ over a neighbourhood of each point with a suitable smoothing function. By taking the neighbourhood small enough one can obtain a function f which has everywhere a timelike gradient in the metric $\mathbf{g}$. Details of this smoothing procedure are given in Seifert (1968). □

The spacelike surfaces $\{f = \text{constant}\}$ may be thought of as surfaces of simultaneity in space–time, though of course they are not unique. If they are all compact they are all diffeomorphic to each other, but this is not necessarily true if some of them are non-compact.

6.5 Cauchy developments

In Newtonian theory there is instantaneous action-at-a-distance and so in order to predict events at future points in space–time one has to know the state of the entire universe at the present time and also to assume some boundary conditions at infinity, such as that the potential goes to zero. In relativity theory, on the other hand, it follows from postulate (a) of §3.2 that events at different points of space–time can be causally related only if they can be joined by a non-spacelike curve. Thus a knowledge of the appropriate data on a closed set $\mathscr{S}$ (if one knew data on an open set, that on its closure would follow by continuity) would determine events in a region $D^+(\mathscr{S})$ to the future of $\mathscr{S}$ called the *future Cauchy development* or *domain of dependence* of $\mathscr{S}$, and defined as the set of all points $p \in \mathscr{M}$ such that every past-inextendible non-spacelike curve through p intersects $\mathscr{S}$ (N.B. $D^+(\mathscr{S}) \supset \mathscr{S}$).

Penrose (1966, 1968) defines the Cauchy development of $\mathscr{S}$ slightly differently, as the set of all points $p \in \mathscr{M}$ such that every past-inextendible timelike curve through p intersects $\mathscr{S}$. We shall denote this set by $\tilde{D}^+(\mathscr{S})$. One has the following result:

Proposition 6.5.1

$\tilde{D}^+(\mathscr{S}) = \overline{D^+(\mathscr{S})}$.

Clearly $\tilde{D}^+(\mathscr{S}) \supset D^+(\mathscr{S})$. If $q \in \mathscr{M} - \tilde{D}^+(\mathscr{S})$ there is a neighbourhood $\mathscr{U}$ of q which does not intersect $\mathscr{S}$. From q there is a past-inextendible curve λ which does not intersect $\mathscr{S}$. If $r \in \lambda \cap I^-(q, \mathscr{U})$ then $I^+(r, \mathscr{U})$ is an open neighbourhood of q in $\mathscr{M} - \tilde{D}^+(\mathscr{S})$. Thus $\mathscr{M} - \tilde{D}^+(\mathscr{S})$ is open and the set $\tilde{D}^+(\mathscr{S})$ is closed. Suppose there were a point $p \in \tilde{D}^+(\mathscr{S})$ which had a neighbourhood $\mathscr{V}$ which did not intersect $D^+(\mathscr{S})$. Choose a point $x \in I^-(p, \mathscr{V})$. From x there would be a past-inextendible non-spacelike curve γ which did not intersect $\mathscr{S}$. Let y_n be a sequence of points on γ which did not converge to any point and which were such that y_{n+1} was to the past of y_n. Let $\mathscr{W}_n$ be convex normal neighbourhoods of the corresponding points y_n such that $\mathscr{W}_{n+1}$ did not intersect $\mathscr{W}_n$. Let z_n be a sequence of points such that

$$z_{n+1} \in I^+(y_{n+1}, \mathscr{W}_{n+1}) \cap I^-(z_n, \mathscr{M} - \mathscr{S}).$$

There would be an inextendible timelike curve from p which passed through each point z_n and which did not intersect $\mathscr{S}$. This would contradict $p \in \tilde{D}^+(\mathscr{S})$. Thus $\tilde{D}^+(\mathscr{S})$ is contained in the closure of $D^+(\mathscr{S})$, and so $\tilde{D}^+(\mathscr{S}) = \overline{D^+(\mathscr{S})}$. $\square$

The future boundary of $D^+(\mathscr{S})$, that is $\overline{D^+(\mathscr{S})} - I^-(D^+(\mathscr{S}))$, marks the limit of the region that can be predicted from knowledge of data on $\mathscr{S}$. We call this closed achronal set the *future Cauchy horizon* of $\mathscr{S}$ and denote it by $H^+(\mathscr{S})$. As figure 43 shows, it will intersect $\mathscr{S}$ if $\mathscr{S}$ is null or if $\mathscr{S}$ has an 'edge'. To make this precise we define edge $(\mathscr{S})$ for an achronal set $\mathscr{S}$ as the set of all points $q \in \overline{\mathscr{S}}$ such that in every neighbourhood $\mathscr{U}$ of q there are points $p \in I^-(q, \mathscr{U})$ and $r \in I^+(q, \mathscr{U})$ which can be joined by a timelike curve in $\mathscr{U}$ which does not intersect $\mathscr{S}$. By an argument similar to that in proposition 6.3.1 it follows that if edge $(\mathscr{S})$ is empty for a non-empty achronal set $\mathscr{S}$, then $\mathscr{S}$ is a three-dimensional imbedded C^{1-} submanifold.

Proposition 6.5.2

For a closed achronal set $\mathscr{S}$,

$$\text{edge}(H^+(\mathscr{S})) = \text{edge}(\mathscr{S}).$$

Let $\mathscr{U}_n$ be a sequence of neighbourhoods of a point $q \in \text{edge}(H^+(\mathscr{S}))$

such that any neighbourhood of q encloses all the $\mathscr{U}_n$ for n sufficiently large. In each $\mathscr{U}_n$ there will be points $p_n \in I^-(q, \mathscr{U}_n)$ and $r_n \in I^+(q, \mathscr{U}_n)$ which can be joined by a timelike curve λ_n which does not intersect $H^+(\mathscr{S})$. This means that λ_n cannot intersect $\overline{D^+(\mathscr{S})}$. By proposition 6.5.1, $q \in \tilde{D}^+(\mathscr{S})$ and so $I^-(q) \subset I^-(\tilde{D}^+(\mathscr{S})) \subset I^-(\mathscr{S}) \cup \tilde{D}^+(\mathscr{S})$. Thus p_n must lie in $I^-(\mathscr{S})$. Also every timelike curve from q which is inextendible in the past direction must intersect $\mathscr{S}$. Therefore for each n, there

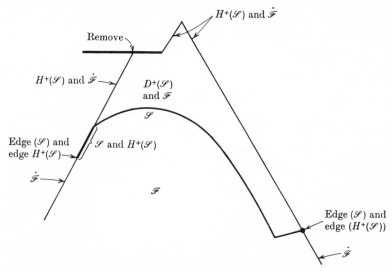

FIGURE 43. The future Cauchy development $D^+(\mathscr{S})$ and future Cauchy horizon $H^+(\mathscr{S})$ of a closed set $\mathscr{S}$ which is partly null and partly spacelike. Note that $H^+(\mathscr{S})$ is not necessarily connected. Null lines are at $\pm 45°$ and a strip has been removed.

must be a point of $\mathscr{S}$ on every timelike curve in $\mathscr{U}_n$ between q and p_n and so q must lie in $\overline{\mathscr{S}}$. As the curves λ_n do not intersect $\mathscr{S}$, q lies in edge $(\mathscr{S})$. The proof the other way round is similar. □

Proposition 6.5.3

Let $\mathscr{S}$ be a closed achronal set. Then $H^+(\mathscr{S})$ is generated by null geodesic segments which either have no past endpoints or have past endpoints at edge $(\mathscr{S})$.

The set $\mathscr{F} \equiv \tilde{D}^+(\mathscr{S}) \cup I^-(\mathscr{S})$ is a past set. Thus by proposition 6.3.1 $\dot{\mathscr{F}}$ is an achronal C^{1-} manifold. $H^+(\mathscr{S})$ is a closed subset of $\dot{\mathscr{F}}$. Let q be a point of $H^+(\mathscr{S})$ − edge $(\mathscr{S})$. If q is not in $\mathscr{S}$ then $q \in I^+(\mathscr{S})$ since $q \in \tilde{D}^+(\mathscr{S})$. As $\mathscr{S}$ is achronal one can find a convex normal neighbourhood $\mathscr{W}$ of q which does not intersect $I^-(\mathscr{S})$. Alternatively if q is in $\mathscr{S}$, let $\mathscr{W}$ be a convex normal neighbourhood of q such that no point of $I^+(q, \mathscr{W})$ can be joined to any point in $I^-(q, \mathscr{W})$ by a timelike curve in $\mathscr{W}$ which does not intersect $\mathscr{S}$.

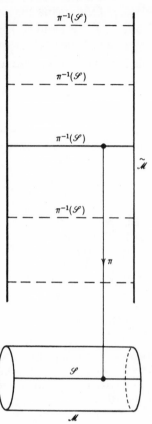

In either case, if p is any point in $I^+(q)$ there must be a past-directed timelike curve from p to some point of $\mathscr{M} - \mathscr{F} - \mathscr{W}$ since otherwise p would be in $D^+(\mathscr{S})$. Therefore by condition (i) of lemma 6.3.2, applied to the future set $\mathscr{M} - \mathscr{F}$, $q \in \dot{\mathscr{F}}_N \cup \dot{\mathscr{F}}_+$.   □

Corollary

If edge $(\mathscr{S})$ vanishes, then $H^+(\mathscr{S})$, if non-empty, is an achronal three-dimensional imbedded C^{1-} manifold which is generated by null geodesic segments which have no past endpoint.

We shall call an acausal set $\mathscr{S}$ with no edge, a *partial Cauchy surface*. That is, a partial Cauchy surface is a spacelike hypersurface which no non-spacelike curve intersects more than once. Suppose there were a connected spacelike hypersurface $\mathscr{S}$ (with no edge) which some non-spacelike curve λ intersected at points p_1 and p_2. Then one could join p_1 and p_2 by a curve μ in $\mathscr{S}$ and $\mu \cup \lambda$ would be a closed curve which crossed $\mathscr{S}$ once only. This curve could not be continuously deformed to zero since such a deformation could change the number of times it crossed $\mathscr{S}$ by an even number only. Thus $\mathscr{M}$ could not be simply connected. This means we could 'unwrap' $\mathscr{M}$ by going to the simply

FIGURE 44. $\mathscr{S}$ is a connected spacelike hypersurface without edge in $\mathscr{M}$. It is not a partial Cauchy surface; however each image $\pi^{-1}(\mathscr{S})$ of $\mathscr{S}$ in the universal covering manifold $\tilde{\mathscr{M}}$ of $\mathscr{M}$, is a partial Cauchy surface in $\tilde{\mathscr{M}}$.

connected universal covering manifold $\tilde{\mathscr{M}}$ in which each connected component of the image of $\mathscr{S}$ is a spacelike hypersurface (with no edge) and is therefore a partial Cauchy surface in $\tilde{\mathscr{M}}$ (figure 44). However going to the universal covering manifold may unwrap $\mathscr{M}$ more than is required to obtain a partial Cauchy surface and may result in the partial Cauchy surface being non-compact even though $\mathscr{S}$ was compact. For the purposes of the following chapters we would like a covering manifold which unwrapped $\mathscr{M}$ sufficiently so that each connected component of the image of $\mathscr{S}$ was a partial Cauchy surface but so that each such component remained homeomorphic to $\mathscr{S}$. Such a covering manifold may be obtained in at least two different ways.

Recall that the universal covering manifold may be defined as the set of all pairs of the form $(p, [\lambda])$ where $p \in \mathscr{M}$ and where $[\lambda]$ is an equivalence class of curves in $\mathscr{M}$ from some fixed point $q \in \mathscr{M}$ to p, which are homotopic modulo q and p. The covering manifold $\mathscr{M}_H$ is defined as the set of all pairs $(p, [\lambda])$ where now $[\lambda]$ is an equivalence class of curves from $\mathscr{S}$ to p homotopic modulo $\mathscr{S}$ and p (i.e. the endpoints on $\mathscr{S}$ can be slid around). $\mathscr{M}_H$ may be characterized as the largest covering manifold such that each connected component of the image of $\mathscr{S}$ is homeomorphic to $\mathscr{S}$. The covering manifold $\mathscr{M}_G$ (Geroch (1967b)) is defined as the set of all pairs $(p, [\lambda])$ where this time $[\lambda]$ is an equivalence class of curves from a fixed point q to p which cross $\mathscr{S}$ the same number of times, crossings in the future direction being counted positive and those in the past direction, negative. $\mathscr{M}_G$ may be characterized as the smallest covering manifold in which each connected component of the image of $\mathscr{S}$ divides the manifold into two parts. In each case the topological and differential structure of the covering manifold is fixed by requiring that the projection which maps $(p, [\lambda])$ to p is locally a diffeomorphism.

Define $D(\mathscr{S}) = D^+(\mathscr{S}) \cup D^-(\mathscr{S})$. A partial Cauchy surface $\mathscr{S}$ is said to be a global Cauchy surface (or simply, a *Cauchy surface*) if $D(\mathscr{S})$ equals $\mathscr{M}$. That is, a Cauchy surface is a spacelike hypersurface which every non-spacelike curve intersects exactly once. The surfaces $\{x^4 = \text{constant}\}$ are examples of Cauchy surfaces in Minkowski space, but the hyperboloids

$$\{(x^4)^2 - (x^3)^2 - (x^2)^2 - (x^1)^2 = \text{constant}\}$$

are only partial Cauchy surfaces since the past or future null cones of the origin are Cauchy horizons for these surfaces (see §5.1 and figure 13). Being a Cauchy surface is a property not only of the surface itself but also of the whole space–time in which it is imbedded. For example, if one cuts a single point out of Minkowski space, the resultant space–time admits no Cauchy surface at all.

If there were a Cauchy surface for $\mathscr{M}$, one could predict the state of the universe at any time in the past or future if one knew the relevant data on the surface. However one could not know the data unless one was to the future of every point in the surface, which would be impossible in most cases. There does not seem to be any physically compelling reason for believing that the universe admits a Cauchy surface; in fact there are a number of known exact solutions of the Einstein field equations which do not, among them the anti-de Sitter space, plane waves, Taub–NUT space and Reissner–Nordström solution, all described in chapter 5. The Reissner–Nordström solution (figure 25) is a specially interesting case: the surface $\mathscr{S}$ shown is adequate for predicting events in the exterior regions I where $r > r_+$ and in the neighbouring region II where $r_- < r < r_+$, but then there is a Cauchy horizon at $r = r_-$. Points in the neighbouring region III are not in $D^+(\mathscr{S})$ since there are non-spacelike curves which are inextendible in the past direction and which do not cross $r = r_-$ but approach the points i^+ (which may be considered to be at infinity) or the singularity at $r = 0$ (which cannot be considered to be in the space–time; see § 8.1). There could be extra information coming in from infinity or from the singularity which would upset any predictions made simply on the basis of data on $\mathscr{S}$. Thus in General Relativity one's ability to predict the future is limited both by the difficulty of knowing data on the whole of a spacelike surface and by the possibility that even if one did it would still be insufficient. Nevertheless despite these limitations one can still predict the occurrence of singularities under certain conditions.

6.6 Global hyperbolicity

Closely related to Cauchy developments is the property of global hyperbolicity (Leray (1952)). A set $\mathcal{N}$ is said to be *globally hyperbolic* if the strong causality assumption holds on $\mathcal{N}$ and if for any two points $p, q \in \mathcal{N}$, $J^+(p) \cap J^-(q)$ is compact and contained in $\mathcal{N}$. In a sense this can be thought of as saying that $J^+(p) \cap J^-(q)$ does not contain any points on the edge of space–time, i.e. at infinity or at a singularity. The reason for the name 'global hyperbolicity' is that on $\mathcal{N}$, the wave equation for a δ-function source at $p \in \mathcal{N}$ has a unique solution which vanishes outside $\mathcal{N} - J^+(p, \mathcal{N})$ (see chapter 7).

Recall that $\mathcal{N}$ is said to be causally simple if for every compact set $\mathcal{K}$ contained in $\mathcal{N}$, $J^+(\mathcal{K}) \cap \mathcal{N}$ and $J^-(\mathcal{K}) \cap \mathcal{N}$ are closed in $\mathcal{N}$.

Proposition 6.6.1

An open globally hyperbolic set $\mathcal{N}$ is causally simple.

Let p be any point of $\mathcal{N}$. Suppose there were a point

$$q \in (\overline{J^+(p)} - J^+(p)) \cap \mathcal{N}.$$

As $\mathcal{N}$ is open, there would be a point $r \in (I^+(q) \cap \mathcal{N})$. But then $q \in \overline{J^+(p) \cap J^-(r)}$, which is impossible as $J^+(p) \cap J^-(r)$ would be compact and therefore closed. Thus $J^+(p) \cap \mathcal{N}$ and $J^-(p) \cap \mathcal{N}$ are closed in $\mathcal{N}$.

Now suppose there exists a point $q \in (\bar{J}^+(\mathcal{K}) - J^+(\mathcal{K})) \cap \mathcal{N}$. Let q_n be an infinite sequence of points in $I^+(q) \cap \mathcal{N}$ converging to q, with $q_{n+1} \in I^-(q_n)$. For each n, $J^-(q_n) \cap \mathcal{K}$ would be a compact non-empty set. Therefore $\bigcap_n \{J^-(q_n) \cap \mathcal{K}\}$ would be a non-empty set. Let p be a point of this set. Then $J^+(p)$ would contain q_n for all n. But $J^+(p)$ is closed. Therefore $J^+(p)$ contains q. □

Corollary

If $\mathcal{K}_1$ and $\mathcal{K}_2$ are compact sets in $\mathcal{N}$, $J^+(\mathcal{K}_1) \cap J^-(\mathcal{K}_2)$ is compact.

One can find a finite number of points $p_i \in \mathcal{N}$ such that

$$\{\bigcup_i J^+(p_i)\} \supset \mathcal{K}_1.$$

Similarly, there will be a finite number of points q_j with $\mathcal{K}_2$ contained in

$$\bigcup_j J^-(q_j).$$

Then $J^+(\mathcal{K}_1) \cap J^-(\mathcal{K}_2)$ will be contained in

$$\bigcup_{i,j} \{J^+(p_i) \cap J^-(q_j)\}$$

and will be closed. □

Leray (1952) did not, in fact, give the above definition of global hyperbolicity but an equivalent one which we shall present: for points $p, q \in \mathcal{M}$ such that strong causality holds on $J^+(p) \cap J^-(q)$, we define $C(p, q)$ to be the space of all (continuous) non-space-like curves from p to q, regarding two curves $\gamma(t)$ and $\lambda(u)$ as representing the same point of $C(p, q)$ if one is a reparametrization of the other, i.e. if there is a continuous monotonic function $f(u)$ such that $\gamma(f(u)) = \lambda(u)$. ($C(p, q)$ can be defined even when the strong causality condition does not hold on $J^+(p) \cap J^-(q)$, but we shall only be interested in the case in which its does hold.) The topology of $C(p, q)$ is defined by saying that a neighbourhood of γ in $C(p,q)$ consists of all the curves in $C(p,q)$ whose points in $\mathcal{M}$ lie in a neighbourhood $\mathcal{W}$ of the points of γ in $\mathcal{M}$ (figure 45). Leray's definition is that an open set $\mathcal{N}$ is globally hyperbolic if $C(p,q)$ is compact for all $p,q \in \mathcal{N}$. These definitions are equivalent, as is shown by the following result.

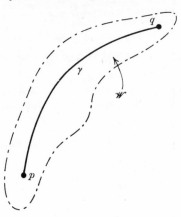

FIGURE 45. A neighbourhood $\mathcal{W}$ of the points of γ in $\mathcal{M}$. A neighbourhood of γ in $C(p, q)$ consists of all non-spacelike curves from p to q whose points lie in $\mathcal{W}$.

Proposition 6.6.2 (*Seifert* (1967), *Geroch* (1970b)).

Let strong causality hold on an open set $\mathcal{N}$ such that

$$\mathcal{N} = J^-(\mathcal{N}) \cap J^+(\mathcal{N}).$$

Then $\mathcal{N}$ is globally hyperbolic if and only if $C(p,q)$ is compact for all $p, q \in \mathcal{N}$.

Suppose first that $C(p,q)$ is compact. Let r_n be an infinite sequence of points in $J^+(p) \cap J^-(q)$ and let λ_n be a sequence of non-spacelike curves from p to q through the corresponding r_n. As $C(p,q)$ is compact, there will be a curve λ to which some subsequence λ'_n converges in the topology on $C(p,q)$. Let $\mathcal{U}$ be a neighbourhood of λ in $\mathcal{M}$ such that $\bar{\mathcal{U}}$ is compact. Then $\mathcal{U}$ will contain all λ'_n and hence all r'_n for n sufficiently large, and so there will be a point $r \in \mathcal{U}$ which is a limit point of the r'_n. Clearly r lies on λ. Thus every infinite sequence in $J^+(p) \cap J^-(q)$ has a limit point in $J^+(p) \cap J^-(q)$. Hence $J^+(p) \cap J^-(q)$ is compact.

Conversely, suppose $J^+(p) \cap J^-(q)$ is compact. Let λ_n be an infinite sequence of non-spacelike curves from p to q. By lemma 6.2.1 applied to the open set $\mathcal{M} - q$, there will be a future-directed non-spacelike curve λ from p which is inextendible in $\mathcal{M} - q$, and is such that there is a subsequence λ'_n which converges to r for every $r \in \lambda$. The curve λ must have a future endpoint at q since by proposition 6.4.7 it cannot be totally future imprisoned in the compact set $J^+(p) \cap J^-(q)$, and it cannot leave the set except at q.

Let $\mathcal{U}$ be any neighbourhood of λ in $\mathcal{M}$ and let r_i $(1 \leqslant i \leqslant k)$ be a finite set of points on λ such that $r_1 = p$, $r_k = q$ and each r_i has a neighbourhood $\mathcal{V}_i$ with $J^+(\mathcal{V}_i) \cap J^-(\mathcal{V}_{i+1})$ contained in $\mathcal{U}$. Then for sufficiently large n, λ'_n will be contained in $\mathcal{U}$. Thus λ'_n converge to λ in the topology on $C(p,q)$ and so $C(p,q)$ is compact. $\qquad\square$

The relation between global hyperbolicity and Cauchy developments is given by the following results.

Proposition 6.6.3

If $\mathscr{S}$ is a closed achronal set, then int $(D(\mathscr{S})) \equiv D(\mathscr{S}) - \dot{D}(\mathscr{S})$, if non-empty, is globally hyperbolic.

We first establish a number of lemmas.

Lemma 6.6.4

If $p \in D^+(\mathscr{S}) - H^+(\mathscr{S})$, then every past-inextendible non-spacelike curve through p intersects $I^-(\mathscr{S})$.

Let p be in $D^+(\mathscr{S}) - H^+(\mathscr{S})$ and let γ be a past-inextendible non-spacelike curve through p. Then one can find a point $q \in D^+(\mathscr{S}) \cap I^+(p)$ and a past-inextendible non-spacelike curve λ through q such that for each point $x \in \lambda$ there is a point $y \in \gamma$ with $y \in I^-(x)$. As λ will intersect $\mathscr{S}$ at some point x_1 there will be a $y_1 \in \gamma \cap I^-(\mathscr{S})$. $\qquad\square$

Corollary

If $p \in$ int $(D(\mathscr{S}))$ then every inextendible non-spacelike curve through p intersects $I^-(\mathscr{S})$ and $I^+(\mathscr{S})$.

int $(D(\mathscr{S})) = D(\mathscr{S}) - \{H^+(\mathscr{S}) \cup H^-(\mathscr{S})\}$. If $p \in I^+(\mathscr{S})$ or $I^-(\mathscr{S})$ the result follows immediately. If $p \in D^+(\mathscr{S}) - I^+(\mathscr{S})$ then $p \in \mathscr{S} \subset D^-(\mathscr{S})$ and the result again follows. $\qquad\square$

Lemma 6.6.5

The strong causality condition holds on int $D(\mathscr{S})$.

Suppose there were a closed non-spacelike curve λ through $p \in$ int $(D(\mathscr{S}))$. By the previous result there would be points $q \in \lambda \cap I^-(\mathscr{S})$ and $r \in \lambda \cap I^+(\mathscr{S})$. As $r \in J^-(q)$, it would also be in $I^-(\mathscr{S})$

which would contradict the fact that $\mathscr{S}$ is achronal. Thus the causality condition holds on int $(D(\mathscr{S}))$. Now suppose that the strong causality condition did not hold at p. Then as in lemma 6.4.6 there would be an infinite sequence of future-directed non-spacelike curves λ_n which converged to an inextendible null geodesic γ through p. There would be points $q \in \gamma \cap I^-(\mathscr{S})$ and $r \in \gamma \cap I^+(\mathscr{S})$ and so there would be some λ_n which intersected $I^+(\mathscr{S})$ and then $I^-(\mathscr{S})$, which would contradict the fact that $\mathscr{S}$ was achronal. $\square$

Proof of proposition 6.6.3. We wish to show that $C(p, q)$ is compact for $p, q \in \text{int}\,(D(\mathscr{S}))$. Consider first the case that $p, q \in I^-(\mathscr{S})$ and suppose $p \in J^-(q)$. Let λ_n be an infinite sequence of non-spacelike curves from q to p. By lemma 6.2.1 there will be a future-directed non-spacelike limit curve from p which is inextendible in $\mathscr{M} - q$. This must have a future endpoint at q since otherwise it would intersect $\mathscr{S}$ which would be impossible as $q \in I^-(\mathscr{S})$. Consider now the case that $p \in J^-(\mathscr{S})$, $q \in J^+(\mathscr{S}) \cap J^+(p)$. If the limit curve λ has an endpoint at q, it is the desired limit point in $C(p, q)$. If it does not have an endpoint at q, it would contain a point $y \in I^+(\mathscr{S})$ since it is inextendible in $\mathscr{M} - q$. Let λ'_n be a subsequence which converges to r for every point r on λ between p and y. Let $\hat{\lambda}$ be a past-directed limit curve from q of the λ'_n. If $\hat{\lambda}$ has a past endpoint at p, it would be the desired limit point in $C(p, q)$. If $\hat{\lambda}$ passed through y, it could be joined up with λ to provide a non-spacelike curve from p to q which would be the desired limit point in $C(p, q)$. Suppose $\hat{\lambda}$ does not have endpoint at p and does not pass through y. Then it would contain some point $z \in I^-(\mathscr{S})$. Let λ''_n be a subsequence of the λ'_n which converges to r for every point r on $\hat{\lambda}$ between q and z. Let $\mathscr{V}$ be an open neighbourhood of $\hat{\lambda}$ which does not contain y. Then for sufficiently large n, all $\lambda''_n \cap J^+(\mathscr{S})$ would be contained in $\mathscr{V}$. This would be impossible as y is a limit point of the λ''_n. Thus there will be a non-spacelike curve from p to q which is a limit point of the λ_n in $C(p, q)$.

The cases $p, q \in I^-(\mathscr{S})$ and $p \in J^-(\mathscr{S})$, $q \in J^+(\mathscr{S})$ together with their duals cover all possible combinations. Thus in all cases we get a non-spacelike curve from p to q which is a limit point of the λ_n in the topology on $C(p, q)$. $\square$

By a similar procedure one can prove:

Proposition 6.6.6

If $q \in \text{int}\,(D(\mathscr{S}))$, then $J^+(\mathscr{S}) \cap J^-(q)$ is compact or empty. $\square$

To show that the whole of $D(\mathscr{S})$ and not merely its interior is globally hyperbolic, one has to impose some extra conditions.

Proposition 6.6.7

If $\mathscr{S}$ is a closed achronal set such that $J^+(\mathscr{S}) \cap J^-(\mathscr{S})$ is both strongly causal and either

(1) acausal (this is the case if and only if $\mathscr{S}$ is acausal), or

(2) compact,

then $D(\mathscr{S})$ is globally hyperbolic.

Suppose that strong causality did not hold at some point $q \in D(\mathscr{S})$. Then by an argument similar to lemma 6.6.5, there would be an inextendible null geodesic through q at each point of which strong causality did not hold. This is impossible, since it would intersect $\mathscr{S}$. Therefore strong causality holds on $D(\mathscr{S})$.

If $p, q \in I^-(\mathscr{S})$, the argument of proposition 6.6.3 holds. If $p \in J^-(\mathscr{S})$, $q \in J^+(\mathscr{S})$ one can as in proposition 6.6.3 construct a future-directed limit curve λ from p and a past-directed limit curve $\hat{\lambda}$ from q, and choose a subsequence λ''_n which converges to r for every point r on λ or $\hat{\lambda}$. In case (1), λ would intersect $\mathscr{S}$ in a single point x. Any neighbourhood of x would contain points of λ''_n for n sufficiently large, and so would contain x''_n, defined as $\lambda''_n \cap \mathscr{S}$, since $\mathscr{S}$ is achronal. Therefore x''_n would converge to x. Similarly x''_n would converge to $\hat{x} \equiv \hat{\lambda} \cap \mathscr{S}$. Thus $\hat{x} = x$ and so one could join λ and $\hat{\lambda}$ to give a non-spacelike limit curve in $C(p, q)$.

In case (2), suppose that λ did not have a future endpoint at q. Then λ would leave $J^-(\mathscr{S})$ since it would intersect $\mathscr{S}$ and by proposition 6.4.7 it would have to leave the compact set $J^+(\mathscr{S}) \cap J^-(\mathscr{S})$. Thus one could find a point x on λ which was not in $J^-(\mathscr{S})$. For each n, choose a point $x''_n \in \mathscr{S} \cap \lambda''_n$. Since $\mathscr{S}$ is compact, there will be some point $y \in \mathscr{S}$ and a subsequence λ'''_n such that the corresponding points x'''_n converge to y. Suppose that y does not lie on λ. Then for sufficiently large n each x'''_n would lie to the future of any neighbourhood $\mathscr{U}$ of x. This would imply $x \in \overline{J^-(\mathscr{S})}$. This is impossible as x is in $J^+(\mathscr{S})$ but is not in the compact set $J^+(\mathscr{S}) \cap J^-(\mathscr{S})$. Therefore λ would pass through y. Similarly $\hat{\lambda}$ would pass through y. One could then join them to obtain a limit curve.　□

Proposition 6.6.3 shows that the existence of a Cauchy surface for an open set $\mathscr{N}$ implies global hyperbolicity of $\mathscr{N}$. The following result shows that the converse is also true:

Proposition 6.6.8 *(Geroch* (1970*b*))

If an open set $\mathcal{N}$ is globally hyperbolic, then $\mathcal{N}$, regarded as a manifold, is homeomorphic to $R^1 \times \mathcal{S}$ where $\mathcal{S}$ is a three-dimensional manifold, and for each $a \in R^1$, $\{a\} \times \mathcal{S}$ is a Cauchy surface for $\mathcal{N}$.

As in proposition 6.4.9, put a measure μ on $\mathcal{N}$ such that the total volume of $\mathcal{N}$ in this measure is one. For $p \in \mathcal{N}$ define $f^+(p)$ to be the volume of $J^+(p, \mathcal{N})$ in the measure μ. Clearly $f^+(p)$ is a bounded function on $\mathcal{N}$ which decreases along every future-directed non-spacelike curve. We shall show that global hyperbolicity implies that $f^+(p)$ is continuous on $\mathcal{N}$ so that we do not need to 'average' the volume of the future as in proposition 6.4.9. To do this it will be sufficient to show that $f^+(p)$ is continuous on any non-spacelike curve λ.

Let $r \in \lambda$ and let x_n be an infinite sequence of points on λ strictly to the past of r. Let $\mathcal{F}$ be $\bigcap_n J^+(x_n, \mathcal{N})$. Suppose that $f^+(p)$ was not upper semi-continuous on λ at r. There would be a point $q \in \mathcal{F} - J^+(r, \mathcal{N})$. Then $r \notin J^-(q, \mathcal{N})$; but each $x_n \in J^-(q, \mathcal{N})$ and so $r \in \overline{J^-}(q, \mathcal{N})$, which is impossible as $J^-(q, \mathcal{N})$ is closed in $\mathcal{N}$ by proposition 6.6.1. The proof that it is lower semi-continuous is similar

As p is moved to the future along an inextendible non-spacelike curve λ in $\mathcal{N}$ the value of $f^+(p)$ must tend to zero. For suppose there were some point q which lay to the future of every point of λ. Then the future-directed curve λ would enter and remain within the compact set $J^+(r) \cap J^-(q)$ for any $r \in \lambda$ which would be impossible by proposition 6.4.7 as the strong causality condition holds on $\mathcal{N}$.

Now consider the function $f(p)$ defined on $\mathcal{N}$ by $f(p) = f^-(p)/f^+(p)$. Any surface of constant f will be an acausal set and, by proposition 6.3.1, will be a three-dimensional C^{1-} manifold imbedded in $\mathcal{N}$. It will also be a Cauchy surface for $\mathcal{N}$ since along any non-spacelike curve, f^- will tend to zero in the past and f^+ will tend to zero in the future. One can put a timelike vector field $\mathbf{V}$ on $\mathcal{N}$ and define a continuous map β which takes points of $\mathcal{N}$ along the integral curves of $\mathbf{V}$ to where they intersect the surface $\mathcal{S}$ ($f = 1$). Then $(\log f(p), \beta(p))$ is a homeomorphism of $\mathcal{N}$ onto $R \times \mathcal{S}$. If one smoothed f as in proposition 6.4.9, one could improve this to a diffeomorphism. $\qquad\square$

Thus if the whole of space–time were globally hyperbolic, i.e. if there were a global Cauchy surface, its topology would be very dull.

6.7   The existence of geodesics

The importance of global hyperbolicity for chapter 8 lies in the following result:

Proposition 6.7.1

Let p and q lie in a globally hyperbolic set $\mathcal{N}$ with $q \in J^+(p)$. Then there is a non-spacelike geodesic from p to q whose length is greater than or equal to that of any other non-spacelike curve from p to q.

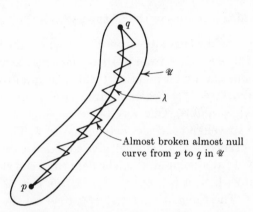

Almost broken almost null curve from p to q in $\mathcal{U}$

FIGURE 46. $\mathcal{U}$ is an open neighbourhood of the timelike curve λ from p to q. There exist in $\mathcal{U}$ timelike curves from p to q which approximate broken null curves and are of arbitrarily small length.

We shall present two proofs of this result: the first, due to Avez (1963) and Seifert (1967), is an argument from the compactness of $C(p,q)$, and the second (applicable only when $\mathcal{N}$ is open) is a procedure whereby the actual geodesic is constructed.

The space $C(p, q)$ contains a dense subset $C'(p, q)$ consisting of all the timelike C^1 curves from p to q. The length of one of these curves λ is defined (cf. § 4.5) as

$$L[\lambda] = \int_p^q (-g(\partial/\partial t, \partial/\partial t))^{\frac{1}{2}} \, dt,$$

where t is a C^1 parameter on λ. The function L is not continuous on $C'(p, q)$ since any neighbourhood of λ contains a zig–zag piecewise almost null curve of arbitrarily small length (figure 46). This lack of continuity arises because we have used the C^0 topology which says that two curves are close if their points in $\mathcal{M}$, but not necessarily their tangent vectors, are close. We could put a C^1 topology on $C'(p,q)$ and so make L continuous but we do not do this because $C'(p,q)$ is not compact; one gets a compact space only when one includes all the continuous non-spacelike curves. Instead, we use the C^0 topology and extend the definition of L to $C(p,q)$.

Because of the signature of the metric, putting wiggles in a timelike curve reduces its length. Thus L is not lower semi-continuous. However one has:

Lemma 6.7.2

L is upper semi-continuous in the C^0 topology on $C'(p,q)$.

Consider a C^1 timelike curve $\lambda(t)$ from p to q, where the parameter t is chosen to be the arc-length from p. In a sufficiently small neighbourhood $\mathcal{U}$ of λ, one can find a function f which is equal to t on λ and is such that the surfaces $\{f = \text{constant}\}$ are spacelike and orthogonal to $\partial/\partial t$ (i.e. $g^{ab}f_{;b}|_\lambda = (\partial/\partial t)^a$). One way to define such an f would be to construct the spacelike geodesics orthogonal to λ. For a sufficiently small neighbourhood $\mathcal{U}$ of λ, they will give a unique mapping of $\mathcal{U}$ to λ, and the value of f at a point in $\mathcal{U}$ can be defined as the value of t at the point on λ into which it is mapped. Any curve μ in $\mathcal{U}$ can be parametrized by f. The tangent vector $(\partial/\partial f)_\mu$ to μ can be expressed as

$$\left(\left(\frac{\partial}{\partial f}\right)_\mu\right)^a = g^{ab}f_{;b} + k^a,$$

where $\mathbf{k}$ is a spacelike vector lying in the surface $\{f = \text{constant}\}$, i.e. $k^a f_{;a} = 0$. Then

$$g\left(\left(\frac{\partial}{\partial f}\right)_\mu, \left(\frac{\partial}{\partial f}\right)_\mu\right) = g^{ab}f_{;a}f_{;b} + g_{ab}k^a k^b$$

$$\geqslant g^{ab}f_{;a}f_{;b}.$$

However on λ, $g^{ab}f_{;a}f_{;b} = -1$. Thus given any $\epsilon > 0$, one can choose $\mathcal{U}' \subset \mathcal{U}$ sufficiently small that on $\mathcal{U}'$, $g^{ab}f_{;a}f_{;b} > -1 + \epsilon$. Therefore for any curve μ in $\mathcal{U}'$, $L[\mu] \leqslant (1+\epsilon)^{\frac{1}{2}} L[\lambda]$.   □

We now define the length of a continuous non-spacelike curve λ from p to q as follows: let $\mathcal{U}$ be a neighbourhood of λ in $\mathcal{M}$ and let $l(\mathcal{U})$ be the least upper bound of the lengths of timelike curves in $\mathcal{U}$ from p to q. Then we define $L[\lambda]$ as the greatest lower bound of $l(\mathcal{U})$ for all neighbourhoods $\mathcal{U}$ of λ in $\mathcal{M}$. This definition of length will work for all curves λ from p to q which have a C^1 timelike curve in every neighbourhood, i.e. it will work for all points in $C(p, q)$ which lie in the closure of $C'(p, q)$. By §4.5, a non-spacelike curve from p to q which is not an unbroken null geodesic curve can be varied to give a piecewise C^1 timelike curve from p to q, and the corners of this curve can be rounded off to give a C^1 timelike curve from p to q. Thus points in $C(p, q) - \overline{C'(p, q)}$ are unbroken null geodesics (containing no conjugate points), and we define their length to be zero.

This definition of L makes it an upper semi-continuous function on the compact space $\overline{C'(p, q)}$. (Actually, as a continuous non-spacelike curve satisfies a local Lipschitz condition, it is differentiable almost everywhere. Thus the length could still be defined as

$$\int (-g(\partial/\partial t, \partial/\partial t))^{\frac{1}{2}} \, dt,$$

and this would agree with the definition above.) If $\overline{C'(p, q)}$ is empty but $C(p, q)$ is non-empty, p and q are joined by an unbroken null geodesic and there are no non-spacelike curves from p to q which are not unbroken null geodesics. If $\overline{C'(p, q)}$ is non-empty, it will contain some point at which L attains its maximum value, i.e. there will be a non-spacelike curve γ from p to q whose length is greater than or equal to that of any other such curve. By proposition 4.5.3, γ must be a geodesic curve as otherwise one could find points $x, y \in \gamma$ which lay in a convex normal coordinate neighbourhood and which could be joined by a geodesic segment of greater length than the portion of γ between x and y. □

For the other, constructive, proof, we first define $d(p, q)$ for $p, q \in \mathcal{M}$ to be zero if $q \notin J^+(p)$ and otherwise to be the least upper bound of the lengths of future-directed piecewise non-spacelike curves from p to q. (Note that $d(p, q)$ may be infinite.) For sets $\mathcal{S}$ and $\mathcal{U}$, we define $d(\mathcal{S}, \mathcal{U})$ to be the least upper bound of $d(p, q)$, $p \in \mathcal{S}$, $q \in \mathcal{U}$.

Suppose $q \in I^+(p)$ and that $d(p, q)$ is finite. Then for any $\delta > 0$ one can find a timelike curve λ of length $d(p, q) - \frac{1}{2}\delta$ from p to q and a neighbourhood $\mathcal{U}$ of q such that λ can be deformed to give a timelike curve of length $d(p, q) - \delta$ from p to any point $r \in \mathcal{U}$. Thus $d(p, q)$, where finite, is lower semi-continuous. In general $d(p, q)$ is not upper semi-continuous but:

Lemma 6.7.3

$d(p, q)$ is finite and continuous in p and q when p and q are contained in a globally hyperbolic set $\mathcal{N}$.

We shall first prove $d(p,q)$ is finite. Since strong causality holds on the compact set $J^+(p) \cap J^-(q)$, one can cover it with a finite number of local causality sets such that each set contains no non-spacelike curve longer than some bound ϵ. Since any non-spacelike curve from p to q can enter each neighbourhood at most once, it must have finite length.

Now suppose that for $p, q \in \mathcal{N}$, there is a $\delta > 0$ such that every neighbourhood of q contains a point $r \in \mathcal{N}$ such that

$$d(p,r) > d(p,q) + \delta.$$

Let x_n be an infinite sequence of points in $\mathcal{N}$ converging to q such that $d(p, x_n) > d(p,q) + \delta$. Then from each x_n one can find a non-spacelike curve λ_n to p of length $> d(p,q) + \delta$. By lemma 6.2.1 there will be a past-directed non-spacelike curve λ through q which is a limit curve of the λ_n. Let $\mathcal{U}$ be a local causality neighbourhood of q. Then λ cannot intersect $I^-(q) \cap \mathcal{U}$ since if it did one of the λ_n could be deformed to give a non-spacelike curve from p to q of length $> d(p,q)$. Thus $\lambda \cap \mathcal{U}$ must be a null geodesic from q and at each point x of $\lambda \cap \mathcal{U}$, $d(p,x)$ will have a discontinuity greater than δ. This argument can be repeated to show that λ is a null geodesic and at each point $x \in \lambda$, $d(p,x)$ has a discontinuity greater than δ. This shows that λ cannot have an endpoint at p, since by proposition 4.5.3, $d(p,x)$ is continuous on a local causality neighbourhood of p. On the other hand, λ would be inextendible in $\mathcal{M} - p$ and so if it did not have an endpoint at p, it would have to leave the compact set $J^+(p) \cap J^-(q)$ by proposition 6.4.7. This shows that $d(p,q)$ is upper semi-continuous on $\mathcal{N}$. $\square$

In the case that $\mathcal{N}$ is open, one can easily construct the geodesic of maximum length from p to q by using the distance function. Let $\mathcal{U} \subset \mathcal{N}$ be a local causality neighbourhood of p which does not contain q and let $x \in J^+(p) \cap J^-(q)$ be such that $d(p,r) + d(r,q)$, $r \in \mathcal{U}$, is maximized for $r = x$. Construct the future-directed geodesic γ from p through x. The relation $d(p,r) + d(r,q) = d(p,q)$ will hold for all points r on γ between p and x. Suppose there were a point $y \in J^-(q) - q$ which was the last point on γ at which this relation held. Let $\mathcal{V} \subset \mathcal{N}$ be a local causality neighbourhood of y which does not contain q and let $z \in J^+(y) \cap J^-(q) \cap \mathcal{V}$ be such that $d(y,r) + d(r,q)$, $r \in \mathcal{V}$, attains its maximum value $d(y,q)$ for $r = z$. If z did not lie on γ, then

$$d(p,z) > d(p,y) + d(y,z) \quad \text{and} \quad d(p,z) + d(z,q) > d(p,q)$$

which is impossible. This shows that the relation

$$d(p,r) + d(r,q) = d(p,q)$$

must hold for all $r \in \gamma \cap J^-(q)$. As $J^+(p) \cap J^-(q)$ is compact, γ must leave $J^-(q)$ at some point y. Suppose $y \neq q$; then y would lie on a past-directed null geodesic λ from q. Joining γ to λ would give a non-spacelike curve from p to q which could be varied to give a curve longer than $d(p,q)$, which is impossible. Thus γ is a geodesic curve from p to q of length $d(p,q)$. □

Corollary

If $\mathscr{S}$ is a C^2 partial Cauchy surface, then to each point $q \in D^+(\mathscr{S})$ there is a future-directed timelike geodesic curve orthogonal to $\mathscr{S}$ of length $d(\mathscr{S}, q)$, which does not contain any point conjugate to $\mathscr{S}$ between $\mathscr{S}$ and q.

By proposition 6.5.2, $H^+(\mathscr{S})$ and $H^-(\mathscr{S})$ do not intersect $\mathscr{S}$ and so are not in $D(\mathscr{S})$. Thus $D(\mathscr{S}) = \operatorname{int} D(\mathscr{S})$ is globally hyperbolic by proposition 6.6.3. By proposition 6.6.6, $\mathscr{S} \cap J^-(q)$ is compact and so $d(p,q)$, $p \in \mathscr{S}$, will attain its maximum value of $d(\mathscr{S}, q)$ at some point $r \in \mathscr{S}$. There will be a geodesic curve γ from r to q of length $d(\mathscr{S}, q)$ which by lemma 4.5.5 and proposition 4.5.9 must be orthogonal to $\mathscr{S}$ and not contain a point conjugate to $\mathscr{S}$ between $\mathscr{S}$ and q. □

6.8 The causal boundary of space–time

In this section we shall give a brief outline of the method of Geroch, Kronheimer and Penrose (1972) for attaching a boundary to space–time. The construction depends only on the causal structure of $(\mathcal{M}, \mathbf{g})$. This means that it does not distinguish between boundary points at a finite distance (singular points) and boundary points at infinity. In § 8.3 we shall describe a different construction which attaches a boundary which represents only singular points. Unfortunately there does not seem to be any obvious relation between the two constructions.

We shall assume that $(\mathcal{M}, \mathbf{g})$ satisfies the strong causality condition. Then any point p in $(\mathcal{M}, \mathbf{g})$ is uniquely determined by its chronological past $I^-(p)$ or its future $I^+(p)$, i.e.

$$I^-(p) = I^-(q) \Leftrightarrow I^+(p) = I^+(q) \Leftrightarrow p = q.$$

The chronological past $\mathscr{W} \equiv I^-(p)$ of any point $p \in \mathcal{M}$ has the properties:

(1) $\mathscr{W}$ is open;

(2) $\mathscr{W}$ is a past set, i.e. $I^-(\mathscr{W}) \subset \mathscr{W}$;

(3) $\mathscr{W}$ cannot be expressed as the union of two proper subsets which have properties (1) and (2).

We shall call a set with properties (1), (2) and (3) an *indecomposable past set*, abbreviated as IP. (The definition given by Geroch, Kronheimer and Penrose does not include property (1). However it is equivalent to the definition given here, since by 'a past set' they mean a set which equals its chronological past, rather than merely containing it.) One can define an IF, or *indecomposable future set*, similarly.

One can divide IPs into two classes: *proper IPs* (*PIPs*) which are the pasts of points in $\mathscr{M}$, and *terminal IPs* (*TIPs*) which are not the past of any point in $\mathscr{M}$. The idea is to regard these TIPs and the similarly defined TIFs as representing points of the causal boundary (*c-boundary*) of $(\mathscr{M}, \mathbf{g})$. For instance, in Minkowski space one would regard the shaded region in figure 47 (i) as representing the point p on $\mathscr{I}^+$. Note that in this example, the whole of $\mathscr{M}$ is itself a TIP and also a TIF. These can be thought of as representing the points i^+ and i^- respectively. In fact all the points of the conformal boundary of Minkowski space, except i^0, can be represented as TIPs or TIFs. In some cases, such as anti-de Sitter space, where the conformal boundary is timelike, points of the boundary will be represented by both a TIP and a TIF (see figure 47 (ii)).

One can also characterize TIPs as the pasts of future-inextendible timelike curves. This means that one can regard the past $I^-(\gamma)$ of a future-inextendible curve γ as representing the future endpoint of γ on the c-boundary. Another curve γ' has the same endpoint if and only if $I^-(\gamma) = I^-(\gamma')$.

Proposition 6.8.1 (*Geroch, Kronheimer and Penrose*)

A set $\mathscr{W}$ is a TIP if and only if there is a future-inextendible timelike curve γ such that $I^-(\gamma) = \mathscr{W}$.

Suppose first that there is a curve γ such that $I^-(\gamma) = \mathscr{W}$. Let $\mathscr{W} = \mathscr{U} \cup \mathscr{V}$ where $\mathscr{U}$ and $\mathscr{V}$ are open past sets. One wants to show that either $\mathscr{U}$ is contained in $\mathscr{V}$, or $\mathscr{V}$ contained in $\mathscr{U}$. Suppose that, on the contrary, $\mathscr{U}$ is not contained in $\mathscr{V}$ and $\mathscr{V}$ not contained in $\mathscr{U}$. Then one could find a point q in $\mathscr{U} - \mathscr{V}$ and a point r in $\mathscr{V} - \mathscr{U}$. Now $q, r \in I^-(\gamma)$, so there would be points $q', r' \in \gamma$ such that $q \in I^-(q')$ and $r \in I^-(r')$. But whichever of $\mathscr{U}$ or $\mathscr{V}$ contained the futuremost of q', r' would also contain both q and r, which contradicts the original definitions of q and r.

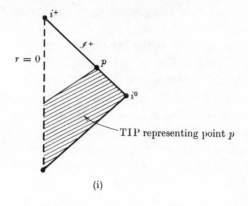

(i)

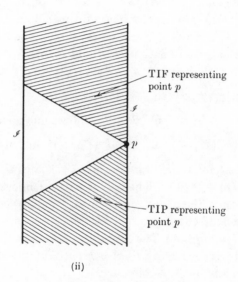

(ii)

FIGURE 47. Penrose diagrams of Minkowski space and anti-de Sitter space (cf. figures 15 and 20), showing (i) the TIP representing a point p on $\mathscr{I}^+$ in Minkowski space, and (ii) the TIP and the TIF representing a point p on $\mathscr{I}$ in anti-de Sitter space.

Conversely, suppose $\mathscr{W}$ is a TIP. Then one must construct a time-like curve γ such that $\mathscr{W} = I^-(\gamma)$. Now if p is any point of $\mathscr{W}$, then $\mathscr{W} = I^-(\mathscr{W} \cap I^+(p)) \cup I^-(\mathscr{W} - I^+(p))$. However $\mathscr{W}$ is indecomposable, so either $\mathscr{W} = I^-(\mathscr{W} \cap I^+(p))$ or $\mathscr{W} = I^-(\mathscr{W} - I^+(p))$. The point p is not contained in $I^-(\mathscr{W} - I^+(p))$, so the second possibility is eliminated. The conclusion may be restated in the following form: given any pair of points of $\mathscr{W}$, then $\mathscr{W}$ contains a point to the future of both of them. Now choose a countable dense family p_n of points of $\mathscr{W}$. Choose a point

q_0 in $\mathscr{W}$ to the future of p_0. Since q_0 and p_1 are in $\mathscr{W}$, one can choose a point q_1 in $\mathscr{W}$ to the future of both of them. Since q_1 and p_2 are in $\mathscr{W}$, one can choose q_2 in $\mathscr{W}$ to the future of both of them, and so on. Since each point q_n obtained in this way lies in the past of its successor, one can find a timelike curve γ in $\mathscr{W}$ through all the points of the sequence. Now for each point $p \in \mathscr{W}$, the set $\mathscr{W} \cap I^+(p)$ is open and non-empty, and so it must contain at least one of the p_n, since these are dense. But for each k, p_k lies in the past of q_k, whence p itself lies in the past of γ. This shows that every point of $\mathscr{W}$ lies to the past of γ, and so since γ is contained in the open past set $\mathscr{W}$, one must have $\mathscr{W} = I^-(\gamma)$. $\square$

We shall denote by $\hat{\mathcal{M}}$ the set of all IPs of the space $(\mathcal{M}, \mathbf{g})$. Then $\hat{\mathcal{M}}$ represents the points of $\mathcal{M}$ plus a future c-boundary; similarly, $\check{\mathcal{M}}$, the set of all IFs of $(\mathcal{M}, \mathbf{g})$, represents $\mathcal{M}$ plus a past c-boundary. One can extend the causal relations I, J and E to $\hat{\mathcal{M}}$ and $\check{\mathcal{M}}$ in the following way. For each $\mathscr{U}$, $\mathscr{V} \subset \hat{\mathcal{M}}$, we shall say

$$\mathscr{U} \in J^-(\mathscr{V}, \hat{\mathcal{M}}) \quad \text{if} \quad \mathscr{U} \subset \mathscr{V},$$

$$\mathscr{U} \in I^-(\mathscr{V}, \hat{\mathcal{M}}) \quad \text{if} \quad \mathscr{U} \subset I^-(q) \text{ for some point } q \in \mathscr{V},$$

$$\mathscr{U} \in E^-(\mathscr{V}, \hat{\mathcal{M}}) \quad \text{if} \quad \mathscr{U} \in J^-(\mathscr{V}, \hat{\mathcal{M}}) \text{ but not } \mathscr{U} \in I^-(\mathscr{V}, \hat{\mathcal{M}}).$$

With these relations, the IP-space $\hat{\mathcal{M}}$ is a causal space (Kronheimer and Penrose (1967)). There is a natural injective map $I^- : \mathcal{M} \to \hat{\mathcal{M}}$ which sends the point $p \in \mathcal{M}$ into $I^-(p) \in \hat{\mathcal{M}}$. This map is an isomorphism of the causality relation J^- as $p \in J^-(q)$ if and only if $I^-(p) \in J^-(I^-(q), \hat{\mathcal{M}})$. The causality relation is preserved by I^- but not by its inverse, i.e. $p \in I^-(q) \Rightarrow I^-(p) \in I^-(I^-(q), \mathcal{M})$. One can define causal relations on $\check{\mathcal{M}}$ similarly.

The idea now is to write $\hat{\mathcal{M}}$ and $\check{\mathcal{M}}$ in some way to form a space $\mathcal{M}^*$ which has the form $\mathcal{M} \cup \Delta$ where Δ will be called the c-boundary of $(\mathcal{M}, \mathbf{g})$. To do so, one needs a method of identifying appropriate IPs and IFs. One starts by forming the space $\mathcal{M}^\#$ which is the union of $\hat{\mathcal{M}}$ and $\check{\mathcal{M}}$, with each PIF identified with the corresponding PIP. In other words, $\mathcal{M}^\#$ corresponds to the points of $\mathcal{M}$ together with the TIPs and TIFs. However as the example of anti-de Sitter space shows, one also wants to identify some TIPs with some TIFs. One way of doing this is to define a topology on $\mathcal{M}^\#$, and then to identify some points of $\mathcal{M}^\#$ to make this topology Hausdorff.

As was mentioned in § 6.4, a basis for the topology of the topological space $\mathcal{M}$ is provided by sets of the form $I^+(p) \cap I^-(q)$. Unfortunately one cannot use a similar method to define a basis for the topology of $\mathcal{M}^{\#}$ as there may be some points of $\mathcal{M}^{\#}$ which are not in the chronological past of any points of $\mathcal{M}^{\#}$. However one can also obtain a topology of $\mathcal{M}$ from a sub-basis consisting of sets of the form $I^{+}(p)$, $I^{-}(p)$, $\mathcal{M} - \overline{I^{+}}(p)$ and $\mathcal{M} - \overline{I^{-}}(p)$. Following this analogy, Geroch, Kromheimer and Penrose have shown how one can define a topology on $\mathcal{M}^{\#}$. For an IF $\mathcal{A} \in \check{\mathcal{M}}$, one defines the sets

$$\mathcal{A}^{\text{int}} \equiv \{\mathcal{V} : \mathcal{V} \in \hat{\mathcal{M}} \text{ and } \mathcal{V} \cap \mathcal{A} \neq \varnothing\},$$

and

$$\mathcal{A}^{\text{ext}} \equiv \{\mathcal{V} : \mathcal{V} \in \hat{\mathcal{M}} \text{ and } \mathcal{V} = I^{-}(\mathcal{W}) \Rightarrow I^{+}(\mathcal{W}) \not\subset \mathcal{A}\}.$$

For an IP $\mathcal{B} \in \hat{\mathcal{M}}$, the sets $\mathcal{B}^{\text{int}}$ and $\mathcal{B}^{\text{ext}}$ are defined similarly. The open sets of $\mathcal{M}^{\#}$ are then defined to be the unions and finite intersections of sets of the form $\mathcal{A}^{\text{int}}$, $\mathcal{A}^{\text{ext}}$, $\mathcal{B}^{\text{int}}$ and $\mathcal{B}^{\text{ext}}$. The sets $\mathcal{A}^{\text{int}}$ and $\mathcal{B}^{\text{int}}$ are the analogues in $\mathcal{M}^{\#}$ of the sets $I^{+}(p)$ and $I^{-}(q)$. If in particular $\mathcal{A} = I^{+}(p)$ and $\mathcal{V} = I^{-}(q)$ then $\mathcal{V} \in \mathcal{A}^{\text{int}}$ if and only if $q \in I^{+}(p)$. However the definitions enable one also to incorporate TIPS into $\mathcal{A}^{\text{int}}$. The sets $\mathcal{A}^{\text{ext}}$ and $\mathcal{B}^{\text{ext}}$ are the analogues of $\mathcal{M} - \overline{I^{+}}(p)$ and $\mathcal{M} - \overline{I^{-}}(q)$.

Finally one obtains $\mathcal{M}^{*}$ by identifying the smallest number of points in the space $\mathcal{M}^{\#}$ necessary to make it a Hausdorff space. More precisely $\mathcal{M}^{*}$ is the quotient space $\mathcal{M}^{\#}/R_{h}$ where R_{h} is the intersection of all equivalence relations $R \subset \mathcal{M}^{\#} \times \mathcal{M}^{\#}$ for which $\mathcal{M}^{\#}/R$ is Hausdorff. The space $\mathcal{M}^{*}$ has a topology induced from $\mathcal{M}^{\#}$ which agrees with the topology of $\mathcal{M}$ on the subset $\mathcal{M}$ of $\mathcal{M}^{*}$. In general one cannot extend the differentiable structure of $\mathcal{M}$ to Δ, though one can on part of Δ in a special case which will be described in the next section.

6.9 Asymptotically simple spaces

In order to study bounded physical systems such as stars, one wants to investigate spaces which are asymptotically flat, i.e. whose metrics approach that of Minkowski space at large distances from the system. The Schwarzschild, Reissner–Nordström and Kerr solutions are examples of spaces which have asymptotically flat regions. As we saw in chapter 5, the conformal structure of null infinity in these spaces is similar to that of Minkowski space. This led Penrose (1964, 1965b, 1968) to adopt this as a definition of a kind of asymptotic flatness. We shall only consider strongly causal spaces. Penrose does not make the requirement of strong causality. However it simplifies matters and implies no loss of generality in the kind of situation we wish to consider.

A time- and space-orientable space $(\mathcal{M}, \mathbf{\mathring{g}})$ is said to be *asymptotically simple* if there exists a strongly causal space $(\tilde{\mathcal{M}}, \tilde{\mathbf{g}})$ and an imbedding $\theta: \mathcal{M} \to \tilde{\mathcal{M}}$ which imbeds $\mathcal{M}$ as a manifold with smooth boundary $\partial\mathcal{M}$ in $\tilde{\mathcal{M}}$, such that:

(1) there is a smooth (say C^3 at least) function Ω on $\tilde{\mathcal{M}}$ such that on $\theta(\mathcal{M})$, Ω is positive and $\Omega^2\mathbf{\mathring{g}} = \theta_*(\tilde{\mathbf{g}})$ (i.e. $\tilde{\mathbf{g}}$ is conformal to $\mathbf{\mathring{g}}$ on $\theta(\mathcal{M})$);

(2) on $\partial\mathcal{M}$, $\Omega = 0$ and $d\Omega \neq 0$;

(3) every null geodesic in $\mathcal{M}$ has two endpoints on $\partial\mathcal{M}$.

We shall write $\mathcal{M} \cup \partial\mathcal{M} \equiv \bar{\mathcal{M}}$.

In fact this definition is rather more general than one wants since it includes cosmological models, such as de Sitter space. In order to restrict it to spaces which are asymptotically flat spaces, we will say that a space $(\mathcal{M}, \mathbf{\mathring{g}})$ is *asymptotically empty and simple* if it satisfies conditions (1), (2), and (3), and

(4) $R_{ab} = 0$ on an open neighbourhood of $\partial\mathcal{M}$ in $\bar{\mathcal{M}}$. (This condition can be modified to allow the existence of electromagnetic radiation near $\partial\mathcal{M}$).

The boundary $\partial\mathcal{M}$ can be thought of as being at infinity, in the sense that any affine parameter in the metric $\mathbf{\mathring{g}}$ on a null geodesic in $\mathcal{M}$ attains unboundedly large values near $\partial\mathcal{M}$. This is because an affine parameter v in the metric g is related to an affine parameter $\tilde{v}$ in the metric $\tilde{\mathbf{g}}$ by $dv/d\tilde{v} = \Omega^{-2}$. Since $\Omega = 0$ at $\partial\mathcal{M}$, $\int dv$ diverges.

From conditions (2) and (4) it follows that the boundary $\partial\mathcal{M}$ is a null hypersurface. This is because the Ricci tensor $\tilde{R}_{ab}$ of the metric $\tilde{g}_{ab}$ is related to the Ricci tensor R_{ab} of g_{ab} by

$$\tilde{R}_a{}^b = \Omega^{-2}R_a{}^b - 2\Omega^{-1}(\Omega)_{|ac}\tilde{g}^{bc} + \{-\Omega^{-1}\Omega_{|cd} + 3\Omega^{-2}\Omega_{|c}\Omega_{|d}\}\tilde{g}^{cd}\delta_a{}^b$$

where $|$ denotes covariant differentiation with respect to $\tilde{g}_{ab}$. Thus

$$\tilde{R} = \Omega^{-2}R - 6\Omega^{-1}\Omega_{|cd}\tilde{g}^{cd} + 3\Omega^{-2}\Omega_{|c}\Omega_{|d}\tilde{g}^{cd}.$$

Since the metric $\tilde{g}_{ab}$ is C^3, $\tilde{R}$ is C^1 at $\partial\mathcal{M}$ where $\Omega = 0$. This implies that $\Omega_{|c}\Omega_{|d}\tilde{g}^{cd} = 0$. However by condition (2), $\Omega_{|c} \neq 0$. Thus $\Omega_{|c}\tilde{g}^{cd}$ is a null vector, and the surface $\partial\mathcal{M}$ ($\Omega = 0$) is a null hypersurface.

In the case of Minkowski space, $\partial\mathcal{M}$ consists of the two null surfaces $\mathscr{I}^+$ and $\mathscr{I}^-$, each of which has the topology $R^1 \times S^2$. (Note that it does not include the points i^0, i^+ and i^- since the conformal boundary is not a smooth manifold at these points.) We shall show that in fact $\partial\mathcal{M}$ has this structure for any asymptotically simple and empty space.

Since $\partial\mathcal{M}$ is a null surface, $\mathcal{M}$ lies locally to the past or future of it. This shows that $\partial\mathcal{M}$ must consist of two disconnected components: $\mathscr{I}^+$ on which null geodesics in $\mathcal{M}$ have their future endpoints, and $\mathscr{I}^-$

on which they have their past endpoints. There cannot be more than two components of $\partial\mathcal{M}$, since there would then be some point $p\in\mathcal{M}$ for which some future-directed null geodesics would go to one component and others to another component. The set of null directions at p going to each component would be open, which is impossible, since the set of future null directions at p is connected.

We next establish an important property.

Lemma 6.9.1

An asymptotically simple and empty space $(\mathcal{M}, \mathbf{g})$ is causally simple.

Let $\mathscr{W}$ be a compact set of $\mathcal{M}$. One wants to show that every null geodesic generator of $\dot{J}^+(\mathscr{W})$ has past endpoint at $\mathscr{W}$. Suppose there were a generator that did not have endpoint there. Then it could not have any endpoint in $\mathcal{M}$, so it would intersect $\mathscr{I}^-$, which is impossible. $\square$

Proposition 6.9.2

An asymptotically simple and empty space $(\mathcal{M}, \mathbf{g})$ is globally hyperbolic.

The proof is similar to that of proposition 6.6.7. One puts a volume element on $\mathcal{M}$ such that the total volume of $\mathcal{M}$ in this measure is unity. Since $(\mathcal{M}, \mathbf{g})$ is causally simple, the functions $f^+(p)$, $f^-(p)$ which are the volumes of $I^+(p)$, $I^-(p)$ are continuous on $\mathcal{M}$. Since strong causality holds on $\mathcal{M}$, $f^+(p)$ will decrease along every future-directed non-spacelike curve. Let λ be a future-inextendible timelike curve. Suppose that $\mathscr{F} = \bigcap_{p\in\lambda} I^+(p)$ was non-empty. Then $\mathscr{F}$ would be a future set and the null generators of the boundary of $\mathscr{F}$ in $\mathcal{M}$ would have no past endpoint in $\mathcal{M}$. Thus they would intersect $\mathscr{I}^-$, which again leads to a contradiction. This shows that $f^+(p)$ goes to zero as p tends to the future on λ. From this it follows that every inextendible non-spacelike curve intersects the surface $\mathscr{H} \equiv \{p: f^+(p) = f^-(p)\}$, which is therefore a Cauchy surface for $\mathcal{M}$. $\square$

Lemma 6.9.3

Let $\mathscr{W}$ be a compact set of an asymptotically empty and simple space $(\mathcal{M}, \mathbf{g})$. Then every null geodesic generator of $\mathscr{I}^+$ intersects $\dot{J}^+(\mathscr{W}, \overline{\mathcal{M}})$ once, where ˙ indicates the boundary in $\overline{\mathcal{M}}$.

Let $p\in\lambda$, where λ is a null geodesic generator of $\mathscr{I}^+$. Then the past set (in $\mathcal{M}$) $J^-(p, \overline{\mathcal{M}})\cap\mathcal{M}$ must be closed in $\mathcal{M}$, since every null geodesic generator of its boundary must have future endpoint on $\mathscr{I}^+$ at p. Since strong causality holds on $\tilde{\mathscr{M}}$, $\mathscr{M} - J^-(p, \bar{\mathscr{M}})$ will be non-empty. Now suppose that λ were contained in $J^+(\mathscr{W}, \bar{\mathscr{M}})$. Then the past set $\bigcap_{p \in \lambda} (J^-(p, \bar{\mathscr{M}}) \cap \mathscr{M})$ would be non-empty. This would be impossible, since the null generators of the boundary of the set would intersect $\mathscr{I}^+$. Suppose on the other hand that λ did not intersect $J^+(\mathscr{W}, \bar{\mathscr{M}})$. Then $\mathscr{M} - \bigcup_{p \in \lambda} (J^-(p, \bar{\mathscr{M}}) \cap \mathscr{M})$ would be non-empty. This would again lead to a contradiction, as the generators of the boundary of the past set $\bigcup_{p \in \lambda} (J^-(p, \bar{\mathscr{M}}) \cap \mathscr{M})$ would intersect $\mathscr{I}^+$.   □

Corollary

$\mathscr{I}^+$ is topologically $R^1 \times (\dot{J}^+(\mathscr{W}, \bar{\mathscr{M}}) \cap \partial\mathscr{M})$.

We shall now show that $\mathscr{I}^+$ (and $\mathscr{I}^-$) and $\mathscr{M}$ are the same topologically as they are for Minkowski space.

Proposition 6.9.4 *(Geroch* (1971))

In an asymptotically simple and empty space $(\mathscr{M}, \mathbf{g})$, $\mathscr{I}^+$ and $\mathscr{I}^-$ are topologically $R^1 \times S^2$, and $\mathscr{M}$ is R^4.

Consider the set N of all null geodesics in $\mathscr{M}$. Since these all intersect the Cauchy surface $\mathscr{H}$, one can define local coordinates on N by the local coordinates and directions of their intersections with $\mathscr{H}$. This makes N into a fibre bundle of directions over $\mathscr{H}$ with fibre S^2. However every null geodesic also intersects $\mathscr{I}^+$. Thus N is also a fibre bundle over $\mathscr{I}^+$. In this case, the fibre is S^2 minus one point which corresponds to the null geodesic generator of $\mathscr{I}^+$ which does not enter $\mathscr{M}$. In other words, the fibre is R^2. Therefore N is topologically $\mathscr{I}^+ \times R^2$. However $\mathscr{I}^+$ is $R^1 \times (\dot{J}^+(\mathscr{W}, \bar{\mathscr{M}}) \cap \partial\mathscr{M})$. This is consistent with $N \approx \mathscr{H} \underset{\sim}{\times} S^2$ only if $\mathscr{H} \approx R^3$ and $\mathscr{I}^+ \approx R^1 \times S^2$.   □

Penrose (1965*b*) has shown that this result implies that the Weyl tensor of the metric $\mathbf{g}$ vanishes on $\mathscr{I}^+$ and $\mathscr{I}^-$. This can be interpreted as saying that the various components of the Weyl tensor of the metric $\mathbf{g}$ 'peel off', that is, they go as different powers of the affine parameter on a null geodesic near $\mathscr{I}^+$ or $\mathscr{I}^-$. Further Penrose (1963), Newman and Penrose (1968) have given conservation laws for the energy–momentum as measured from $\mathscr{I}^+$, in terms of integrals on $\mathscr{I}^+$.

The null surfaces $\mathscr{I}^+$ and $\mathscr{I}^-$ form nearly all the *c*-boundary Δ of $(\mathscr{M}, \mathbf{g})$ defined in the previous section. To see this, note first that any point $p \in \mathscr{I}^+$ defines a TIP $I^-(p, \bar{\mathscr{M}}) \cap \mathscr{M}$. Suppose λ is a future-inextendible curve in $\mathcal{M}$. If λ has a future endpoint at $p \in \mathscr{I}^+$, then the TIP $I^-(\lambda)$ is the same as the TIP defined by p. If λ does not have a future endpoint on $\mathscr{I}^+$, then $\mathcal{M} - I^-(\lambda)$ must be empty, since if it were not, the null geodesic generators of $\dot{I}^-(\lambda)$ would intersect $\mathscr{I}^+$ which is impossible as λ does not intersect $\mathscr{I}^+$. The TIPs therefore consist of one for each point of $\mathscr{I}^+$, and one extra TIP, denoted by i^+, which is $\mathcal{M}$ itself. Similarly, the TIFs consist of one for each point of $\mathscr{I}^-$, and one, denoted by i^-, which again is $\mathcal{M}$ itself.

One now wants to verify that one does not have to identify any TIPs or TIFs, i.e. that $\mathcal{M}^\#$ is Hausdorff. It is clear that no two TIPs or TIFs corresponding to $\mathscr{I}^+$ or $\mathscr{I}^-$ are non-Hausdorff separated. If $p \in \mathscr{I}^+$ then one can find $q \in \mathcal{M}$ such that $p \notin I^+(q, \bar{\mathcal{M}})$. Then $(I^+(q, \bar{\mathcal{M}}))^{\text{ext}}$ is a neighbourhood in $\mathcal{M}^\#$ of the TIP $I^-(p, \bar{\mathcal{M}}) \cap \mathcal{M}$, and $(I^+(q, \bar{\mathcal{M}}))^{\text{int}}$ is a disjoint neighbourhood of the TIP i^+. Thus i^+ is Hausdorff separated from every point of $\mathscr{I}^+$. Similarly it is Hausdorff separated from every point of $\mathscr{I}^-$. Thus the c-boundary of any asymptotically simple and empty space $(\mathcal{M}, \mathbf{g})$ is the same as that of Minkowski space–time, consisting of $\mathscr{I}^+$, $\mathscr{I}^-$ and the two points i^+, i^-.

Asymptotically simple and empty spaces include Minkowski space and the asymptotically flat spaces containing bounded objects such as stars which do not undergo gravitational collapse. However they do not include the Schwarzschild, Reissner–Nordström or Kerr solutions, because in these spaces there are null geodesics which do not have endpoints on $\mathscr{I}^+$ or $\mathscr{I}^-$. Nevertheless these spaces do have asymptotically flat regions which are similar to those of asymptotically empty and simple spaces. This suggests that one should define a space $(\mathcal{M}, \mathbf{g})$ to be *weakly asymptotically simple and empty* if there is an asymptotically simple and empty space $(\mathcal{M}', \mathbf{g}')$ and a neighbourhood $\mathcal{U}'$ of $\partial\mathcal{M}'$ in $\mathcal{M}'$ such that $\mathcal{U}' \cap \mathcal{M}'$ is isometric to an open set $\mathcal{U}$ of $\mathcal{M}$. This definition covers all the spaces mentioned above. In the Reissner–Nordström and Kerr solutions there is an infinite sequence of asymptotically flat regions $\mathcal{U}$ which are isometric to neighbourhoods $\mathcal{U}'$ of asymptotically simple spaces. There is thus an infinite sequence of null infinities $\mathscr{I}^+$ and $\mathscr{I}^-$. However we shall consider only one asymptotically flat region in these spaces. One can then regard $(\mathcal{M}, \mathbf{g})$ as being conformally imbedded in a space $(\tilde{\mathcal{M}}, \tilde{\mathbf{g}})$ such that a neighbourhood $\mathcal{U}$ of $\partial\mathcal{M}$ in $\tilde{\mathcal{M}}$ is isometric to $\mathcal{U}'$. The boundary $\partial\mathcal{M}$ consists of a single pair of null surfaces $\mathscr{I}^+$ and $\mathscr{I}^-$.

We shall discuss weakly asymptotically simple and empty spaces in § 9.2 and § 9.3.

7

The Cauchy problem in General Relativity

In this chapter we shall give an outline of the Cauchy problem in General Relativity. We shall show that, given certain data on a space-like three-surface $\mathscr{S}$, there is a unique maximal future Cauchy development $D^+(\mathscr{S})$ and that the metric on a subset $\mathscr{U}$ of $D^+(\mathscr{S})$ depends only on the initial data on $J^-(\mathscr{U}) \cap \mathscr{S}$. We shall also show that this dependence is continuous if $\mathscr{U}$ has a compact closure in $D^+(\mathscr{S})$. This discussion is included here because of its intrinsic interest, because it uses some of the results of the previous chapter, and because it demonstrates that the Einstein field equations do indeed satisfy postulate (a) of § 3.2 that signals can only be sent between points that can be joined by a non-spacelike curve. However it is not really needed for the remaining three chapters, and so could be skipped by the reader more interested in singularities.

In § 7.1, we discuss the various difficulties and give a precise formulation of the problem. In § 7.2 we introduce a global background metric $\hat{\mathbf{g}}$ to generalize the relation which holds between the Ricci tensor and the metric in each coordinate patch to a single relation which holds over the whole manifold. We impose four gauge conditions on the covariant derivatives of the physical metric $\mathbf{g}$ with respect to the background metric $\hat{\mathbf{g}}$. These remove the four degrees of freedom to make diffeomorphisms of a solution of Einstein's equations, and lead to the second order hyperbolic reduced Einstein equations for $\mathbf{g}$ in the background metric $\hat{\mathbf{g}}$. Because of the conservation equations, these gauge conditions hold at all times if they and their first derivatives hold initially.

In § 7.3 we show that the essential part of the initial data for $\mathbf{g}$ on the three-dimensional manifold $\mathscr{S}$ can be expressed as two three-dimensional tensor fields h^{ab}, χ^{ab} on $\mathscr{S}$. The three-dimensional manifold $\mathscr{S}$ is then imbedded in a four-dimensional manifold $\mathscr{M}$ and a metric $\mathbf{g}$ is defined on $\mathscr{S}$ such that h^{ab} and χ^{ab} become respectively the first and second fundamental forms of $\mathscr{S}$ in $\mathbf{g}$. This can be done in such a way that the gauge conditions hold on $\mathscr{S}$. In § 7.4 we establish some basic inequalities for second order hyperbolic equations. These relate integrals of squared derivatives of solutions of such equations to their initial values. These inequalities are used to prove the existence and uniqueness of solutions of second order hyperbolic equations. In § 7.5 the existence and uniqueness of solutions of the reduced empty space Einstein equations is proved for small perturbations of an empty space solution. The local existence and uniqueness of empty space solutions for arbitrary initial data is then proved by dividing the initial surface up into small regions which are nearly flat, and then joining the resulting solutions together. In § 7.6 we show there is a unique maximal empty space solution for given initial data and that in a certain sense this solution depends continuously on the initial data. Finally in §7.7 we indicate how these results may be extended to solutions with matter.

7.1 The nature of the problem

The Cauchy problem for the gravitational field differs in several important respects from that for other physical fields.

(1) The Einstein equations are *non-linear*. Actually in this respect they are not so different from other fields, for while the electromagnetic field, the scalar field, etc., *by themselves* obey linear equations in a given space–time, they form a non-linear system when their mutual interactions are taken into account. The distinctive feature of the gravitational field is that it is *self-interacting*: it is non-linear even in the absence of other fields. This is because it defines the space–time over which it propagates. To obtain a solution of the non-linear equations one employs an iterative method on approximate linear equations whose solutions are shown to converge in a certain neighbourhood of the initial surface.

(2) Two metrics $\mathbf{g}_1$ and $\mathbf{g}_2$ on a manifold $\mathcal{M}$ are physically equivalent if there is a diffeomorphism $\phi \colon \mathcal{M} \to \mathcal{M}$ which takes $\mathbf{g}_1$ into $\mathbf{g}_2$ ($\phi_* \mathbf{g}_1 = \mathbf{g}_2$), and clearly $\mathbf{g}_1$ satisfies the field equations if and only if $\mathbf{g}_2$ does. Thus the solutions of the field equations can be unique only up to a diffeomorphism. In order to obtain a definite member of the equivalence class of metrics which represents a space–time, one introduces a fixed 'background' metric and imposes four 'gauge conditions' on the covariant derivatives of the physical metric with respect to the background metric. These conditions remove the four degrees of freedom to make diffeomorphisms and lead to a unique solution for the metric components. They are analogous to the Lorentz condition

which is imposed to remove the gauge freedom for the electromagnetic field.

(3) Since the metric defines the space–time structure, one does not know in advance what the domain of dependence of the initial surface is and hence what the region is on which the solution is to be determined. One is simply given a three-dimensional manifold $\mathscr{S}$ with certain initial data $\boldsymbol{\omega}$ on it, and is required to find a four-dimensional manifold $\mathscr{M}$, an imbedding $\theta\colon\mathscr{S}\to\mathscr{M}$ and a metric $\mathbf{g}$ on $\mathscr{M}$ which satisfies the Einstein equations, agrees with the initial values on $\theta(\mathscr{S})$ and is such that $\theta(\mathscr{S})$ is a Cauchy surface for $\mathscr{M}$. We shall say that $(\mathscr{M},\theta,\mathbf{g})$, or simply $\mathscr{M}$, is a *development* of $(\mathscr{S},\boldsymbol{\omega})$. Another development $(\mathscr{M}',\theta',\mathbf{g}')$ of $(\mathscr{S},\boldsymbol{\omega})$ will be called an *extension* of $\mathscr{M}$ if there is a diffeomorphism α of $\mathscr{M}$ into $\mathscr{M}'$ which leaves the image of $\mathscr{S}$ pointwise fixed and takes $\mathbf{g}'$ into $\mathbf{g}$ (i.e. $\theta^{-1}\alpha^{-1}\theta' = \mathrm{id}$ on $\mathscr{S}$, and $\alpha_*\mathbf{g}' = \mathbf{g}$). We shall show that provided the initial data $\boldsymbol{\omega}$ satisfies certain *constraint equations* on $\mathscr{S}$, there will exist developments of $(\mathscr{S},\boldsymbol{\omega})$ and further, there will be a development which is maximal in the sense that it is an extension of any development of $(\mathscr{S},\boldsymbol{\omega})$. Note that by formulating the Cauchy problem in these terms we have included the freedom to make diffeomorphisms, since any development is an extension of any diffeomorphism of itself which leaves the image of $\mathscr{S}$ pointwise fixed.

7·2   The reduced Einstein equations

In chapter 2, the Ricci tensor was obtained in terms of coordinate partial derivatives of the components of the metric tensor. For the purposes of this chapter it will be convenient to obtain an expression that applies to the whole manifold $\mathscr{M}$ and not just to each coordinate neighbourhood separately. To this end we introduce a *background metric* $\hat{\mathbf{g}}$ as well as the physical metric $\mathbf{g}$. With two metrics one has to be careful to maintain the distinction between covariant and contravariant indices. (To avoid confusion, we shall suspend the usual conventions for raising and lowering indices.) The covariant and contravariant forms of $\mathbf{g}$ and $\hat{\mathbf{g}}$ are related by

$$g^{ab}g_{bc} = \delta^a{}_c = \hat{g}^{ab}\hat{g}_{bc}. \tag{7.1}$$

It will be convenient to take the contravariant form g^{ab} of the metric to be more fundamental and the covariant form g_{ab} as derived from it by (7.1). Using the alternating tensor $\hat{\eta}_{abcd}$ defined by the background metric, this relation can be expressed explicitly as

$$g_{ab} = \frac{1}{3!} g^{cd}g^{ef}g^{ij}(\det \mathbf{g}) \, \hat{\eta}_{acei}\hat{\eta}_{bdfj}, \tag{7.2}$$

where

$$(\det \mathbf{g})^{-1} \equiv \frac{1}{4!} g^{ab}g^{cd}g^{ef}g^{ij}\hat{\eta}_{acei}\hat{\eta}_{bdfj}$$

is the determinant of the components of g^{ab} in a basis which is orthonormal with respect to the metric $\hat{\mathbf{g}}$.

The difference between the connection $\mathbf{\Gamma}$ defined by $\mathbf{g}$ and the connection $\hat{\mathbf{\Gamma}}$ defined by $\hat{\mathbf{g}}$ is a tensor, and can be expressed in terms of the covariant derivative of $\mathbf{g}$ with respect to $\hat{\mathbf{\Gamma}}$ (cf § 3.3):

$$\delta\Gamma^a{}_{bc} \equiv \Gamma^a{}_{bc} - \hat{\Gamma}^a{}_{bc}$$

$$= \tfrac{1}{2}g^{ij}{}_{|k}(g_{bi}g_{cj}g^{ak} - g_{bi}\delta^k{}_c\delta^a{}_j - g_{ci}\delta^k{}_b\delta^a{}_j), \tag{7.3}$$

where we have used a stroke to denote covariant differentiation with respect to $\hat{\mathbf{\Gamma}}$ and the symbol δ to denote the difference between quantities defined from $\mathbf{g}$ and $\hat{\mathbf{g}}$. Then from (2.20),

$$\delta R_{ab} = \delta\Gamma^d{}_{ab|d} - \delta\Gamma^d{}_{ad|b} + \delta\Gamma^d{}_{ab}\,\delta\Gamma^e{}_{de} - \delta\Gamma^d{}_{ae}\,\delta\Gamma^e{}_{bd}. \tag{7.4}$$

Thus

$$\delta(R^{ab} - \tfrac{1}{2}g^{ab}R) = g^{ai}g^{bj}\delta R_{ij} + 2\delta g^{i(a}g^{b)j}\hat{R}_{ij} - \delta g^{ai}\delta g^{bj}\hat{R}_{ij}$$

$$- \tfrac{1}{2}\delta g^{ab}\hat{R} - \tfrac{1}{2}g^{ab}(\delta g^{ij}\hat{R}_{ij} + g^{ij}\delta R_{ij})$$

$$= \tfrac{1}{2}g^{ij}\delta g^{ab}{}_{|ij} - g^{i(a}\psi^{b)}{}_{|i} + \tfrac{1}{2}g^{ab}(\psi^i{}_{|i} - g_{cd}g^{ij}\delta g^{cd}{}_{|ij})$$

$$+ (\text{terms in } \delta g^{cd}{}_{|i} \text{ and } \delta g^{ef}), \tag{7.5}$$

$$\psi^b \equiv g^{bc}{}_{|c} - \tfrac{1}{2}g^{bc}g_{de}g^{de}{}_{|c} = (\det \mathbf{g})^{-1}((\det \mathbf{g})g^{bc})_{|c} = (\det \mathbf{g})^{-1}\phi^{bc}{}_{|c} \tag{7.6}$$

and

$$\phi^{bc} \equiv (\det \mathbf{g})\,\delta g^{bc}.$$

The plan is now as follows. We choose some suitable background metric $\hat{\mathbf{g}}$ and express the Einstein equations in the form

$$R^{ab} - \tfrac{1}{2}Rg^{ab} = \delta(R^{ab} - \tfrac{1}{2}Rg^{ab}) + \hat{R}^{ab} - \tfrac{1}{2}\hat{g}^{ab}\hat{R} = 8\pi T^{ab}. \tag{7.7}$$

One regards this as a second order non-linear set of differential equations to determine $\mathbf{g}$ in terms of the values of it and its first derivatives on some initial surface. Of course to complete the system one has to specify the equations governing the physical fields which make up the energy–momentum tensor T^{ab}. However even when this is done one does not have a system of equations which uniquely determines the time development in terms of the initial values and first derivatives. The reason for this is, as was mentioned above, that a solution of the Einstein equations can be unique only up to a diffeomorphism. In order to obtain a definite solution one removes this freedom to make diffeomorphisms by imposing four *gauge conditions* on the covariant derivatives of $\mathbf{g}$ with respect to the background metric $\mathbf{\hat{g}}$. We shall use the so-called 'harmonic' conditions

$$\psi^b = \phi^{bc}{}_{|c} = 0$$

which are analogous to the Lorentz gauge conditions $A^i{}_{;i} = 0$ in electrodynamics. With this condition one obtains the *reduced Einstein equations*

$$g^{ij}\phi^{ab}{}_{|ij} + (\text{terms in } \phi^{cd}{}_{|e} \text{ and } \phi^{ab}) = 16\pi T^{ab} - 2\hat{R}^{ab} + \hat{g}^{ab}\hat{R}. \quad (7.8)$$

We shall denote the left-hand side of (7.8) by $E^{ab}{}_{cd}(\phi^{cd})$, where $E^{ab}{}_{cd}$ is the *Einstein operator*. For suitable forms of the energy–momentum tensor T^{ab} these are second order hyperbolic equations for which we shall demonstrate the existence and uniqueness of solutions in §7.5. We still have to check that the harmonic conditions are consistent with the Einstein equations. That is to say: we derived (7.8) from the Einstein equations by assuming that $\phi^{bc}{}_{|c}$ was zero. We now have to verify that the solution that (7.8) gives rise to does indeed have this property. To do this, differentiate (7.8) and contract. This gives an equation of the form

$$g^{ij}\psi^b{}_{|ij} + B_c{}^{bi}\psi^c{}_{|i} + C_c{}^b \psi^c = 16\pi T^{ab}{}_{;a}, \quad (7.9)$$

where a semi-colon denotes differentiation with respect to g, and the tensors $B_c{}^{bi}$ and $C_c{}^b$ depend on $\hat{g}^{ab}$, $\hat{R}^a{}_{bcd}$, g^{ab} and $g^{ab}{}_{|c}$. Equations (7.9) may be regarded as second order linear hyperbolic equations for ψ^b. Since the right-hand side vanishes, one can use the uniqueness theorem for such equations (proposition 7.4.5) to show that ψ^b will vanish everywhere if it and its first derivatives are zero on the initial surface. We shall see in the next section that this can be arranged by a suitable diffeomorphism.

We still have to show that the unique solution obtained by imposing the harmonic gauge condition is related by a diffeomorphism to any other solution of the Einstein equations with the same initial data. This will be done in §7.4 by making a special choice of the background metric.

7.3 The initial data

As (7.8) is a second order hyperbolic system it seems that to determine the solution one should prescribe the values of g^{ab} and $g^{ab}{}_{|c}u^c$ on the initial surface $\theta(\mathscr{S})$, where u^c is some vector field which is not tangent to $\theta(\mathscr{S})$. However not all these twenty components are significant or independent: some can be given arbitrary initial values without changing the solution by more than a diffeomorphism, and others have to obey certain consistency conditions.

Consider a diffeomorphism $\mu \colon \mathscr{M} \to \mathscr{M}$ which leaves $\theta(\mathscr{S})$ pointwise fixed. This will induce a map μ_* which takes g^{ab} at $p \in \theta(\mathscr{S})$ into a new tensor $\mu_* g^{ab}$ at p. If $n_a \in T^*{}_p$ is orthogonal to $\theta(\mathscr{S})$ (i.e. $n_a V^a = 0$ for any $V^a \in T_p$ tangent to $\theta(\mathscr{S})$) and normalized so that $n_a \hat{g}^{ab} n_b = -1$ then, by suitable choice of μ, $n_a \mu_* g^{ab}$ can be made equal to any vector at p which is not tangent to $\theta(\mathscr{S})$. Thus the components $n_a g^{ab}$ are not significant. On the other hand as μ leaves $\theta(\mathscr{S})$ pointwise fixed, the induced metric $h_{ab} = \theta^* g_{ab}$ on $\mathscr{S}$ will remain unchanged. It is therefore only this part of $\mathbf{g}$ which lies in $\theta(\mathscr{S})$ which need be given to determine the solution. The other components $n_a g^{ab}$ can be prescribed arbitrarily without changing the solution by more than a diffeomorphism. Another way of seeing this is to recall that we formulated the Cauchy problem in terms of certain data on a disembodied three-manifold $\mathscr{S}$ and then looked for an imbedding into some four-manifold $\mathscr{M}$. Now on $\mathscr{S}$ itself one cannot define a four-dimensional tensor field like $\mathbf{g}$ but only a three-dimensional metric $\mathbf{h}$, which we shall take to be positive definite. The contravariant and covariant forms of $\mathbf{h}$ are related by

$$h^{ab} h_{bc} = \delta^a{}_c, \tag{7.10}$$

where now $\delta^a{}_c$ is a three-dimensional tensor in $\mathscr{S}$. The imbedding θ will carry h_{ab} into a contravariant tensor field $\theta_* h^{ab}$ on $\theta(\mathscr{S})$ which has the property

$$n_a \theta_* h^{ab} = 0. \tag{7.11}$$

As $n_a g^{ab}$ is arbitrary, one may now define $\mathbf{g}$ on $\theta(\mathscr{S})$ by

$$g^{ab} = \theta_* h^{ab} - u^a u^b, \tag{7.12}$$

where u^a is any vector field on $\theta(\mathscr{S})$ which is nowhere zero or tangent to $\theta(\mathscr{S})$. Defining g_{ab} by (7.1), one has:

$$h_{ab} = \theta^* g_{ab}, \quad n_a g^{ab} = -n_a u^a u^b, \quad g_{ab} u^a u^b = -1. \tag{7.13}$$

Thus h_{ab} is the metric induced on $\mathscr{S}$ by $\mathbf{g}$ and u^a is the unit vector orthogonal to $\theta(\mathscr{S})$ in the metric $\mathbf{g}$.

The situation with the first derivatives $g^{ab}{}_{|c} u^c$ is similar: $n_a g^{ab}{}_{|c} u^c$ can be given any value by suitable diffeomorphisms. However there is now an additional complication in that $g^{ab}{}_{|c}$ depends not only on $\mathbf{g}$ but also on the background metric $\hat{\mathbf{g}}$ on $\mathcal{M}$. In order to give a description of the significant part of the first derivative of $\mathbf{g}$ in terms only of tensor fields defined on $\mathcal{S}$, we proceed as follows. We prescribe a symmetric contravariant tensor field χ^{ab} on $\mathcal{S}$. Under the imbedding χ^{ab} is mapped into a tensor field $\theta_* \chi^{ab}$ on $\theta(\mathcal{S})$. We require that this is equal to the second fundamental form (see §2.7) of the submanifold $\theta(\mathcal{S})$ in the metric $\mathbf{g}$. This gives

$$\theta_* \chi^{ab} = \theta_* h^{ac} \theta_* h^{bd} (u^e g_{ec})_{;d}$$
$$= \theta_* h^{ac} \theta_* h^{bd} ((u^e g_{ec})_{|d} - \delta \Gamma^f_{cd} u^e g_{ef}). \tag{7.14}$$

Using (7.3), one has

$$\theta_* \chi^{ab} = \tfrac{1}{2} \theta_* h^{ac} \theta_* h^{bd} (- g_{ci} g_{dj} g^{ij}{}_{|k} u^k + g_{bi} u^i{}_{|c} + g_{ci} u^i{}_{|b}). \tag{7.15}$$

This may be inverted to give $g^{ab}{}_{|c} u^c$ in terms of $\theta_* \chi^{ab}$:

$$\tfrac{1}{2} g^{ab}{}_{|c} u^c = - \theta_* \chi^{ab} + \theta_* h^{ac} \theta_* h^{bd} g_{i(c} u^i{}_{|d)} + u^{(a} W^{b)}, \tag{7.16}$$

where W^b is some vector field on $\theta(\mathcal{S})$. It can be given any required value by a suitable diffeomorphism μ.

The tensor fields h^{ab} and χ^{ab} cannot be prescribed completely independently on $\mathcal{S}$. For multiplying the Einstein equations (7.7) by n_a, one obtains four equations which do not contain $g^{ab}{}_{|cd} u^c u^d$, the second derivatives of $\mathbf{g}$ out of $\mathcal{S}$. Thus there must be four relations between g^{ab}, $g^{ab}{}_{|c} u^c$ and $n_a T^{ab}$. Using (2.36) and (2.35), they can be expressed as equations in the three-manifold $\mathcal{S}$:

$$\chi^{cd}{}_{||d} h_{ce} - \chi^{cd}{}_{||e} h_{cd} = 8\pi \theta^* (T_{de} u^d), \tag{7.17}$$

$$\tfrac{1}{2} (R' + (\chi^{dc} h_{dc})^2 - \chi^{ab} \chi^{cd} h_{ac} h_{bd}) = 8\pi \theta^* (T_{de} u^d u^e), \tag{7.18}$$

where a double stroke $\|$ denotes covariant differentiation in $\mathcal{S}$ with respect to the metric $\mathbf{h}$, and R' is the curvature scalar of $\mathbf{h}$.

The data $\boldsymbol{\omega}$ on $\mathcal{S}$ that is required to determine the solution therefore consists of the initial data for the matter fields (in the case of a scalar field ϕ for example, this would consist of two functions on $\mathcal{S}$ representing the value of ϕ and its normal derivative) and two tensor fields h^{ab} and χ^{ab} on $\mathcal{S}$ which obey the *constraint equations* (7.17–18). These contraint equations are elliptic equations on the surface $\mathcal{S}$ which impose four constraints on the twelve independent components of (h^{ab}, χ^{ab}). In such situations, one can show one can prescribe eight of these components independently and then solve the constraint equations to find the other four, see e.g. Bruhat (1962). We shall call a pair $(\mathcal{S}, \boldsymbol{\omega})$ satisfying these conditions, an *initial data set*. We then imbed $\mathcal{S}$ in some suitable four-manifold $\mathcal{M}$ with metric $\mathbf{g}$ and define g^{ab} on $\theta(\mathcal{S})$ by (7.12) for some suitable choice of u^a. We shall take u^a to be $g^{ab}n_b$. Thus it will be the unit vector orthogonal to $\theta(\mathcal{S})$ in both the metric $\mathbf{g}$ and $\hat{\mathbf{g}}$. We shall also exploit our freedom of choice of W^a in the definition of $g^{ab}{}_{|c}u^c$ by (7.16) to make ψ^b zero on $\theta(\mathcal{S})$. This requires

$$W^b = -g^{bc}{}_{|d}g_{ce}\theta_*h^{ed} + \tfrac{1}{2}g_{cd}g^{cd}{}_{|e}\theta_*h^{eb}$$
$$+ u^b(g_{cd}\theta_*\chi^{cd} - g_{ic}u^i{}_{|d}\theta_*h^{cd}). \quad (7.19)$$

(Note that all the derivatives in (7.19) are tangent to $\theta(\mathcal{S})$ as is required by the fact that the fields involved have been defined only on $\theta(\mathcal{S})$.) To ensure that ψ^b vanishes everywhere one also needs $\psi^b{}_{|c}u^c$ to be zero on $\theta(\mathcal{S})$. However this now follows from the constraint equations providing the reduced Einstein equations (7.8) hold on $\theta(\mathcal{S})$. One may therefore proceed to solve (7.8) as a second order non-linear hyperbolic system on the manifold $\mathcal{M}$ with metric $\hat{\mathbf{g}}$.

(Note that there are 10 such equations for the ϕ's; in proving the existence of solutions of these 10 equations we do not split them into a set of constraint equations and a set of evolution equations, and so the question as to whether the constraint equations are conserved does not arise.)

7.4 Second order hyperbolic equations

In this section we shall reproduce some results on second order hyperbolic equations given in Dionne (1962). They will be generalized to apply to a whole manifold, not just one coordinate neighbourhood. These results will be used in the following sections to prove the existence and uniqueness of developments for an initial data set $(\mathcal{S}, \boldsymbol{\omega})$.

We first introduce a number of definitions. We use Latin letters to denote multiple contravariant or covariant indices; thus a tensor of type (r, s) will be written as $K^I{}_J$, and we denote by $|I| = r$ the number of indices that the multiple index I represents. We introduce a positive definite metric e_{ab} on $\mathcal{M}$ and define

$$e_{IJ} = \underbrace{e_{ab}e_{cd} \ldots e_{pq}}_{r \text{ times}}, \quad e^{IJ} = \underbrace{e^{ab}e^{cd} \ldots e^{pq}}_{r \text{ times}},$$

where $|I| = |J| = r$. We then define the magnitude $|K^I{}_J|$ (or simply, $|\mathbf{K}|$) as $(K^I{}_J K^L{}_M e_{IL} e^{JM})^{\frac{1}{2}}$ where repeated multiple indices imply contraction over all the indices they represent. We define $| D^m K^I{}_J |$ (or simply, $|D^m\mathbf{K}|$) to be $|K^I{}_{J|L}|$ where $|L| = m$ and as before, $|$ indicates covariant differentiation with respect to $\mathbf{\hat{g}}$.

Let $\mathcal{N}$ be an imbedded submanifold of $\mathcal{M}$ with compact closure in $\mathcal{M}$. Then $\|K^I{}_J, \mathcal{N}\|_m$ is defined to be

$$\left\{ \sum_{p=0}^{m} \int_{\mathcal{N}} (|D^p K^I{}_J|)^2 \, d\sigma \right\}^{\frac{1}{2}},$$

where $d\sigma$ is the volume element on $\mathcal{N}$ induced by $\mathbf{e}$. We also define $\|\mathbf{K}, \widetilde{\mathcal{N}}\|_m$ to be the same expression where the derivatives are taken only in directions tangent to $\mathcal{N}$. Clearly, $\|\mathbf{K}, \mathcal{N}\|_m \geqslant \|\mathbf{K}, \widetilde{\mathcal{N}}\|_m$.

The *Sobolev spaces* $W^m(r, s, \mathcal{N})$ (or simply $W^m(\mathcal{N})$) are then defined to be the vector spaces of tensor fields $K^I{}_J$ of type (r, s) whose values and derivatives (in the sense of distributions) are defined almost everywhere on $\mathcal{N}$ (i.e. except, possibly, on a set of measure zero; for the rest of this section 'almost everywhere' is to be understood almost everywhere) and for which $\|K^I{}_J, \mathcal{N}\|_m$ is finite. With the norms $\| \quad , \mathcal{N}\|_m$ the Sobolev spaces are Banach spaces in which the C^m tensor fields of type (r, s) form dense subsets. If $\mathbf{e}'$ is another continuous positive definite metric on $\mathcal{M}$ then there will be positive constants C_1 and C_2 such that

$$C_1 |K^I{}_J| \leqslant |K^I{}_J|' \leqslant C_2 |K^I{}_J| \quad \text{on} \quad \mathcal{N},$$

and $C_1 \|K^I{}_J, \mathcal{N}\|_m \leqslant \|K^I{}_J, \mathcal{N}\|_m' \leqslant C_2 \|K^I{}_J, \mathcal{N}\|_m.$

Thus $\| \quad , \mathcal{N}\|_m'$ will be an equivalent norm. Similarly another C^m background metric $\mathbf{\hat{g}}'$ will give an equivalent norm. In fact it follows from two lemmas given below that if $\mathbf{\hat{g}}'' \in W^m(\mathcal{N})$ and $2m$ is greater than the dimension of $\mathcal{N}$, then the norm obtained using the covariant derivatives defined by $\mathbf{\hat{g}}''$ is again equivalent.

We now quote three fundamental results on Sobolev spaces. The proofs can be derived from results given in Sobolev (1963). They require a mild restriction on the shape of $\mathcal{N}$. A sufficient condition will be that for each point p of the boundary $\partial\mathcal{N}$ it should be possible to imbed an n-dimensional half cone in $\mathcal{N}$ with vertex at p, where n is the dimension of $\mathcal{N}$. In particular this condition will be satisfied if the boundary $\partial\mathcal{N}$ is smooth.

Lemma 7.4.1

There is a positive constant P_1 (depending on $\mathcal{N}$, **e** and $\hat{\mathbf{g}}$) such that for any field $K^I_J \in W^m(\mathcal{N})$ with $2m > n$, where n is the dimension of $\mathcal{N}$, $|\mathbf{K}| \leqslant P_1 \|\mathbf{K}, \widetilde{\mathcal{N}}\|_m$ on $\mathcal{N}$.

From this and the fact that the vector space of all continuous fields K^I_J on $\mathcal{N}$ is a Banach space with norm $\sup_{\mathcal{N}} |\mathbf{K}|$, it follows that if $K^I_J \in W^m(\mathcal{N})$ where $2m > n$, then K^I_J is continuous on $\mathcal{N}$. Similarly if $K^I_J \in W^{m+p}(\mathcal{N})$, then K^I_J is C^p on $\mathcal{N}$.

Lemma 7.4.2

There is a positive constant P_2 (depending on $\mathcal{N}$, **e** and $\hat{\mathbf{g}}$) such that for any fields K^I_J, $L^P_Q \in W^m(\mathcal{N})$ with $4m \geqslant n$,

$$\|K^I_J L^P_Q, \mathcal{N}\|_0 \leqslant P_2 \|\mathbf{K}, \mathcal{N}\|_m \|\mathbf{L}, \mathcal{N})\|_m.$$

From this and the previous lemma it follows that if $n \leqslant 4$ and $2m > n$, then for any two fields K^I_J, $L^P_Q \in W^m(\mathcal{N})$, the product $K^I_J L^P_Q$ is also in $W^m(\mathcal{N})$.

Lemma 7.4.3

If $\mathcal{N}'$ is an $(n-1)$-dimensional submanifold smoothly imbedded in $\mathcal{N}$, there is a positive constant P_3 (depending on $\mathcal{N}$, $\mathcal{N}'$, **e** and $\hat{\mathbf{g}}$) such that for any field $K^I_J \in W^{m+1}(\mathcal{N})$,

$$\|\mathbf{K}, \mathcal{N}'\|_m \leqslant P_3 \|\mathbf{K}, \mathcal{N}\|_{m+1}.$$

We shall prove the existence and uniqueness of developments for $(\mathcal{S}, \boldsymbol{\omega})$ when $h^{ab} \in W^{4+a}(\mathcal{S})$ and $\chi^{ab} \in W^{3+a}(\mathcal{S})$ where a is any non-negative integer. (If $\mathcal{S}$ is non-compact, we mean by $h^{ab} \in W^m(\mathcal{S})$ that $h^{ab} \in W^m(\mathcal{N})$ for any open subset $\mathcal{N}$ of $\mathcal{S}$ with compact closure.) A sufficient condition for this is that h^{ab} be C^{4+a} and χ^{ab} be C^{3+a} on $\mathcal{S}$; by lemma 7.4.1, a necessary condition is that h^{ab} be C^{2+a} and χ^{ab} be C^{1+a}. The solution obtained for g^{ab} will belong to $W^{4+a}(\mathcal{H})$ for each smooth spacelike surface $\mathcal{H}$ and so the $(2+a)$th derivatives will be bounded, i.e. g^{ab} will be $C^{(2+a)-}$ on $\mathcal{M}$.

These differentiability conditions can be weakened to cases such as shock waves where the solution departs from W^4 behaviour on well-behaved hypersurfaces; see Choquet–Bruhat (1968), Papapetrou and Hamoui (1967), Israel (1966), and Penrose (1972a). However no proof is known for cases in which such departures occur generally. The W^4 condition for the existence and uniqueness of developments is an improvement on previous work (Choquet-Bruhat (1968)) but it is somewhat stronger than one would like since the Einstein equations can be defined in a distributional sense if the metric is continuous and its generalized derivatives are locally square integrable (i.e. if $\mathbf{g}$ is C^0 and W^1). On the other hand any W^p conditions for p less than 4 would

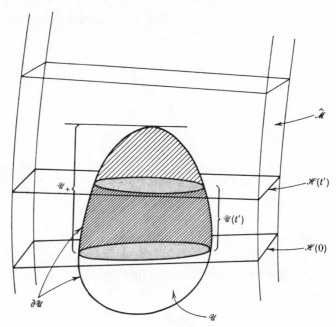

FIGURE 48. $\mathcal{U}$ is an open set with compact closure in the manifold $\hat{\mathcal{M}} = \mathcal{H} \times R^1$. $\mathcal{U}_+$ is the region of $\mathcal{U}$ for which $t \geqslant 0$ and $\mathcal{U}(t')$ is the region of $\mathcal{U}$ between $t = 0$ and $t = t' > 0$.

not guarantee the uniqueness of geodesics, or, for p less than 3, their existence. Our own view is that these differences of differentiability conditions are not important since as explained in § 3.1, the model for space–time may as well be taken to be C^∞.

In order to prove the existence and uniqueness of developments we now establish some fundamental inequalities (lemmas 7.4.4 and 7.4.6) for second order hyperbolic equations, in a manner similar to that of the conservation theorem in § 4.3.

Consider a manifold $\hat{\mathcal{M}}$ of the form $\mathcal{H} \times R^1$ where $\mathcal{H}$ is a three-dimensional manifold. Let $\mathcal{U}$ be an open set of $\hat{\mathcal{M}}$ with compact closure which has boundary $\partial\mathscr{U}$ and which intersects $\mathscr{H}(0)$, where $\mathscr{H}(t)$ denotes the surface $\mathscr{H} \times \{t\}$, $t \in R^1$. Let $\mathscr{U}_+$ and $\mathscr{U}(t')$ denote the parts of $\mathscr{U}$ for which $t \geqslant 0$ and $t' \geqslant t \geqslant 0$ respectively (figure 48). On $\mathscr{U}_+$ let $\hat{\mathbf{g}}$ be a C^{2-} background metric and let $\mathbf{e}$ be a C^{1-} positive definite metric. We shall consider tensor fields $K^I{}_J$ which obey second order hyperbolic equations of the form

$$L(K) \equiv A^{ab}K^I{}_{J|ab} + B^{aPI}{}_{QJ}K^Q{}_{P|a} + C^{PI}{}_{QJ}K^Q{}_P = F^I{}_J, \quad (7.20)$$

where $\mathbf{A}$ is a Lorentz metric on $\mathscr{U}_+$ (i.e. a symmetric tensor field of signature $+2$), $\mathbf{B}$, $\mathbf{C}$ and $\mathbf{F}$ are tensor fields of type indicated by their indices, and $|$ denotes covariant differentiation with respect to the metric $\hat{\mathbf{g}}$.

Lemma 7.4.4

If (1) $\partial\mathscr{U} \cap \overline{\mathscr{U}}_+$ is achronal with respect to $\mathbf{A}$,
  (2) there exists some $Q_1 > 0$ such that on $\overline{\mathscr{U}}_+$

$$A^{ab}t_{|a}t_{|b} \leqslant -Q_1$$

and $\qquad\qquad A^{ab}W_a W_b \geqslant Q_1 e^{ab}W_a W_b$

for any form $\mathbf{W}$ which satisfies $A^{ab}t_{|a}W_b = 0$,
  (3) there exists some Q_2 such that on $\overline{\mathscr{U}}_+$

$$|\mathbf{A}| \leqslant Q_2, \quad |\mathbf{DA}| \leqslant Q_2, \quad |\mathbf{B}| \leqslant Q_2, \quad |\mathbf{C}| \leqslant Q_2,$$

then there exists some positive constant P_4 (depending on $\mathscr{U}$, $\mathbf{e}$, $\hat{\mathbf{g}}$, Q_1 and Q_2) such that for all solutions $K^I{}_J$ of (7.20),

$$\|\mathbf{K}, \mathscr{H}(t) \cap \mathscr{U}_+\|_1 \leqslant P_4\{\|\mathbf{K}, \mathscr{H}(0) \cap \mathscr{U}_+\|_1 + \|\mathbf{F}, \mathscr{U}(t)\|_0\}.$$

One forms the 'energy tensor' S^{ab} for the field $K^I{}_J$ in analogy to the energy–momentum tensor of a scalar field of unit mass (§ 3.2):

$$S^{ab} = \{(A^{ac}A^{bd} - \tfrac{1}{2}A^{ab}A^{cd})K^I{}_{J|c}K^P{}_{Q|d} - \tfrac{1}{2}A^{ab}K^I{}_J K^P{}_Q\}e^{JQ}e_{IP}. \quad (7.21)$$

The tensor S^{ab} obeys the dominant energy condition (§4.3) with respect to the metric $\mathbf{A}$ (i.e. if W_a is timelike with respect to $\mathbf{A}$ then $S^{ab}W_a W_b \geqslant 0$ and $S^{ab}W_a$ is non-spacelike with respect to $\mathbf{A}$). Moreover by conditions (2) and (3) there will be positive constants Q_3 and Q_4 such that

$$Q_3(|\mathbf{K}|^2 + |\mathbf{DK}|^2) \leqslant S^{ab}t_{|a}t_{|b} \leqslant Q_4(|\mathbf{K}|^2 + |\mathbf{DK}|^2). \quad (7.22)$$

We now apply lemma 4.3.1 to S^{ab}, taking $\mathscr{U}_+$ as the compact region $\mathscr{F}$

and using the volume element $\mathrm{d}\hat{v}$ and covariant differentiation defined by the metric $\hat{\mathbf{g}}$:

$$\int_{\mathscr{H}(t)\,\cap\,\bar{\mathscr{U}}_+} S^{ab}t_{|a}\,\mathrm{d}\hat{\sigma}_b \leqslant \int_{\mathscr{H}(0)\,\cap\,\bar{\mathscr{U}}_+} S^{ab}t_{|a}\,\mathrm{d}\hat{\sigma}_b$$
$$+ \int_0^t \left\{ \int_{\mathscr{H}(t')\,\cap\,\bar{\mathscr{U}}_+} (PS^{ab}t_{|a} + S^{ab}{}_{|a})\,\mathrm{d}\hat{\sigma}_b \right\} \mathrm{d}t' \quad (7.23)$$

where P is a positive constant independent of S^{ab}. (The sign has been changed in the first term on the right-hand side since the surface element $\mathrm{d}\hat{\sigma}_b$ of the surface $\mathscr{H}(t)$ is taken to have the same orientation as $t_{|b}$, i.e. $\mathrm{d}\hat{\sigma}_b = t_{|b}\,\mathrm{d}\tilde{\sigma}$ where $\mathrm{d}\tilde{\sigma}$ is a positive definite measure on $\mathscr{H}(t)$.) Since $\mathbf{e}$ and $\hat{\mathbf{g}}$ are continuous there will be positive constants Q_5 and Q_6 such that on $\bar{\mathscr{U}}_+$

$$Q_5\,\mathrm{d}\sigma \leqslant \mathrm{d}\tilde{\sigma} \leqslant Q_6\,\mathrm{d}\sigma, \quad (7.24)$$

where $\mathrm{d}\sigma$ is the area element on $\mathscr{H}(t)$ induced by $\mathbf{e}$. Thus by (7.22) and (7.23) there is some Q_7 such that

$$\|\mathbf{K}, \mathscr{H}(t)\cap\mathscr{U}_+\|_1^2 \leqslant Q_7 \Big\{ \|\mathbf{K}, \mathscr{H}(0)\cap\mathscr{U}_+\|_1^2$$
$$+ \int_0^t \|\mathbf{K}, \mathscr{H}(t')\cap\mathscr{U}_+\|_1^2\,\mathrm{d}t' + \int_0^t (S^{ab}{}_{|b}t_{|a}\,\mathrm{d}\sigma)\,\mathrm{d}t' \Big\}. \quad (7.25)$$

By (7.20),

$$S^{ab}{}_{|b} = A^{ac}K^I{}_{J|c}F^P{}_Q\,e^{JQ}e_{IP} + \text{(terms quadratic in } K^I{}_J \text{ and}$$
$$K^P{}_{Q|c} \text{ with coefficients involving } A^{cd}, A^{cd}{}_{|e},$$
$$\hat{R}^c{}_{def}, B^{cPI}{}_{QJ} \text{ and } C^{PI}{}_{QJ}). \quad (7.26)$$

Since the coefficients are all bounded on $\mathscr{U}_+$, there is some Q_8 such that

$$S^{ab}{}_{|b}t_{|a} \leqslant Q_8\{|\mathbf{F}|^2 + |\mathbf{K}|^2 + |\mathbf{DK}|^2\}. \quad (7.27)$$

Thus there is some Q_9 such that, from (7.25) and (7.27),

$$\|\mathbf{K}, \mathscr{H}(t)\cap\mathscr{U}_+\|_1^2 \leqslant Q_9 \Big\{ \|\mathbf{K}, \mathscr{H}(0)\cap\mathscr{U}_+\|_1^2$$
$$+ \int_0^t \|\mathbf{K}, \mathscr{H}(t')\cap\mathscr{U}_+\|_1^2\,\mathrm{d}t' + \|\mathbf{F}, \mathscr{U}(t)\|_0^2 \Big\}.$$

This is of the form
$$\mathrm{d}x/\mathrm{d}t \leqslant Q_9\{x+y\}, \quad (7.28)$$

where
$$x(t) = \int_0^t \|\mathbf{K}, \mathscr{H}(t')\cap\mathscr{U}_+\|_1^2\,\mathrm{d}t'.$$

Therefore
$$x \leqslant e^{Q_9 t}\int_0^t e^{-Q_9 t'}y(t')\,\mathrm{d}t'. \quad (7.29)$$

Since y is a monotonically increasing function of t and since t is bounded on $\overline{\mathscr{U}}_+$, there is some Q_{10} such that

$$x \leqslant Q_{10} y.$$

Thus $\|\mathbf{K}, \mathscr{H}(t) \cap \mathscr{U}_+\|_1 \leqslant P_4\{\|\mathbf{K}, \mathscr{H}(0) \cap \mathscr{U}_+\|_1 + \|\mathbf{F}, \mathscr{U}(t)\|_0\}$, where

$$P_4 = (Q_9 + Q_{10})^{\frac{1}{2}}. \qquad \square$$

With this inequality one can immediately prove the uniqueness of solutions of second order hyperbolic equations which are linear, i.e. for which $\mathbf{A}$, $\mathbf{B}$, $\mathbf{C}$ and $\mathbf{F}$ do not depend on $\mathbf{K}$. For suppose $K^{1I}{}_J$ and $K^{2I}{}_J$ were solutions of the equation $L(\mathbf{K}) = \mathbf{F}$ which had the same initial values and first derivatives on $\mathscr{H}(0) \cap \mathscr{U}$. Then one can apply the above result to the equation $L(\mathbf{K}^1 - \mathbf{K}^2) = 0$ and obtain

$$\|\mathbf{K}^1 - \mathbf{K}^2, \mathscr{H}(t) \cap \mathscr{U}_+\|_1 = 0.$$

Therefore $\mathbf{K}^1 = \mathbf{K}^2$ on $\overline{\mathscr{U}}_+$. One has thus

Proposition 7.4.5

Let $\mathbf{A}$ be a C^{1-} Lorentz metric on $\hat{\mathscr{M}}$ and let $\mathbf{B}$, $\mathbf{C}$, and $\mathbf{F}$ be locally bounded. Let $\mathscr{H} \subset \hat{\mathscr{M}}$ be a three-surface which is spacelike and acausal with respect to $\mathbf{A}$. Then if $\mathscr{V}$ is a set in $D^+(\mathscr{H}, \mathbf{A})$, the solution on $\mathscr{V}$ of the linear equation (7.20) is uniquely determined by its values and the values of its first derivatives on $\mathscr{H} \cap J^-(\mathscr{V}, \mathbf{A})$.

By proposition 6.6.7, $D^+(\mathscr{H}, \mathbf{A})$ is of the form $\mathscr{H} \times R^1$. If $q \in \mathscr{V}$, then by proposition 6.6.6, $J^-(q) \cap J^+(\mathscr{H})$ is compact and so may be taken for $\overline{\mathscr{U}}_+$. $\qquad \square$

Thus a physical field obeying a linear equation of the form (7.20) will satisfy the causality postulate (a) of §3.2 provided the null cone of $\mathbf{A}$ coincides with or lies within the null cone of the space–time metric $\mathbf{g}$.

In order to prove the existence of solutions of the equations (7.20) we shall need inequalities for higher order derivatives of $\mathbf{K}$. We shall now take the background metric $\hat{\mathbf{g}}$ to be at least C^{5+a} where a is a non-negative integer and we shall take $\mathscr{U}$ to be such that $\mathscr{H}(0) \cap \overline{\mathscr{U}}$ has a smooth boundary and such that there is a diffeomorphism

$$\lambda : (\mathscr{H}(0) \cap \overline{\mathscr{U}}) \times [0, t_1] \to \overline{\mathscr{U}}_+$$

which has the property that for each $t \in [0, t_1]$,

$$\lambda\{(\mathscr{H}(0) \cap \overline{\mathscr{U}}), t\} = \mathscr{H}(t) \cap \overline{\mathscr{U}}_+.$$

We do this so that there shall be upper bounds $\tilde{P}_1$, $\tilde{P}_2$ and $\tilde{P}_3$ to the constants P_1, P_2 and P_3 in lemmas 7.4.1–7.4.3 for the surface $\mathscr{H}(t) \cap \mathscr{U}_+$.

Lemma 7.4.6

If conditions (1) and (2) of lemma 7.4.4 hold and if
  (4) there is some Q_3 such that

$$\|A, \mathcal{U}_+\|_{4+a} < Q_3, \quad \|B, \mathcal{U}_+\|_{3+a} < Q_3, \quad \|C, \mathcal{U}_+\|_{3+a} < Q_3$$

(by lemma 7.4.1, this implies condition (3)), then there exist positive constants $P_{5,\,a}$ (depending on $\mathcal{U}$, e, $\hat{g}$, a, Q_1 and Q_3) such that

$$\|K, \mathcal{H}(t) \cap \mathcal{U}_+\|_{4+a} \leqslant P_{5,\,a}\{\|K, \mathcal{H}(0) \cap \mathcal{U}_+\|_{4+a} + \|F, \mathcal{U}(t)\|_{3+a}\}. \quad (7.30)$$

From lemma 7.4.4 one has an inequality for $\|K, \mathcal{H}(t) \cap \mathcal{U}_+\|_1$. To obtain an inequality for $\|K, \mathcal{H}(t) \cap \mathcal{U}_+\|_2$ one forms the 'energy' tensor S^{ab} for the first derivatives $K^I_{J|c}$ and proceeds as before. The divergence $S^{ab}_{\ |b}$ can now be evaluated by differentiating equations (7.20):

$$S^{ab}_{\ |b} = A^{ad}K^I_{J|cd}F^P_{Q|e}e^{ec}e^{JQ}e_{IP} + \text{(terms quadratic in } K^I_J,$$

$$K^I_{J|c} \text{ and } K^I_{J|cd} \text{ with coefficients involving } A^{cd},$$

$$A^{cd}_{\ |e}, \hat{R}^c_{\ def}, \hat{R}^c_{\ def|g}, B^{cPI}_{QJ}, B^{cPI}_{QJ|d}, C^{PI}_{QJ}$$

$$\text{and } C^{PI}_{QJ|d}). \quad (7.31)$$

With the possible exceptions of $B^{cPI}_{QJ|d}$ and $C^{PI}_{QJ|d}$, these coefficients are all bounded on $\overline{\mathcal{U}}_+$ in the case $a = 0$. When integrated over the surface $\mathcal{H}(t') \cap \mathcal{U}_+$, the term in (7.31) involving $B^{cPI}_{QJ|d}$ is

$$-\int_{\mathcal{H}(t') \cap \mathcal{U}_+} A^{ab}K^I_{J|cb}B^{dPR}_{QS|e}K^S_{R|d}e^{ce}e^{QJ}e_{PI}\,d\hat{\sigma}_a. \quad (7.32)$$

There is some Q_4 such that for all t', (7.32) is less than or equal to

$$Q_4\int_{\mathcal{H}(t') \cap \mathcal{U}_+} |DB|\,|DK|\,|D^2K|\,d\sigma$$

$$\leqslant \tfrac{1}{2}Q_4\int_{\mathcal{H}(t') \cap \mathcal{U}_+} (|D^2K|^2 + |DB|^2\,|DK|^2)\,d\sigma. \quad (7.33)$$

By lemma 7.4.2,

$$\int_{\mathcal{H}(t') \cap \mathcal{U}_+} |DB|^2\,|DK|^2\,d\sigma \leqslant \tilde{P}_2^2\|B, \mathcal{H}(t') \cap \mathcal{U}_+\|_2^2\,\|K, \mathcal{H}(t') \cap \mathcal{U}_+\|_2^2,$$

where, by condition (4) and lemma 7.4.3, $\|B, \mathcal{H}(t') \cap \mathcal{U}_+\|_2 < \tilde{P}_3Q_3$. The term involving $C^{PI}_{QJ|d}$ can be bounded similarly. Thus by lemma 4.3.1 there is some constant Q_5 such that

$$\int_{\mathcal{H}(t) \cap \mathcal{U}_+} (|D^2K| + |DK|^2)\,d\sigma \leqslant Q_5\bigg\{\int_{\mathcal{H}(0) \cap \mathcal{U}_+} (|D^2K|^2 + |DK|^2)\,d\sigma$$

$$+ \int_0^t \|K, \mathcal{H}(t') \cap \mathcal{U}_+\|_2^2\,dt' + \int_{\mathcal{U}(t)} |DF|^2\,d\sigma\bigg\}. \quad (7.34)$$

By lemma 7.4.4,

$$\int_{\mathscr{H}(t)\cap\mathscr{U}_+} |\mathbf{K}|^2 \, d\sigma \leqslant \|\mathbf{K}, \mathscr{H}(t)\cap\mathscr{U}_+\|_1^2$$
$$\leqslant 2P_4^2\{\|\mathbf{K}, \mathscr{H}(0)\cap\mathscr{U}\|_1^2 + \|\mathbf{F}, \mathscr{U}(t)\|_0^2\}. \quad (7.35)$$

Adding this to (7.34), one obtains

$$\|\mathbf{K}, \mathscr{H}(t)\cap\mathscr{U}_+\|_2^2 \leqslant Q_6\Big\{\|\mathbf{K}, \mathscr{H}(0)\cap\mathscr{U}\|_2^2$$
$$+ \int_0^t \|\mathbf{K}, \mathscr{H}(t')\cap\mathscr{U}_+\|_2^2 \, dt' + \|\mathbf{F}, \mathscr{U}(t)\|_1^2\Big\}, \quad (7.36)$$

where $Q_6 = Q_5 + 2P_4$. By a similar argument to that in lemma 7.4.4, there is some constant Q_7 such that

$$\|\mathbf{K}, \mathscr{H}(t)\cap\mathscr{U}_+\|_2 \leqslant Q_7\{\|\mathbf{K}, \mathscr{H}(0)\cap\mathscr{U}\|_2 + \|\mathbf{F}, \mathscr{U}(t)\|_1\}. \quad (7.37)$$

From lemma 7.4.1 it now follows that on $\mathscr{U}_+$,

$$|\mathbf{K}| \leqslant \tilde{P}_1 Q_7\{\|\mathbf{K}, \mathscr{H}(0)\cap\mathscr{U}\|_2 + \|\mathbf{F}, \mathscr{U}(t)\|_0\}. \quad (7.38)$$

Using this one may proceed in a similar way to establish an inequality for $\|\mathbf{K}, \mathscr{H}(t)\cap\mathscr{U}_+\|_3$. The divergence of the 'energy' tensor now gives a term of the form

$$Q_8 \int_{\mathscr{H}(t')\cap\mathscr{U}_+} (|\mathrm{D}^3\mathbf{K}|^2 + |\mathrm{D}^2\mathbf{B}|^2 |\mathrm{D}\mathbf{K}|^2) \, d\sigma. \quad (7.39)$$

By lemma 7.4.2 the second term above is bounded by

$$Q_8 \tilde{P}_2^2 \|\mathbf{B}, \mathscr{H}(t')\cap\mathscr{U}_+\|_3^2 \|\mathbf{K}, \mathscr{H}(t')\cap\mathscr{U}_+\|_2^2,$$

where by condition (4), $\|\mathbf{B}, \mathscr{H}(t)\cap\mathscr{U}_+\|_3$ is defined for almost all values of t' and is square integrable with respect to t'. Thus one can obtain an inequality for $\|\mathbf{K}, \mathscr{H}(t)\cap\mathscr{U}_+\|_3$ in the same manner as for $\|\mathbf{K}, \mathscr{H}(t)\cap\mathscr{U}_+\|_2$. The procedure for higher order derivatives is similar. $\qquad\square$

Corollary

There exist constants $P_{6,a}$ and $P_{7,a}$ such that

$$\|\mathbf{K}, \mathscr{H}(t)\cap\mathscr{U}_+\|_{4+a} \leqslant P_{6,a}\{\|\mathbf{K}, \mathscr{H}(0)\cap\widetilde{\mathscr{U}}\|_{4+a}$$
$$+ \|K^I{}_{J|a}u^a, \mathscr{H}(0)\cap\widetilde{\mathscr{U}}\|_{3+a} + \|\mathbf{F}, \mathscr{U}_+\|_{3+a}\},$$

and
$$\|\mathbf{K}, \mathscr{U}_+\|_{4+a} \leqslant P_{7\ a}\{\text{ditto}\},$$

where u^a is some C^{3+a} vector field on $\mathscr{H}(0)$ which is nowhere tangent to $\mathscr{H}(0)$.

By (7.20), the second and higher derivatives of $\mathbf{K}$ out of the surface $\mathscr{H}(0)$ may be expressed in terms of $\mathbf{F}$ and its derivatives out of $\mathscr{H}(0)$, $K^I{}_{J|a}u^a$ and derivatives of $\mathbf{K}$ in the surface $\mathscr{H}(0)$. By lemma 7.4.3,

$$
\left.
\begin{aligned}
\|\mathbf{A}, \mathscr{H}(0) \cap \mathscr{U}\|_{3+a} &< \tilde{P}_3 Q_3, \\
\|\mathbf{B}, \mathscr{H}(0) \cap \mathscr{U}\|_{2+a} &< \tilde{P}_3 Q_3, \\
\|\mathbf{C}, \mathscr{H}(0) \cap \mathscr{U}\|_{2+a} &< \tilde{P}_3 Q_3, \\
\|\mathbf{F}, \mathscr{H}(0) \cap \mathscr{U}\|_{2+a} &< \tilde{P}_3 \|\mathbf{F}, \mathscr{U}_+\|_{3+a}.
\end{aligned}
\right\} \tag{7.40}
$$

Thus there will be some constant Q_4 such that

$$
\|\mathbf{K}, \mathscr{H}(0) \cap \mathscr{U}\|_{4+a} \leqslant Q_4 \{ \|\mathbf{K}, \mathscr{H}(0) \cap \widetilde{\mathscr{U}}\|_{4+a}
$$
$$
+ \|K^I{}_{J|a}u^a, \mathscr{H}(0) \cap \widetilde{\mathscr{U}}\|_{3+a} + \|\mathbf{F}, \widetilde{\mathscr{U}}_+\|_{3+a} \}. \tag{7.41}
$$

The second result follows immediately, since t is bounded on $\mathscr{U}_+$. $\square$

We can now proceed to prove the existence of solutions of linear equations of the form (7.20). We first suppose that the components of $\mathbf{A}$, $\mathbf{B}$, $\mathbf{C}$, $\mathbf{F}$, $\mathbf{u}$ and $\hat{\mathbf{g}}$ are analytic functions of the local coordinates x^1, x^2, x^3 and x^4 ($x^4 = t$) on a coordinate neighbourhood $\mathscr{V}$ and take the initial data $K^I{}_J = {}_0 K^I{}_J$ and $K^I{}_{J|a}u^a = {}_1 K^I{}_J$ to be analytic functions of the coordinates x^1, x^2 and x^3 on $\mathscr{H}(0) \cap \mathscr{V}$. Then from (7.20) one can calculate the partial derivatives $\partial^2(K^I{}_J)/\partial t^2$, $\partial^3(K^I{}_J)/\partial t^2\, \partial x^i$, $\partial^3(K^I{}_J)/\partial t^3$, etc. of the components of $\mathbf{K}$ out of the surface $\mathscr{H}(0)$ in terms of derivatives of ${}_0\mathbf{K}$ and ${}_1\mathbf{K}$ in $\mathscr{H}(0)$. One can then express $K^I{}_J$ as a formal power series in x^1, x^2, x^3 and t about the origin of coordinates p. By the Cauchy–Kowaleski theorem (Courant and Hilbert (1962), p. 39) this series will converge in some ball $\mathscr{V}(r)$ of coordinate radius r to give a solution of (7.20) with the given initial conditions. One now selects an analytic atlas from the C^∞ atlas of $\mathscr{M}$, covers $\mathscr{H}(0) \cap \widetilde{\mathscr{U}}$ with co-ordinate neighbourhoods of the form $\mathscr{V}(r)$ from this atlas, and in each coordinate neighbourhood constructs a solution as above. One thus obtains a solution on a region $\mathscr{U}(t_2)$ for some $t_2 > 0$. One then repeats the process using $\mathscr{H}(t_2)$. By the Cauchy–Kowaleski theorem, the ratio of successive intervals of t for which the power series converges is independent of the initial data and so the solution can be extended to the whole of $\mathscr{U}_+$ in a finite number of steps. This proves the existence of solutions of linear equations of the form (7.20) when the coefficients, the source term and the initial data are all analytic. We shall now remove the requirement of analyticity.

Proposition 7.4.7

If conditions (1), (2) and (4) hold and if

(5) $\mathbf{F} \in W^{3+a}(\mathcal{U}_+)$,

(6) $_0\mathbf{K} \in W^{4+a}(\mathcal{H}(0) \cap \overline{\mathcal{U}})$, $\quad _1\mathbf{K} \in W^{3+a}(\mathcal{H}(0) \cap \overline{\mathcal{U}})$,

then there exists a unique solution $\mathbf{K} \in W^{4+a}(\mathcal{U}_+)$ of the linear equation (7.20) such that on $\mathcal{H}(0)$, $K^I{}_J = {}_0K^I{}_J$ and $K^I{}_{J|a} u^a = {}_1K^I{}_J$.

We prove this result by approximating the coefficients and initial data by analytic fields and showing that the analytic solutions obtained converge to a field which is a solution of the given equations with the given initial conditions. Let $\mathbf{A}_n$ ($n = 1, 2, 3, \ldots$) be a sequence of analytic fields on $\overline{\mathcal{U}}_+$ which converge strongly to $\mathbf{A}$ in $W^{4+a}(\mathcal{U}_+)$. ($\mathbf{A}_n$ is said to converge strongly to $\mathbf{A}$ in W^m if $\|\mathbf{A}_n - \mathbf{A}\|_m$ converges to zero.) Let $\mathbf{B}_n$, $\mathbf{C}_n$ and $\mathbf{F}_n$ be analytic fields on $\overline{\mathcal{U}}_+$ which converge strongly to $\mathbf{B}$, $\mathbf{C}$ and $\mathbf{F}$ respectively in $W^{3+a}(\mathcal{U}_+)$, and let $_0\mathbf{K}_n$ and $_1\mathbf{K}_n$ be analytic fields on $\mathcal{H}(0) \cap \overline{\mathcal{U}}$ which converge strongly to $_0\mathbf{K}$ and $_1\mathbf{K}$ in $W^{4+a}(\mathcal{H}(0) \cap \mathcal{U})$ and $W^{3+a}(\mathcal{H}(0) \cap \mathcal{U})$ respectively. For each value of n there will be an analytic solution $\mathbf{K}_n$ to (7.20) with the initial values $K_n{}^I{}_J = {}_0K_n{}^I{}_J$, $K_n{}^I{}_{J|a} u^a = {}_1K_n{}^I{}_J$. By the corollary to lemma 7.4.6, $\|\mathbf{K}_n, \mathcal{U}_+\|_{4+a}$ will be bounded as $n \to \infty$. Therefore by a theorem of Riesz (1955) there will be a field $\mathbf{K} \in W^{4+a}(\mathcal{U}_+)$ and a subsequence $\mathbf{K}_{n'}$ of the $\mathbf{K}_n$ such that for each b, $0 \leqslant b \leqslant 4 + a$, $D^b\mathbf{K}_{n'}$ converges weakly to $D^b\mathbf{K}$. (A sequence of fields $I_n{}^I{}_J$ on $\mathcal{N}$ is said to converge weakly to $I^I{}_J$ if for each C^∞ field $J^I{}_J$,

$$\int_{\mathcal{N}} I_n{}^I{}_J J^J{}_I \, d\sigma \to \int_{\mathcal{N}} I^I{}_J J^J{}_I \, d\sigma.\Big)$$

Since $\mathbf{A}_n$, $\mathbf{B}_n$ and $\mathbf{C}_n$ converge strongly to $\mathbf{A}$, $\mathbf{B}$ and $\mathbf{C}$ in $W^3(\mathcal{U}_+)$, $\sup|\mathbf{A} - \mathbf{A}_n|$, $\sup|\mathbf{B} - \mathbf{B}_n|$ and $\sup|\mathbf{C} - \mathbf{C}_n|$ will converge to zero. Thus $L_{n'}(\mathbf{K}_{n'})$ will converge weakly to $L(\mathbf{K})$. But $L_{n'}(\mathbf{K}_{n'})$ is equal to $\mathbf{F}_{n'}$ which converges strongly to $\mathbf{F}$. Therefore $L(\mathbf{K}) = \mathbf{F}$. On $\mathcal{H}(0) \cap \overline{\mathcal{U}}$ $K_{n'}{}^I{}_J$ and $K_{n'}{}^I{}_{J|a} u^a$ will converge weakly to $K^I{}_J$ and $K^I{}_{J|a} u^a$ which must therefore be equal to $_0K^I{}_J$ and $_1K^I{}_J$ respectively. Thus $\mathbf{K}$ is a solution of the given equation with the given initial conditions. By proposition 7.4.5 it is unique. Since each $\mathbf{K}_n$ satisfies the inequality in lemma 7.4.6, $\mathbf{K}$ will satisfy it also. $\qquad\square$

7.5 The existence and uniqueness of developments for the empty space Einstein equations

We shall now apply the results of the previous section to the Cauchy problem in General Relativity. We shall first deal with the Einstein equations for empty space ($T^{ab} = 0$), and shall discuss the effect of matter in §7.7.

The reduced Einstein equations

$$E^{ab}{}_{cd}(\phi^{cd}) = 8\pi T^{ab} - (\hat{R}^{ab} - \tfrac{1}{2}\hat{R}\hat{g}^{ab}) \qquad (7.42)$$

are *quasi-linear* second order hyperbolic equations. That is, they have the form (7.20) where the coefficients **A**, **B** and **C** are functions of **K** and **DK** (actually, in this case $A^{ab} = g^{ab}$ is a function of ϕ^{ab} and not of $\phi^{ab}{}_{|c}$). To prove the existence of solutions of these equations we proceed as follows. We take some suitable trial field ϕ'^{ab} and use this to determine the values of the coefficients **A**, **B** and **C** in the operator E. Using these values we then solve (7.42) as a *linear* equation with the prescribed initial data and obtain a new field ϕ''^{ab}. We thus have a map α which takes $\boldsymbol{\phi}'$ into $\boldsymbol{\phi}''$, and we show that under suitable conditions this map has a fixed point (i.e. there is some ϕ such that $\alpha(\boldsymbol{\phi}) = \boldsymbol{\phi}$). This fixed point will be the desired solution of the quasi-linear equation.

We shall take the background metric $\hat{\mathbf{g}}$ to be a solution of the empty space Einstein equations and choose the surfaces $\mathscr{H}(t) \cap \overline{\mathscr{U}}_+$ and $\partial \mathscr{U} \cap \overline{\mathscr{U}}_+$ to be spacelike in $\hat{\mathbf{g}}$. Then by lemma 7.4.1 there will be some positive constants $\tilde{Q}_a$ such that if for some value of $a \geqslant 0$

$$\|\boldsymbol{\phi}', \mathscr{U}_+\|_{4+a} < \tilde{Q}_a, \qquad (7.43)$$

then the coefficients **A**′, **B**′ and **C**′ determined by $\boldsymbol{\phi}'$ satisfy conditions (1), (2) and (4) of lemma 7.4.6 for given values of Q_1 and Q_3. From (7.41) one then has

$$\|\boldsymbol{\phi}'', \mathscr{U}_+\|_{4+a} \leqslant P_{7,\,a}\{\|_0\boldsymbol{\phi}, \mathscr{H}(0) \cap \overline{\mathscr{U}}\|_{4+a} + \|_1\boldsymbol{\phi}, \mathscr{H}(0) \cap \overline{\mathscr{U}}\|_{3+a}\}.$$

Thus the map $\alpha\colon W^{4+a}(\mathscr{U}_+) \to W^{4+a}(\mathscr{U}_+)$ will take the closed ball $W(r)$ of radius r ($r < \tilde{Q}_a$) in $W^{4+a}(\mathscr{U}_+)$ into itself provided that

$$\|_0\boldsymbol{\phi}, \mathscr{H}(0) \cap \overline{\mathscr{U}}\|_{4+a} \leqslant \tfrac{1}{2} r P_{7,\,a}{}^{-1}$$

and $\qquad\qquad \|_1\boldsymbol{\phi}, \mathscr{H}(0) \cap \overline{\mathscr{U}}\|_{3+a} \leqslant \tfrac{1}{2} r P_{7,\,a}{}^{-1}. \qquad (7.44)$

We shall show that α has a fixed point if (7.44) holds and if r is sufficiently small.

Suppose $\boldsymbol{\phi_1}'$ and $\boldsymbol{\phi_2}'$ are in $W(r)$. The fields $\boldsymbol{\phi_1}'' = \alpha(\boldsymbol{\phi_1}')$ and $\boldsymbol{\phi_2}'' = \alpha(\boldsymbol{\phi_2}')$ satisfy $E_1'(\boldsymbol{\phi_1}'') = 0$, $E_2'(\boldsymbol{\phi_2}'') = 0$ where E_1' is the Einstein operator with coefficients $\mathbf{A_1}'$, $\mathbf{B_1}'$ and $\mathbf{C_1}'$ determined by $\boldsymbol{\phi_1}'$. Thus

$$E_1'(\boldsymbol{\phi_1}'' - \boldsymbol{\phi_2}'') = -(E_1' - E_2')(\boldsymbol{\phi_2}''). \tag{7.45}$$

Since the coefficients $\mathbf{A_1}'$, $\mathbf{B_1}'$ and $\mathbf{C_1}'$ depend differentiably on $\boldsymbol{\phi_1}'$ and $\mathrm{D}\boldsymbol{\phi_1}'$ for $\boldsymbol{\phi_1}'$ in $W(r)$, there will be some constant Q_4 such that on $\overline{\mathscr{U}}_+$

$$\left.\begin{aligned}
|\mathbf{A}'_1 - \mathbf{A}'_2| &\leqslant Q_4 |\boldsymbol{\phi}'_1 - \boldsymbol{\phi}'_2|, \\
|\mathbf{B}'_1 - \mathbf{B}'_2| &\leqslant Q_4(|\boldsymbol{\phi}'_1 - \boldsymbol{\phi}'_2| + |\mathrm{D}\boldsymbol{\phi}'_1 - \mathrm{D}\boldsymbol{\phi}'_2|), \\
|\mathbf{C}'_1 - \mathbf{C}'_2| &\leqslant Q_4(|\boldsymbol{\phi}'_1 - \boldsymbol{\phi}'_2| + |\mathrm{D}\boldsymbol{\phi}'_1 - \mathrm{D}\boldsymbol{\phi}'_2|).
\end{aligned}\right\} \tag{7.46}$$

Therefore by lemmas 7.4.1 and 7.4.6,

$$|(E'_1 - E'_2)(\boldsymbol{\phi}''_2)| \leqslant 3rQ_4 \tilde{P}_1 P_{7,a}^{-1} P_{6,a}(|\boldsymbol{\phi}'_1 - \boldsymbol{\phi}'_2| + |\mathrm{D}\boldsymbol{\phi}'_1 - \mathrm{D}\boldsymbol{\phi}'_2|).$$

We now apply lemma 7.4.4 to (7.45) to obtain the result

$$\|\boldsymbol{\phi}''_1 - \boldsymbol{\phi}''_2, \mathscr{U}_+\|_1 \leqslant rQ_5 \|\boldsymbol{\phi}'_1 - \boldsymbol{\phi}'_2, \mathscr{U}_+\|_1, \tag{7.47}$$

where Q_5 is some constant independent of r. Thus for sufficiently small r, the map α will be contracting in the $\|\ \|_1$ norm (i.e. $\|\alpha(\boldsymbol{\phi_1}) - \alpha(\boldsymbol{\phi_2})\|_1 < \|\boldsymbol{\phi_1} - \boldsymbol{\phi_2}\|_1$) and the sequence $\alpha^n(\boldsymbol{\phi}'_1)$ will converge strongly in $W^1(\mathscr{U}_+)$ to some field $\boldsymbol{\phi}$. But by the theorem of Riesz some subsequence of the $\alpha^n(\boldsymbol{\phi}'_1)$ will converge weakly to some field $\tilde{\boldsymbol{\phi}} \in W(r)$. Thus $\boldsymbol{\phi}$ must equal $\tilde{\boldsymbol{\phi}}$ and so be in $W(r)$. Therefore $\alpha(\boldsymbol{\phi})$ will be defined. Now

$$\|\alpha(\boldsymbol{\phi}) - \alpha^{n+1}(\boldsymbol{\phi}'_1), \mathscr{U}_+\|_1 \leqslant rQ_5 \|\boldsymbol{\phi} - \alpha^n(\boldsymbol{\phi}'_1), \mathscr{U}_+\|_1.$$

As $n \to \infty$, the right-hand side tends to zero. This implies that $\|\alpha(\boldsymbol{\phi}) - \boldsymbol{\phi}, \mathscr{U}_+\|_1 = 0$ and so that $\alpha(\boldsymbol{\phi}) = \boldsymbol{\phi}$. Since the map α is contracting the fixed point is unique in $W(r)$. We have therefore proved:

Proposition 7.5.1

If $\hat{\mathbf{g}}$ is a solution of the empty space Einstein equations, the reduced empty space Einstein equations have a solution $\boldsymbol{\phi} \in W^{4+a}(\mathscr{U}_+)$ if $\|_0\boldsymbol{\phi}, \mathscr{H}(0) \cap \overline{\widetilde{\mathscr{U}}}\|_{4+a}$ and $\|_1\boldsymbol{\phi}, \mathscr{H}(0) \cap \overline{\widetilde{\mathscr{U}}}\|_{3+a}$ are sufficiently small. $\|\boldsymbol{\phi}, \mathscr{H}(0) \cap \overline{\widetilde{\mathscr{U}}}_+\|_{4+a}$ will be bounded and so $\boldsymbol{\phi}$ will be at least $C^{(2+a)-}$. $\square$

This solution will be locally unique even among solutions which are not in $W^4(\mathscr{U}_+)$.

Proposition 7.5.2

Let $\tilde{\phi}$ be a C^{1-} solution of the reduced empty space Einstein equations with the same initial data on an open set $\mathscr{V} \subset \mathscr{H}(0) \cap \mathscr{U}$. Then $\tilde{\phi} = \phi$ on a neighbourhood of $\mathscr{V}$ in $\mathscr{U}_+$.

Since $\tilde{\phi}$ is continuous one can find a neighbourhood $\mathscr{U}'$ of $\mathscr{V}$ in $\mathscr{U}$ such that the conditions of lemma 7.4.4 hold for **A**, **B** and **C**. As before one has

$$\tilde{E}(\tilde{\phi} - \phi) = -(\tilde{E} - E)(\phi). \tag{7.48}$$

Similarly there will be some Q_6 such that

$$\|(\tilde{E} - E)(\phi), \mathscr{H}(t) \cap \mathscr{U}'_+\|_0 \leqslant Q_6 \|\tilde{\phi} - \phi, \mathscr{H}(t) \cap \mathscr{U}'_+\|_1.$$

Applying lemma 7.4.4 to (7.48) one obtains an inequality of the form

$$\mathrm{d}x/\mathrm{d}t \leqslant Q_7 x,$$

where $\qquad x = \displaystyle\int_0^t \|\tilde{\phi} - \phi, \mathscr{H}(t') \cap \mathscr{U}'_+\|_1 \, \mathrm{d}t'.$

Therefore $\tilde{\phi} = \phi$ on $\overline{\mathscr{U}}'_+$. $\qquad\qquad\qquad\square$

Proposition 7.5.1 shows that if one makes a sufficiently small perturbation in the initial data of an empty space solution of the Einstein equations one obtains a solution in a region $\mathscr{U}_+$. What one wants however is to prove the existence of developments for any initial data h^{ab} and χ^{ab} which satisfy the constraint equations on a three-manifold $\mathscr{S}$. To do this we proceed as follows. We take $\mathscr{M}$ to be R^4, **e** to be the Euclidean metric and $\hat{\mathbf{g}}$ to be the flat, Minkowski metric (this is a solution of the empty space Einstein equations). In the usual Minkowski coordinates x^1, x^2, x^3 and x^4 ($x^4 = t$) we take $\mathscr{U}$ to be such that $\partial\mathscr{U} \cap \overline{\mathscr{U}}_+$ is spacelike and $\mathscr{H}(0) \cap \overline{\mathscr{U}}$ consists of the points for which $(x^1)^2 + (x^2)^2 + (x^3)^2 \leqslant 1$, $x^4 = 0$. The idea now is that any metric appears nearly flat if looked at on a fine enough scale. Therefore if one maps a sufficiently small region of $\mathscr{S}$ onto $\mathscr{H}(0) \cap \overline{\mathscr{U}}$, one can use proposition 7.5.1 and obtain a solution on $\mathscr{U}_+$. We then repeat this for other portions of $\mathscr{S}$ and join up the resulting solutions to form a manifold $\mathscr{M}$ with metric $\mathbf{g}$ which is a development of $(\mathscr{S}, \boldsymbol{\omega})$.

Let $\mathscr{V}_1$ be a coordinate neighbourhood in $\mathscr{S}$ with coordinates y^1, y^2 and y^3 such that at p, the origin of the coordinates, the coordinate components of h^{ab} equal δ^{ab}. Let $\mathscr{V}_1(f_1)$ be the open ball of coordinate radius f_1 about p. Define an imbedding $\theta_1 \colon \mathscr{V}_1(f_1) \to \mathscr{U}$ by $x^i = f_1^{-1} y^i$ ($i = 1, 2, 3$), $x^4 = 0$. By the usual law of transformation of a basis, the components of $\theta_* h^{ab}$ and $\theta_* \chi^{ab}$ with respect to the coordinates $\{x\}$ are f_1^{-2} times the components of h^{ab} and χ^{ab} with respect to the co-ordinates $\{y\}$. We define new fields h'^{ab} and χ'^{ab} on $\mathcal{V}_1$ by $h'^{ab} = f_1^2 h^{ab}$ and $\chi'^{ab} = f_1^3 \chi^{ab}$. Then since $\mathbf{h}$ is continuous (in fact C^{2+a}) on $\mathcal{S}$ one can make $g'^{ab} - \hat{g}^{ab}$ and $g'^{ab}{}_{|c} u^c$ arbitrarily small on $\mathcal{H}(0) \cap \mathcal{U}$ by taking f_1 sufficiently small, where g'^{ab} and $g'^{ab}{}_{|c} u^c$ are defined from h'^{ab} and χ'^{ab} in the manner of §7.3. The derivatives of g'^{ab} and $g'^{ab}{}_{|c} u^c$ in the surface $\mathcal{H}(0)$ will also become smaller as f_1 is made smaller. Thus $\|_0 \boldsymbol{\phi}', \mathcal{H}(0) \cap \widetilde{\overline{\mathcal{U}}}\|_{4+a}$ and $\|_1 \boldsymbol{\phi}', \mathcal{H}(0) \cap \widetilde{\overline{\mathcal{U}}}\|_{3+a}$ can be made small enough that proposition 7.5.1 can be applied and a solution for $\boldsymbol{\phi}'$ obtained on $\mathcal{U}_+$. Then $g_1^{ab} = f_1^{-2} g'^{ab}$ will be a solution of the reduced Einstein equations with the initial data determined by h^{ab} and χ^{ab}. Similarly one can obtain a solution on $\mathcal{U}_-$, the part of $\mathcal{U}$ on which $t \leqslant 0$.

One can now cover $\mathcal{S}$ by coordinate neighbourhoods $\mathcal{V}_\alpha(f_\alpha)$ of the form $\mathcal{V}_1(f_1)$, map them by imbeddings θ_α to neighbourhoods $\mathcal{U}_\alpha$ of the form $\mathcal{U}$ and obtain solutions $g_\alpha{}^{ab}$ on $\mathcal{U}_\alpha$. The problem now is to identify suitable points in the overlaps to make the collection of the $\mathcal{U}_\alpha$ into a manifold with a metric $\mathbf{g}$. To do this we make use of the harmonic gauge condition

$$\phi^{bc}{}_{|c} = g^{bc}{}_{|c} - \tfrac{1}{2} g^{bc} g_{de} g^{dc}{}_{|c} = 0. \tag{7.49}$$

By the definition (7.3) of $\delta\Gamma^a{}_{bc}$, this is equivalent to $g^{de} \delta\Gamma^b{}_{de} = 0$. Therefore for any function z,

$$z_{;\,ab} g^{ab} = z_{|ab} g^{ab} - \delta\Gamma^c{}_{ab} z_{|c} g^{ab} = z_{|ab} g^{ab}. \tag{7.50}$$

If the background metric is the Minkowski metric and z is one of the Minkowski coordinates x^1, x^2, x^3 and x^4, the right-hand side of (7.50) will vanish. Suppose now one has an arbitrary W^{4+a} Lorentz metric $\mathbf{g}$ on a manifold $\mathcal{M}$. In some neighbourhood $\mathcal{Y} \subset \mathcal{M}$ one can find four solutions z^1, z^2, z^3 and z^4 of the linear equation

$$z_{;\,ab} g^{ab} = 0 \tag{7.51}$$

which are such that their gradients are linearly independent at each point of $\mathcal{Y}$. We may then define a diffeomorphism $\mu: \mathcal{Y} \to \hat{\mathcal{M}}$ by $x^a = z^a$ $(a = 1, 2, 3, 4)$. This diffeomorphism will have the property that the metric $\mu_* g^{ab}$ on $\hat{\mathcal{M}}$ will satisfy the harmonic gauge condition with respect to the Minkowski metric $\hat{\mathbf{g}}$ on $\hat{\mathcal{M}}$. Thus if the metric $\mathbf{g}$ is a solution of the Einstein equations on $\mathcal{M}$, the metric $\mu_* \mathbf{g}$ will be a solution of the reduced Einstein equations on $\hat{\mathcal{M}}$ with the background metric $\hat{\mathbf{g}}$.

The procedure to identify points in the overlap between two neighbourhoods $\mathscr{U}_\alpha$ and $\mathscr{U}_\beta$ is therefore to solve (7.51) on $\mathscr{U}_\alpha$ for the coordinates $x_\beta{}^1$, $x_\beta{}^2$, $x_\beta{}^3$ and $x_\beta{}^4$ using the initial values for $x_\beta{}^a$ and $x_\beta{}^a{}_{|b}u^b$ determined by the overlap of the coordinate neighbourhoods $\mathscr{V}_\alpha$ and $\mathscr{V}_\beta$ on $\mathscr{S}$. In fact $x_\beta{}^i{}_{|a}u^a = 0$ ($i = 1$, 2, 3) and $x_\beta{}^4{}_{|a}u^a = 1$ where $u^a = \partial/\partial x_\alpha{}^a$ is the unit vector in $\mathscr{U}_\alpha$ orthogonal to $\mathscr{H}(0)$ in the metric $\hat{\mathbf{g}}$. Thus $x_\beta{}^4 = x_\alpha{}^4$ though $x_\beta{}^i$ will not in general be equal to $x_\alpha{}^i$. By proposition 7.4.7. the coordinates $x_\beta{}^a$ will be $C^{(2+a)-}$ functions on $\mathscr{U}_\alpha$. (In proposition 7.4.7 the background metric with respect to which the covariant derivatives are taken has to be $C^{(5+a)-}$. Thus it cannot be applied directly to (7.51), since the covariant derivatives are taken with respect to $\mathbf{g}$, which is only W^{4+a}. However one can introduce a C^{5+a} background metric $\tilde{\mathbf{g}}$ and express (7.51) in the form

$$z_{\|ab}g^{ab} + z_{\|a}B^a = 0,$$

where $\|$ indicates covariant differentiation with respect to $\tilde{\mathbf{g}}$. Proposition 7.4.7 can then be applied to this equation.)

Since the gradients of $x_\beta{}^a$ are linearly independent on $\mathscr{H}(0) \cap \mathscr{U}_\alpha$, they will be linearly independent on some neighbourhood $\mathscr{U}''_\alpha$ of $\mathscr{H}(0)$ in $\mathscr{U}_\alpha$. The metric $\mu_* g^{ab}_\alpha$ will be at least C^{1-} on $\mu(\mathscr{U}''_\alpha)$ in $\mathscr{U}_\beta$. Since it will obey the reduced empty space Einstein equations on $\mathscr{U}_\beta$ in the background metric $\hat{\mathbf{g}}$ and since it has the same initial data on $\theta_\beta(\mathscr{V}_\alpha \cap \mathscr{V}_\beta)$, it must coincide with $\mathbf{g}_\beta$ on some neighbourhood $\mathscr{U}'_\beta$ of $\theta_\beta(\mathscr{V}_\alpha \cap \mathscr{V}_\beta)$ in $\mathscr{U}_\beta$. This shows that one may join together $\mathscr{U}''_\alpha$ and $\mathscr{U}'_\beta$ to obtain a development of the region $\mathscr{V}_\alpha \cup \mathscr{V}_\beta$ of $\mathscr{S}$. Taking the covering $\{\mathscr{V}_\alpha\}$ of $\mathscr{S}$ to be locally finite, one may proceed in a similar fashion to join together the subsets of the other neighbourhoods $\{\mathscr{U}_\alpha\}$ to obtain a development of $\mathscr{S}$, i.e. a manifold $\mathscr{M}$ with a metric $\mathbf{g}$ and an imbedding $\theta: \mathscr{S} \to \mathscr{M}$ such that $\mathbf{g}$ satisfies the empty space Einstein equations and agrees with the prescribed initial data $\boldsymbol{\omega}$ on $\theta(\mathscr{S})$, which is a Cauchy surface for $\mathscr{M}$. If $(\mathscr{M}', \mathbf{g}')$ is another development of $(\mathscr{S}, \boldsymbol{\omega})$ one can by a similar procedure establish a diffeomorphism μ between some neighbourhood of $\theta'(\mathscr{S}')$ in $\mathscr{M}'$ and some neighbourhood of $\theta(\mathscr{S})$ in $\mathscr{M}$ such that $\mu_* g'^{ab} = g^{ab}$. We have therefore proved:

The local Cauchy development theorem

If $h^{ab} \in W^{4+a}(\mathscr{S})$ and $\chi^{ab} \in W^{3+a}(\mathscr{S})$ satisfy the empty space constraint equations there exist developments $(\mathscr{M}, \mathbf{g})$ for the empty space Einstein equations such that $\mathbf{g} \in W^{4+a}(\mathscr{M})$ and $\mathbf{g} \in W^{4+a}(\mathscr{H})$ for any smooth spacelike surface $\mathscr{H}$. These developments are locally unique in that if $(\mathcal{M}', \mathbf{g}')$ is another W^{4+a} development of $(\mathcal{S}, \omega)$ then $(\mathcal{M}, \mathbf{g})$ and $(\mathcal{M}', \mathbf{g}')$ are both extensions of some common development of $(\mathcal{S}, \omega)$.

That $\mathbf{g} \in W^{4+a}(\mathcal{H})$ follows from lemma 7.4.6 since the surfaces of constant t can be chosen arbitrarily. $\square$

7.6 The maximal development and stability

We have shown that if the initial data satisfied the empty space constraint equations one can find a development, i.e. one can construct a solution some distance into the future and past of the initial surface. In general, this development can be extended further into the future and past to give a larger development of $(\mathcal{S}, \omega)$. However we shall show by an argument similar to that of Choquet-Bruhat and Geroch (1969) that there is a unique (up to a diffeomorphism) development $(\mathcal{M}, \mathbf{g})$ of $(\mathcal{S}, \omega)$ which is an extension of any other development of $(\mathcal{S}, \omega)$.

Recall that $(\mathcal{M}_1, \mathbf{g}_1)$ is an extension of $(\mathcal{M}_2, \mathbf{g}_2)$ if there is an imbedding $\mu : \mathcal{M}_2 \to \mathcal{M}_1$ such that $\mu_* \mathbf{g}_2 = \mathbf{g}_1$, and such that $\theta_1^{-1} \mu \theta_2$ is the identity map on $\mathcal{S}$. Given a point $q \in \mathcal{S}$, and a distance s one can uniquely determine points $p_1 \in \mathcal{M}_1$ and $p_2 \in \mathcal{M}_2$ by going a distance s along the geodesics orthogonal to $\theta_1(\mathcal{S})$ and $\theta_2(\mathcal{S})$ through $\theta_1(q)$ and $\theta_2(q)$ respectively. Since $\mu(p_2)$ must equal p_1, the imbedding μ must be unique. One can therefore partially order the set of all developments of $(\mathcal{S}, \omega)$, writing $(\mathcal{M}_2, \mathbf{g}_2) \leqslant (\mathcal{M}_1, \mathbf{g}_1)$ if $(\mathcal{M}_1, \mathbf{g}_1)$ is an extension of $(\mathcal{M}_2, \mathbf{g}_2)$. If now $\{(\mathcal{M}_\alpha, \mathbf{g}_\alpha)\}$ is a totally ordered set (a set $\mathcal{A}$ is said to be totally ordered if for every pair a, b of distinct elements of $\mathcal{A}$, either $a \leqslant b$ or $b \leqslant a$) of developments of $(\mathcal{S}, \omega)$, one can form the manifold $\mathcal{M}'$ as the union of all the $\mathcal{M}_\alpha$ where for $(\mathcal{M}_\alpha, \mathbf{g}_\alpha) \leqslant (\mathcal{M}_\beta, \mathbf{g}_\beta)$ each $p_\alpha \in \mathcal{M}_\alpha$ is identified with $\mu_{\alpha\beta}(p_\alpha) \in \mathcal{M}_\beta$, where $\mu_{\alpha\beta} : \mathcal{M}_\alpha \to \mathcal{M}_\beta$ is the imbedding. The manifold $\mathcal{M}'$ will have an induced metric $\mathbf{g}'$ equal to $\mu_{\alpha*} \mathbf{g}_\alpha$ on each $\mu_\alpha(\mathcal{M}_\alpha)$ where $\mu_\alpha : \mathcal{M}_\alpha \to \mathcal{M}'$ is the natural imbedding. Clearly $(\mathcal{M}', \mathbf{g}')$ will also be a development of $(\mathcal{S}, \omega)$; therefore every totally ordered set has an upper bound, and so by Zorn's lemma (see, for example, Kelley (1965), p. 33) there is a maximal development $(\tilde{\mathcal{M}}, \tilde{\mathbf{g}})$ of $(\mathcal{S}, \omega)$ whose only extension is itself.

We shall now show that $(\tilde{\mathcal{M}}, \tilde{\mathbf{g}})$ is an extension of every development of $(\mathcal{S}, \omega)$. Suppose $(\mathcal{M}', \mathbf{g}')$ is another development of $(\mathcal{S}, \omega)$. By the local Cauchy theorem, there exist developments of $(\mathcal{S}, \omega)$ of which $(\tilde{\mathcal{M}}, \tilde{\mathbf{g}})$ and $(\mathcal{M}', \mathbf{g}')$ are both extensions. The set of all such common developments is likewise partially ordered and so again by Zorn's lemma there will be a maximal development $(\mathcal{M}'', \mathbf{g}'')$ with the imbeddings $\tilde{\mu}\colon \mathcal{M}'' \to \tilde{\mathcal{M}}$ and $\mu'\colon \mathcal{M}'' \to \mathcal{M}'$, etc. Let $\mathcal{M}^+$ be the union of $\tilde{\mathcal{M}}$, $\mathcal{M}'$ and $\mathcal{M}''$, where each $p'' \in \mathcal{M}''$ is identified with $\tilde{\mu}(p'') \in \tilde{\mathcal{M}}$ and $\mu'(p'') \in \mathcal{M}'$. If one can show that the manifold $\mathcal{M}^+$ is Hausdorff, the pair $(\mathcal{M}^+, \mathbf{g}^+)$ will be a development of $(\mathcal{S}, \boldsymbol{\omega})$. It will be an extension of both $(\tilde{\mathcal{M}}, \tilde{\mathbf{g}})$ and $(\mathcal{M}', \mathbf{g}')$. However the only extension of $(\tilde{\mathcal{M}}, \tilde{\mathbf{g}})$ is $(\tilde{\mathcal{M}}, \tilde{\mathbf{g}})$ itself, and so $(\tilde{\mathcal{M}}, \tilde{\mathbf{g}})$ must equal $(\mathcal{M}^+, \mathbf{g}^+)$ and be an extension of $(\mathcal{M}', \mathbf{g}')$.

Suppose that $\mathcal{M}^+$ were not Hausdorff. Then there exist points $\tilde{p} \in (\tilde{\mu}(\mathcal{M}''))^{\textstyle\cdot} \subset \tilde{\mathcal{M}}$ and $p' \in (\mu'(\mathcal{M}''))^{\textstyle\cdot} \subset \mathcal{M}'$ such that every neighbourhood $\mathcal{U}$ of $\tilde{p}$ has the property that $\overline{\mu'(\tilde{\mu}^{-1}(\mathcal{U}))}$ contains p'. Now since $(\mathcal{M}'', \mathbf{g}'')$ is a development, it will be globally hyperbolic as will its image $\tilde{\mu}(\mathcal{M}'')$ in $\tilde{\mathcal{M}}$. Therefore the boundary of $\tilde{\mu}(\mathcal{M}'')$ in $\tilde{\mathcal{M}}$ must be achronal. Let γ be a timelike curve in $\tilde{\mathcal{M}}$ with future endpoint at $\tilde{p}$. Then p' must be a limit point in $\mathcal{M}'$ of the curve $\mu'\tilde{\mu}^{-1}(\gamma)$. In fact it must be a future endpoint, since strong causality holds in $(\mathcal{M}', \mathbf{g}')$. Thus the point p' is unique, given $\tilde{p}$. Further, by continuity vectors at p' can be uniquely associated with vectors at $\tilde{p}$. Thus one can find normal coordinate neighbourhoods $\tilde{\mathcal{U}}$ of $\tilde{p}$ in $\tilde{\mathcal{M}}$ and $\mathcal{U}'$ of p' in $\mathcal{M}'$ such that under the map $\mu'\tilde{\mu}^{-1}$ points of $\tilde{\mathcal{U}} \cap \tilde{\mu}(\mathcal{M}'')$ are mapped into points of $\mathcal{U}' \cap \mu'(\mathcal{M}'')$ with the same coordinate values. This shows that the set $\mathcal{F}$ of all 'non-Hausdorff' points of $(\tilde{\mu}(\mathcal{M}''))^{\textstyle\cdot}$ is open in $(\tilde{\mu}(\mathcal{M}''))^{\textstyle\cdot}$. We shall suppose that $\mathcal{F}$ is non-empty, and so obtain a contradiction.

If $\tilde{\lambda}$ is a past-directed null geodesic in $\tilde{\mathcal{M}}$ through $\tilde{p} \in \mathcal{F}$, then since one can associate directions at p with directions at p', one can construct a past-directed null geodesic λ' through p' in $\mathcal{M}'$ in the corresponding direction. To each point of $\tilde{\lambda} \cap (\tilde{\mu}(\mathcal{M}''))^{\textstyle\cdot}$ there will correspond a point of $\lambda' \cap (\mu'(\mathcal{M}''))^{\textstyle\cdot}$ and so every point of $\tilde{\lambda} \cap (\tilde{\mu}(\mathcal{M}''))^{\textstyle\cdot}$ will be in $\mathcal{F}$. Since $\partial(\mathcal{S})$ is a Cauchy surface for $\tilde{\mathcal{M}}$, $\tilde{\lambda}$ must leave $(\tilde{\mu}(\mathcal{M}''))^{\textstyle\cdot}$ at some point $\tilde{q}$. There will be some point $\tilde{r} \in \mathcal{F}$ in a neighbourhood of $\tilde{q}$ such that there is a spacelike surface $\tilde{\mathcal{H}}$ through $\tilde{r}$ which has the property that $(\tilde{\mathcal{H}} - \tilde{r}) \subset \tilde{\mu}(\mathcal{M}'')$. There will be a corresponding spacelike surface $\mathcal{H}' = (\mu'\tilde{\mu}^{-1}(\tilde{\mathcal{H}} - \tilde{r})) \cup r'$ in $\mathcal{M}'$ through the corresponding point r'. The surfaces $\tilde{\mathcal{H}}$ and $\mathcal{H}'$ may be regarded as images of a three-dimensional manifold $\mathcal{H}$ under imbeddings $\tilde{\psi}\colon \mathcal{H} \to \tilde{\mathcal{M}}$ and $\psi'\colon \mathcal{H} \to \mathcal{M}'$ such that $\tilde{\psi}^{-1}\tilde{\mu}\mu'^{-1}\psi'$ is the identity map on $\mathcal{H} - \tilde{\psi}^{-1}(\tilde{p})$. The induced metrics $\tilde{\psi}_*(\tilde{\mathbf{g}})$ and $\psi'_*(\mathbf{g}')$ on $\mathcal{H}$ will agree since $\tilde{\mathcal{H}} - \tilde{p}$ and $\mathcal{H}' - p'$ are isometric. By the local Cauchy theorem, they will be in $W^{4+a}(\mathcal{H})$. Similarly the second fundamental forms will agree and be in $W^{3+a}(\mathscr{H})$. Neighbourhoods of $\mathscr{H}$ in $\tilde{\mathscr{M}}$ and $\mathscr{H}'$ in $\mathscr{M}'$ would be W^{4+a} developments of $\mathscr{H}$. By the local Cauchy theorem they must be extensions of the same common development $(\mathscr{M}^*, \mathbf{g}^*)$. Joining $(\mathscr{M}^*, \mathbf{g}^*)$ to $(\mathscr{M}'', \mathbf{g}'')$ one would obtain a larger development of $(\mathscr{S}, \omega)$, of which $(\tilde{\mathscr{M}}, \tilde{\mathbf{g}})$ and $(\mathscr{M}', \mathbf{g}')$ would be extensions. This is impossible, since $(\mathscr{M}'', \mathbf{g}'')$ was the largest such common development. This shows that $\mathscr{M}^+$ must be Hausdorff, and so that $(\tilde{\mathscr{M}}, \tilde{\mathbf{g}})$ must be an extension of $(\mathscr{M}', \mathbf{g}')$.

We have therefore proved:

The global Cauchy development theorem

If $h^{ab} \in W^{4+a}(\mathscr{S})$ and $\chi^{ab} \in W^{3+a}(\mathscr{S})$ satisfy the empty space constraint equations, there exists a maximal development $(\mathscr{M}, \mathbf{g})$ of the empty space Einstein equations with $\mathbf{g} \in W^{4+a}(\mathscr{M})$ and $\mathbf{g} \in W^{4+a}(\mathscr{H})$ for any smooth spacelike surface $\mathscr{H}$. This development is an extension of any other such development.

We have so far only proved that this development is maximal among W^{4+a} developments. If a is greater than zero, there will also be $W^{4+a-1}, W^{4+a-2}, \ldots, W^4$ developments which are extensions of the W^{4+a} development. However, Choquet-Bruhat (1971) has pointed out that these developments must all coincide with the W^4 development. This is because one can differentiate the reduced Einstein equations and then regard them as *linear* equations on the W^4 development, for the first derivatives of g^{ab}. Then using proposition 7.4.7 one can show that g^{ab} is W^5 on the W^4 development, if the initial data is W^5. By continuing in this way, one can show that if the initial data is C^∞, there will be a C^∞ development which will in fact coincide with the W^4 development.

We have proved the existence and uniqueness of maximal developments only for W^4 or higher metrics. In fact, it is possible to prove the existence of developments for W^3 initial data, but we have not been able to prove the uniqueness in this case. It may be possible to extend the W^4 maximal development either so that the metric does not remain in W^4, or so that $\theta(\mathscr{S})$ does not remain a Cauchy surface. In the latter case, a Cauchy horizon occurs; examples of this were given in chapter 6. On the other hand it may be that some sort of singularity occurs, in which case the development cannot be extended with a metric which is sufficiently differentiable to be interpreted physically. In fact, theorem 4 of the next chapter will show that if $\mathscr{S}$ is compact and $\chi^{ab}h_{ab}$ is negative everywhere on $\mathscr{S}$, then the development cannot be extended to be geodesically complete with a C^{2-} metric, i.e. with locally bounded curvature.

We have shown there is a map from the space of pairs of tensors (h^{ab}, χ^{ab}) on $\mathscr{S}$ which satisfy the constraint equations to the space of equivalence classes of metrics $\hat{\mathbf{g}}$ on a manifold $\mathscr{M}$, which, by proposition 6.6.8, is diffeomorphic to $\mathscr{S} \times R^1$. If two pairs (h^{ab}, χ^{ab}) and (h'^{ab}, χ'^{ab}) are equivalent under a diffeomorphism $\lambda: \mathscr{S} \rightarrow \mathscr{S}$ (i.e. $\lambda_* h^{ab} = h'^{ab}$ and $\lambda_* \chi^{ab} = \chi'^{ab}$) they will produce equivalent metrics $\hat{\mathbf{g}}$. We thus have a map from equivalence classes of pairs (h^{ab}, χ^{ab}) to equivalence classes of metrics $\hat{\mathbf{g}}$. Now h^{ab} and χ^{ab} together have twelve independent components. The constraint equations impose four relations between these, and the equivalence under diffeomorphisms may be regarded as removing a further three arbitrary functions, leaving five independent functions. One of these functions may be regarded as specifying the position of $\theta(\mathscr{S})$ within the development $(\mathscr{M}, \hat{\mathbf{g}})$. Therefore maximal developments of the empty space Einstein equations are specified by four functions of three variables.

One would like to show that the map from equivalence classes of (h^{ab}, χ^{ab}) to equivalence classes of $\hat{\mathbf{g}}$ is continuous in some sense. The appropriate topology on the equivalence classes for this is the W^r compact-open topology (cf. §6.4). Let $\hat{\mathbf{g}}$ be a C^r Lorentz metric on $\mathscr{M}$ and $\mathscr{U}$ be an open set with compact closure. Let V be an open set in $W^r(\mathscr{U})$ and let $O(\mathscr{U}, V)$ be the set of all Lorentz metrics on $\mathscr{M}$ whose restrictions to $\mathscr{U}$ lie in V. The open sets of the W^r compact open topology on the space $\mathscr{L}_r(\mathscr{M})$ of all W^r Lorentz metrics on $\mathscr{M}$ are defined to be the unions and finite intersections of sets of the form $O(U, V)$. The topology of the space $\mathscr{L}_r^*(\mathscr{M})$ of equivalence classes of W^r metrics on $\mathscr{M}$ is then that induced by the projection

$$\pi: \mathscr{L}_r(\mathscr{M}) \rightarrow \mathscr{L}_r^*(\mathscr{M})$$

which assigns a metric to its equivalence class (i.e. the open sets of $\mathscr{L}_r^*(\mathscr{M})$ are of the form $\pi(Q)$ where Q is open in $\mathscr{L}_r(\mathscr{M})$). Similarly the W^r compact open topology on the space $\Omega_r(\mathscr{S})$ of all pairs (h^{ab}, χ^{ab}) which satisfy the constraint equations is defined by sets of the form $O(\mathscr{U}, V, V')$ consisting of the pairs for which $h^{ab} \in V$ and $\chi^{ab} \in V'$ where V and V' are open sets in $W^r(\mathscr{S})$ and $W^{r-1}(\mathscr{S})$ respectively. The C^∞ metrics on $\mathscr{M}$ form a subspace $\mathscr{L}_\infty(\mathscr{M})$ of the space $\mathscr{L}(\mathscr{M})$ of all Lorentz metrics on $\mathscr{M}$. Since a C^∞ metric is W^r for any r, one has the W^r topology on $\mathscr{L}_\infty(\mathscr{M})$. One can then define the C^∞ or W^∞ topology on $\mathcal{L}_\infty(\mathcal{M})$ as that given by all the open sets in the W^r topologies on $\mathcal{L}_\infty(\mathcal{M})$ for every r. The C^∞ topology on $\mathcal{L}_\infty^*(\mathcal{M})$ and on $\Omega_\infty(\mathcal{S})$ are defined similarly.

One would like to show that the map Δ_r from the space $\Omega_r^*(\mathcal{S})$ of equivalence classes of pairs (h^{ab}, χ^{ab}) to the space $\mathcal{L}_r^*(\mathcal{M})$ of equivalence classes of metrics is continuous with the W^r compact open topology on both spaces. In other words, suppose one has initial data $h^{ab} \in W^r(\mathcal{S})$ and $\chi^{ab} \in W^{r-1}(\mathcal{S})$ which gives rise to a solution $\mathbf{g} \in W^r(\mathcal{M})$ on $\mathcal{M}$. Then if $\mathcal{V}$ is a region of $\mathcal{M}$ with compact closure, and $\epsilon > 0$, one would like to show there was some region $\mathcal{Y}$ of $\mathcal{S}$ with compact closure and some $\delta > 0$ such that $\|\mathbf{g}' - \mathbf{g}, \mathcal{V}\|_r < \epsilon$ for all initial data (h'^{ab}, χ'^{ab}) such that $\|\mathbf{h}' - \mathbf{h}, \mathcal{Y}\|_r < \tfrac{1}{2}\delta$ and $\|\mathbf{\chi}' - \mathbf{\chi}, \mathcal{Y}\|_{r-1} < \tfrac{1}{2}\delta$. This result may be true, but we have been unable to prove it. What we can prove is that this result holds if the metric is $C^{(r+1)-}$. This follows immediately from proposition 7.5.1, taking $\mathbf{g}$ to be the background metric and $\mathcal{U}$ to be some suitable neighbourhood of $J^-(\mathcal{V}) \cap J^+(\theta(\mathcal{S}))$. In fact if one examines lemma 7.4.6, one sees that the condition on the background metric can be weakened from $C^{(r+1)-}$ to $W^{(r+1)}$, but not to W^r, since the $(r-1)$th derivatives of the Riemann tensor of the background metric appear. (By the background metric being W^{r+1} we mean that it is W^{r+1} with respect to a further C^{r+1} background metric.) Thus the map $\Delta_r: \Omega_r^*(\mathcal{S}) \to \mathcal{L}_r^*(\mathcal{M})$ from the equivalence classes of initial data to the equivalence classes of metrics will be continuous in the W^r compact open topology at every W^{r+1} metric. Although the W^{r+1} metrics form a dense set in the W^r metrics, there is a possibility that the map might not be continuous at a W^r metric which was not also a W^{r+1} metric. However $\infty + 1 = \infty$ and so the map $\Delta_\infty: \Omega^*_\infty(\mathcal{S}) \to \mathcal{L}^*_\infty(\mathcal{M})$ will be continuous in the C^∞ topology on both spaces.

One can express this result as:

The Cauchy stability theorem

Let $(\mathcal{M}, g)$ be the W^{5+a} $(0 \leqslant a \leqslant \infty)$ maximal development of initial data $\mathbf{h} \in W^{5+a}(\mathcal{S})$ and $\mathbf{\chi} \in W^{4+a}(\mathcal{S})$, and let $\mathcal{V}$ be a region of $J^+(\theta(\mathcal{S}))$ with compact closure. Let Z be a neighbourhood of $\mathbf{g}$ in $\mathcal{L}_{5+a}(\mathcal{V})$ and $\mathcal{U}$ be an open neighbourhood in $\theta(\mathcal{S})$ of $J^-(\mathcal{V}) \cap \theta(\mathcal{S})$ with compact closure. Then there is some neighbourhood Y of $(\mathbf{h}, \mathbf{\chi})$ in $\Omega_{5+a}(\mathcal{U})$ such that for all initial data $(\mathbf{h}', \mathbf{\chi}') \in Y$ satisfying the constraint equations, there is a diffeomorphism $\mu: \mathcal{M}' \to \mathcal{M}$ with the properties

(1) $\theta^{-1}\mu\theta'$ is the identity on $\theta^{-1}(\mathcal{U})$,

(2) $\mu_* \mathbf{g}' \in Z$,

where $(\mathcal{M}', \mathbf{g}')$ is the maximal development of $(\mathbf{h}', \chi')$.   □

Roughly speaking what this theorem says is that if the perturbation of initial data on the Cauchy surface $\theta(\mathcal{S})$ is small on $J^-(\bar{\mathcal{V}}) \cap \theta(\mathcal{S})$, then one gets a new solution which is near the old solution in $\mathcal{V}$. In fact the perturbation of the initial data has to be small on a slightly larger region of the Cauchy surface than $J^-(\bar{\mathcal{V}}) \cap \theta(\mathcal{S})$, since the null cones will be slightly different in the new solution and so $\mathcal{V}$ may not lie in the Cauchy development of $J^-(\bar{\mathcal{V}}) \cap \theta(\mathcal{S})$.

7.7 The Einstein equations with matter

For simplicity we have so far considered the Einstein equations only for empty space. However similar results hold when matter is present providing that the equations governing the matter fields $\Psi_{(i)}{}^I{}_J$ obey certain physically reasonable conditions. The idea is to solve the matter equations with the prescribed initial conditions in a given space–time metric $\mathbf{g}'$. One then solves the reduced Einstein equations (7.42) as *linear* equations with the coefficients determined by $\mathbf{g}'$ and with the source term T'^{ab} determined by $\mathbf{g}'$ and by the solution for the matter fields. One thus obtains a new metric $\mathbf{g}''$ and repeats the procedure with $\mathbf{g}''$ in place of $\mathbf{g}'$. To show that this converges to a solution of the combined Einstein and matter equations one has to impose certain conditions on the matter equations. We shall require:

(a) if $\{{}_0\Psi_{(i)}\} \in W^{4+a}(\mathcal{H})$ and $\{{}_1\Psi_{(i)}\} \in W^{3+a}(\mathcal{H})$ are the initial data on an achronal spacelike surface $\mathcal{H}$ in a W^{4+a} metric $\mathbf{g}$, there exists a unique solution of the matter equations in a neighbourhood of $\mathcal{H}$ in $D^+(\mathcal{H})$ with $\{\Psi_{(i)}\} \in W^{4+a}(\mathcal{H}')$ for any smooth spacelike surface $\mathcal{H}'$, and $\Psi_{(i)} = {}_0\Psi_{(i)}, \quad \Psi_{(i)}{}^I{}_{J|a} u^a = {}_1\Psi_{(i)}{}^I{}_J \quad$ on $\mathcal{H}$;

(b) if $\{\Psi_{(i)}\}$ is a W^{5+a} solution in the W^{5+a} metric $\mathbf{g}$ on the set $\mathcal{U}_+$, then there exist positive constants $\tilde{Q}_1$ and $\tilde{Q}_2$ such that

$$\sum_{(i)} \|\Psi'_{(i)} - \Psi_{(i)}, \mathcal{U}_+\|_{4+a} \leqslant \tilde{Q}_2 \{\|\mathbf{g}' - \mathbf{g}, \mathcal{U}_+\|_{4+a}$$
$$+ \sum_{(i)} \|{}_0\Psi'_{(i)} - {}_0\Psi_{(i)}, \mathcal{H}(0) \cap \tilde{\mathcal{U}}\|_{4+a} + \sum_{(i)} \|{}_1\Psi'_{(i)} - {}_1\Psi_{(i)}, \mathcal{H}(0) \cap \tilde{\mathcal{U}}\|_{3+a}\}$$

for any W^{4+a} solution $\{\Psi'_{(i)}\}$ in the metric $\mathbf{g}'$ such that

$$\|\mathbf{g}' - \mathbf{g}, \mathcal{U}_+\|_{4+a} < \tilde{Q}_1$$

and

$$\sum_{(i)} \{\|{}_0\Psi'_{(i)} - {}_0\Psi_{(i)}, \mathcal{H}(0) \cap \tilde{\mathcal{U}}\|_{4+a} + \|{}_1\Psi'_{(i)} - {}_1\Psi_{(i)}, \mathcal{H}(0) \cap \tilde{\mathcal{U}}\|_{3+a}\} < \tilde{Q}_1;$$

(c) the energy–momentum tensor T_{ab} is polynomial in

$$\Psi_{(i)}{}^I{}_J, \quad \Psi_{(i)}{}^I{}_{J;a} \quad \text{and} \quad g^{ab}.$$

Condition (a) is the local Cauchy theorem for the matter field in a given space–time metric. Condition (b) is the Cauchy stability theorem for the matter field under a variation of the initial conditions and under a variation of the space–time metric $\mathfrak{g}$. If the matter equations are quasi-linear second order hyperbolic equations, these conditions may be established in a similar manner to that for the reduced Einstein equations, providing that the null cones of the matter equations coincide with or lie within the null cone of the space–time metric $\mathfrak{g}$. In the case of the scalar field or the electromagnetic potential which obey linear equations, these conditions follow from proposition 7.4.7. One can also deal with a scalar field coupled to the electromagnetic potential; one fixes the metric and the electromagnetic potential, solves the scalar field as a linear equation in that metric and potential, and then solves the electromagnetic field in the given metric with the scalar field as the source. Iterating this procedure one can show that one converges on a set of the form $\mathcal{U}_+$ to a solution of the coupled scalar and electromagnetic equations in the given metric, providing that the initial data are sufficiently small. One then shows, by rescaling the metric and the fields, that for $\mathcal{U}_+$ sufficiently small (as measured by the space–time metric $\mathfrak{g}$) one can obtain a solution for any suitable initial data. The same procedure will work for any finite number of coupled quasi-linear second order hyperbolic equations, where the coupling does not involve derivatives higher than the first.

The equations of a perfect fluid are not second order hyperbolic, but form a quasi-linear first order *system*. (For the definition of a first order hyperbolic system, see Courant and Hilbert (1962), p. 577.) Similar results can be obtained for such systems providing that the ray cone coincides with or lies within the null cone of the space–time with metric $\mathfrak{g}$. The requirement that the matter equations should be second order hyperbolic equations or first order hyperbolic systems with their cones coinciding with or lying within that of the space–time metric $\mathfrak{g}$, may be thought of as a more rigorous form of the local causality postulate of chapter 3.

With the conditions (a), (b) and (c) one can establish propositions 7.5.1 and 7.5.2 for the combined reduced Einstein's equations and the matter equations; from these, the local and global Cauchy development theorems and the Cauchy stability theorem follow.

8

Space-time singularities

In this chapter, we use the results of chapters 4 and 6 to establish some basic results about space–time singularities. The astrophysical and cosmological implications of these results are considered in the next chapters.

In §8.1, we discuss the problem of defining singularities in space–time. We adopt b-incompleteness, a generalization of the idea of geodesic incompleteness, as an indication that singular points have been cut out of space–time, and characterize two possible ways in which b-incompleteness can be associated with some form of curvature singularity. In §8.2, four theorems are given which prove the existence of incompleteness under a wide variety of situations. In §8.3 we give Schmidt's construction of the b-boundary which represents the singular points of space–time. In §8.4 we prove that the singularities predicted by at least one of the the the theorems cannot be just a discontinuity in the curvature tensor. We also show that there is not only one incomplete geodesic, but a three-parameter family of them. In §8.5 we discuss the situation in which the incomplete curves are totally or partially imprisoned in a compact region of space–time. This is shown to be related to non-Hausdorff behaviour of the b-boundary. We show that in a generic space–time, an observer travelling on one of these incomplete curves would experience infinite curvature forces. We also show that the kind of behaviour which occurs in Taub–NUT space cannot happen if there is some matter present.

8·1 The definition of singularities

By analogy with electrodynamics one might think it reasonable to define a space–time singularity as a point where the metric tensor was undefined or was not suitably differentiable. However the trouble with this is that one could simply cut out such points and say that the remaining manifold represented the whole of space–time, which would then be non-singular according to this definition. Indeed, it would seem inappropriate to regard such singular points as being part of space–time, for the normal equations of physics would not hold at them and it would be impossible to make any measurements. We therefore defined space–time in §3.1 as a pair $(\mathcal{M}, \mathbf{g})$ where the metric $\mathbf{g}$ is Lorentzian and suitably differentiable and we ensured that no regular points were omitted from the manifold $\mathcal{M}$ along with the singular points by requiring that $(\mathcal{M}, \mathbf{g})$ could not be extended with the required differentiability.

The problem of defining whether space–time has a singularity now becomes one of determining whether any singular points have been cut out. One would hope to recognize this by the fact that space–time was incomplete in some sense.

In the case of a manifold $\mathcal{M}$ with a positive definite metric $\mathbf{g}$, one can define a distance function $\rho(x, y)$ which is the greatest lower bound of the length of curves from x to y. The distance function $\rho(x, y)$ is a metric in the topological sense; that is, a basis for the open sets of $\mathcal{M}$ is provided by the sets $\mathcal{B}(x, r)$ consisting of all points $y \in \mathcal{M}$ such that $\rho(x, y) < r$. The pair $(\mathcal{M}, \mathbf{g})$ is said to be *metrically complete* (*m-complete*) if every Cauchy sequence with respect to the distance function ρ converges to a point in $\mathcal{M}$. (A *Cauchy sequence* is an infinite sequence of points x_n such that for any $\epsilon > 0$ there is a number N such that $\rho(x_n, x_m) < \epsilon$ whenever n and m are greater than N.) An alternative formulation is that $(\mathcal{M}, \mathbf{g})$ is m-complete if every C^1 curve of finite length has an endpoint in the sense of §6.2 (note that the curve need not be C^1 at the endpoint). It therefore follows that m-completeness implies *geodesic completeness* (*g-completeness*), that is every geodesic can be extended to arbitrary values of its affine parameter. In fact it can be shown (see Kobayashi and Nomizu (1963)) that g-completeness and m-completeness are equivalent for a positive definite metric.

A Lorentz metric, on the other hand, does not define a topological metric and so one is left only with g-completeness. One can distinguish three kinds of g-incompleteness: that of timelike, null and spacelike geodesics. If one cuts a regular point out of space–time, the resulting manifold is incomplete in all three ways and so one might hope that a space–time which was complete in one of the above senses would also be complete in the other two. Unfortunately this is not necessarily so (Kundt (1963)), as is shown by the following example given by Geroch (1968 b). Consider two-dimensional Minkowski space with coordinates x and t and metric g_{ab}. Define a new metric $\hat{g}_{ab} = \Omega^2 g_{ab}$ where the positive function Ω has the properties:

(1) $\Omega = 1$ outside the region between the vertical lines $x = -1$ and $x = +1$;

(2) Ω is symmetric about the t-axis, that is, $\Omega(t, x) = \Omega(t, -x)$;

(3) on the t-axis, $t^2\Omega \to 0$ as $t \to \infty$.

By (2) the t-axis is a timelike geodesic which by (3) is incomplete as $t \to \infty$. However every null and spacelike geodesic must leave and not re-enter the region between $x = -1$ and $x = +1$. Therefore by (1) the space is null and spacelike complete. In fact one can construct examples which are incomplete in any of the three possible ways and complete in the remaining two.

Timelike geodesic incompleteness has an immediate physical significance in that it presents the possibility that there could be freely moving observers or particles whose histories did not exist after (or before) a finite interval of proper time. This would appear to be an even more objectionable feature than infinite curvature and so it seems appropriate to regard such a space as singular. Although the affine parameter on a null geodesic does not have quite the same physical significance as proper time does on timelike geodesics, one should probably also regard a null geodesically incomplete space–time as singular both because null geodesics are the histories of zero rest-mass particles and because there are some examples (such as the Reissner–Nordström solution, §5.5) which one would think of as singular but which are timelike but not null geodesically complete. As nothing moves on spacelike curves, the significance of spacelike geodesic incompleteness is not so clear. We shall therefore adopt the view that *timelike and null geodesic completeness are minimum conditions for space–time to be considered singularity-free*. Therefore if a space–time is timelike or null geodesically incomplete, we shall say that it has a singularity.

The advantage of taking timelike and/or null incompleteness as being indicative of the presence of a singularity is that on this basis one can establish a number of theorems about their occurrence. However, the class of timelike and/or null incomplete space–times does not include all those one might wish to consider as singular in some sense. For example Geroch (1968b) has constructed a space–time which is geodesically complete but which contains an inextendible timelike curve of bounded acceleration and finite length. An observer with a suitable rocketship and a finite amount of fuel could traverse this curve. After a finite interval of time he would no longer be represented by a point of the space–time manifold. If one is going to say that there is a singularity in a space–time in which a freely falling observer comes to an untimely end, one should presumably do the same for an observer in a rocketship. What one needs is some generalization of the concept of an affine parameter to all C^1 curves, geodesic or non-geodesic. One could then define a notion of completeness by requiring that every C^1 curve of finite length as measured by such a parameter had an endpoint. The idea we are going to use seems to have been first suggested by Ehresman (1957), and has been reformulated in an elegant manner by Schmidt (1971).

Let $\lambda(t)$ be a C^1 curve through $p \in \mathcal{M}$ and let $\{\mathbf{E}_i\}$ ($i = 1, 2, 3, 4$) be a basis for T_p. One can parallelly propagate $\{\mathbf{E}_i\}$ along $\lambda(t)$ to obtain a basis for $T_{\lambda(t)}$ for each value of t. Then the tangent vector $\mathbf{V} = (\partial/\partial t)_{\lambda(t)}$ can be expressed in terms of the basis as $\mathbf{V} = V^i(t)\,\mathbf{E}_i$, and one can define a *generalized affine parameter* u on λ by

$$u = \int_p (\sum_i V^i V^i)^{\frac{1}{2}}\, dt.$$

The parameter u depends on the point p and the basis $\{\mathbf{E}_i\}$ at p. If $\{\mathbf{E}_{i'}\}$ is another basis at p, then there is some non-singular matrix $A_i{}^{j}$ such that

$$\mathbf{E}_i = \sum_{j'} A_i{}^{j'} \mathbf{E}_{j'}.$$

As $\{\mathbf{E}_{i'}\}$ and $\{\mathbf{E}_i\}$ are parallelly transported along $\lambda(t)$, this relation is maintained with constant $A_i{}^{j'}$. Thus

$$V^{i'}(t) = \sum_j A_j{}^{i'}\, V^j(t).$$

Since $A_i{}^{j'}$ is a non-singular matrix, there is some constant $C > 0$ such that

$$C \sum_i V^i V^i \leqslant \sum_{i'} V^{i'} V^{i'} \leqslant C^{-1} \sum_i V^i V^i.$$

Thus the length of a curve λ is finite in the parameter u if and only if it is finite in the parameter u'. If λ is a geodesic curve then u is an affine parameter on λ, but the beauty of the definition is that u can be defined on any C^1 curve. We shall say that $(\mathcal{M}, \mathbf{g})$ is *b-complete* (short for bundle complete, see §8.3) if there is an endpoint for every C^1 curve of finite length as measured by a generalized affine parameter. If the length is finite in one such parameter it will be finite in all such parameters, so one loses nothing by restricting the bases to be ortho-normal bases. If the metric $\mathbf{g}$ is positive definite, the generalized affine parameter defined by an orthonormal basis is arc-length and so b-completeness coincides with m-completeness. However b-complete-ness can be defined even if the metric is not positive definite; in fact it can be defined providing there is a connection on $\mathcal{M}$. Clearly b-completeness implies g-completeness, but the example quoted shows that the converse is not true.

We shall therefore define a space–time to be *singularity-free* if it is b-complete. This definition conforms with the requirement made above, that timelike and null geodesic completeness are minimum conditions for a space–time to be considered singularity-free. One might possibly wish to weaken this condition slightly, to say that space–time is singularity-free it it is only *non-spacelike b-complete*, i.e. if there is an endpoint for all non-spacelike C^1 curves with finite length as measured by a generalized affine parameter. However this definition would appear rather awkward in the bundle formulation of b-completeness which we shall give in § 8.3. In fact each of the theorems we give in § 8.2 implies that $(\mathcal{M}, \mathbf{g})$ is timelike or null g-incomplete and hence has a singularity by both the above definitions.

One feels intuitively that a singularity ought to involve the curvature becoming unboundedly large near a singular point. However since we have excluded singular points from our definition of space–time, difficulty arises in defining both 'near' and 'unboundedly large'. One can say that points on a b-incomplete curve are near the singularity if they correspond to values of a generalized affine parameter which is near the upper bound of that parameter. 'Unboundedly large' is more difficult, since the size of components of the curvature tensor depend on the basis in which it is measured. One possibility is to look at scalar polynomials in g_{ab}, η_{abcd}, and R_{abcd}. We shall say that a b-incomplete curve corresponds to a scalar polynomial curvature singularity (*s.p. curvature singularity*) if any of these scalar polynomials is unbounded on the incomplete curve. However, with a Lorentz metric these polynomials do not fully characterize the Riemann tensor since, as Penrose has pointed out, in plane-wave solutions the scalar polynomials are all zero but the Riemann tensor does not vanish. (This is similar to the fact that a non-zero vector may have zero length.) Thus the curvature might become very large in some sense even though the scalar polynomials remained small. Alternatively one might measure the components of the curvature tensor in a basis that was parallelly propagated along a curve. We shall say that a b-incomplete curve corresponds to a curvature singularity with respect to a parallelly propagated basis (a *p.p. curvature singularity*) if any of these components is unbounded on the curve. Clearly an s.p. curvature singularity implies a p.p. curvature singularity.

One might expect that in any physically realistic solution, a b-incomplete curve would correspond both to an s.p. and a p.p. curvature singularity. However an example of a solution where this does not seem to be true is provided by Taub–NUT space (§5.8). Here the incomplete geodesics are totally imprisoned in a compact neighbourhood of the horizon. As the metric is perfectly regular on this compact neighbourhood, the scalar polynomials in the curvature remain finite. Because of the special nature of this solution, the components of the curvature in a parallelly propagated basis along the imprisoned geodesics remains bounded. Since the imprisoned geodesics are contained in a compact set, one could not extend the manifold $\mathcal{M}$ to a larger four-dimensional Hausdorff paracompact manifold $\mathcal{M}'$, in which the incomplete geodesics could be continued. Thus there is no possibility of the incompleteness having arisen from the cutting out of singular points. Nevertheless it would be unpleasant to be moving on one of the incomplete timelike geodesics for although one's world-line never comes to an end and would continue to wind round and round inside the compact set, one would never get beyond a certain time in one's life. It would, therefore, seem reasonable to say that such a space–time was singular even though there is no p.p. or s.p. curvature singularity. By lemma 6.4.8, such totally imprisoned incompleteness can only occur if strong causality is violated. In §8.5 we shall show that in a generic space–time, a partially or totally imprisoned b-incomplete curve will correspond to a p.p. curvature singularity. We shall also show that the Taub–NUT kind of totally imprisoned incompleteness cannot occur if there is some matter present.

8.2 Singularity theorems

In §5.4 it was shown that there would be singularities in spatially homogeneous solutions under certain reasonable conditions. Similar theorems can be obtained for a number of other types of exact symmetry. Such results, although suggestive, do not necessarily have any physical significance because they depend on the symmetry being exact and clearly in any physical situation this will not be the case. It was therefore suggested by a number of authors that singularities were simply the result of symmetries and that they would not occur in general solutions. This view was supported by Lifshitz, Khalatnikov and co-workers who showed that certain classes of solutions with spacelike singularities did not have the full number of arbitrary functions expected in a general solution of the field equations (see Lifshitz and Khalatnikov (1963) for an account of this work). This presumably indicates that the Cauchy data which gave rise to such singularities is of measure zero in the set of all possible Cauchy data and so should not occur in the real universe. However more recently Belinskii, Khalatnikov and Lifshitz (1970) have found other classes of solutions which seem to have the full number of arbitrary functions and to contain singularities. They have therefore withdrawn the claim that singularities do not occur in general solutions. Their methods are interesting for the light they shed on the possible structure of singularities but it is not clear whether the power series which are used will converge. Neither does one obtain general conditions which imply that a singularity is inevitable. Nevertheless we may take their results as supporting our view that the singularities implied by the theorems of this section involve infinite curvature in general.

The first theorem about singularities which did not involve any assumption of symmetry was given by Penrose (1965c). It was designed to prove the occurrence of a singularity in a star which collapsed inside its Schwarzschild radius. If the collapse were exactly spherical, the solution could be integrated explicitly and a singularity would always occur. However it is not obvious that this would be the case if there were irregularities or a small amount of angular momentum. Indeed in Newtonian theory the smallest amount of angular momentum could prevent the occurrence of infinite density and cause the star to re-expand. However Penrose showed that the situation was very different in General Relativity: once the star had passed inside the Schwarzschild surface (the surface $r = 2m$) it could not come out again. In fact the Schwarzschild surface is defined only for an exactly spherically symmetric solution but the more general criterion used by Penrose is equivalent for such a solution and is applicable also to solutions without exact symmetry. It is that there should exist a *closed trapped surface* $\mathscr{T}$. By this is meant a C^2 closed (i.e. compact, without boundary) spacelike two-surface (normally, S^2) such that the two families of null geodesics orthogonal to $\mathscr{T}$ are converging at $\mathscr{T}$ (i.e. $_1\hat{\chi}_{ab}g^{ab}$ and $_2\hat{\chi}_{ab}g^{ab}$ are negative, where $_1\hat{\chi}_{ab}$ and $_2\hat{\chi}_{ab}$ are the two null second fundamental forms of $\mathscr{T}$. In the following chapters we shall discuss the circumstances under which such a surface would arise.) One may think of $\mathscr{T}$ as being in such a strong gravitational field that even the 'outgoing' light rays are dragged back and are, in fact, converging. Since nothing can travel faster than light, the matter within $\mathcal{T}$ is trapped inside a succession of two-surfaces of smaller and smaller area and so it seems that something must go wrong. That this is so is shown rigorously by Penrose's theorem:

Theorem 1

Space–time $(\mathcal{M}, \mathbf{g})$ cannot be null geodesically complete if:

(1) $R_{ab} K^a K^b \geqslant 0$ for all null vectors K^a (cf. §4.3);

(2) there is a non-compact Cauchy surface $\mathcal{H}$ in $\mathcal{M}$;

(3) there is a closed trapped surface $\mathcal{T}$ in $\mathcal{M}$.

Note: the method of proof is to show that the boundary of the future of $\mathcal{T}$ would be compact if $\mathcal{M}$ were null geodesically complete. This is then shown to be incompatible with $\mathcal{H}$ being non-compact.

Proof. The existence of a Cauchy surface implies that $\mathcal{M}$ is globally hyperbolic (proposition 6.6.3) and therefore causally simple (proposition 6.6.1). This means that the boundary of $J^+(\mathcal{T})$ will be $E^+(\mathcal{T})$ and will be generated by null geodesic segments which have past endpoints on $\mathcal{T}$ and which are orthogonal to $\mathcal{T}$. Suppose $\mathcal{M}$ were null geodesically complete. Then by conditions (1) and (3) and proposition 4.4.6 there would be a point conjugate to $\mathcal{T}$ along every future-directed null geodesic orthogonal to $\mathcal{T}$ within an affine distance $2c^{-1}$ where c is the value of $_n\hat{\chi}_{ab} g^{ab}$ at the point where the null geodesic intersects $\mathcal{T}$. By proposition 4.5.14, points on such a null geodesic beyond the point conjugate to $\mathcal{T}$ would lie in $I^+(\mathcal{T})$. Thus each generating segment of $\dot{J}^+(\mathcal{T})$ would have a future endpoint at or before the point conjugate to $\mathcal{T}$. At $\mathcal{T}$ one could assign, in a continuous manner, an affine parameter on each null geodesic orthogonal to $\mathcal{T}$. Consider the continuous map $\beta \colon \mathcal{T} \times [0, b] \times Q \to \mathcal{M}$ (Q is the discrete set 1, 2) defined by taking a point $p \in \mathcal{T}$ an affine distance $v \in [0, b]$ along one or other of the two future-directed null geodesics through p orthogonal to $\mathcal{T}$. Since $\mathcal{T}$ is compact, there will be some minimum value c_0 of $(-_1\hat{\chi}_{ab} g^{ab})$ and $(-_2\hat{\chi}_{ab} g^{ab})$. Then if $b_0 = 2c_0^{-1}$, $\beta(\mathcal{T} \times [0, b_0] \times Q)$ would contain $\dot{J}^+(\mathcal{T})$. Thus $\dot{J}^+(\mathcal{T})$ would be compact being a closed subset of a compact set. This would be possible if the Cauchy surface $\mathcal{H}$ were compact because then $\dot{J}^+(\mathcal{T})$ could meet up round the back and form a compact Cauchy surface homeomorphic to $\mathcal{H}$ (figure 49). However there is clearly going to be trouble if one demands that $\mathcal{H}$ is non-compact. To show this rigorously one can use the fact (see §2.6) that $\mathcal{M}$ admits a past-directed C^1 timelike

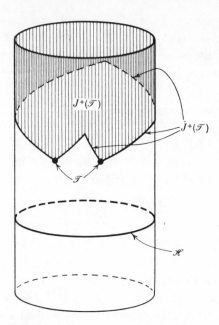

FIGURE 49. A two-dimensional section of a geodesically complete space with a compact Cauchy surface $\mathscr{H}$. The two-sphere $\mathscr{T}$ has a compact boundary $\dot{J}^+(\mathscr{T})$ to its future $J^+(\mathscr{T})$, as the outgoing null geodesics from $\mathscr{T}$ meet up round the back of the cylinder.

vector field. Each integral curve of this field will intersect $\mathscr{H}$ (as it is a Cauchy surface) and will intersect $\dot{J}^+(\mathscr{T})$ at most once. Thus they will define a continuous one-to-one map $\alpha \colon \dot{J}^+(\mathscr{T}) \to \mathscr{H}$. If $\dot{J}^+(\mathscr{T})$ were compact, its image $\alpha(\dot{J}^+(\mathscr{T}))$ would also be compact and would be homeomorphic to $\dot{J}^+(\mathscr{T})$. However as $\mathscr{H}$ is non-compact, $\alpha(\dot{J}^+(\mathscr{T}))$ could not contain the whole of $\mathscr{H}$ and would therefore have to have a boundary in $\mathscr{H}$. This would be impossible since by proposition 6.3.1, $\dot{J}^+(\mathscr{T})$, and therefore $\alpha(\dot{J}^+(\mathscr{T}))$, would be a three-dimensional manifold (without boundary). This shows that the assumption that $\mathscr{M}$ is null geodesically complete (which we made in order to prove $\dot{J}^+(\mathscr{T})$ compact) is incorrect.   □

Condition (1) of this theorem (that $R_{ab}K^aK^b \geqslant 0$ for any null vector $\mathbf{K}$) was discussed in §4.3. It will hold no matter what value the value of the constant Λ, provided that the energy density is positive for every observer. It will be shown in chapter 9 that condition (3) (that there is a closed trapped surface) should be satisfied in at least some region of space–time. This leaves condition (2) (that there is a non-compact spacelike surface $\mathcal{H}$ which is a Cauchy surface) to be discussed. By proposition 6.4.9, the existence of spacelike surfaces is guaranteed provided one assumes stable causality. That the spacelike surface $\mathcal{H}$ be non-compact is not too serious a restriction since the only place it was used was to show that $\alpha(\dot{J}^+(\mathcal{T}))$ could not be the whole of $\mathcal{H}$. This could also be shown if, instead of taking $\mathcal{H}$ to be non-compact, one required that there exist a future-directed inextendible curve from $\mathcal{H}$ which did not intersect $\dot{J}^+(\mathcal{T})$. In other words, the theorem would still hold even if $\mathcal{H}$ were compact, provided there was some observer who could avoid falling into the collapsing star. This might not be possible if the whole universe were collapsing also, but in such a case one would expect singularities anyway as will be shown presently. The real weakness of the theorem is the requirement that $\mathcal{H}$ be a Cauchy surface. This was used in two places: first, to show that $\mathcal{M}$ was causally simple which implied that the generators of $\dot{J}^+(\mathcal{T})$ had past endpoints on $\mathcal{T}$, and second, to ensure that under the map α every point of $\dot{J}^+(\mathcal{T})$ was mapped into a point of $\mathcal{H}$. That the Cauchy surface condition is necessary is shown by an example due to Bardeen. This has the same global structure as the Reissner–Nordström solution except that the real singularities at $r = 0$ have been smoothed out so that they are just the origins of polar coordinates. The space–time obeys the condition $R_{ab}K^aK^b \geqslant 0$ for any null but not timelike vector **K**, and contains closed trapped surfaces. The only way in which it fails to satisfy the conditions of the theorem is that it does not have a Cauchy surface.

It therefore seems that what the theorem tells us is that in a collapsing star there will occur either a singularity or a Cauchy horizon. This is a very important result since in either case our ability to predict the future breaks down. However it does not answer the question of whether singularities occur in physically realistic solutions. To decide this we need a theorem which does not assume the existence of Cauchy surfaces. One of the conditions of such a theorem must be that $R_{ab}K^aK^b \geqslant 0$ for all *timelike* as well as null vectors, since failure to obey this condition is the only way in which Bardeen's example is unreasonable. The theorem we shall give below requires this condition and also the chronology condition that there be no closed timelike curves. On the other hand it is applicable to a wider class of situations since the existence of a closed trapped surface is now only one of three possible conditions. One of these alternative conditions is that there should be a compact partial Cauchy surface, and the other is that there

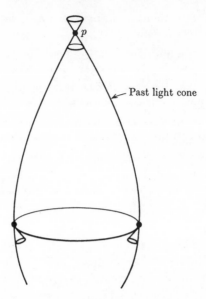

FIGURE 50. A point p whose past light cone starts reconverging.

should be a point whose past (or future) light cone starts converging again (figure 50). The first of these other conditions is satisfied in a spatially closed solution while the second is closely related to the existence of a closed trapped surface but is in a form which is more convenient for some purposes; for in the case in which the light cone is our own past light cone, one can directly determine whether this condition is satisfied. In the last chapter it will be shown that recent observations of the microwave background indicate that it is.

The precise statement is:

Theorem 2 (Hawking and Penrose (1970))

Space–time $(\mathcal{M}, \mathbf{g})$ is not timelike and null geodesically complete if:

(1) $R_{ab}K^aK^b \geqslant 0$ for every non-spacelike vector **K** (cf. §4.3).

(2) The generic condition is satisfied (§4.4), i.e. every non-spacelike geodesic contains a point at which $K_{[a}R_{b]cd[e}K_{f]}K^cK^d \neq 0$, where **K** is the tangent vector to the geodesic.

(3) The chronology condition holds on $\mathcal{M}$ (i.e. there are no closed timelike curves).

(4) There exists at least one of the following:

(i) a compact achronal set without edge,

(ii) a closed trapped surface,

(iii) a point p such that on every past (or every future) null geodesic from p the divergence $\hat{\theta}$ of the null geodesics from p becomes negative (i.e. the null geodesics from p are focussed by the matter or curvature and start to reconverge).

Remark. An alternative version of the theorem is that the following three conditions cannot all hold:

(a) every inextendible non-spacelike geodesic contains a pair of conjugate points;

(b) the chronology condition holds on $\mathcal{M}$;

(c) there is an achronal set $\mathcal{S}$ such that $E^+(\mathcal{S})$ or $E^-(\mathcal{S})$ is compact. (We shall say that such a set is, respectively, *future trapped* or *past trapped*).

In fact it is this form of the theorem that we shall prove. The other version will then follow since if $\mathcal{M}$ were timelike and null geodesically complete, (1) and (2) would imply (a) by propositions 4.4.2 and 4.4.5, (3) is the same as (b), and (1) and (4) would imply (c), since in case (i) $\mathcal{S}$ would be the compact achronal set without edge and

$$E^+(\mathcal{S}) = E^-(\mathcal{S}) = \mathcal{S};$$

in cases (ii) and (iii) $\mathcal{S}$ would be the closed trapped surface and the point p respectively, and by propositions 4.4.4, 4.4.6, 4.5.12 and 4.5.14 $E^+(\mathcal{S})$ and $E^-(\mathcal{S})$ would be compact respectively, being the intersections of the closed sets $\dot{J}^+(\mathcal{S})$ and $\dot{J}^-(\mathcal{S})$ with compact sets consisting of all the null geodesics of some finite length from $\mathcal{S}$.

Proof. As the proof is rather long, we shall break it up by first establishing a lemma and corollary. We note that by an argument similar to that of proposition 6.4.6, (a) and (b) imply that strong causality holds on $\mathcal{M}$.

Lemma 8.2.1

If $\mathcal{S}$ is a closed set and if the strong causality condition holds on $\bar{J}^+(\mathcal{S})$ then $H^+(\bar{E}^+(\mathcal{S}))$ is non-compact or empty (figure 51).

By lemma 6.3.2, through every point $q \in \dot{J}^+(\mathcal{S}) - \mathcal{S}$ there is a past-directed null geodesic segment lying in $\dot{J}^+(\mathcal{S})$ which has a past endpoint if and only if $q \in E^+(\mathcal{S})$. (Note that as we no longer assume the existence of a Cauchy surface, $\mathcal{M}$ may not be causally simple and so $\dot{J}^+(\mathcal{S}) - E^+(\mathcal{S})$ may be non-empty.) Therefore if $q \in \dot{J}^+(\mathcal{S}) - E^+(\mathcal{S})$, there is a past-inextendible null geodesic through q which lies in $\dot{J}^+(\mathcal{S})$

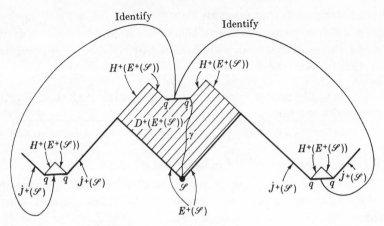

FIGURE 51. A future trapped set $\mathscr{S}$; null lines are at $\pm\,45°$, three lines have been identified and the points q are at infinity. The achronal sets $E^+(\mathscr{S}), J^+(\mathscr{S})$ and $H^+(E^+(\mathscr{S}))$ are shown. A future-inextendible timelike curve $\gamma \in D^+(E^+(\mathscr{S}))$ is shown.

and so does not intersect $I^-(J^+(\mathscr{S}))$. From lemma 6.6.4 it then follows that q is not in $D^+(J^+(\mathscr{S})) - H^+(J^+(\mathscr{S}))$. Hence

$$D^+(\bar{E}^+(\mathscr{S})) - H^+(\bar{E}^+(\mathscr{S})) = D^+(J^+(\mathscr{S})) - H^+(J^+(\mathscr{S}))$$

and $$H^+(\bar{E}^+(\mathscr{S})) \subset H^+(J^+(\mathscr{S})).$$

Now suppose that $H^+(\bar{E}^+(\mathscr{S}))$ was non-empty and compact. Then it could be covered by a finite number of local causality neighbourhoods $\mathscr{U}_i$. Let p_1 be a point of $J^+(\mathscr{S}) \cap [\mathscr{U}_1 - D^+(J^+(\mathscr{S}))]$. Then from p_1 there would be a past-inextendible non-spacelike curve λ_1 which did not intersect either $J^+(\mathscr{S})$ or $D^+(\bar{E}^+(\mathscr{S}))$. Since the $\mathscr{U}_i$ have compact closure, λ_1 would leave $\mathscr{U}_1$. Let q_1 be a point on λ_1 not in $\mathscr{U}_1$. Then since $q_1 \in J^+(\mathscr{S})$ there would be a non-spacelike curve μ_1 from q_1 to $\mathscr{S}$. This curve would intersect $D^+(\bar{E}^+(\mathscr{S}))$ and hence would intersect some $\mathscr{U}_i$ other than $\mathscr{U}_1$ (say, $\mathscr{U}_2$). Then let p_2 be a point of $\mu_1 \cap [\mathscr{U}_2 - D^+(J^+(\mathscr{S}))]$ and continue as before.

This leads to a contradiction since there were only a finite number of the local causality neighbourhoods $\mathscr{U}_i$, and one could not return to an earlier $\mathscr{U}_j$ because no non-spacelike curve can intersect a $\mathscr{U}_i$ more than once. Thus $H^+(\bar{E}^+(\mathscr{S}))$ must be non-compact or empty. $\square$

Corollary

If $\mathscr{S}$ is a future trapped set, there is a future-inextendible timelike curve γ contained in $D^+(E^+(\mathscr{S}))$.

Put a timelike vector field on $\mathcal{M}$. If every integral curve of this field which intersected $E^+(\mathcal{S})$ also intersected $H^+(E^+(\mathcal{S}))$ they would define a continuous one–one mapping of $E^+(\mathcal{S})$ onto $H^+(E^+(\mathcal{S}))$ and hence $H^+(E^+(\mathcal{S}))$ would be compact. The intersection of $I^+(\mathcal{S})$ with a curve which does not intersect $H^+(E^+(\mathcal{S}))$ gives the desired curve γ (figure 51 indicates one possible situation).   □

Now consider the compact set $\mathcal{F}$ defined as $E^+(\mathcal{S}) \cap \overline{J^-(\gamma)}$. Since γ was contained in $\operatorname{int} I^+(E^+(\mathcal{S}))$, $E^-(\mathcal{F})$ would consist of $\mathcal{F}$ and a portion of $\dot{J}^-(\gamma)$. Since γ was future inextendible, the null geodesic segments generating $\dot{J}^-(\gamma)$ could have no future endpoints. But by (a) every inextendible non-spacelike geodesic contains a pair of conjugate points. Thus by proposition 4.5.12, the past-inextendible extension ν' of each generating segment ν of $\dot{J}^-(\gamma)$ would enter $I^-(\gamma)$. There would be a past endpoint for ν at or before the first point p of $\overline{\nu' \cap I^-(\gamma)}$. As $I^-(\gamma)$ would be an open set, a neighbourhood of p would contain points in $I^-(\gamma)$ on neighbouring null geodesics. Thus the affine distance of the points p from $\mathcal{F}$ would be upper semi-continuous, and $E^-(\mathcal{F})$ would be compact being the intersection of the closed set $\dot{J}^-(\gamma)$ with a compact set generated by null geodesic segments from $\mathcal{F}$ of some bounded affine length. It would then follow from the lemma that there would be a past-inextendible timelike curve λ contained in $\operatorname{int} D^-(E^-(\mathcal{F}))$ (figure 52). Let a_n be an infinite sequence of points on λ such that:

(I) $a_{n+1} \in I^-(a_n)$,

(II) no compact segment of λ contains more than a finite number of the a_n.

Let b_n be a similar sequence on γ but with I^+ instead of I^- in (I) and with $b_1 \in I^+(a_1)$.

As γ and λ were contained in the globally hyperbolic set $\operatorname{int} D(E^-(\mathcal{F}))$ (proposition 6.6.3), there would be a non-spacelike geodesic μ_n of maximum length between each a_n and the corresponding b_n (proposition 6.7.1). Each would intersect the compact set $E^+(\mathcal{S})$. Thus there would be a $q \in E^+(\mathcal{S})$ which was a limit point of the $\mu_n \cap E^+(\mathcal{S})$ and a non-spacelike direction at q which is a limit of the directions of the μ_n. (The point q and the direction at q define a point of the bundle of directions over $\mathcal{M}$. Such a limit point exists because the portion of the bundle over $E^+(\mathcal{S})$ is compact.) Let μ'_n be a subsequence of the μ_n such that $\mu'_n \cap E^+(\mathcal{S})$ converges to q and such that the directions of the μ'_n at $E^+(\mathcal{S})$ converge to the limit direction.

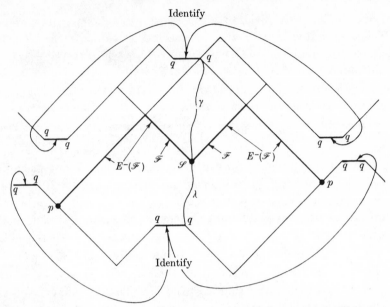

FIGURE 52. As figure 51, but with three further lines identified. $\mathscr{F}$ is the set $E^+(\mathscr{S}) \cap \bar{J}^-(\gamma)$; the points p are past endpoints of null geodesic generating segments of $E^-(\mathscr{F})$. The curve λ is a past-inextendible timelike curve contained in $\operatorname{int} D^-(E^-(\mathscr{F}))$.

(More precisely, the points defined by the μ'_n in the bundle of directions over $E^+(\mathscr{S})$ converge to the limit point.) Let μ be the inextendible geodesic through q in the limit direction. By (a) there would be conjugate points x and y on μ with $y \in I^+(x)$. Let x' and y' be on μ to the past and future of x and y respectively. By proposition 4.5.8, there is some $\epsilon > 0$ and some timelike curve α from x' to y' whose length is ϵ plus the length of μ from x' to y'. Let $\mathscr{U}$ and $\mathscr{V}$ be convex normal coordinate neighbourhoods of x' and y' respectively, each of which contains no curve of length $\frac{1}{4}\epsilon$. Let x'' and y'' be $\dot{\mathscr{U}} \cap \alpha$ and $\dot{\mathscr{V}} \cap \alpha$ respectively. Let x'_n and y'_n be points on μ'_n converging to x' and y' respectively. For n sufficiently large, the length μ'_n from x'_n to y'_n will be less than $\frac{1}{4}\epsilon$ plus the length of μ from x' to y'. Also for n sufficiently large, x'_n and y'_n would be in $I^-(x'', \mathscr{U})$ and $I^+(y'', \mathscr{V})$ respectively. Then going from x'_n to x'', along α to y'', and from y'' to y'_n would give a longer non-spacelike curve than μ'_n from x'_n to y'_n. But by property (II), a'_n would lie to the past of x'_n on μ'_n and b'_n would lie to the future of y'_n on μ'_n, for n large enough. Therefore μ'_n ought to be the longest non-spacelike curve from x'_n to y'_n. This establishes the desired contradiction.   □

While this theorem establishes the existence of singularities under very general conditions, it has the disadvantage of not showing whether the singularity is in the future or the past. In case (ii) of condition (4), when there is a compact spacelike surface, one has no reason to believe that it should be in the future rather than in the past, but in case (i) when there is a closed trapped surface, one would expect the singularity to be in the future, and in case (iii) when the past null cone starts reconverging, one would expect the singularity to be in the past. One can show that there is a singularity in the past if condition (iii) is strengthened somewhat to say that all past-directed timelike as well as null geodesics from p start to reconverge within a compact region in $J^-(p)$.

Theorem 3 (Hawking (1967))

If (1) $R_{ab}K^aK^b \geqslant 0$ for every non-spacelike vector **K** (cf. §4.3);

(2) the strong causality condition holds on $(\mathcal{M}, \mathbf{g})$;

(3) there is some past-directed unit timelike vector **W** at a point p and a positive constant b such that if **V** is the unit tangent vector to the past-directed timelike geodesics through p, then on each such geodesic the expansion $\theta \equiv V^a{}_{;a}$ of these geodesics becomes less than $-3c/b$ within a distance b/c from p, where $c \equiv -W^aV_a$,

then there is a past incomplete non-spacelike geodesic through p.

Let K^a be the parallelly propagated tangent vector to the past-directed non-spacelike geodesics through p, normalized by $K^aW_a = -1$. Then for the timelike geodesics through p, $K^a = c^{-1}V^a$ and so $K^a{}_{;a} = c^{-1}V^a{}_{;a}$. Since $K^a{}_{;a}$ is continuous on the non-spacelike geodesics, it will become less than $-3/b$ on the null geodesics through p within an affine distance b. If $\mathbf{Y}_1$, $\mathbf{Y}_2$, $\mathbf{Y}_3$ and $\mathbf{Y}_4$ are a pseudo-orthonormal tetrad on these null geodesics with $\mathbf{Y}_1$ and $\mathbf{Y}_2$ spacelike unit vectors and $\mathbf{Y}_3$ and $\mathbf{Y}_4$ null with $Y_3{}^aY_{4a} = -1$ and $\mathbf{Y}_4 = \mathbf{K}$, the expansion $\hat{\theta}$ of the null geodesics through p is defined as

$$\hat{\theta} = K_{a;b}(Y_1{}^aY_1{}^b + Y_2{}^aY_2{}^b)$$
$$= K^a{}_{;a} + K_{a;b}(Y_3{}^aY_4{}^b + Y_4{}^aY_3{}^b).$$

The second term is zero because K^a is parallelly propagated. The third term can be expressed as $\frac{1}{2}(K_aK^a)_{;b}Y_3{}^b$, which is less than zero as K_aK^a is zero on the null geodesics and negative for timelike geodesics. This shows that $\hat{\theta}$ will become less than $-3/b$ within an affine distance b along each null geodesic from p. Thus if all past-directed null geodesics from p were complete, $E^-(p)$ would be compact. Any point $q \in J^-(E^-(p)) - E^-(p)$ would be in $I^-(p)$. Thus it could not be in $J^+(E^-(p))$ since $E^-(p)$ is achronal. Therefore

$$J^+(E^-(p)) \cap J^-(E^-(p)) = E^-(p)$$

and so would be compact. Then by proposition 6.6.7, $D^-(E^-(p))$ would be globally hyperbolic. By proposition 6.7.1, each point $r \in D^-(E^-(p))$ would be joined to p by a non-spacelike geodesic which did not contain any point conjugate to p between r and p. Thus by proposition 4.4.1, $D^-(E^-(p))$ would be contained in $\exp_p(F)$ where F is the compact region of T_p consisting of all past-directed non-spacelike vectors K^a such that $K^a W_a \leqslant -2b$. If all past non-spacelike geodesics from p were complete, $\exp_p(K^a)$ would be defined for every $K^a \in F$, and so $\exp_p(F)$ would be compact being the image of a compact set under a continuous map. However by the corollary to lemma 8.2.1, $D^-(E^-(p))$ contains a past-inextendible timelike curve. By proposition 6.4.7 this could not be totally imprisoned in the compact set $\exp_p(F)$, therefore the assumption that all past-directed non-spacelike geodesics from p are complete must be false.   □

Theorems 2 and 3 are the most useful theorems on singularities since it can be shown that their conditions are satisfied in a number of physical situations (see next chapter). However it might be that what occurred was not a singularity but a closed timelike curve, violating the causality conditions. This would be much worse than the mere breakdown of prediction which was the alternative after theorem 1, and it is our personal opinion that it would be physically more objectionable than a singularity. Nevertheless one would like to know whether such causality violations would prevent the occurrence of singularities. The following theorem shows that they cannot in certain situations. This means that we have to take singularities seriously and it gives us confidence that, in general, causality breakdowns are not the way out.

Theorem 4 (Hawking (1967))

Space–time is not timelike geodesically complete if:

   (1) $R_{ab} K^a K^b \geqslant 0$ for every non-spacelike vector $\mathbf{K}$ (cf. §4.3);

   (2) there exists a compact spacelike three-surface $\mathscr{S}$ (without edge);

   (3) the unit normals to $\mathscr{S}$ are everywhere converging (or everywhere diverging) on $\mathscr{S}$.

Remarks. Condition (2) may be interpreted as saying that the universe is spatially closed and condition (3) as saying that it is contracting (or expanding). As explained in § 6.5 one may take a covering manifold $\hat{\mathcal{M}}$ in which each connected component of the image of $\mathcal{S}$ is diffeomorphic to $\mathcal{S}$ and is a partial Cauchy surface in $\hat{\mathcal{M}}$. We shall work in $\hat{\mathcal{M}}$ and shall denote by $\hat{\mathcal{S}}$ one connected component of the image of $\mathcal{S}$. Considering the Cauchy evolution problem in $\hat{\mathcal{M}}$ one sees that the occurrence of singularities (though not necessarily their nature) is a stable property of the Cauchy data on $\hat{\mathcal{S}}$ since a sufficiently small variation of the data on $\hat{\mathcal{S}}$ will not violate condition (3). This is a counterexample to the conjecture by Lifshitz and Khalatnikov that singularities occur only for a set of Cauchy data of measure zero, though it must be remembered that the definition of a singularity adopted here is not that used by Lifshitz and Khalatnikov

Proof. By conditions (2) and (3) the contraction $\chi^a{}_a$ of the second fundamental form of $\hat{\mathcal{S}}$ has a negative upper bound on $\mathcal{S}$. Thus if $\mathcal{M}$ (and hence $\hat{\mathcal{M}}$) was timelike geodesically complete there would be a point conjugate to $\hat{\mathcal{S}}$ on every future-directed geodesic orthogonal to $\hat{\mathcal{S}}$ within a finite upper bound b of distance from $\hat{\mathcal{S}}$ (proposition 4.4.3). But by the corollary to proposition 6.7.1, to every point $q \in D^+(\hat{\mathcal{S}})$ there is a future-directed geodesic orthogonal to $\hat{\mathcal{S}}$ which does not contain any point conjugate to $\hat{\mathcal{S}}$ between $\hat{\mathcal{S}}$ and q. Let $\beta: \hat{\mathcal{S}} \times [0, b] \to \hat{\mathcal{M}}$ be the differentiable map which takes a point $p \in \hat{\mathcal{S}}$ a distance $s \in [0, b]$ up the future-directed geodesic through p orthogonal to $\hat{\mathcal{S}}$. Then $\beta(\hat{\mathcal{S}} \times [0, b])$ would be compact and would contain $D^+(\hat{\mathcal{S}})$. Thus $\bar{D}^+(\hat{\mathcal{S}})$ and hence $H^+(\hat{\mathcal{S}})$ would be compact. If one assumed the strong causality condition the desired contradiction would follow from lemma 8.2.1. However even without strong causality one can obtain a contradiction. Consider a point $q \in H^+(\hat{\mathcal{S}})$. Since every past-directed non-spacelike curve from q to $\hat{\mathcal{S}}$ would consist of a (possibly zero) null geodesic segment in $H^+(\hat{\mathcal{S}})$ and then a non-spacelike curve in $D^+(\hat{\mathcal{S}})$, it follows that $d(\hat{\mathcal{S}}, q)$ would be less than or equal to b. Thus, as d is lower semi-continuous, one could find an infinite sequence of points $r_n \in D^+(\hat{\mathcal{S}})$ converging to q such that $d(\hat{\mathcal{S}}, r_n)$ converged to $d(\hat{\mathcal{S}}, q)$. To each r_n there would correspond at least one element $\beta^{-1}(r_n)$ of $\hat{\mathcal{S}} \times [0, b]$ Since $\hat{\mathcal{S}} \times [0, b]$ is compact there would be an element (p, s) which was a limit point of the $\beta^{-1}(r_n)$. By continuity $s = d(\hat{\mathcal{S}}, q)$ and $\beta(p, s) = q$. Thus to every point $q \in H^+(\hat{\mathcal{S}})$ there would be a timelike geodesic of length $d(\hat{\mathcal{S}}, q)$ from $\hat{\mathcal{S}}$. Now let

$q_1 \in H^+(\mathscr{S})$ lie to the past of q on the same null geodesic generator λ of $H^+(\mathscr{S})$. Joining the geodesic of length $d(\mathscr{S}, q_1)$ from $\mathscr{S}$ to q_1 to the segment of λ between q_1 and q, one would obtain a non-spacelike curve of length $d(\mathscr{S}, q_1)$ from $\mathscr{S}$ to q which could be varied to give a longer curve between these endpoints (proposition 4.5.10). Thus $d(\mathscr{S}, q)$, $q \in H^+(\mathscr{S})$, would strictly decrease along every past-directed generator of $H^+(\mathscr{S})$. But by proposition 6.5.2, such generators could have no past endpoints. This leads to a contradiction since as $d(\mathscr{S}, q)$ is lower semi-continuous in q, it would have a minimum on the compact set $H^+(\mathscr{S})$. □

Condition (2) that $\mathscr{S}$ is compact is necessary, since in Minkowski space $(\mathscr{M}, \boldsymbol{\eta})$ the non-compact surface $\mathscr{S} : (x^1)^2 + (x^2)^2 + (x^3)^2 - (x^4)^2 = -1$, $x^4 < 0$, is a partial Cauchy surface with $\chi^a{}_a = -3$ at all points. If one took the region of Minkowski space defined by

$$x^4 < 0, \quad (x^1)^2 + (x^2)^2 + (x^3)^2 - (x^4)^2 < 0,$$

one could identify points under a discrete group of isometries G such that $\mathscr{S}/G$ was compact (Löbell (1931)). As required by theorem 4, the space $(\mathscr{M}/G, \boldsymbol{\eta})$ would be timelike geodesically incomplete because one could not extend the identification under G to the whole of $\mathscr{M}$ (neither conditions (1) nor (2) of §5.8 would hold at the origin). In this case the incompleteness singularity arises from bad global properties and is not accompanied by a curvature singularity. This example was suggested by Penrose.

Conditions (2) and (3) can be replaced by:

(2′) $\mathscr{S}$ is a Cauchy surface for $\mathscr{M}$;
(3′) $\chi^a{}_a$ is bounded away from zero on $\mathscr{S}$;

since in this case there cannot be a Cauchy horizon, yet all the future-directed timelike curves from $\mathscr{S}$ must have lengths less than some finite upper bound.

Geroch (1966) has shown that if condition (2) holds, and if conditions (1) and (3) are replaced by:

(1″) $R_{ab} K^a K^b \geqslant 0$ for every non-spacelike vector, equality holding only if $R_{ab} = 0$;
(3″) there is a point $p \in \mathscr{S}$ such that any inextendible non-spacelike curve which intersects $\mathscr{S}$ also intersects both $J^+(p)$ and $J^-(p)$;

then either the Cauchy development of $\mathscr{S}$ is flat, or $\mathscr{M}$ is timelike geodesically incomplete.

Condition (3″) requires that an observer at p can see, and be seen by, every particle that intersects $\mathscr{S}$. The method of proof is to consider all spacelike surfaces without edge which contain p. One can form a topological space $S(p)$ out of all these surfaces, in a manner analogous to that in which one forms a topological space out of all the non-spacelike curves between two points. Conditions (2) and (3″) then imply that $S(p)$ is compact. One can show that the area of the surfaces is an upper semi-continuous function on $S(p)$ and so there will be some surface $\mathscr{S}'$ through p which has an area greater than or equal to that of any other surface. By a variation argument similar to that used for non-spacelike curves, one can show that $\chi^a{}_a$ vanishes everywhere on $\mathscr{S}'$ except possibly at p, where the surface may not be differentiable.

Consider a one-parameter family of spacelike surfaces $\mathscr{S}(u)$ where $\mathscr{S}(0) = \mathscr{S}'$. The variation vector $\mathbf{W} \equiv \partial/\partial u$ can be expressed as $f\mathbf{n}$ where $\mathbf{n}$ is the unit normal to the surfaces and f is some function. One can apply the Raychaudhuri equation to the congruence of integral curves of $\mathbf{W}$ to show

$$\partial\theta/\partial u = f\{-\tfrac{1}{3}\theta^2 - 2\sigma^2 - R_{ab}n^an^b + f^{-1}f_{;ab}h^{ab}\},$$

where $\theta \equiv \chi^a{}_a, \quad \sigma_{ab} \equiv \chi_{ab} - \tfrac{1}{3}\theta h_{ab}, \quad h_{ab} \equiv g_{ab} + n_an_b,$

and $$\sigma^2 = \tfrac{1}{2}\sigma_{ab}\sigma^{ab}.$$

If there is some point $q \in \mathscr{S}'$ at which $R_{ab}n^an^b \neq 0$ or $\chi_{ab} \neq 0$ one can find an f such that $\partial\theta/\partial u$ is negative everywhere on S'. If $R_{ab}n^an^b$ and χ_{ab} were zero everywhere on $\mathscr{S}'$, but there was some point q on $\mathscr{S}'$ at which $C_{abcd}n^bn^d$ was not equal to zero, then $\partial\sigma/\partial u \neq 0$ and one could find an f such that $\partial\theta/\partial u = 0$ and $\partial^2\theta/\partial u^2 < 0$ everywhere on $\mathscr{S}'$. In either case, one would obtain a surface $\mathscr{S}''$ on which $\chi^a{}_a < 0$ everywhere, and so $\hat{\mathscr{M}}$ would be timelike geodesically incomplete by theorem 4. If R_{ab}, χ_{ab} and $C_{abcd}n^bn^d$ were zero everywhere on $\mathscr{S}'$, then the Ricci identities for n^a show that $C_{abcd} = 0$ on $\mathscr{S}'$. Hence space–time is flat in $D(\mathscr{S})$. An example in which conditions (1″), (2) and (3″) hold and in which $D(\mathscr{S})$ is flat is Minkowski space with $\{x^1, x^2, x^3, x^4\}$ identified with $\{x^1 + 1, x^2, x^3, x^4\}, \{x^1, x^2 + 1, x^3, x^4\}$, and $\{x^1, x^2, x^3 + 1, x^4\}$. This is geodesically complete. However the example given previously also satisfies these conditions and shows that $D(\mathscr{S})$ can be both geodesically incomplete and flat.

8.3 The description of singularities

The preceding theorems prove the occurrence of singularities in a large
class of solutions but give little information as to their nature. To
investigate this in more detail, one would need to define what one
meant by the size, shape, location and so on of a singularity. This would
be fairly easy if the singular points were included in the space–time
manifold. However it would be impossible to determine the manifold
structure at such points by physical measurements. In fact there
would be many manifold structures which agreed for the non-singular
regions but which differed for the singular points. For example, the
manifold at the $t = 0$ singularity in the Robertson–Walker solutions
could be that described by the coordinates

$$\{t, r \cos \theta, r \sin \theta \cos \phi, \, r \sin \theta \sin \phi\}$$

or that described by

$$\{t, Sr \cos \theta, \, Sr \sin \theta \cos \phi, \, Sr \sin \theta \sin \phi\}.$$

In the first case the singularity would be a three-surface, in the second
case a single point.

What is needed is a prescription for attaching some sort of boundary
∂ to $\mathcal{M}$ which is uniquely determined by measurements at non-
singular points, i.e. by the structure of $(\mathcal{M}, \mathbf{g})$. One would then like to
define at least a topology, and possibly a differentiable structure and
metric, on the space $\mathcal{M}^+ \equiv \mathcal{M} \cup \partial$. One possibility would be to use the
method of indecomposable infinity sets described in §6.8. However
since this depends only on the conformal metric, it does not distinguish
between infinity and singular points at a finite distance. To make this
distinction it would seem one should base one's construction for $\mathcal{M}^+$
on the criterion that has been adopted for the existence of a singularity:
namely b-incompleteness. An elegant way of doing this has been
developed by Schmidt. This supersedes earlier constructions by
Hawking (1966b) and Geroch (1968a) which defined the singular
points as equivalence classes of incomplete geodesics. These constructions did not necessarily provide endpoints for all b-incomplete curves,
such as incomplete timelike curves of bounded acceleration. There was
also a certain ambiguity in their definition of equivalence classes.
Schmidt's construction does not suffer from these weaknesses.

Schmidt's procedure is to define a positive definite metric $\mathbf{e}$ on the
bundle of orthonormal frames $\pi \colon O(\mathcal{M}) \to \mathcal{M}$. Here $O(\mathcal{M})$ is the set of
all orthonormal four-tuples of vectors $\{\mathbf{E}_a\}$, $\mathbf{E}_a \in T_p$ for each $p \in \mathcal{M}$

(a ranges from 1 to 4), and π is the projection which maps a basis at a point p to the point p. It turns out that $O(\mathcal{M})$ is m-incomplete in the metric **e** if and only if $\mathcal{M}$ is b-incomplete. If $O(\mathcal{M})$ is m-incomplete, one can form the metric space completion $\overline{O(\mathcal{M})}$ of $O(\mathcal{M})$ by Cauchy sequences. The projection π can be extended to $\overline{O(\mathcal{M})}$, and the quotient of $\overline{O(\mathcal{M})}$ by π is defined to be $\mathcal{M}^+$ which is the union of $\mathcal{M}$ with a set of additional points ∂. The set ∂ consists of the singular points of $\mathcal{M}$ in the sense that it is the set of endpoints for every b-incomplete curve in $\mathcal{M}$.

To perform this construction, we recall (§ 2.9) that the connection on $\mathcal{M}$ given by the metric $\mathbf{g}$ defines a four-dimensional *horizontal subspace* H_u of the ten-dimensional tangent space T_u at the point $u \in O(\mathcal{M})$. Then T_u is the direct sum of H_u and the vertical subspace V_u consisting of all the vectors in T_u which are tangent to the fibre $\pi^{-1}(\pi(u))$. We now construct a basis $\{\mathbf{G}_A\} = \{\mathbf{E}_a, \mathbf{F}_i\}$ for T_u where A runs from 1 to 10, a runs from 1 to 4 and i runs from 1 to 6; $\{\overline{\mathbf{E}}_a\}$ is a basis for H_u, and $\{\mathbf{F}_i\}$ is a basis for V_u.

Given any vector $\mathbf{X} \in T_{\pi(u)}(\mathcal{M})$ there is a unique vector $\overline{\mathbf{X}} \in H_u(O(\mathcal{M}))$ such that $\pi_* \overline{\mathbf{X}} = \mathbf{X}$. Thus on $O(\mathcal{M})$ there are four uniquely defined horizontal vector fields $\overline{\mathbf{E}}_a$ which are the horizontal lifts of the orthonormal basis vectors $\mathbf{E}_a$ for each point $u \in O(\mathcal{M})$. The integral curves of the field $\overline{\mathbf{E}}_a$ in $O(\mathcal{M})$ represent parallel propagation of the basis $\{\mathbf{E}_a\}$ along the geodesic in $\mathcal{M}$ in the direction of the vector $\mathbf{E}_a$.

The group $O(3, 1)$, the multiplicative group of all non-singular 4×4 real Lorentz matrices A_{ab}, acts in the fibres of $O(\mathcal{M})$ sending a point $u = \{p, \mathbf{E}_a\} \in O(\mathcal{M})$ to the point $A(u) = \{p, A_{ab} \mathbf{E}_b\} \in O(\mathcal{M})$. One can regard $O(3, 1)$ as a six-dimensional manifold and represent the tangent space $T_I(O(3, 1))$ to $O(3, 1)$ at the unit matrix I by the vector space of all 4×4 matrices a such that $a_{ab} G_{bc} = -a_{cb} G_{ba}$. Then if $a \in T_I(O(3, 1))$, one can define a curve in $O(3, 1)$ by $A_t = \exp(ta)$ where

$$\exp(b) = \sum_{n=0}^{\infty} \frac{b^n}{n!}.$$

Thus if $u \in O(\mathcal{M})$ one can define a curve through u in $\pi^{-1}(\pi(u))$ by $\lambda_{au}(t) = A_t(u)$. As the curve $\lambda_{au}(t)$ lies in the fibre, its tangent vector $(\partial/\partial t)_{\lambda_{au}}$ is vertical. For each $a \in T_I$, one can therefore define a vertical vector field $\mathbf{F}(a)$ by $\mathbf{F}(a)|_u = (\partial/\partial t)_{\lambda_{au}}|_u$ for each $u \in O(\mathcal{M})$. If $\{a_i\}$ ($i = 1, 2, \ldots, 6$) are a basis for T_I, then $\mathbf{F}_i \equiv \mathbf{F}(a_i)$ will be six vertical vector fields on $O(\mathcal{M})$ which will provide a basis for V_u at each point $u \in O(\mathcal{M})$.

A matrix $B \in O(3, 1)$ defines a mapping $O(\mathcal{M}) \to O(\mathcal{M})$ by $u \to B(u)$. Under the induced map $B_* : T_u \to T_{B(u)}$, the vertical and horizontal vector fields transform as follows:

$$B_*(\overline{\mathbf{E}}_a) = B_{ab}^{-1} \overline{\mathbf{E}}_b,$$

$$B_*(\mathbf{F}_i) = C_i{}^j \, \mathbf{F}_j,$$

where $C_i{}^j = B_{ab} a_{ibc} B^{-1}{}_{cd} a^j{}_{da}$ and $\{a^j\}$ are the basis for $T^*{}_I$ dual to the basis $\{a_i\}$ for T_I (thus $a^i{}_{ab} a_{jab} = \delta^i{}_j$, $a^j{}_{ab} a_{jcd} = \frac{1}{4}\delta_{ac}\delta_{bd}$). The property of these induced maps which will be important for what follows is not their actual form but the fact that they are constant over $O(\mathcal{M})$.

One now has a basis $\{\mathbf{G}_A\} = \{\overline{\mathbf{E}}_a, \mathbf{F}_i\}$ ($A = 1, \ldots, 10$) for T_u at each point $u \in O(\mathcal{M})$. One can thus define a positive definite metric $\mathbf{e}$ on $O(\mathcal{M})$ by $e(\mathbf{X}, \mathbf{Y}) = \sum_A X^A Y^A$ where $\mathbf{X}, \mathbf{Y} \in T(u)$ and X^A, Y^A are the components of $\mathbf{X}, \mathbf{Y}$ respectively in the basis $\{\mathbf{G}_A\}$.

Using the metric $\mathbf{e}$, one can define a distance function $\rho(u, v)$, $u, v \in O(\mathcal{M})$, as the greatest lower bound of lengths (measured by $\mathbf{e}$) of curves from u to v. One can then ask whether $O(\mathcal{M})$ is m-complete with the distance function ρ.

Proposition 8.3.1

$(O(\mathcal{M}), \mathbf{e})$ is m-complete if and only if $(\mathcal{M}, \mathbf{g})$ is b-complete.

Suppose $\gamma(t)$ is a curve in $\mathcal{M}$. Then given a point $u \in \pi^{-1}(p)$ where $p \in \gamma$ one can construct a horizontal curve $\overline{\gamma}(t)$ through u such that $\pi(\overline{\gamma}(t)) = \gamma(t)$. From the definition of the positive definite metric $\mathbf{e}$, it follows that the arc-length of $\overline{\gamma}(t)$ as measured in this metric is equal to the generalized affine parameter of $\gamma(t)$, defined by the basis at p represented by the point u. If therefore $\gamma(t)$ has no endpoint but has finite length as measured by the generalized affine parameter, then $\overline{\gamma}(t)$ will also have no endpoint but will have finite length in the metric $\mathbf{e}$. Thus m-completeness in $O(\mathcal{M})$ implies b-completeness in $\mathcal{M}$.

To prove the converse, one needs to show that if $\lambda(t)$ is a C^1 curve in $O(\mathcal{M})$ of finite length without endpoint, then $\pi(\lambda(t))$ is a C^1 curve in $\mathcal{M}$ with

(1) finite affine length,

(2) no endpoint in $\mathcal{M}$.

To prove (1), one proceeds as follows. Let $u \in \lambda(t)$. Then one can construct a horizontal curve $\overline{\lambda}(t)$ through u such that $\pi(\overline{\lambda}(t)) = \pi(\lambda(t))$. For each value of t, $\lambda(t)$ and $\overline{\lambda}(t)$ will lie in the same fibre, so there will be a unique curve $B(t)$ in $O(3, 1)$ such that $\lambda(t) = B(t)\bar{\lambda}(t)$. This implies

$$\left(\frac{\partial}{\partial t}\right)_\lambda = B_* \left(\frac{\partial}{\partial t}\right)_{\bar{\lambda}} + F(B^{\cdot}B^{-1}),$$

where $B^{\cdot} \equiv \mathrm{d}B/\mathrm{d}t$. Therefore

$$e\left(\left(\frac{\partial}{\partial t}\right)_\lambda, \left(\frac{\partial}{\partial t}\right)_\lambda\right) = \sum_b \left(\left\langle \bar{\mathbf{E}}^a, \left(\frac{\partial}{\partial t}\right)_{\bar{\lambda}}\right\rangle B^{-1}{}_{ab}\right)^2 + \sum_i (B^{\cdot}{}_{ab} B^{-1}{}_{bc} a^i{}_{ca})^2,$$

where $\{\bar{\mathbf{E}}^a\}$ is the basis of $H^*{}_u$ dual to the basis $\{\bar{\mathbf{E}}_a\}$ (i.e. $\langle \bar{\mathbf{E}}^a, \bar{\mathbf{E}}_b \rangle = \delta^a{}_b$) and $a^i{}_{ab}$ is the basis of $T_I{}^*$ dual to the basis $a_{i\,ab}$ (i.e. $a_{i\,ab}\, a^j{}_{ab} = \delta_i{}^j$). The matrix B_{ab} satisfies $B_{ab}\, G_{bc}\, B_{dc} = G_{ad}$. Therefore

$$B_{ab}\, G_{ac}\, B_{cd} = G_{bd}$$

as $G_{ab} = G^{-1}{}_{ab}$. Differentiating with respect to t, one has

$$B^{\cdot}{}_{ab}\, B^{-1}{}_{bc}\, G_{cd} = - G_{ac}\, B^{\cdot}{}_{db}\, B^{-1}{}_{bc}.$$

Thus $B^{\cdot}{}_{ab}\, B^{-1}{}_{bc} \in T_I(O(3, 1))$. Since the $a^i{}_{ab}$ are a basis for $T^*{}_I$, there is some constant C such that

$$\sum_i (B^{\cdot}{}_{ab}\, B^{-1}{}_{bc}\, a^i{}_{ca})^2 \geqslant C(B^{\cdot}{}_{ab}\, B^{-1}{}_{bc}\, B^{\cdot}{}_{ad}\, B^{-1}{}_{dc}).$$

Any matrix $B \in O(3, 1)$ can be expressed in the form $B = \bar{\Omega}\Delta\Omega$, where (i) $\bar{\Omega}$ and Ω are orthogonal matrices of the form

$$\left(\begin{array}{c|c} \bar{O} & \\ \hline & 1 \end{array}\right) \quad \text{and} \quad \left(\begin{array}{c|c} O & \\ \hline & 1 \end{array}\right)$$

where $\bar{O}$ and O are 3×3 orthogonal matrices, and the basis $\{\mathbf{E}_a\}$ has been numbered so that $\mathbf{E}_4$ is the timelike vector; these matrices represent rotations; and (ii) Δ is the matrix

$$\begin{pmatrix} \cosh\xi & 0 & 0 & \sinh\xi \\ 0 & 1 & 0 & 0 \\ 0 & 0 & 1 & 0 \\ \sinh\xi & 0 & 0 & \cosh\xi \end{pmatrix}$$

which represents a change of velocity in the 1-direction. With this decomposition, $\qquad B^{\cdot}{}_{ab}\, B^{-1}{}_{bc}\, B^{\cdot}{}_{ad}\, B^{-1}{}_{dc} \geqslant 2(\xi^{\cdot})^2.$

For any vector $\mathbf{X} \in T_u$,

$$\sum_b (\langle \bar{\mathbf{E}}^a, \mathbf{X} \rangle\, \Omega_{ab})^2 = \sum_a (\langle \bar{\mathbf{E}}^a, \mathbf{X} \rangle)^2.$$

Thus
$$\sum_b \left(\left\langle \mathbf{E}^a, \left(\frac{\partial}{\partial t} \right)_{\bar{\lambda}} \right\rangle B^{-1}{}_{ab} \right)^2 \geqslant \sum_a \left(\left\langle \mathbf{E}^a, \left(\frac{\partial}{\partial t} \right)_{\bar{\lambda}} \right\rangle \right)^2 e^{-2|\xi|}$$

$$= e \left(\left(\frac{\partial}{\partial t} \right)_{\bar{\lambda}}, \left(\frac{\partial}{\partial t} \right)_{\bar{\lambda}} \right) e^{-2|\xi|}.$$

Therefore

$$e \left(\left(\frac{\partial}{\partial t} \right)_{\lambda}, \left(\frac{\partial}{\partial t} \right)_{\lambda} \right) \geqslant e \left(\left(\frac{\partial}{\partial t} \right)_{\bar{\lambda}}, \left(\frac{\partial}{\partial t} \right)_{\bar{\lambda}} \right) e^{-2|\xi|} + 2C(\xi^{\cdot})^2,$$

and so

$$\left[e \left(\left(\frac{\partial}{\partial t} \right)_{\lambda}, \left(\frac{\partial}{\partial t} \right)_{\lambda} \right) \right]^{\frac{1}{2}} \geqslant \frac{1}{2} \left[e \left(\left(\frac{\partial}{\partial t} \right)_{\bar{\lambda}}, \left(\frac{\partial}{\partial t} \right)_{\bar{\lambda}} \right) \right]^{\frac{1}{2}} e^{-|\xi|} + C^{\frac{1}{2}} |\xi^{\cdot}|.$$

Let $\xi_0 \leqslant \infty$ be the least upper bound for $|\xi|$ on $\lambda(t)$. Then

$$L(\lambda) \geqslant \tfrac{1}{2} L(\bar{\lambda}) e^{-\xi_0} + C^{\frac{1}{2}} \xi_0,$$

where $L(\lambda)$ is the length of the curve λ in the metric $\mathbf{e}$. Since this is finite, ξ_0 and $L(\bar{\lambda})$ must be finite. Thus the affine length of the curve $\pi(\lambda(t))$ in $\mathcal{M}$, which is equal to $L(\bar{\lambda})$, will be finite.

To complete the proof of proposition 8.3.1, we have to show that the curve $\pi(\lambda(t))$ in $\mathcal{M}$ has no endpoint, that is, we have to show that there is no point $p \in \mathcal{M}$ such that $\pi(\lambda(t))$ enters and remains within every neighbourhood $\mathcal{U}$ of p. Because of the existence of normal neighbourhoods $\mathcal{U}$ of p, this is a consequence of the following result:

Proposition 8.3.2 (*Schmidt* (1972))

Let $\mathcal{N}$ be a compact subset of $\mathcal{M}$. Suppose there is a curve $\lambda(t)$ in $O(\mathcal{M})$ without endpoint and of finite length, which enters and remains within $\pi^{-1}(\mathcal{N})$. Then there is an inextendible null geodesic γ contained in $\mathcal{N}$.

Let $\bar{\lambda}(t)$ be the horizontal curve through some point $u \in \lambda(t)$ such that $\pi(\bar{\lambda}(t)) = \pi(\lambda(t))$. The curve $\lambda(t)$ has no endpoint. Suppose there were a point $v \in O(\mathcal{M})$ which was an endpoint of the horizontal curve $\bar{\lambda}(t)$. Then there would be an open neighbourhood $\mathcal{W}$ of v with compact closure such that $\bar{\lambda}(t)$ entered and remained within $\mathcal{W}$. Let $\mathcal{W}'$ be the set $\{x \in O(\mathcal{M}): Bx \in \mathcal{W} \text{ for all matrices } B \text{ with } |\xi| \leqslant \xi_0\}$. Since $\bar{\mathcal{W}}$ was compact and ξ_0 is finite, $\bar{\mathcal{W}'}$ would be compact. The curve $\lambda(t)$ would enter and remain within $\bar{\mathcal{W}'}$. But any compact set is m-complete with respect to the positive definite metric $\mathbf{e}$. Thus $\lambda(t)$, having finite length, would have an endpoint in $\bar{\mathcal{W}'}$. This shows that $\bar{\lambda}(t)$ has no endpoint.

Let $\{x_n\}$ be a sequence of points on $\bar{\lambda}(t)$ without any limit point. Since $\mathcal{N}$ is compact, there will be a point $x \in \mathcal{N}$ which is a limit point of $\pi(x_n)$. Let $\mathcal{U}$ be a normal neighbourhood of x with compact closure, and let $\sigma: \mathcal{U} \to O(\mathcal{M})$ be a cross-section of $O(\mathcal{M})$ over $\mathcal{U}$, i.e. $\sigma(p)$, $p \in \mathcal{U}$, is an orthonormal basis at p. Let $\tilde{\lambda}(t) \equiv \sigma(\pi(\lambda(t)))$ for $\lambda(t) \in \pi^{-1}(\mathcal{U})$. Then as in the previous proposition, there will be a unique family of matrices $A(t) \in O(3, 1)$ such that $\bar{\lambda}(t) = A(t) \tilde{\lambda}(t)$, and one can express the matrix A in the form $A = \bar{\Omega}\Delta\Omega$. Suppose that $|\xi(t_{n'})|$ had a finite upper bound ξ_1, where $x_{n'} = \bar{\lambda}(t_{n'})$ is a subsequence of the x_n which converges to x. Then the points $x_{n'}$ would be contained in the set $\mathcal{U}' = \{v \in O(\mathcal{M}): A^{-1}v \subset \sigma(\mathcal{U})$ for some $A \in O(3, 1)$ with $|\xi| < \xi_1\}$. However $\bar{\mathcal{U}}'$ would be compact and so would contain a limit point of the $\{x_{n'}\}$, which is contrary to our choice of the $\{x_n\}$. Thus $|\xi(t_{n'})|$ has no finite upper bound. Since the orthogonal group is compact, one can choose a subsequence $\{x_{n''}\}$ such that $\bar{\Omega}_{n''}$ converges to some $\bar{\Omega}'$, $\Omega_{n''}$ converges to some Ω', $\xi_{n''} \to \infty$, and

$$\xi_{n''+1} - \xi_{n''} > a > 0 \tag{8.1}$$

for some constant a (here $\bar{\Omega}_{n''} = \bar{\Omega}(t_{n''})$, etc.).

Let $\lambda'(t) = (\bar{\Omega}')^{-1}\bar{\lambda}(t)$, and let $\hat{\lambda}_{n''}(t) \equiv \Delta_{n''}^{-1}(\bar{\Omega}')^{-1}\bar{\lambda}(t)$. Then $\hat{\lambda}_{n''}(t_{n''})$ tends to $\hat{x} \equiv \Omega'\sigma(x)$. Since the length of the curve $\bar{\lambda}(t)$ is finite, the curve $\lambda'(t)$ also has finite length. This means that

$$\int_{t_{n''}}^{t_{n''+1}} ((X^u)^2 + (X^v)^2 + (X^2)^2 + (X^3)^2)^{\frac{1}{2}} \, dt$$

tends to zero, where

$$X^A \equiv \langle \mathbf{E}^A, (\partial/\partial t)_{\lambda'} \rangle, \quad A = u, v, 2, 3,$$

and

$$\mathbf{E}^u = \frac{1}{\sqrt{2}}(\mathbf{E}^4 + \mathbf{E}^1), \quad \mathbf{E}^v = \frac{1}{\sqrt{2}}(\mathbf{E}^4 - \mathbf{E}^1).$$

Thus

$$\int_{t_{n''}}^{t_{n''+1}} |X^A| \, dt$$

tends to zero, for each A. The components $Y_{n''}{}^A$ of the tangent vector of the horizontal curve $\hat{\lambda}_{n''}(t)$ are

$$Y_{n''}{}^u = e^{-\xi_{n''}} X^u, \quad Y_{n''}{}^v = e^{\xi_{n''}} X^v, \quad Y_{n''}{}^2 = X^2, \quad Y_{n''}{}^3 = X^3.$$

Thus

$$\int_{t_{n''}}^{t_{n''+1}} |Y_{n''}{}^A| \, dt \quad (A = u, 2, 3), \tag{8.2}$$

tend to zero.

Let μ be the integral curve of the horizontal vector field $\overline{E}^v$ through $\hat{x}$. Then $\pi(\mu)$ will be a null geodesic in $\mathcal{M}$. Suppose that $\pi(\mu)$ left $\mathcal{N}$ in both the past and future directions. Then there would be some neighbourhood $\mathcal{V}$ of $\hat{x}$ with compact closure and with the property that in each direction μ left and did not re-enter the set $\overline{\mathcal{V}'}$, where $\mathcal{V}' \equiv \{v \in O(\mathcal{M})$: there is a Δ with Δv contained in $\mathcal{V}\}$. One could choose $\mathcal{V}$ sufficiently small that it had this property for any integral curve of $\overline{E}^v$ which intersected $\overline{\mathcal{V}}$ and so that any such curve would leave $\pi^{-1}(\mathcal{N})$ in both directions. Let $\mathcal{Y}$ be the tube consisting of all points on integral curves of $\overline{E}^v$ which intersect $\overline{\mathcal{V}}$. Then $\mathcal{Y} \cap \pi^{-1}(\mathcal{N})$ would be compact. For sufficiently large n'', $\hat{\lambda}_{n''}(t_{n''})$ would be contained in $\mathcal{V}$. By (8.2) the components of the tangent vector to $\hat{\lambda}_{n''}$ transverse to the direction $\overline{E}^v$ are so small that for large n'' and $t > t_{n''}$, the curve $\hat{\lambda}_{n''}(t)$ could not leave the tube $\mathcal{Y} \cap \pi^{-1}(\mathcal{N})$ except at its ends where $\mathcal{Y}$ left $\pi^{-1}(\mathcal{N})$. However $\hat{\lambda}_{n''}(t)$ cannot leave $\pi^{-1}(\mathcal{N})$, as $\lambda(t)$ does not leave $\pi^{-1}(\mathcal{N})$. Thus $\hat{\lambda}_{n''}(t)$ would be contained in $\mathcal{Y} \cap \pi^{-1}(\mathcal{N})$ for $t \geqslant t_{n''}$. This leads to a contradiction as follows: $\hat{\lambda}_{n''+1}(t_{n''+1})$ is contained in $\mathcal{V}$. However by (8.1), $\mathcal{V}$ can be chosen sufficiently small that

$$\hat{\lambda}_{n''}(t_{n''+1}) = \Delta_{n''+1} \Delta_{n''}{}^{-1} \hat{\lambda}_{n''+1}(t_{n''+1})$$

is not contained in $\mathcal{V}$, though it is contained in $\mathcal{V}'$. This shows that our assumption that the null geodesic $\pi(\mu)$ left $\mathcal{N}$ in both directions is false. Thus there will be some point $p \in \mathcal{N}$ which is a limit point of $\pi(\mu)$. By lemma 6.2.1 there will be an inextendible null geodesic γ through p which is contained in $\mathcal{N}$ and which is a limit curve of $\pi(\mu)$. ☐

If $O(\mathcal{M})$ is m-incomplete, one can form the metric space completion $\overline{O(\mathcal{M})}$. This is defined to be the set of equivalence classes of Cauchy sequences of points in $O(\mathcal{M})$. If $x \equiv \{x_n\}$ and $y \equiv \{y_m\}$ are Cauchy sequences in $O(\mathcal{M})$, the distance $\bar{\rho}(x, y)$ between x and y is defined to be $\lim_{n \to \infty} \rho(x_n, y_n)$ where ρ is the distance function on $O(\mathcal{M})$ defined by the positive definite metric $\mathbf{e}$; x and y are said to be equivalent if $\bar{\rho}(x, y) = 0$. One can decompose $\overline{O(\mathcal{M})}$ into a part homeomorphic to $O(\mathcal{M})$ and a set of boundary points $\bar{\partial}$ (i.e. $\overline{O(\mathcal{M})} = O(\mathcal{M}) \cup \bar{\partial}$). The distance function $\bar{\rho}$ defines a topology on $\overline{O(\mathcal{M})}$. From (8.1), it follows that the topology on $\overline{O(\mathcal{M})}$ is independent of the choice of basis $\{a_i\}$ of T_I.

One can extend the action of $O(3, 1)$ to $\overline{O(\mathcal{M})}$. For under the action of $A \in O(3, 1)$, the transformation of the basis $\{\mathbf{G}_A\}$ is independent of position in $O(\mathcal{M})$. Thus there are positive constants C_1 and C_2 (depending only on A) such that $C_1\rho(u, v) \leqslant \rho(A(u), A(v)) \leqslant C_2\rho(u, v)$. This means that under the action of A, Cauchy sequences will map to Cauchy sequences and equivalence classes of Cauchy sequences are mapped to equivalence classes of Cauchy sequences. Therefore the action of $O(3, 1)$ extends to $\overline{O(\mathcal{M})}$ in a unique way. One can then define $\mathcal{M}^+$ to be the quotient of $\overline{O(\mathcal{M})}$ by the action of $O(3, 1)$. Since the quotient of $O(\mathcal{M})$ by $O(3, 1)$ is $\mathcal{M}$, and since $O(3, 1)$ maps incomplete Cauchy sequences to incomplete Cauchy sequences, one can express $\mathcal{M}^+$ as the union of $\mathcal{M}$ and a set ∂ of points called the *b-boundary* of $\mathcal{M}$. One can regard points of ∂ as representing the endpoint of equivalence classes of b-incomplete curves in $\mathcal{M}$.

The projection $\bar{\pi}\colon \overline{O(\mathcal{M})} \to \mathcal{M}^+$, which assigns a point in $\overline{O(\mathcal{M})}$ to its equivalence class under $O(3, 1)$, induces a topology on $\mathcal{M}^+$ from the topology on $O(\mathcal{M})$. However $\bar{\pi}$ does not induce a distance function on $\mathcal{M}^+$ because $\bar{\rho}$ is not invariant under $O(3, 1)$. Thus although the topology of $\overline{O(\mathcal{M})}$ is a metric topology, and so Hausdorff, that of $\mathcal{M}^+$ need not be Hausdorff. This means that there may be a point $p \in \mathcal{M}$ and a point $q \in \partial$ such that every neighbourhood of p in $\mathcal{M}^+$ intersects every neighbourhood of q. This happens when the point q corresponds to an incomplete curve which is totally or partially imprisoned in $\mathcal{M}$. We shall discuss imprisoned incompleteness further in §8.5.

If $\mathbf{g}$ is a positive definite metric on $\mathcal{M}$, then $\mathcal{M}^+$ is homeomorphic to the completion of $(\mathcal{M}, \mathbf{g})$ by Cauchy sequences. Schmidt's construction also has the desirable property that if one cuts a closed set $\mathscr{A}$ out of a space, then one gets at least one point of the b-boundary for every point of $\mathscr{A}^{\cdot}$ that is the endpoint of a curve in $\mathcal{M} - \mathscr{A}$. An example where one gets more than one b-boundary point for a point of $\mathscr{A}^{\cdot}$ is provided by two-dimensional Minkowski space in which the set $\mathscr{A}$ is taken to be the t-axis between -1 and $+1$. Then there will be two b-boundary points for each point $(0, t)$ where $-1 < t < 1$. An example where a point in $\mathscr{A}^{\cdot}$ cannot be reached by a curve in $\mathcal{M} - \mathscr{A}$ is given by the set

$$\mathscr{A} = \left\{t = \sin \frac{1}{x}, t \neq 0\right\} \cup \{-1 \leqslant t \leqslant 1, x = 0\}.$$

There is no curve in $\mathcal{M} - \mathscr{A}$ which has an endpoint at the origin, and hence this point will not be in $(\mathcal{M} - \mathscr{A})^+$, although it is in $\mathscr{A}^{\cdot}$.

Although Schmidt's construction has an elegant formulation, it is unfortunately very difficult to apply in practice. The only solutions for which $\mathcal{M}^+$ has been found, apart from spaces of constant curvature, are the two-dimensional Robertson–Walker solutions with normal matter. In these ∂ turns out to be a spacelike one-surface as might be expected from the conformal picture. In this case, one can define a natural differential structure on ∂ and make $\mathcal{M}^+$ into a manifold with boundary. However there does not seem to be any general way of defining a manifold structure on ∂. Indeed one might expect that in generic situations ∂ would be highly irregular and could not be given a smooth structure.

8.4   The character of the singularities

In this and the following section we shall discuss the character of the singularities predicted by theorem 4. We consider this theorem rather than the others because more information about the singularity can be obtained. We expect however that the singularities predicted by the other theorems will have similar properties.

First there is the question of how bad the breakdown of differentiability of the metric must be. The theorems of the previous section showed that space–time must be geodesically incomplete if the metric was C^2. The C^2 condition was necessary in order that the conjugate points and variation of arc-length should be well-defined; in other words, in order that solutions of the geodesic equation should depend *differentiably* on their initial position and direction. However one can talk about geodesic incompleteness provided that solutions of the geodesic equation are defined. They will exist if the metric is C^1 and will be unique and depend *continuously* on initial position and direction if the metric is C^{2-} (i.e. if the connection is locally Lipschitz). In fact one can discuss b-incompleteness provided merely that the positive definite metric **e** on the bundle of frames $O(\mathcal{M})$ is defined almost every where and is locally bounded. This will be the case if the components $\Gamma^a{}_{bc}$ of the connection are defined almost everywhere and are locally bounded, i.e. if the metric is C^{1-}.

It thus might appear that what the theorems indicate is not that the curvature becomes unboundedly large but merely that it has a discontinuity (i.e. the metric is C^{2-} rather than C^2). We shall show that this is not the case: under the conditions of theorem 4 space–time must be timelike geodesically incomplete (and hence b-incomplete) even if the metric is only required to be C^{2-}. The method of proof is to approximate the C^{2-} metric by a C^2 metric and to perform variation of arc-length in this metric.

Suppose that space–time is defined to be inextendible with a C^{2-} metric and that the conditions of theorem 4 are satisfied. The timelike convergence condition, $R_{ab}K^aK^b \geqslant 0$, is now required to hold 'almost everywhere' with the Ricci tensor defined by generalized derivatives. The only part of the proof of theorem 4 that does not hold in a C^{2-} metric is where variation of arc-length is used to show that there can be no point $p \in D^+(\mathscr{S})$ such that $d(\mathscr{S}, p) > -3/\theta_0$, where θ_0 is the maximum value of $\chi^a{}_a$ on $\mathscr{S}$. Thus if $\mathscr{M}$ were timelike geodesically complete there would be some such point p and a geodesic orthogonal to $\mathscr{S}$ of length $d(\mathscr{S}, p)$ from $\mathscr{S}$ to p. Let $\mathscr{U}$ be an open set with compact closure which contains $J^-(p) \cap J^+(\mathscr{S})$ and let $\mathbf{e}$ and $\hat{\mathbf{g}}$ be C^∞ positive definite and Lorentz metrics respectively. For any $\epsilon > 0$ one could find a C^∞ Lorentz metric $g_\epsilon{}^{ab}$ such that on $\overline{\mathscr{U}}$

(1) $|g_\epsilon{}^{ab} - g^{ab}| < \epsilon$,

(2) $|g_\epsilon{}^{ab}{}_{|c} - g^{ab}{}_{|c}| < \epsilon$,

(3) $|g_\epsilon{}^{ab}{}_{|cd}| < C$, where C is a constant depending on $\mathscr{U}$, $\mathbf{e}$, $\hat{\mathbf{g}}$ and $\mathbf{g}$,

(4) $R_{\epsilon ab}K^aK^b > -\epsilon|K^a|^2$ for any vector $\mathbf{K}$ such that $g_{\epsilon ab}K^aK^b \geqslant 0$.

(The $g_\epsilon{}^{ab}$ may be constructed by covering $\overline{\mathscr{U}}$ by a finite number of local coordinate neighbourhoods $(\mathscr{V}_\alpha, \phi_\alpha)$, integrating the coordinate components of g^{ab} with a suitable smoothing function $\rho_\epsilon(x)$ and summing with a partition of unity $\{\psi_\alpha\}$, i.e.

$$g_\epsilon{}^{ab}(q) = \sum_\alpha \psi_\alpha(q) \int_{\phi_\alpha(\mathscr{V}_\alpha)} g^{ab}(x) \rho_\epsilon(x - \phi_\alpha(q)) \, \mathrm{d}^4x,$$

where $\int \rho_\epsilon(x) \, \mathrm{d}^4x = 1$.)

Property (1) implies that for sufficiently small values of ϵ, p would be in $D^+(\mathscr{S}, \mathbf{g}_\epsilon)$ and $J^-(p, \mathbf{g}_\epsilon) \cap J^+(\mathscr{S}, \mathbf{g}_\epsilon)$ would be contained in $\mathscr{U}$. There would therefore be a geodesic γ_ϵ in the metric $\mathbf{g}_\epsilon$ from $\mathscr{S}$ to p of length $d_\epsilon(\mathscr{S}, p)$. Also $|d_\epsilon(\mathscr{S}, p) - d(\mathscr{S}, p)|$ would tend to zero as $\epsilon \to 0$.

By properties (1), (2) and (3), and the standard theorems on ordinary differential equations, as $\epsilon \to 0$ the tangent vector to a geodesic in the metric $\mathbf{g}_\epsilon$ would tend to that of the geodesic in the metric $\mathbf{g}$ with the same initial position and direction. There would be some upper bound to $|V^a|$ on $\overline{\mathscr{U}} \cap \beta(\mathscr{S} \times [0, 2d(\mathscr{S}, p)])$, where V^a is the unit tangent vector to the geodesic orthogonal to $\mathscr{S}$ in the metric $\mathbf{g}$. Thus for any $\delta > 0$ there would be an $\epsilon_1 > 0$ such that for any $\epsilon < \epsilon_1$, $R_{\epsilon ab}V_\epsilon^a V_\epsilon^b > -\delta$. We can now establish a contradiction by showing that a sufficiently small variation of the energy condition will not prevent the occurrence of conjugate points in the metric $\hat{g}_\epsilon$ within a distance less than $d_\epsilon(\mathcal{S}, p)$. For the expansion θ_ϵ of the geodesics in the metric $\hat{g}_\epsilon$ obeys the Raychaudhuri equation:

$$d\theta_\epsilon/ds = -\tfrac{1}{3}\theta_\epsilon{}^2 - 2\sigma_\epsilon{}^2 - R_{\epsilon ab} V_\epsilon{}^a V_\epsilon{}^b.$$

Thus $d(\theta_\epsilon{}^{-1})/ds \geqslant \tfrac{1}{3} + R_{\epsilon ab} V^a V^b \theta_\epsilon{}^{-2}$. Therefore if the initial value θ_{ϵ_0} were negative and $3\delta\theta_{\epsilon_0}{}^{-2}$ were less than one, $\theta_\epsilon{}^{-1}$ would become zero within a distance $3/\theta_0(1 - 3\delta\theta_0{}^{-2})$ from $\mathcal{S}$. But $\theta_{\epsilon_0} \to \theta_0$ as $\epsilon \to 0$. This shows that for sufficiently small values of ϵ there would be a conjugate point on every geodesic in the metric $\hat{g}_\epsilon$ orthogonal to $\mathcal{S}$ within a distance less than $d_\epsilon(\mathcal{S}, p)$. Therefore $\mathcal{M}$ must be timelike geodesically incomplete even if the metric is required only to be C^{2-}.

This result implies that if space–time is extended to try to continue the incomplete geodesics, the metric must fail to be Lorentzian or the curvature must be locally unbounded, i.e. there would be a curvature singularity. However even though the curvature were locally unbounded, the metric might still be able to be interpreted as a distributional solution of the Einstein equations provided that the volume integrals of the components of the curvature tensor over any compact region were finite. This would be the case if the metric were Lorentz, continuous and had square integrable first derivatives. In particular this would be true if the metric were Lorentz and C^{1-} (i.e. locally Lipschitz). Examples of such C^{1-} solutions include gravitational shock waves (where the curvature has a δ-function behaviour on a null three-surface, see, for example, Choquet-Bruhat (1968) and Penrose (1972a)); thin mass shells (where the curvature has a δ-function behaviour on a timelike three-surface, see, for example, Israel (1966)); and solutions containing pressure-free matter where the geodesic flow lines have two- or three-dimensional caustics (see Papapetrou and Hamoui (1967), Grischuk (1967)). Because of the non-linear dependence of the curvature on the metric one cannot necessarily approximate a C^{1-} distributional solution by a C^2 metric which obeys the convergence condition at every point, or at least does not violate it by more than a small amount as in the case above (property (4)). However in all the examples given above one can. Indeed this is their physical justification: they are regarded as mathematical idealizations of C^2 or C^∞ solutions which obey the convergence condition and in which the curvature is very large in a small region. One could apply the theorems of §8.2 to these C^2 solutions and prove the existence of incomplete geodesics in them. This shows that the singularities predicted cannot be just gravitational impulse waves or caustics of flow lines but must be more serious breakdowns of the metric. (Ordinary hydrodynamic shock waves involve only discontinuities of density and pressure and so can exist with a C^{2-} metric.) Although we are not quite able to prove it we believe that the singularities must be such that the metric cannot be extended to be even a distributional solution of the Einstein equations, i.e. as well as the components of the curvature being unbounded at a singular point, their volume integral over any neighbourhood of such a point must also be unbounded. This is so in all known examples of singularities other than the exceptional case of the Taub–NUT solution, which will be dealt with in the next section. If this conjecture is correct for 'generic' singularities (i.e. except for those arising from a set of initial conditions of measure zero), then one can regard a singularity as a point where the Einstein equations (and presumably the other presently known laws of physics) break down.

Another question one would like to answer is: how many incomplete geodesics are there? If there were only one, one might be tempted to feel that the singularity could be ignored. From the proof of theorem 4 one can see that if there is no Cauchy horizon, i.e. if $\mathscr{S}$ is a Cauchy surface, then no timelike curve from $\mathscr{S}$ (geodesic or not) can be extended to a length greater than $-3/\theta_0$ where θ_0 is the maximum value of $\chi^a{}_a$ on $\mathscr{S}$. In fact this result is true even if $\mathscr{S}$ is non-compact provided that $\chi^a{}_a$ still has a negative upper bound. However this does not necessarily indicate that what happens is that every timelike curve hits the singularity. Rather it suggests that a singularity will be accompanied by a Cauchy horizon and so our ability to predict the future will break down. An example of this is shown in figure 53. Here the metric is singular at the point p and so this point has been removed from the space–time manifold. Spreading out from this hole there is a Cauchy horizon. This example shows that the most one can hope to prove is that there is a three-dimensional family of geodesics which are incomplete and which remain within the Cauchy development of $\mathscr{S}$ (in the example these are the geodesics which would pass through p). There may be other geodesics which leave the Cauchy development of $\mathscr{S}$ and which are incomplete but one cannot predict their behaviour from knowledge of conditions on $\mathscr{S}$.

It is clear that there must be more than one incomplete geodesic in $D^+(\mathscr{S})$. For from theorem 4 it follows that there must be a geodesic γ, orthogonal to $\mathscr{S}$, which remains in $D^+(\mathscr{S})$ but which is incomplete.

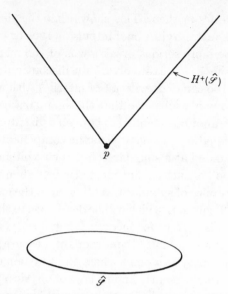

FIGURE 53. The point p has been removed from space–time because a singularity occurs there. Consequently there is a Cauchy horizon $H^+(\hat{\mathscr{S}})$ for the surface $\hat{\mathscr{S}}$.

Let p be the point where γ intersects $\mathscr{S}$. Then one can make a small variation of $\mathscr{S}$ in a neighbourhood of p to obtain a new surface $\mathscr{S}'$ for which $\chi^a{}_a$ is still negative, but which is not orthogonal to γ. Then by theorem 4 there must be some other timelike geodesic γ' orthogonal to $\mathscr{S}'$ which is incomplete and which does not cross $H^+(\mathscr{S}')$, which is the same as $H^+(\mathscr{S})$.

One can in fact prove that there is at least a three-dimensional family of timelike geodesics (one through each point of some achronal surface) which remain within $D^+(\mathscr{S})$ and which are incomplete. These geodesics all correspond to the same boundary point in the sense of the indecomposable past sets of §6.8, that is, they all have the same past. They may not, however, all correspond to the same points as defined by the construction of the previous section. An outline of the proof is as follows: in theorem 4 it was shown that there must be a future-directed timelike geodesic orthogonal to $\mathscr{S}$ which cannot be extended to length $3/\theta_0$. One can say more than this: there must be such a geodesic γ which remains within $D^+(\mathscr{S})$ and is at each point a curve of maximum length from $\mathscr{S}$, i.e. for each $q \in \gamma$, the length of γ from $\mathscr{S}$ to q equals $d(\mathscr{S}, q)$. The idea is now to consider the function $d(r, \gamma)$ for $r \in J^-(\gamma)$. Clearly this is bounded on $J^+(\mathscr{S}) \cap J^-(\gamma)$. From the fact that γ is a curve of maximum length from $\mathscr{S}$, it follows that in a neighbourhood of γ, $d(r, \gamma)$ is continuous and the surfaces of constant $d(r, \gamma)$ are spacelike surfaces which intersect γ orthogonally. The timelike geodesics orthogonal to these surfaces will then remain within $J^-(\gamma)$ and so will be incomplete.

8.5 Imprisoned incompleteness

In §8.1 we proposed b-incompleteness as a definition of a singularity. The idea was that a b-incomplete curve corresponded to a singular point which had been left out of space–time. However suppose that there is a b-incomplete curve λ which has a limit point $p \in \mathcal{M}$, i.e. λ is partially or totally imprisoned in a compact neighbourhood of p. Then one cannot imbed $\mathcal{M}$ in a larger four-dimensional Hausdorff paracompact manifold $\mathcal{M}'$ such that λ can be continued in $\mathcal{M}'$. For if q were the point where λ intersected the boundary of $\mathcal{M}$ in $\mathcal{M}'$, then any neighbourhood of q would intersect any neighbourhood of p, which would be impossible as $\mathcal{M}'$ is Hausdorff and $q \neq p$. In fact, one can characterize imprisoned incompleteness of $\mathcal{M}$ by non-Hausdorff behaviour of the Schmidt completion $\mathcal{M}^+$.

Proposition 8.5.1

A point $p \in \mathcal{M}$ is not Hausdorff separated in $\mathcal{M}^+$ from a point $r \in \partial$ if there is an incomplete curve λ in $\mathcal{M}$ which has p as a limit point and which has r as an endpoint in $\mathcal{M}^+$.

Suppose that $p \in \mathcal{M}$ is a limit point of a b-incomplete curve λ. One can construct a horizontal lift $\bar{\lambda}$ of λ in the bundle of orthonormal frames $O(\mathcal{M})$. This will have an endpoint at some point

$$x \in \pi^{-1}(r) \subset \bar{\partial} \equiv \overline{O(\mathcal{M})} - O(\mathcal{M}).$$

If $\mathscr{V}$ is a neighbourhood of r in $\mathcal{M}^+$ then $\pi^{-1}(\mathscr{V})$ is an open neighbourhood of x in $\overline{O(\mathcal{M})}$. Thus it contains all points on $\bar{\lambda}$ beyond some point y. Therefore all points on λ beyond $\pi(y)$ will lie in $\mathscr{V}$ and hence $\mathscr{V}$ will intersect any neighbourhood of p since p is a limit point of λ. $\qquad\square$

Taub–NUT space (§5.8) is an example where there are incomplete geodesics which are all totally imprisoned in compact neighbourhoods of the past and future horizons $U(t) = 0$. As the metric is perfectly regular on these compact neighbourhoods, the incomplete geodesis. do not correspond to s.p. (scalar polynomial) curvature singularities. Consider a future incomplete closed null geodesic $\lambda(v)$ in the future horizon $U(t) = 0$. Let $p = \lambda(0)$ and let v_1 be the first positive value of v for which $\lambda(v) = p$. Then as in §6.4, the parallelly propagated tangent vector to λ will satisfy

$$(\partial/\partial v)|_{v=v_1} = a(\partial/\partial v)|_{v=0},$$

where $a > 1$. For each n, the point $\lambda(v_n) = p$, where

$$v_n = v_1 \sum_{r=1}^{n} a^{1-r} = v_1 \frac{1 - a^{-n}}{1 - a^{-1}},$$

and

$$(\partial/\partial v)|_{v=v_n} = a^n(\partial/\partial v)|_{v=0}.$$

Thus if one takes a pseudo-orthonormal parallelly propagated basis $\{\mathbf{E}_a\}$ on $\lambda(v)$, where $\mathbf{E}_4 = \partial/\partial v$, then the other null basis vector $\mathbf{E}_3$ obeys $\mathbf{E}_3|_{v=v_n} = a^{-n}\mathbf{E}_3|_{v=0}$. Each time one goes round the closed null geodesic λ, the vector $\mathbf{E}_4$ gets bigger and the vector $\mathbf{E}_3$ gets smaller. The vectors $\mathbf{E}_1$ and $\mathbf{E}_2$ remain the same. If therefore there were some non-zero component of the Riemann tensor which involved $\mathbf{E}_4$ and possibly $\mathbf{E}_1$ and $\mathbf{E}_2$, it would appear bigger and bigger each time one went round λ and so there would be a p.p. (parallelly propagated) curvature singularity. However in Taub–NUT space it turns out that the vector $\mathbf{E}_3$ can be chosen so that there is only one independent non-zero component of the Riemann tensor, which is $R(\mathbf{E}_3, \mathbf{E}_4, \mathbf{E}_3, \mathbf{E}_4)$. This involves $\mathbf{E}_3$ and $\mathbf{E}_4$ equally, and so has the same value each time round. Since a similar argument will probably hold for any imprisoned curve, it seems there is no p.p. curvature singularity in Taub–NUT space, although this space is singular by our definition. One would like to know whether this kind of behaviour would occur in physically realistic solutions containing matter, or whether Taub–NUT space is an isolated pathological example. This question is important because, as we shall argue in the next chapter, we interpret the preceding theorems as indicating not that geodesic incompleteness necessarily occurs, but that General Relativity breaks down in very strong gravitational fields. Such fields do not occur in the Taub–NUT kind of situation. This conclusion is a result of the very special nature of the Riemann tensor in Taub–NUT space. In general, one would expect some other components of the Riemann tensor to be non-zero on the imprisoned curve, and so there would be a p.p. curvature singularity even though there might be no s.p. curvature singularity. In fact one can prove:

Proposition 8.5.2

If $p \in \mathcal{M}$ is a limit point of a b-incomplete curve λ and if at p, $R_{ab}K^aK^b \neq 0$ for all non-spacelike vectors $\mathbf{K}$, then λ corresponds to a p.p. curvature singularity. (This condition can be replaced by the condition that there do not exist any null directions K^a such that $K^a K^c C_{abcd} K_e] = 0$.)

Let $\mathscr{U}$ be a convex normal coordinate neighbourhood of p with compact closure, and let $\{\mathbf{Y}_i\}$, $\{\mathbf{Y}^i\}$ be a field of dual orthonormal bases on $\mathscr{U}$. Let $\{\mathbf{E}_a\}$, $\{\mathbf{E}^a\}$ be a parallelly propagated dual orthonormal basis on the curve $\lambda(t)$. Let $\tilde{t}$ be a parameter on λ such that in $\mathscr{U}$,

$$d\tilde{t}/dt = (\sum_i X^i X^i)^{\frac{1}{2}},$$

where X^i are the components of the tangent vector $\partial/\partial t$ in the basis $\{\mathbf{Y}_i\}$. Then $\tilde{t}$ measures arc-length in the positive definite metric on $\mathscr{U}$ in which the bases $\{\mathbf{Y}_i\}$, $\{\mathbf{Y}^i\}$ are orthonormal.

Since $R_{ab}K^a K^b \neq 0$ at p for any non-spacelike vector K^a, there is a neighbourhood $\mathscr{V} \subset \mathscr{U}$ such that $R_{ab} = CZ_a Z_b + R'_{ab}$, where $C \neq 0$ is a constant, Z_a is a unit timelike vector, and R'_{ab} is such that $CR'_{ab}K^a K^b > 0$ for any non-spacelike vector K^a. Suppose that after some value $\tilde{t}_0$ of $\tilde{t}$ the curve λ intersects $\mathscr{V}$. Since λ has no endpoint and since p is a limit point of λ, the part of λ in $\mathscr{V}$ will have infinite length as measured by $\tilde{t}$. However, the generalized affine parameter is given by

$$du/d\tilde{t} = \{\sum_a (E^a{}_i \tilde{X}^i)^2\}^{\frac{1}{2}},$$

where $\tilde{X}^i$ are the components of the tangent vector $(\partial/\partial\tilde{t})_\lambda$, so $\sum_i \tilde{X}^i \tilde{X}^i = 1$, and $E^a{}_i$ are the components of the basis $\{\mathbf{E}^a\}$ in the basis $\{\mathbf{Y}^i\}$. Since u is finite on the curve, the modulus of the column vector $E^a{}_i \tilde{X}^i$ must go to zero, and so the Lorentz transformation represented by the components $E^a{}_i$ must become unboundedly large. Since $\mathbf{Z}$ is a unit timelike vector, the components of $\mathbf{Z}$ in the basis $\{\mathbf{E}_a\}$ will therefore become unboundedly large and hence some component of the Ricci tensor in the basis $\{\mathbf{E}_a\}$ will become unboundedly large. $\square$

This result shows that an observer whose history was a b-incomplete imprisoned non-spacelike curve in a generic space–time would be torn apart by unboundedly large curvature forces in a finite time. However another observer could travel through the same region without experiencing any such effects. An interesting example in this connection is provided by Taub–NUT space in which the metric has been altered by a conformal factor Ω which differs from one only in a small neighbourhood of a point p on the horizon. This conformal transformation would not alter the causal structure of the space and would not affect the incompleteness of the closed null curve through the point p. However in general $R_{ab}K^a K^b \neq 0$ where K^a is the tangent vector to the closed null geodesic. After each cycle, $R_{ab}K^a K^b$ increases by a factor a^2 and so there is a p.p. curvature singularity. Yet the metric is perfectly regular on a compact neighbourhood of the horizon and so there is no s.p. curvature singularity associated with the incompleteness.

One would like to rule out this kind of situation in which the incomplete curves are totally imprisoned in a compact region. This kind of behaviour might occur in a countably infinite number of different regions of space–time. Thus one cannot describe it by saying that *all* the incomplete curves are totally imprisoned in one compact set. Instead one wants to describe it by saying that a set of incomplete curves which are compact in some sense are totally imprisoned in a compact region of $\mathcal{M}$. To make this concept precise, we define b-boundedness as follows.

We define the space $B(\mathcal{M})$ to be the set of all pairs (λ, u), where u is a point in the bundle of linear frames $L(\mathcal{M})$ and λ is a C^1 curve in $\mathcal{M}$ which has only one endpoint, which is at $\pi(u)$. Let $\mathcal{U}$ be an open set in $\mathcal{M}$ and $\mathcal{V}$ be an open set in $L(\mathcal{M})$. We define the open set $O(\mathcal{U}, \mathcal{V})$ to be the set of all elements of $B(\mathcal{M})$ such that λ intersects $\mathcal{U}$ and $u \in \mathcal{V}$. The sets of the form $O(\mathcal{U}, \mathcal{V})$ for all $\mathcal{U}, \mathcal{V}$ form a sub-basis for the topology of $B(\mathcal{M})$. Recall that the map exp: $T(\mathcal{M}) \to \mathcal{M}$ is defined by taking a vector $\mathbf{X}$ at a point p and proceeding along the geodesic from p in the direction of $\mathbf{X}$ a unit distance as measured in the affine parameter defined by $\mathbf{X}$. Similarly we may define a map Exp: $B(\mathcal{M}) \to \mathcal{M}$ by proceeding from $\pi(u)$ along the curve λ a unit distance as measured in the generalized affine parameter on λ defined by u. The map Exp is continuous and will be defined for all of $B(\mathcal{M})$ if $\mathcal{M}$ is b-complete. We shall say that $(\mathcal{M}, \mathbf{g})$ is *b-bounded* if for every compact set $W \subset B(\mathcal{M})$, Exp (W) has a compact closure in $\mathcal{M}$. Since Exp is continuous, $(\mathcal{M}, \mathbf{g})$ is b-bounded if it is b-complete. However, Taub–NUT space is an example which is b-bounded but not b-complete. We shall show that this can be possible only because Taub–NUT space is completely empty. The presence of any matter on the surface $\mathcal{S}$ in theorem 4 will mean that the space is both b-incomplete and b-unbounded.

Theorem 5

Space–time is not b-bounded if conditions (1)–(3) of theorem 4 hold, and
    (4) the energy–momentum tensor is non-zero somewhere on $\mathcal{S}$,

(5) the energy–momentum tensor obeys a slightly stronger form of the dominant energy condition (§4.3): if K^a is a non-spacelike vector, then $T^{ab}K_a$ is zero or non-spacelike and $T_{ab}K^aK^b \geqslant 0$, equality holding only if $T^{ab}K_b = 0$.

Remark. Condition (4) could be replaced by the generic condition (see Theorem 2).

Proof. Consider the covering space $\mathcal{M}_G$ (§6.5) defined as the set of all pairs $(p, i[\lambda])$, where λ is a curve from q to p, $p, q \in M$, and $i[\lambda]$ is the number of times λ cuts $\mathscr{S}$ in the future direction minus the number of times it cuts it in the past direction. For each integer a,

$$\mathscr{S}_a \equiv \{(p, i[\lambda]): p \in \mathscr{S}, i[\lambda] = a\}$$

is diffeomorphic to $\mathscr{S}$ and is a partial Cauchy surface in $\mathcal{M}_G$. In general $\mathcal{M}_G$ need not be b-bounded if $\mathcal{M}$ is, but in the situation under consideration we have the following result:

Lemma 8.5.3

Let conditions (1)–(3) hold and let $D^+(\mathscr{S}_0)$ not have compact closure in $\mathcal{M}_G$; then if ψ is the covering projection $\psi: \mathcal{M}_G \to \mathcal{M}$, $\psi(D^+(\mathscr{S}_0))$ will not have compact closure in $\mathcal{M}$.

$\mathcal{M}$ is either diffeomorphic to $\mathcal{M}_G$ or to $\mathcal{M}_a$, the portion of $\mathcal{M}_G$ between $\mathscr{S}_a$ and $\mathscr{S}_{a+1}$ with $\mathscr{S}_a$ and $\mathscr{S}_{a+1}$ identified. If for any $a \geqslant 0$, $\mathcal{M}_a \cap D^+(\mathscr{S}_0)$ does not have compact closure in $\mathcal{M}_G$, then $\psi(D^+(\mathscr{S}_0))$ will not have compact closure in $\mathcal{M}$. If however $\mathcal{M}_a \cap D^+(\mathscr{S}_0)$ had compact closure for all $a \geqslant 0$ it would also have to be non-empty for all $a \geqslant 0$ since $\bar{D}^+(\mathscr{S}_0)$ is non-compact. But for $p \in \mathscr{S}_a$, the proper volume of $I^-(p) \cap \mathcal{M}_{a-1}$ has some lower bound c. Thus for every $a \geqslant 0$ the proper volume of $\mathcal{M}_a \cap D^+(\mathscr{S}_0)$ could not be less than c. But this is impossible since by conditions (1)–(3) and proposition 6.7.1, the proper volume of $D^+(\mathscr{S}_0)$ is less than $3/(-\theta_0) \times$ (area of $\mathscr{S}$), where θ_0 is the negative upper bound of $\chi^a{}_a$ on $\mathscr{S}$. □

Using this result, one can prove:

Lemma 8.5.4

If $D^+(\mathscr{S}_0)$ does not have compact closure, $\mathcal{M}$ is not b-bounded.

Let $\mathcal{W}$ be the subset of $B(\mathcal{M}_G)$ consisting of all pairs (λ, u) where λ is any future-inextendible timelike geodesic curve in $\mathcal{M}_G$ orthogonal to

$\mathscr{S}_0$ with endpoint $r \in \mathscr{S}_0$, and $u \in \pi^{-1}(r)$ is any basis at r, one of whose vectors is tangent to λ and of length $-3/\theta_0$, the remaining vectors being an orthonormal basis in $\mathscr{S}_0$.

Let $\{\mathscr{P}_\alpha\}$ be a collection of open sets which cover $\mathscr{W}$. Each $\mathscr{P}_\alpha$ will be the union of finite intersections of sets of the form $O(\mathscr{U}, \mathscr{V})$. It is sufficient to consider the case when the $\mathscr{P}$ can be represented as

$$\mathscr{P}_\alpha = \bigcap_\beta O(\mathscr{U}_{\alpha\beta}, \mathscr{V}_\alpha),$$

where for each α the $\mathscr{U}_{\alpha\beta}$ are a finite number of open sets in $\mathscr{M}_G$, and $\mathscr{V}_\alpha$ is an open set in $L(\mathscr{M}_G)$. Let $(\mu, v) \in \mathscr{W}$. Then there is some α such that $(\mu, v) \in \mathscr{P}_\alpha$. This means that the geodesic μ intersects the open set $\mathscr{U}_{\alpha\beta}$ for each value of β and that $v \in \mathscr{V}_\alpha$. Since geodesics depend continuously on their initial conditions there will be some neighbourhood $\mathscr{Y}_\alpha$ of $\pi(v)$ such that every future-inextendible geodesic through $\mathscr{Y}_\alpha$ orthogonal to $\mathscr{S}_0$ will intersect $\mathscr{U}_{\alpha\beta}$ for each value of β. Let $\mathscr{V}'_\alpha$ be an open set contained in $\mathscr{V}_\alpha$ such that $\pi(\mathscr{V}'_\alpha) \subset \mathscr{Y}_\alpha$. Then

$$(\mu, v) \in O(\pi(\mathscr{V}'_\alpha), \mathscr{V}'_\alpha)$$

is contained in $\mathscr{P}_\alpha$. Thus the sets $\{O(\pi(\mathscr{V}'_\alpha), \mathscr{V}'_\alpha)\}$ form a refinement of the covering $\mathscr{P}_\alpha$.

Consider the subset $\mathscr{Q}$ of $L(\mathscr{M}_G)$ consisting of all bases over $\mathscr{S}_0$ where one of the basis vectors is orthogonal to $\mathscr{S}_0$ and of length $-3/\theta_0$, and the remaining vectors are an orthonormal basis of $\mathscr{S}_0$. Since $\mathscr{Q}$ is compact, it can be covered by a finite number of the sets $\mathscr{V}'_\alpha$. Thus $\mathscr{W}$ is compact since it can be covered by a finite number of the sets $O(\pi(\mathscr{V}'_\alpha), \mathscr{V}'_\alpha)$.

By proposition 6.7.1 each point of $D^+(\mathscr{S}_0)$ lies within a proper distance $-3/\theta_0$ along the future-directed geodesic orthogonal to $\mathscr{S}_0$. This means that $\mathrm{Exp}(\mathscr{W})$ contains $D^+(\mathscr{S}_0)$. Let $\psi_* : B(\mathscr{M}_G) \to B(\mathscr{M})$ be the map which takes $(\lambda, u) \in B(\mathscr{M}_G)$ to $(\psi(\lambda), \psi_* u) \in B(\mathscr{M})$. Then $\psi_* \mathscr{W}$ will be a compact subset of $B(\mathscr{M})$ such that

$$\mathrm{Exp}(\psi_* W) \supset \psi(D^+(\mathscr{S}_0)).$$

Thus if $\overline{D^+(\mathscr{S}_0)}$ is not compact, $\overline{\psi}(D^+(\mathscr{S}_0))$ is not compact, so $(\mathscr{M}, \mathbf{g})$ is not b-bounded.   □

This shows that it is sufficient to prove $\overline{D^+}(\mathscr{S}_0)$ non-compact. Suppose it were compact. Then $H^+(\mathscr{S}_0)$ would also be compact. We show below that this would imply that the divergence of the null geodesic generators would have to be zero everywhere on $H^+(\mathscr{S}_0)$. This would be impossible if the matter density were non zero somewhere on $H^+(\mathscr{S}_0)$.

Lemma 8.5.5

If $H^+(\mathscr{D})$ is compact for a partial Cauchy surface $\mathscr{D}$, then the null geodesic generating segments of $H^+(\mathscr{D})$ are geodesically complete in the past direction.

From proposition 6.5.2 it follows that the generating segments have no past endpoints. They must therefore form 'almost closed' curves in the compact set $H^+(\mathscr{D})$. If they formed actual closed curves, one could use proposition 6.4.4 to show that if they were incomplete in the past direction, they could be varied towards the past to give closed timelike curves. This however would be impossible since such curves would lie in $D^+(\mathscr{D})$. The proof in the case when the null geodesic generators of $H^+(\mathscr{D})$ are only 'almost closed' is similar though a little more delicate.

Introduce a future-directed timelike unit vector field $\mathbf{V}$ which is geodesic in a neighbourhood $\mathscr{U}$ of $H^+(\mathscr{D})$ with compact closure. Define the positive definite metric $\mathbf{g}'$ as in proposition 6.4.4 by

$$g'(\mathbf{X}, \mathbf{Y}) = g(\mathbf{X}, \mathbf{Y}) + 2g(\mathbf{X}, \mathbf{V})\,g(\mathbf{Y}, \mathbf{V})$$

and let t be a parameter which measures proper distance in the metric $\mathbf{g}'$ along a null geodesic generating segment γ of $H^+(\mathscr{D})$, and which is zero at some point $q \in \gamma$. Then $g(\mathbf{V}, \partial/\partial t) = -2^{-\frac{1}{2}}$. As γ has no past endpoint, t will have no lower bound. Let f and h be given by

$$f\frac{\partial}{\partial t} = \frac{\mathrm{D}}{\partial t}\left(\frac{\partial}{\partial t}\right), \quad \frac{\partial}{\partial v} = h\frac{\partial}{\partial t},$$

where v is an affine parameter. Suppose γ were geodesically incomplete in the past, then the affine parameter

$$v = \int_0^t h^{-1}\,\mathrm{d}t'$$

would have a lower bound v_0 as $t \to -\infty$. Now consider a variation α of γ whose variation vector $\partial/\partial u$ is equal to $-x\mathbf{V}$. Then

$$\frac{\partial}{\partial u}g\left(\frac{\partial}{\partial t}, \frac{\partial}{\partial t}\right)\bigg|_{u=0} = 2^{-\frac{1}{2}}\left(\frac{\mathrm{d}x}{\mathrm{d}t} + xh^{-1}\frac{\mathrm{d}h}{\mathrm{d}t}\right). \tag{8.3}$$

Since $h \to \infty$ as $t \to -\infty$, one could find a bounded function $x(t)$ such that (8.3) was negative for all $t \leqslant 0$. However this would not be sufficient to ensure that the variation gave an everywhere timelike curve since it could be that the range of u for which (8.3) remained negative tended to zero as $t \to -\infty$. To deal with this we shall consider the second derivative under the variation:

$$\frac{\partial^2}{\partial u^2} g\left(\frac{\partial}{\partial t}, \frac{\partial}{\partial t}\right) = \frac{\partial}{\partial u}\left(g\left(\frac{\partial}{\partial t}, \frac{D}{\partial t}\frac{\partial}{\partial u}\right)\right)$$

$$= g\left(\frac{D}{\partial t}\frac{\partial}{\partial u}, \frac{D}{\partial t}\frac{\partial}{\partial u}\right) + g\left(\frac{\partial}{\partial t}, \frac{D}{\partial t}\frac{D}{\partial u}\frac{\partial}{\partial u}\right) + g\left(\frac{\partial}{\partial t}, R\left(\frac{\partial}{\partial u}, \frac{\partial}{\partial t}\right)\frac{\partial}{\partial u}\right).$$

Choosing $\partial x/\partial u$ to be zero and using the fact that $\mathbf{V}$ is a geodesic in a neighbourhood $\mathscr{U}$ of $H^+(\mathscr{Q})$ this reduces to

$$-\left(\frac{\mathrm{d}x}{\mathrm{d}t}\right)^2 + x^2\left[g\left(\frac{D\mathbf{V}}{\partial t}, \frac{D\mathbf{V}}{\partial t}\right) + g\left(\frac{\partial}{\partial t}, R\left(\mathbf{V}, \frac{\partial}{\partial t}\right)\mathbf{V}\right)\right]$$

for $0 \leqslant u \leqslant \epsilon$. In any basis orthonormal with respect to the metric $\mathbf{g}'$, the components of the Riemann tensor and of the covariant derivative of $\mathbf{V}$ (with respect to $\mathbf{g}$) will be bounded on $\mathscr{U}$. Thus there is some $C > 0$ such that

$$\frac{\partial^2}{\partial u^2} g\left(\frac{\partial}{\partial t}, \frac{\partial}{\partial t}\right) \leqslant C^2 x^2 g'\left(\frac{\partial}{\partial t}, \frac{\partial}{\partial t}\right).$$

Now

$$\frac{\partial}{\partial u}\left(g\left(\mathbf{V}, \frac{\partial}{\partial t}\right)\right) = -\frac{\mathrm{d}x}{\mathrm{d}t},$$

so

$$g\left(\mathbf{V}, \frac{\partial}{\partial t}\right) = -2^{-\frac{1}{2}} - u\frac{\mathrm{d}x}{\mathrm{d}t}.$$

Therefore

$$g'\left(\frac{\partial}{\partial t}, \frac{\partial}{\partial t}\right) = g\left(\frac{\partial}{\partial t}, \frac{\partial}{\partial t}\right) + 1 - (2\sqrt{2})\,u\frac{\mathrm{d}x}{\mathrm{d}t} + 2u^2\left(\frac{\mathrm{d}x}{\mathrm{d}t}\right)^2$$

$$\leqslant g\left(\frac{\partial}{\partial t}, \frac{\partial}{\partial t}\right) + d$$

for $0 \leqslant u \leqslant \epsilon$, where $d = (2\sqrt{2})\,\epsilon C_1 + 2\epsilon^2 C_1{}^2 + 1$, and C_1 is an upper bound to $|\mathrm{d}x/\mathrm{d}t|$. Thus we have

$$\frac{\partial^2 y}{\partial u^2} \leqslant C^2 x^2 (y + d)$$

and

$$\left.\frac{\partial y}{\partial u}\right|_{u=0} = 2^{-\frac{1}{2}} h^{-1}\frac{\mathrm{d}}{\mathrm{d}t}(hx), \quad y|_{u=0} = 0,$$

where $y = g(\partial/\partial t, \partial/\partial t)$. Therefore

$$y \leqslant d\,(\cosh Cxu - 1) + a \sinh Cxu$$

$$\leqslant \sinh Cxu(d \tanh \tfrac{1}{2}Cxu + a),$$

where $a = 2^{-\frac{1}{2}} C^{-1}\,\mathrm{d}\,(\log hx)/\mathrm{d}t$.

Now take
$$x = h^{-1}\left[-\int_t^0 h^{-1}\,\mathrm{d}t' + K \right]^{-1},$$

where
$$K = 2\int_{-\infty}^0 h^{-1}\,\mathrm{d}t';$$

then $a = -2^{-\frac{1}{2}}C^{-1}hx$. Since $f = -h^{-1}(\mathrm{d}h/\mathrm{d}t)$ is bounded on the compact set $H^+(\mathscr{Q})$ and since
$$\int_t^0 h^{-1}\,\mathrm{d}t' = -v$$

was assumed to converge as $t \to -\infty$, there would be upper bounds for x and $|\mathrm{d}x/\mathrm{d}t|$ and a positive lower bound C_2 for h when $-\infty < t \leqslant 0$. Then for $0 < u < \min(\epsilon, 2C^{-2}d^{-1}C_2)$, y would be negative when $-\infty < t \leqslant 0$.

In other words, the variation α would give a past-inextendible time-like curve which lay in int $D^+(\mathscr{Q})$ and which was totally imprisoned in the compact set $\overline{\mathscr{U}}$. But this is impossible, since by lemma 6.6.5 the strong causality condition holds on int $D^+(\mathscr{Q})$. Thus γ must be geodesically complete in the past direction. $\square$

Consider the expansion $\hat{\theta}$ of the tangent vectors $\partial/\partial t$ to the null geodesic generators of $H^+(\mathscr{S}_0)$. Suppose that $\hat{\theta} > 0$ at some point q on a generator γ and let $\mathscr{T}$ be a spacelike two-surface through q in a neighbourhood of q in $H^+(\mathscr{S}_0)$. The generators of $H^+(\mathscr{S}_0)$ will be orthogonal to $\mathscr{T}$ and would be converging into the past. Then by condition (1) and the above lemma there would be a point $r \in \gamma$ conjugate to $\mathscr{T}$ along γ (proposition 4.4.6). Points on γ beyond r could be joined to $\mathscr{T}$ by timelike curves (proposition 4.5.14). But this would be impossible since $H^+(\mathscr{S}_0)$ is an achronal set. Therefore $\hat{\theta} \leqslant 0$ on $H^+(\mathscr{S}_0)$.

Now consider the family of differentiable maps $\beta_z : H^+(\mathscr{S}_0) \to H^+(\mathscr{S}_0)$ defined by taking a point $q \in H^+(\mathscr{S}_0)$ a distance z (measured in the metric $\mathbf{g}'$) to the past along the null geodesic generator through q. Let $\mathrm{d}A$ be the area measured in the metric $\mathbf{g}'$ of a small element of $H^+(\mathscr{S}_0)$. Under the map β_z,

$$\frac{\mathrm{d}}{\mathrm{d}z}\mathrm{d}A = -\hat{\theta}\,\mathrm{d}A.$$

Thus
$$\frac{\mathrm{d}}{\mathrm{d}z}\int_{\beta_z(H^+(\mathscr{S}_0))} \mathrm{d}A = -\int_{\beta_z(H^+(\mathscr{S}_0))} \hat{\theta}\,\mathrm{d}A. \qquad (8.4)$$

But β_z maps $H^+(\mathscr{S}_0)$ into $H^+(\mathscr{S}_0)$ (and onto if the generating segments have no future endpoints). Thus (8.4) must be less than or equal to zero. Together with the previous result this would imply $\theta = 0$ on $H^+(\mathcal{S}_0)$. By the propagation equation (4.35) this is possible only if $R_{ab}K^aK^b = 0$ everywhere on $H^+(\mathcal{S}_0)$, where $\mathbf{K}$ is the tangent vector to the null geodesic generator. However by the conservation theorem of §4.3 condition (5) implies that $T_{ab}K^aK^b$ is non-zero somewhere on $H^+(\mathcal{S})$ and by the Einstein equations (with or without Λ), $T_{ab}K^aK^b$ equals $R_{ab}K^aK^b$. (Strictly, the form of the conservation theorem required is slightly different from that in §4.3. Since there are no suitable spacelike surfaces which intersect $H^+(\mathcal{S}_0)$, one uses instead a family of surfaces one of which is $H^+(\mathcal{S}_0)$, the others being spacelike. These surfaces can be defined by taking the value of the function t at the point $p \in \overline{D^+}(\mathcal{S}_0)$ to be minus the proper volume of $J^+(p) \cap D^+(\mathcal{S}_0)$. Since $t_{;a}$ becomes null on $H^+(\mathcal{S}_0)$, it is no longer necessarily true that there is a constant $C > 0$ such that on $\overline{D^+}(\mathcal{S}_0)$,

$$T^{ab}t_{;ab} \leqslant CT^{ab}t_{;a}t_{;b}.$$

However if V^a is a timelike vector field on $\overline{D^+}(\mathcal{S}_0)$, there is a constant C such that

$$T^{ab}t_{;ab} \leqslant CT^{ab}(t_{;a}t_{;b} + t_{;a}V_b)$$

and

$$T^{ab}V_{a;b} \leqslant CT^{ab}(t_{;a}t_{;b} + t_{;a}V_b).$$

One can then proceed as in §4.3 using $T^{ab}(t_{;ab} + V_{a;b})$ in place of $T^{ab}t_{;ab}$, and proving that $T^{ab}(t_{;a}t_{;b} + t_{;a}V_b)$ cannot be zero on $H^+(\mathcal{S}_0)$ if it is non-zero on $\mathcal{S}_0$. The result then follows from (5).)   □

9

Gravitational collapse and black holes

In this chapter, we shall show that stars of more than about $1\frac{1}{2}$ times the solar mass should collapse when they have exhausted their nuclear fuel. If the initial conditions are not too asymmetric, the conditions of theorem 2 should be satisfied and so there should be a singularity. This singularity is however probably hidden from the view of an external observer who sees only a 'black hole' where the star once was. We derive a number of properties of such black holes, and show that they probably settle down finally to a Kerr solution.

In §9.1 we discuss stellar collapse, showing how one would expect a closed trapped surface to form around any sufficiently large spherical star at a late stage in its evolution. In §9.2 we discuss the event horizon which seems likely to form around such a collapsing body. In §9.3 we consider the final stationary state to which the solution outside the horizon settles down. This seems to be likely to be one of the Kerr family of solutions. Assuming that this is the case, one can place certain limits on the amount of energy which can be extracted from such solutions.

For further reading on black holes, see the 1972 Les Houches summer school proceedings, edited by B. S. de Witt, to be published by Gordon and Breach.

9.1 Stellar collapse

Outside a static spherically symmetric body such as a star, the solution of Einstein's equations is necessarily that part of one of the asymptotically flat regions of the Schwarzschild solution for which r is greater than some value r_0 corresponding to the surface of the star. This will be joined, for $r < r_0$, onto a solution which depends in detail on the radial distribution of density and pressure in the star. In fact even if the star is not static, providing it remains spherically symmetric the solution outside will still be part of the Schwarzschild solution cut off by the surface of the star. (This is Birkhoff's theorem, proof of which is given in appendix B.) If the star is static then r_0 must be

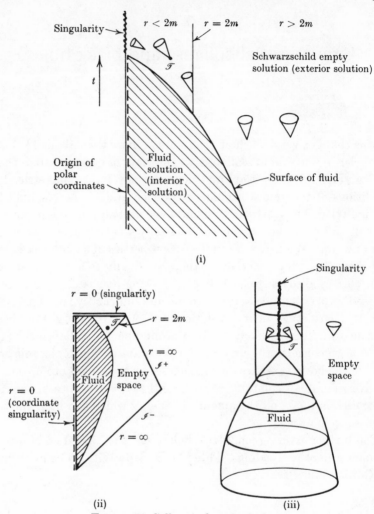

FIGURE 54. Collapse of a spherical star.

(i) Finkelstein diagram ((r, t) plane) of a collapsing spherically symmetric fluid ball. Each point represents a two-sphere.

(ii) Penrose diagram of the collapsing fluid ball.

(iii) Diagram of the collapse with only one spatial dimension suppressed.

greater than $2m$ (the 'Schwarzschild radius'). This follows because the surface of a static star must correspond to the orbit of a timelike Killing vector, and in the Schwarzschild solution there is a timelike Killing vector only where $r > 2m$. If r_0 were less than $2m$, the surface of the star would be expanding or contracting. To get an idea of the magnitude of the Schwarzschild radius, we note that the Schwarzschild radius of the earth is 1.0 cm and that of the sun is 3.0 Km;

the ratios of the Schwarzschild radius to the radius of the earth and the sun are 7×10^{-10} and 2×10^{-6} respectively. Thus normal stars are a long way from their Schwarzschild radii.

The life of a typical star will consist of a long ($\sim 10^9$ years) quasi-static phase in which it is burning nuclear fuel and supporting itself against gravity by thermal and radiation pressure. However when the nuclear fuel is exhausted, the star will cool, the pressure will be reduced, and so it will contract. Now suppose that this contraction cannot be halted by the pressure before the radius becomes less than the Schwarzschild radius (we shall see below that this seems likely for stars of greater than a certain mass). Then since the solution outside the star is the Schwarzschild solution, there will be a closed trapped surface $\mathcal{T}$ around the star (see figure 54), and so, by theorem 2, a singularity will occur provided that causality is not violated and the appropriate energy condition holds. Of course in this case, because the exterior solution is the Schwarzschild solution, it is obvious (see figure 54) that there must be a singularity. However the point is that even if the star is not exactly spherically symmetric, a closed trapped surface will still occur providing the departures from spherical symmetry are not too great. This follows from the stability of the Cauchy development proved in § 7.5; for one can regard the solution as developing from a partial Cauchy surface $\mathcal{H}$ (figure 55). Now if one changes the initial data by a sufficiently small amount on the compact region $J^-(\mathcal{T}) \cap \mathcal{H}$, the new development of $\mathcal{H}$ will still be sufficiently near the old in the compact region $J^+(\mathcal{H}) \cap J^-(\mathcal{T})$ that there will still be a closed trapped surface around the star in the perturbed solution. Thus we have shown that there is a non-zero measure set of initial conditions which lead to a closed trapped surface and hence to a singularity by theorem 2.

The two principal reasons why a star may depart from spherical symmetry are that it may be rotating or may have a magnetic field. One may get some idea of how large the rotation may be without preventing the occurrence of a trapped surface by considering the Kerr solution. This solution can be thought of as representing the exterior solution for a body with mass m and angular momentum $L = am$. If a is less than m there are closed trapped surfaces, but if a is greater than m they do not occur. Thus one might expect that if the angular momentum of the star were greater than the square of its mass, it would be able to halt the contraction of the star before a closed trapped surface developed. Another way of seeing this is that if $L = m^2$ and angular momentum is conserved during the collapse, then the velocity

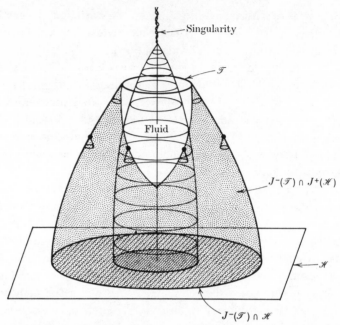

FIGURE 55. Collapse of a spherical star as in figure 54 (iii) showing a partial Cauchy surface $\mathscr{H}$. It is the initial data on the compact region $J^-(\mathscr{T}) \cap \mathscr{H}$ of $\mathscr{H}$ which leads to the occurrence of the closed trapped surface $\mathscr{T}$ in the compact region $J^-(\mathscr{T}) \cap J^+(\mathscr{H})$.

of the surface of the star would be about the velocity of light when the star was at its Schwarzschild radius. Now many stars have an angular momentum greater than the square of their mass (for the sun, $L \sim m^2$). However it seems reasonable to expect some loss of angular momentum during the collapse because of braking by magnetic fields and because of gravitational radiation. The situation is therefore that in some stars, and probably most, angular momentum would not prevent occurrence of closed trapped surfaces, and hence a singularity.

In a nearly spherical collapse a magnetic field **B** which is frozen into a star will increase as the matter density ρ to the $\frac{2}{3}$ power. Thus the magnetic pressure is proportional to $\rho^{\frac{4}{3}}$. This rate of increase is so slow that if the magnetic pressure is not important initially in supporting the star, then it will never be strong enough to have a significant effect on the collapse.

To see why a burnt-out star of more than a certain mass cannot support itself against gravity, we shall give a qualitative discussion (based on unpublished work by Carter) of the zero temperature equation of state for matter.

In hot matter there is pressure produced by the thermal motions of the atoms and by the radiation present. However in cold matter at densities lower than that of nuclear matter ($\sim 10^{14}\,\mathrm{gm\,cm^{-3}}$), the only significant pressure will arise from the quantum mechanical exclusion principle. To estimate this, consider a number density n of fermions of mass m. By the exclusion principle, each fermion will effectively occupy a volume of n^{-1}. Thus by the uncertainty principle, it will have a spatial component of momentum of order $\hbar n^{\frac{1}{3}}$. If the fermions are non-relativistic, i.e. if $\hbar n^{\frac{1}{3}}$ is less than m, the velocity of the fermions will be of order $\hbar n^{\frac{1}{3}}/m$, while if the fermions are relativistic (i.e. $\hbar n^{\frac{1}{3}}$ is greater than m) then the velocity will be practically one (the speed of light). The pressure will be of order (momentum) × (velocity) × (number density), and so will be $\sim \hbar^2 n^{\frac{5}{3}} m^{-1}$ if $\hbar n^{\frac{1}{3}} < m$, and will be $\sim \hbar n^{\frac{4}{3}}$ if $\hbar n^{\frac{1}{3}} > m$. When the matter is non-relativistic, the principal contribution to the degeneracy pressure comes from the electrons, since m^{-1} for them is bigger than it is for baryons. However at high densities, when the particles become relativistic, the pressure is independent of the mass of the particles producing it and depends simply on their number density.

For small cold bodies, self-gravity can be neglected and the degeneracy pressure will be balanced by attractive electrostatic forces between nearest neighbour particles arranged in some sort of lattice. (We assume that there are equal numbers of positive and negative charges and approximately equal numbers of electrons and baryons.) These forces will produce a negative pressure of order $e^2 n^{\frac{4}{3}}$. Thus the mass density of a small cold body will be of order

$$e^6 m_e{}^3 m_n \hbar^{-6} \quad (\sim 1\,\mathrm{gm\,cm^{-3}}), \qquad (9.1)$$

where m_e is the electron rest-mass and m_n is the nucleon rest-mass.

For larger bodies self-gravity will be important, and will compress the matter against the degeneracy pressure. To obtain an exact solution would involve a detailed integration of Einstein's equations. However the important qualitative features can be seen more easily from a simple Newtonian order of magnitude argument. In a star of mass M and radius r_0, the gravitational force on a typical unit volume is of the order $(M/r_0{}^2)\,n m_n$, where $n m_n \simeq M/r_0{}^3$ is the mass density. The gravitational force will be balanced by a pressure gradient of order P/r_0, where P is the average pressure in the star. Thus

$$P = M^2/r_0{}^4 \simeq M^{\frac{2}{3}} n^{\frac{4}{3}} m_n{}^{\frac{4}{3}}.$$

If the density is sufficiently low that the main contribution to the pressure is from the degeneracy of non-relativistic electrons,

$$P = \hbar^2 n^{\frac{5}{3}} m_e^{-1} = M^{\frac{2}{3}} n^{\frac{5}{3}} m_n^{\frac{4}{3}},$$

so
$$n = M^2 m_n{}^4 m_e{}^3 \hbar^{-6}.$$

This will be the correct formula for bodies for which it yields a value of n greater than (9.1) and less than $m_e{}^3 \hbar^{-3}$, i.e. for $e^3 m_n{}^{-2} < M < \hbar^{\frac{3}{2}} m_n{}^{-2}$. Such stars are known as white dwarfs.

If the density is so high that the electrons are relativistic, i.e. $n > m_e{}^3 \hbar^{-3}$, then the pressure will be given by the relativistic formula; so $P = \hbar n^{\frac{4}{3}} = M^{\frac{2}{3}} n^{\frac{4}{3}} m_n{}^{\frac{4}{3}}$. Now n cancels out of this equation. Thus apparently one obtains a star of mass

$$M_{\rm L} = \hbar^{\frac{3}{2}} m_n{}^{-2} \simeq 1.5 \, M_\odot,$$

which can have any density greater than $m_e{}^3 m_n \hbar^{-3}$, i.e. any radius less than $\hbar^{\frac{3}{2}} m_n{}^{-1} m_e{}^{-1}$. Stars of mass greater than $M_{\rm L}$ simply cannot be supported by the degeneracy pressure of electrons.

In fact, when the electrons become relativistic they tend to induce inverse beta decay with the protons, producing neutrons:

$$e^- + p \to \nu_e + n.$$

This denudes the electrons and hence reduces their degeneracy pressure, thereby causing the star to contract and making the electrons more relativistic. This is an unstable situation, and the process will continue until nearly all the electrons and protons have been converted into neutrons. At this stage, equilibrium is again possible with the star supported by the degeneracy pressure of the neutrons. Such a body is called a neutron star. If the neutrons are non-relativistic, one finds

$$n = M^2 m_n{}^7 \hbar^{-6}.$$

If the neutrons are relativistic, the star must again have a mass $M_{\rm L}$ and a radius less than or equal to $\hbar^{\frac{3}{2}} m_n{}^{-2}$. However $M_{\rm L}/\hbar^{\frac{3}{2}} m_n{}^{-2} = 1$ and so such a star is near the General Relativity limit $M_{\rm L}/R \approx 2$.

The conclusion is that a cold star of mass greater than $M_{\rm L}$ cannot be supported by either electron or neutron degeneracy pressure. To show this rigorously, consider the Newtonian equation of support:

$$\mathrm{d}p/\mathrm{d}r = -\rho M(r) r^{-2}, \tag{9.2}$$

where
$$M(r) = 4\pi \int_0^r \rho r^2 \, \mathrm{d}r$$

is the mass within radius r. Multiply both sides of (9.2) by r^4 and integrate by parts from 0 to r_0. This gives

$$4\int_0^{r_0} pr^3 \, \mathrm{d}r = (M(r_0))^2/8\pi,$$

since $p = 0$ at $r = r_0$. On the other hand,

$$\frac{\mathrm{d}}{\mathrm{d}r}\left(\int_0^r pr'^3 \, \mathrm{d}r'\right)^{\frac{3}{4}} = \frac{3}{4}\left(\int_0^r pr'^3 \, \mathrm{d}r'\right)^{-\frac{1}{4}} pr^3$$

$$= \frac{3}{4}\left(\tfrac{1}{4}pr^4 - \frac{1}{4}\int_0^r \frac{\mathrm{d}p}{\mathrm{d}r'}r'^4 \, \mathrm{d}r'\right)^{-\frac{1}{4}} pr^3 < \frac{3\sqrt{2}}{4}p^{\frac{3}{4}}r^2,$$

since $\mathrm{d}p/\mathrm{d}r$ is never positive. As p is never greater than $\hbar n^{\frac{4}{3}}$, this shows that

$$\int_0^{r_0} pr^3 \, \mathrm{d}r < \hbar\left(\int_0^{r_0} nr^2 \, \mathrm{d}r\right)^{\frac{4}{3}} = \hbar(M(r_0))^{\frac{4}{3}} (4\pi m_{\mathrm{n}})^{-\frac{4}{3}}.$$

Therefore $M(r_0)$ must be less than $(8\hbar)^{\frac{3}{2}} (4\pi)^{-\frac{1}{2}} m_{\mathrm{n}}^{-2}$, i.e.

$$M(r_0) < 8\hbar^{\frac{3}{2}}m_{\mathrm{n}}^{-2}.$$

We summarize these results in figure 56. In this diagram we plot the average nucleon density n against the total mass M of the body. The solid line shows the approximate equilibrium configuration of a cold body. In a hot body there will be thermal and radiation pressure in addition to degeneracy pressure and so such bodies may be in equilibrium above the solid line. The heavy dashed line on the right indicates where M/r_0 (which is $M^{\frac{2}{3}}n^{\frac{1}{3}}m_{\mathrm{p}}^{\frac{1}{3}}$) is equal to two. The region to the right of this line contains no equilibrium states, and corresponds to a star being within its Schwarzschild radius. Far away from this line to the left, the difference between Newtonian theory and General Relativity may be neglected. Near this line, one has to take into account General Relativistic effects. For a static spherically symmetric body composed of a perfect fluid, the Einstein field equations can be reduced to (see appendix B)

$$\frac{\mathrm{d}p}{\mathrm{d}r} = -\frac{(\mu + p)(\hat{M}(r) + 4\pi r^3 p)}{r(r - 2\hat{M}(r))}, \tag{9.3}$$

where the radial coordinate is such that the area of the two-surface $\{r = \text{constant}, t = \text{constant}\}$ is $4\pi r^2$. $\hat{M}(r)$ is now defined as

$$\int_0^r 4\pi r^2 \mu \, \mathrm{d}r,$$

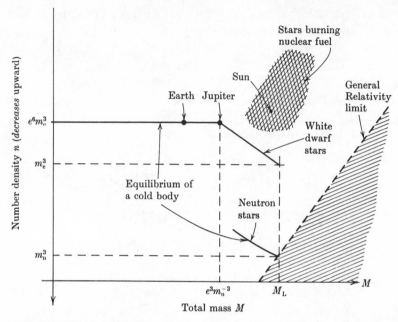

FIGURE 56. Nucleon number density n plotted against total mass of a static body M. The heavy line shows the equilibrium of cold bodies; hot bodies at suitable temperatures can be in equilibrium above this line. General Relativity forbids any bodies in the shaded region from being static.

where $\mu = \rho(1 + \epsilon)$ is the total energy density, ρ is nm_n, and ϵ is the relativistic increase of mass associated with the momentum of the fermions. $\hat{M}(r_0)$ is equal to the Schwarzschild mass $\hat{M}$ of the exterior Schwarzschild solution for $r > r_0$. For a bound star this will be less than the conserved mass

$$\tilde{M} = \int_0^{r_0} \frac{4\pi\rho r^2 \, dr}{(1 - 2M/r)^{\frac{1}{2}}} = Nm_n,$$

where N is the total number of nucleons in the star, because the difference $(\tilde{M} - \hat{M})$ represents the amount of energy radiated to infinity since the formation of the star from dispersed matter initially at rest. In practice this difference is never more than a few percent and in no case can it exceed $2\hat{M}$, since Bondi (1964) has shown that $(1 - 2\hat{M}/r)^{\frac{1}{2}}$ cannot be less than $\frac{1}{3}$ provided μ and p are positive and that μ decreases outwards, and cannot be less than $\frac{1}{2}$ if p is less than or equal to μ. Therefore $\hat{M} < \tilde{M} < 3\hat{M}$.

Comparing (9.3) with (9.2), with μ in place of ρ and $\hat{M}$ in place of M, one sees that the extra terms on the right-hand side of (9.3) are all negative provided $\epsilon \geqslant 0$ and $p \geqslant 0$. Thus since in Newtonian theory a cold star of mass $M > M_L$ cannot support itself, neither can a cold star of Schwarzschild mass $\hat{M} > M_L$ in General Relativity. This means that a cold star which contains more than $3M_L/m_n$ nucleons cannot support itself. In practice, the extra terms in (9.3) mean that the limiting nucleon number is less than M_L/m_n.

In our discussion of neutron stars, we ignored the effects of nuclear forces. These will somewhat modify the position of the equilibrium line in figure 56 for such stars. For details, see Harrison, Thorne, Wakano and Wheeler (1965), Thorne (1966), Cameron (1970), and Tsuruta (1971). However they will not affect the important point that a star containing slightly more than M_L/m_n nucleons will not have any zero temperature equilibrium. This is because the point at which neutrons become relativistic in a star of mass M_L almost coincides with the General Relativity limit $M/R \approx 2$. Thus a star containing somewhat more than M_L/m_n nucleons will not reach nuclear densities until it is already inside its Schwarzschild radius.

The life history of a star will lie in a vertical line on figure 56, unless it manages to lose a significant amount of material by some process. The star will condense out of a cloud of gas. As it contracts, the temperature will rise due to the compression of the gas. If the mass is less than about $10^{-2}M_L$, the temperature will never rise sufficiently high to start nuclear reactions and the star will eventually radiate away its heat and settle down to a state in which gravity is balanced by degeneracy pressure of non-relativistic electrons. If the mass is greater than about $10^{-2}M_L$, the temperature will rise high enough to start the nuclear reaction which converts hydrogen to helium. The energy produced by this reaction will balance the energy lost by radiation and the star will spend a long period ($\sim 10^{10}(M_L/M)^2$ years) in quasi-static equilibrium. When the hydrogen in the core is exhausted, the core will contract and the temperature will rise. Further nuclear reactions may now take place, converting helium in the core into heavier elements. However the energy available from this conversion is not very great, and so the core cannot remain in this phase very long. If the mass is less than M_L, the star can settle down eventually to a white dwarf state in which it is supported by degeneracy pressure of non-relativistic electrons, or possibly to a neutron star state in which it is supported by neutron degeneracy pressure. However if the mass is more than slightly greater than M_L, there is no low temperature equilibrium state. Therefore the star must either pass within its Schwarzschild radius, or manage to eject sufficient matter that its mass is reduced to less than M_L.

Ejection of matter has been observed in supernovae and planetary nebulae, but the theory is not yet very well understood. What calculations there have been suggest that stars up to $20M_L$ may possibly be able to throw off most of their mass and leave a white dwarf or neutron star of mass less than M_L (see Weymann (1963), Colgate and White (1966), Arnett (1966), Le Blanc and Wilson (1970), and Zel'dovich and Novikov (1971)). However it is not really credible that a star of more than $20M_L$ could manage to lose more than 95 % of its matter, and so one would expect that the inner part of the star at any rate would collapse within its Schwarzschild radius. (Present calculations in fact indicate that stars of mass $M > 5M_L$ would not be able to eject sufficient mass to avoid a relativistic collapse.)

Going to larger masses, consider a body of about $10^8 M_L$. If this collapsed to its Schwarzschild radius, the density would only be of the order of 10^{-4} gm cm^{-3} (less than the density of air). If the matter were fairly cold initially, the temperature would not have risen sufficiently either to support the body or to ignite the nuclear fuel; thus there would be no possibility of mass loss, or uncertainty about the equation of state. This example also shows that the conditions when a body passes through its Schwarzschild radius need not be in any way extreme.

To summarize, it seems that certainly some, and probably most, bodies of mass $> M_L$ will eventually collapse within their Schwarzschild radius, and so give rise to a closed trapped surface. There are at least 10^9 stars more massive than M_L in our galaxy. Thus there are a large number of situations in which theorem 2 predicts the existence of singularities. We discuss the observable consequences of stellar collapse in the next sections.

9.2   Black holes

What would a collapsing body look like to an observer O who remained at a large distance from it? One can answer this if the collapse is exactly spherically symmetric, since then the solution outside the body will be the Schwarzschild solution. In this case, an observer O' on the surface of the star would pass within $r = 2m$ at some time, say 1 o'clock, as measured by his watch. He would not notice anything special at that time. However after he passes $r = 2m$ he will not be

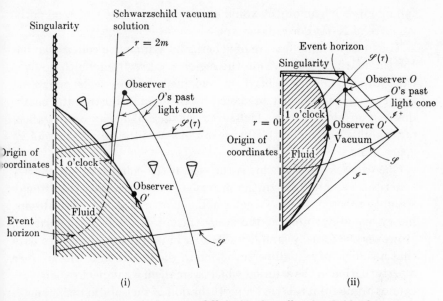

FIGURE 57. An observer O who never falls inside the collapsing fluid sphere never sees beyond a certain time (say, 1 o'clock) in the history of an observer O' on the surface of the collapsing fluid sphere.

(i) Finkelstein diagram; (ii) Penrose diagram.

visible to the observer O who remains outside $r = 2m$ (figure 57). However long the observer O waits, he will never see O' at a time later than 1 o'clock as measured by O''s watch. Instead he will see O''s watch apparently slow down and asymptotically approach 1 o'clock. This means that the light he receives from O' will have a greater and greater shift of frequency to the red and as a consequence a greater and greater decrease of intensity. Thus although the surface of the star never actually disappears from O's sight, it soon becomes so faint as to be invisible in practice. In fact O would first see the centre of the disc of the star become faint, and then this faint region would spread outwards to the limb (Ames and Thorne (1968)). The time scale for this diminution of intensity is of the order for light to travel a distance $2m$.

One would be left with an object which, for all practical purposes, is invisible. However it would still have the same Schwarzschild mass, and would still produce the same gravitational field, as it did before it collapsed. One might be able to detect its presence by its gravitational effects, for instance its effects on the orbits of nearby objects, or by the deflection of light passing near it. It is also possible that gas falling into such an object would set up a shock wave which might be a source of X-rays or radio waves.

The most striking feature of spherically symmetric collapse is that the singularity occurs within the region $r < 2m$, from which no light can escape to infinity. Thus if one remained outside $r = 2m$ one would never see the singularity predicted by theorem 2. Further the breakdown of physical theory which occurs at the singularity cannot affect one's ability to predict the future in the asymptotically flat region of space–time.

One can ask whether this is the case if the collapse is not exactly spherically symmetric. In the previous section we used the Cauchy stability theorem to show that small departures from spherical symmetry would not prevent the occurrence of closed trapped surfaces. However the Cauchy stability theorem in its present form says only that a sufficiently small perturbation in the initial data will produce a perturbation in the solution which is small on a compact region. One cannot argue from this that a perturbation of the solution will remain small at arbitrarily large times.

We expect that in general the occurrence of singularities will lead to Cauchy horizons (as in the Reissner–Nordström and Kerr solutions) and hence to a breakdown of one's ability to predict the future. However if the singularities are not visible from outside, one would still be able to predict in the exterior asymptotically flat region.

To make this precise, we shall suppose that $(\mathcal{M}, \mathbf{g})$ has a region which is asymptotically flat in the sense of being weakly asymptotically simple and empty (§ 6.9). There is then a space $(\tilde{\mathcal{M}}, \tilde{\mathbf{g}})$ into which $(\mathcal{M}, \mathbf{g})$ is conformally imbedded as a manifold with boundary $\bar{\mathcal{M}} = \mathcal{M} \cup \partial \mathcal{M}$, where the boundary $\partial \mathcal{M}$ of $\mathcal{M}$ in $\tilde{\mathcal{M}}$ consists of two null surfaces $\mathcal{I}^+$ and $\mathcal{I}^-$ which represent future and past null infinity respectively. Let $\mathcal{S}$ be a partial Cauchy surface in $\mathcal{M}$. We shall say that the space $(\mathcal{M}, \mathbf{g})$ is (future) asymptotically predictable from $\mathcal{S}$ if $\mathcal{I}^+$ is contained in the closure of $D^+(\mathcal{S})$ in the conformal manifold $\tilde{\mathcal{M}}$. Examples of spaces which are future asymptotically predictable from some surface $\mathcal{S}$ include Minkowski space, the Schwarzschild solution for $m \geqslant 0$, the Kerr solution for $m \geqslant 0$, $|a| \leqslant m$, and the Reissner–Nordström solution for $m \geqslant 0$, $|e| \leqslant m$. The Kerr solution with $|a| > m$ and the Reissner–Nordström solution with $|e| > m$ are not future asymptotically predictable, since for any partial Cauchy surface $\mathcal{S}$, there are past-inextendible non-spacelike curves from $\mathcal{I}^+$ which do not intersect $\mathcal{S}$ but approach a singularity. One can regard future asymptotic predictability as the condition that there should be no singularities to the future of $\mathscr{S}$ which are 'naked', i.e. which are visible from $\mathscr{I}^+$.

In a spherical collapse, one gets a space which is future asymptotically predictable. The question is whether this would still be the case for non-spherical collapse. We cannot answer this completely. Perturbation calculations by Doroshkevich, Zel'dovich and Novikov (1966) and Price (1971) seem to indicate that small perturbations from spherical symmetry do not give rise to naked singularities. In addition, Gibbons and Penrose (1972) have tried, and failed, to obtain contradictions which would show that in some situations the development of a future asymptotically predictable space was inconsistent. Their failure does not of course prove that asymptotic predictability will hold, but it does make it more plausible. If it does not hold, one cannot say anything definite about the evolution of any region of a space containing a collapsing star, as new information could come out of the singularity. We shall therefore proceed on the assumption that future asymptotic predictability holds at least for sufficiently small departures from spherical symmetry.

One would expect a particle on a closed trapped surface to be unable to escape to $\mathscr{I}^+$. However if one allowed arbitrary singularities one could always make suitable cuts and identifications to form an escape route for the particle. The following result shows that this is not possible in a future asymptotically predictable space.

Proposition 9.2.1

If

(a) $(\mathcal{M}, \mathbf{g})$ is future asymptotically predictable from a partial Cauchy surface $\mathscr{S}$,

(b) $R_{ab}K^aK^b \geqslant 0$ for all null vectors K^a,

then a closed trapped surface $\mathscr{T}$ in $D^+(\mathscr{S})$ cannot intersect $J^-(\mathscr{I}^+, \overline{\mathcal{M}})$, i.e. cannot be seen from $\mathscr{I}^+$.

For suppose $\mathscr{T} \cap J^-(\mathscr{I}^+, \overline{\mathcal{M}})$ is non-empty. Then there would be a point $p \in \mathscr{I}^+$ in $J^+(\mathscr{T}, \overline{\mathcal{M}})$. Let $\mathscr{U}$ be the neighbourhood of $\mathcal{M}$ which is isometric to the neighbourhood $\mathscr{U}'$ of $\partial\mathcal{M}'$ in the conformal manifold $\tilde{\mathcal{M}}'$ of an asymptotically simple and empty space $(\mathcal{M}', \mathbf{g}')$. Let $\mathscr{S}'$ be a Cauchy surface in $\mathcal{M}'$, which coincides with $\mathscr{S}$ on $\mathscr{U}' \cap \mathcal{M}'$. Then $\mathscr{S}' - \mathscr{U}'$ is compact and so by lemma 6.9.3, every generator of $\mathscr{I}^+$ leaves $J^+(\mathscr{S}' - \mathscr{U}', \overline{\mathcal{M}}')$. This shows that if $\mathscr{W}$ is any compact set of $\mathscr{S}$, every generator of $\mathscr{I}^+$ leaves $J^+(\mathscr{W},\overline{\mathscr{M}})$. From this it follows that every generator of $\mathscr{I}^+$ would leave $J^+(\mathscr{T},\overline{\mathscr{M}})$, since this is contained in $J^+(J^-(\mathscr{T})\cap\mathscr{S},\overline{\mathscr{M}})$. Therefore a null geodesic generator μ of $\dot{J}^+(\mathscr{T},\overline{\mathscr{M}})$ would intersect $\mathscr{I}^+$. The generator μ must have past end-point at $\mathscr{T}$, since otherwise it would intersect $I^-(\mathscr{S})$. Since μ meets $\mathscr{I}^+$ it would have infinite affine length. However by the condition (b) every null geodesic orthogonal to $\mathscr{T}$ would contain a point conjugate to $\mathscr{T}$ within a finite affine length. Thus it could not remain in $\dot{J}^+(\mathscr{T},\overline{\mathscr{M}})$ all the way out to $\mathscr{I}^+$. This shows that $\mathscr{T}$ cannot intersect $J^-(\mathscr{I}^+,\overline{\mathscr{M}})$. $\square$

From the above it follows that a closed trapped surface in $D^+(\mathscr{S})$ in a future asymptotically predictable space must be contained in $\mathscr{M}-J^-(\mathscr{I}^+,\overline{\mathscr{M}})$. Therefore there must be a non-trivial (future) *event horizon* $\dot{J}^-(\mathscr{I}^+,\overline{\mathscr{M}})$. This is the boundary of the region from which particles or photons can escape to infinity in the future direction. By § 6.3 the event horizon is an achronal boundary which is generated by null geodesic segments which may have past endpoints but which can have no future endpoints.

Lemma 9.2.2

If conditions (a), (b) of proposition 9.2.1 are satisfied and if there is a non-empty event horizon $\dot{J}^-(\mathscr{I}^+,\overline{\mathscr{M}})$, then the expansion θ of the null geodesic generators of $\dot{J}^-(\mathscr{I}^+,\overline{\mathscr{M}})$ is non-negative in

$$\dot{J}^-(\mathscr{I}^+,\overline{\mathscr{M}})\cap D^+(\mathscr{S}).$$

Suppose there was an open set $\mathscr{U}$ such that $\theta < 0$ in $\mathscr{U}\cap\dot{J}^-(\mathscr{I}^+,\overline{\mathscr{M}})$. Let $\mathscr{T}$ be a spacelike two-surface in $\mathscr{U}\cap\dot{J}^-(\mathscr{I}^+,\overline{\mathscr{M}})$. Then $\theta = \chi_2{}^a{}_a < 0$. Let $\mathscr{V}$ be an open subset of $\mathscr{U}$ which intersects $\mathscr{T}$ and has compact closure contained in $\mathscr{U}$. One can vary $\mathscr{T}$ by a small amount in $\mathscr{V}$ so that $\chi_2{}^a{}_a$ is still negative but such that in $\mathscr{U}$, $\mathscr{T}$ intersects $J^-(\mathscr{I}^+,\overline{\mathscr{M}})$. As before, this leads to a contradiction since any generator of $\dot{J}^+(\mathscr{T},\overline{\mathscr{M}})$ in $J^-(\mathscr{I}^+,\overline{\mathscr{M}})$ would have past endpoint at $\mathscr{T}$ in $\mathscr{V}$, where it would be orthogonal to $\mathscr{T}$. However as $\chi_2{}^a{}_a < 0$ in $\mathscr{V}$, every outgoing null geodesic orthogonal to $\mathscr{T}$ in $\mathscr{V}$ would contain a point conjugate to $\mathscr{T}$ within a finite affine distance, and so could not remain in $\dot{J}^+(\mathscr{T},\overline{\mathscr{M}})$ all the way out to $\mathscr{I}^+$. $\square$

In a future asymptotically predictable space, $J^+(\mathscr{S})\cap J^-(\mathscr{I}^+,\overline{\mathscr{M}})$ is contained in $D^+(\mathscr{S})$. If there were a point p on the event horizon in $J^+(\mathscr{S})$ which was not in $D^+(\mathscr{S})$, the smallest perturbation could lead to p being in $J^-(\mathscr{I}^+,\overline{\mathscr{M}})$, i.e. being visible from infinity, which would mean that the space was no longer asymptotically predictable. It therefore seems reasonable to slightly extend the definition of future asymptotically predictable, to say that space–time is *strongly future asymptotically predictable* from a partial Cauchy surface $\mathscr{S}$ if $\mathscr{I}^+$ is contained in the closure of $D^+(\mathscr{S})$ in $\bar{\mathscr{M}}$, and $J^+(\mathscr{S}) \cap \bar{J}^-(\mathscr{I}^+, \bar{\mathscr{M}})$ is contained in $D^+(\mathscr{S})$. In other words, one can also predict a neighbourhood of the event horizon from $\mathscr{S}$.

Proposition 9.2.3

If $(\mathscr{M}, \mathbf{g})$ is strongly future asymptotically predictable from a partial Cauchy surface $\mathscr{S}$, there is a homeomorphism

$$\alpha\colon (0, \infty) \times \mathscr{S} \to D^+(\mathscr{S}) - \mathscr{S}$$

such that for each $\tau \in (0, \infty)$, $\mathscr{S}(\tau) \equiv (\{\tau\} \times \mathscr{S})$ is a partial Cauchy surface such that:

(a) for $\tau_2 > \tau_1$, $\mathscr{S}(\tau_2) \subset I^+(\mathscr{S}(\tau_1))$;

(b) for each τ, the edge of $\mathscr{S}(\tau)$ in the conformal manifold $\bar{\mathscr{M}}$ is a spacelike two-sphere $\mathscr{Q}(\tau)$ in $\mathscr{I}^+$ such that for $\tau_2 > \tau_1$, $\mathscr{Q}(\tau_2)$ is strictly to the future of $\mathscr{Q}(\tau_1)$,

(c) for each τ, $\mathscr{S}(\tau) \cup \{\mathscr{I}^+ \cap J^-(\mathscr{Q}(\tau), \bar{\mathscr{M}})\}$ is a Cauchy surface in $\bar{\mathscr{M}}$ for $D(\mathscr{S})$.

In other words, $\mathscr{S}(\tau)$ is a family of spacelike surfaces homeomorphic to $\mathscr{S}$ which cover $D^+(\mathscr{S}) - \mathscr{S}$ and intersect $\mathscr{I}^+$ (see figure 58). One could regard them as surfaces of constant time in the asymptotically predictable region. We choose them to intersect $\mathscr{I}^+$ so that the mass measured on them at infinity will decrease when the emission of gravitational or other forms of radiation takes place.

The construction for $\mathscr{S}(\tau)$ is rather similar to that of proposition 6.4.9. Choose a continuous family $\mathscr{Q}(\tau)$ ($\infty > \tau > 0$) of spacelike two-spheres which cover $\mathscr{I}^+$, such that for $\tau_2 > \tau_1$, $\mathscr{Q}(\tau_2)$ is strictly to the future of $\mathscr{Q}(\tau_1)$. Put a volume measure on $\mathscr{M}$ such that the total volume of $\mathscr{M}$ in this measure is finite. We first prove:

Lemma 9.2.4

$k(\tau)$, the volume of the set $I^-(\mathscr{Q}(\tau), \bar{\mathscr{M}}) \cap D^+(\mathscr{S})$ is a continuous function of τ.

Let $\mathscr{V}$ be any open set with compact closure contained in

$$I^-(\mathscr{Q}(\tau), \bar{\mathscr{M}}) \cap D^+(\mathscr{S}).$$

Then there are timelike curves from every point of $\mathscr{V}$ to $\mathscr{Q}(\tau)$, which can be deformed to give timelike curves to $\mathscr{Q}(\tau - \delta)$ for some $\delta > 0$.

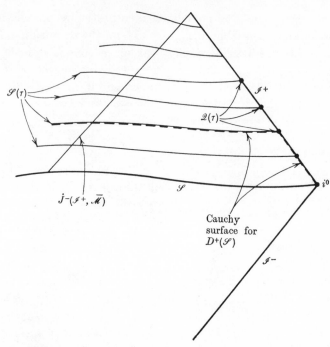

FIGURE 58. A space $(\mathcal{M}, \mathbf{g})$ which is strongly future asymptotically predictable from a partial Cauchy surface $\mathcal{S}$, showing a family $\mathcal{S}(\tau)$ of spacelike surfaces which cover $D^+(\mathcal{S}) - \mathcal{S}$ and intersect $\mathscr{I}^+$ in a family of two-spheres $\mathcal{Q}(\tau)$.

Given any $\epsilon > 0$, one can find a $\mathcal{V}$ whose volume is $> k(\tau) - \epsilon$. Thus there is a $\delta > 0$ such that $k(\tau - \delta) > k(\tau) - \epsilon$. On the other hand, suppose there were an open set $\mathcal{W}$ which did not intersect $I^-(\mathcal{Q}(\tau), \bar{\mathcal{M}}) \cap D^+(\mathcal{S})$ but which was contained in $I^-(\mathcal{Q}(\tau'), \bar{\mathcal{M}}) \cap D^+(\mathcal{S})$ for any $\tau' > \tau$. Then if $p \in \mathcal{W}$, there would be past-directed timelike curves $\lambda_{\tau'}$ from each $\mathcal{Q}(\tau')$ to p. As the region of $\mathscr{I}^+$ between $\mathcal{Q}(\tau)$ and $\mathcal{Q}(\tau_1)$ is compact for any $\tau_1 > \tau$, there would be a past-directed non-spacelike curve λ from $\mathcal{Q}(\tau)$ which was the limit curve of the $\{\lambda_{\tau'}\}$. Since the $\{\lambda_{\tau'}\}$ did not intersect $I^-(\mathcal{Q}(\tau), \bar{\mathcal{M}})$, λ would not either, and so it would be a null geodesic and would lie in $\dot{I}^-(\mathcal{Q}(\tau), \bar{\mathcal{M}})$. It would enter $\mathcal{M}$ and so it would either have a past endpoint at p, or would intersect $\mathcal{S}$. The former is impossible as it would imply that $\mathcal{W}$ intersected $I^-(\mathcal{Q}(\tau), \bar{\mathcal{M}})$, and the latter is impossible as $p \in I^+(\mathcal{S})$. This shows that there is no open set which is in $I^-(\mathcal{Q}(\tau'), \bar{\mathcal{M}})$ for every $\tau' > \tau$, but which is not in $I^-(\mathcal{Q}(\tau), \bar{\mathcal{M}}) \cap D^+(\mathcal{S})$. Thus given ϵ, there is a δ such that

$$k(\tau + \delta) < k(\tau) + \epsilon.$$

Therefore $k(\tau)$ is continuous.   □

Proof of proposition 9.2.3. Define functions $f(p)$ and $h(p, \tau)$, $p \in D^+(\mathscr{S})$, which are volumes of $I^+(p)$ and $I^-(p) - \overline{I^-}(\mathcal{Q}(\tau), \tilde{\mathcal{M}})$. As in proposition 6.4.9, the function $f(p)$ is continuous on the globally hyperbolic region $D^+(\mathscr{S}) - \mathscr{S}$, and goes to zero on every future-inextendible non-spacelike curve. Since $\overline{I^-}(\mathcal{Q}(\tau), \tilde{\mathcal{M}}) \cap \mathcal{M}$ is a past set,

$$D^+(\mathscr{S}) - \overline{I^-}(\mathcal{Q}(\tau), \tilde{\mathcal{M}}) - \mathscr{S}$$

is globally hyperbolic. Thus for each τ, $h(p, \tau)$ is continuous on $D^+(\mathscr{S}) - \mathscr{S}$. This means that given any $\epsilon > 0$, one can find a neighbourhood $\mathcal{U}$ of p such that $|h(q, \tau) - h(p, \tau)| < \tfrac{1}{2}\epsilon$ for any $q \in \mathcal{U}$. By lemma 9.2.4, one can find a $\delta > 0$ such that $|k(\tau') - k(\tau)| < \tfrac{1}{2}\epsilon$ for $|\tau' - \tau| < \delta$. Then $|h(q, \tau') - h(p, \tau)| < \epsilon$, which shows that $h(p, \tau)$ is continuous on $(D^+(\mathscr{S}) - \mathscr{S}) \times (0, \infty)$. The surfaces $\mathscr{S}(\tau)$ can then be defined as the set of points $p \in D^+(\mathscr{S}) - \mathscr{S}$ such that $h(p, \tau) = \tau f(p)$. Clearly these are spacelike surfaces which cover $D^+(\mathscr{S}) - \mathscr{S}$ and satisfy properties (a)–(c).

To define the homeomorphism α, one needs a timelike vector field on $D^+(\mathscr{S}) - \mathscr{S}$ which intersects each surface $\mathscr{S}(\tau)$. We construct such a vector field as follows. Let $\mathscr{V}$ be a neighbourhood of $\mathscr{I}^+$ in the conformal manifold $\tilde{\mathcal{M}}$, let $\mathbf{X}_1$ be a non-spacelike vector field on $\mathscr{V}$ which is tangent to the generators of $\mathscr{I}^+$ on $\mathscr{I}^+$, and let $x_1 \geqslant 0$ be a C^2 function which vanishes outside $\mathscr{V}$ and is non-zero on $\mathscr{I}^+$. Let $\mathbf{X}_2$ be a timelike vector field on $\mathcal{M}$, and let $x_2 \geqslant 0$ be a C^2 function on $\tilde{\mathcal{M}}$ which is non-zero on $\mathcal{M}$ and is zero on $\mathscr{I}^+$. Then the vector field $\mathbf{X} = x_1 \mathbf{X}_1 + x_2 \mathbf{X}_2$ has the required property. The homeomorphism $\alpha : D^+(\mathscr{S}) - \mathscr{S} \to (0, \infty) \times \mathscr{S}$ then maps a point $p \in D^+(\mathscr{S}) - \mathscr{S}$ to (τ, q) where τ is such that $p \in \mathscr{S}(\tau)$, and the integral curve of $\mathbf{X}$ through p intersects $\mathscr{S}$ at q. $\qquad\Box$

If there is an event horizon $\dot{J}^-(\mathscr{I}^+, \tilde{\mathcal{M}})$ in the region $D^+(\mathscr{S})$ of a future asymptotically predictable space, then it follows from property (b) of proposition 9.2.3 that for sufficiently large τ, the surfaces $\mathscr{S}(\tau)$ will intersect it. We define a *black hole* on the surface $\mathscr{S}(\tau)$ to be a connected component of the set $\mathscr{B}(\tau) \equiv \mathscr{S}(\tau) - J^-(\mathscr{I}^+, \tilde{\mathcal{M}})$. In other words, it is a region of $\mathscr{S}(\tau)$ from which particles or photons cannot escape to $\mathscr{I}^+$.

As τ increases, black holes can merge together, and new black holes can form as the result of further bodies collapsing. However, the following result shows that black holes can never bifurcate.

Proposition 9.2.5

Let $\mathscr{B}_1(\tau_1)$ be a black hole on $\mathscr{S}(\tau_1)$. Let $\mathscr{B}_2(\tau_2)$ and $\mathscr{B}_3(\tau_2)$ be black holes on a later surface $\mathscr{S}(\tau_2)$. If $\mathscr{B}_2(\tau_2)$ and $\mathscr{B}_3(\tau_2)$ both intersect $J^+(\mathscr{B}_1(\tau_1))$, then $\mathscr{B}_2(\tau_2) = \mathscr{B}_3(\tau_2)$.

By property (c) of proposition 9.2.3, every future-directed inextendible timelike curve from $\mathscr{B}_1(\tau_1)$ will intersect $\mathscr{S}(\tau_2)$. Thus

$$J^+(\mathscr{B}_1(\tau_1)) \cap \mathscr{S}(\tau_2)$$

is connected, and will be contained in a connected component of $\mathscr{B}(\tau_2)$.   □

For physical applications, one is interested primarily in black holes which form as the result of gravitational collapse from an initially non-singular state. To make this notion precise, we shall say that the partial Cauchy surface $\mathscr{S}$ has an *asymptotically simple past* if $J^-(\mathscr{S})$ is isometric to the region $J^-(\mathscr{S}')$ of some asymptotically simple and empty space–time $(\mathscr{M}', \mathbf{g}')$, where $\mathscr{S}'$ is a Cauchy surface for $(\mathscr{M}', \mathbf{g}')$. By proposition 6.9.4, the surface $\mathscr{S}'$ has the topology R^3 and so $\mathscr{S}$ also has this topology. Proposition 9.2.3 therefore shows that if $(\mathscr{M}, \mathbf{g})$ is strongly future asymptotically predictable from a surface $\mathscr{S}$ with an asymptotically simple past, then each surface $\mathscr{S}(\tau)$ has the topology R^3, and the union of $\mathscr{S}(\tau)$ with the boundary two-sphere $\mathscr{Q}(\tau)$ on $\mathscr{I}^+$ is homeomorphic to the unit cube I^3.

Although one is primarily interested in spaces which have asymptotically simple pasts it will in the next section be convenient to consider future asymptotically predictable spaces which do not have this property, but which at large times may closely approximate to spaces which do. An example of this is the spherically symmetric collapse we considered at the beginning of this section. Once the surface of the star has passed inside the event horizon, the metric of the exterior region is that of the Schwarzschild solution, and is unaffected by the fate of the star. When studying the asymptotic behaviour it is therefore convenient simply to forget about the star, and consider the empty Schwarzschild solution as a space which is strongly future asymptotically predictable from a surface $\mathscr{S}$ such as that shown in figure 24 on p. 154. This surface does not have an asymptotically simple past, and its topology is $S^2 \times R^1$ instead of R^3. However the portion of $\mathscr{S}$ outside the event horizon in region I has the same topology as the region outside the event horizon of the surface $\mathscr{S}(\tau)$ in figure 57. We want to consider spaces which are strongly future asymptotically predictable from a surface $\mathscr{S}$, and are such that the portion of $\mathscr{S}$ outside the event horizon has the same topology as some surface $\mathscr{S}(\tau)$ in a space with an asymptotically simple past. Of course in more complicated cases there may be several components of $\mathscr{B}(\tau)$, corresponding to the collapse of several bodies. We shall therefore consider spaces which are strongly future asymptotically predictable from a surface $\mathscr{S}$, and with the property:

(α) $\mathscr{S} \cap \overline{J^-}(\mathscr{I}^+, \overline{\mathscr{M}})$ is homeomorphic to $R^3 - ($an open set with compact closure).

(Note that this open set may not be connected.) It will also be convenient to demand the property:

(β) $\mathscr{S}$ is simply connected.

Proposition 9.2.6

Let $(\mathscr{M}, \mathbf{g})$ be a space which is strongly future asymptotically predictable from a partial Cauchy surface $\mathscr{S}$ which satisfies (α), (β). Then:

(1) the surfaces $\mathscr{S}(\tau)$ also satisfy (α), (β);

(2) for each τ, $\partial \mathscr{B}_1(\tau)$, the boundary in $\mathscr{S}(\tau)$ of a black hole $\mathscr{B}_1(\tau)$, is compact and connected.

Since the surfaces $\mathscr{S}(\tau)$ are homeomorphic to $\mathscr{S}$, they satisfy property (β). One can define an injective map

$$\gamma: \mathscr{S}(\tau) \cap \overline{J^-}(\mathscr{I}^+, \overline{\mathscr{M}}) \to \mathscr{S} \cap \overline{J^-}(\mathscr{I}^+, \overline{\mathscr{M}})$$

by mapping each point of $\mathscr{S}(\tau)$ down the integral curves of the vector field of **X** proposition 9.2.3. Since $(\mathscr{M}, \mathbf{g})$ is weakly asymptotically simple, one can find a two-sphere $\mathscr{P}$ near $\mathscr{I}^+$ in $\mathscr{S}(\tau) \cap \overline{J^-}(\mathscr{I}^+, \overline{\mathscr{M}})$. The portion of $\mathscr{S}(\tau)$ outside $\mathscr{P}$ will map into the region of $\mathscr{S}$ outside the two-sphere $\gamma(\mathscr{P})$. This shows that the region of $\mathscr{S} \cap \overline{J^-}(\mathscr{I}^+, \overline{\mathscr{M}})$ which is not in $\gamma(\mathscr{S}(\tau) \cap \overline{J^-}(\mathscr{I}^+, \overline{\mathscr{M}}))$ must have compact closure. Therefore $\gamma(\mathscr{S}(\tau) \cap \overline{J^-}(\mathscr{I}^+, \overline{\mathscr{M}}))$ will be homeomorphic to $R^3 - ($an open set with compact closure). Since $\mathscr{S}(\tau)$ is homeomorphic to $R^3 - \mathscr{V}$ where $\mathscr{V}$ is an open subset of R^3 with compact closure, $\partial \mathscr{B}(\tau)$ will be homeomorphic to $\partial \mathscr{V}$ and so will be compact. $\partial \mathscr{B}_1(\tau)$ being a closed subset of $\partial \mathscr{B}(\tau)$ will be compact.

Suppose that $\partial \mathscr{B}_1(\tau)$ consisted of two disconnected components $\partial \mathscr{B}_1{}^1(\tau)$ and $\partial \mathscr{B}_1{}^2(\tau)$. One could find curves λ_1 and λ_2 in $\mathscr{S}(\tau) - \mathscr{B}(\tau)$ from $\mathscr{Q}(\tau)$ to $\partial \mathscr{B}_1{}^1(\tau)$ and $\partial \mathscr{B}_1{}^2(\tau)$ respectively. One could also find a curve μ in int $\mathscr{B}_1(\tau)$ from $\partial \mathscr{B}_1{}^1(\tau)$ to $\partial \mathscr{B}_1{}^2(\tau)$. Joining these together one would obtain a closed curve in $\mathscr{S}(\tau)$ which crossed $\partial\mathscr{B}_1{}^1(\tau)$ only once. This cannot be deformed to zero in $\mathscr{S}(\tau)$, contradicting the fact that $\mathscr{S}(\tau)$ is simply connected. $\square$

We are only interested in black holes that one can actually fall into, i.e. ones in which the boundary $\partial\mathscr{B}(\tau)$ is contained in $J^+(\mathscr{I}^-, \bar{\mathscr{M}})$. We shall therefore add to properties (α), (β) the requirement:

$\quad(\gamma)$ for sufficiently large τ, $\mathscr{S}(\tau) \cap \overline{J^-}(\mathscr{I}^+, \bar{\mathscr{M}})$ is contained in $\overline{J^+}(\mathscr{I}^-, \bar{\mathscr{M}})$.

We shall say that $(\mathscr{M}, \mathbf{g})$ is a *regular predictable space* if it is strongly future asymptotically predictable from a partial Cauchy surface $\mathscr{S}$ and if properties (α), (β), (γ) are satisfied. All the spaces mentioned at the beginning of this section as being future asymptotically predictable are in fact also regular predictable spaces. Proposition 9.2.6 shows that when one is dealing with regular predictable spaces developing from a partial Cauchy surface $\mathscr{S}$, there is a one–one correspondence between black holes $\mathscr{B}_i(\tau)$ and their boundaries $\partial\mathscr{B}_i(\tau)$ in $\mathscr{S}(\tau)$. One could therefore in such a situation give an equivalent definition of a black hole as a connected component of $\mathscr{S}(\tau) \cap \dot{J}^-(\mathscr{I}^+, \bar{\mathscr{M}})$.

The next result gives a property of the boundaries of black holes which will be important in the next section.

Proposition 9.2.7

Let $(\mathscr{M}, \mathbf{g})$ be a regular predictable space developing from a partial Cauchy surface $\mathscr{S}$, in which $R_{ab}K^aK^b \geqslant 0$ for every null vector K^a. Let $\mathscr{B}_1(\tau)$ be a black hole on the surface $\mathscr{S}(\tau)$, and let $\{\mathscr{B}_i(\tau')\}$ $(i = 1$ to $N)$ be the black holes on an earlier surface $\mathscr{S}(\tau')$ which are such that $J^+(\mathscr{B}_i(\tau')) \cap \mathscr{B}_1(\tau) \neq \varnothing$. Then the area $A_1(\tau)$ of $\partial\mathscr{B}_1(\tau)$ is greater than or equal to the sum of the areas $A_i(\tau')$ of $\partial\mathscr{B}_i(\tau')$; the equality can hold only if $N = 1$.

In other words, the area of the boundary of a black hole cannot decrease with time, and if two or more black holes merge to form a single black hole, the area of its boundary will be greater than the areas of the boundaries of the original black holes.

Since the event horizon is the boundary of the past of $\mathscr{I}^+$, its null geodesic generators would have future endpoints only if they intersected $\mathscr{I}^+$. However this is impossible, as the null geodesic generators of $\mathscr{I}^+$ have no future endpoints. Thus the null generators of the event horizon have no future endpoints. By lemma 9.2.2, their expansion θ is non-negative. Thus the area of a two-dimensional cross-section of the generators cannot decrease with τ. By property (c) of proposition 9.2.3, and by proposition 9.2.5, all the null geodesic generators of $J^-(\mathscr{I}^+, \bar{\mathscr{M}})$ which intersect $\mathscr{S}(\tau')$ in any of the $\partial\mathscr{B}_i(\tau')$ must intersect $\mathscr{S}(\tau)$ in $\partial\mathscr{B}_1(\tau)$. Thus the area of $\partial\mathscr{B}_1(\tau)$ is greater than or equal to the sum of the areas of the $\{\mathscr{B}_i(\tau')\}$. When $N > 1$, $\partial\mathscr{B}_1(\tau)$ will contain N disjoint closed subsets which correspond to the generators of $J^-(\mathscr{I}^+, \bar{\mathscr{M}})$ which intersect each $\partial\mathscr{B}_i(\tau')$. Since $\partial\mathscr{B}_1(\tau)$ is connected, it must contain an open set of generators which do not intersect any $\partial\mathscr{B}_i(\tau')$, but have past endpoints between $\mathscr{S}(\tau)$ and $\mathscr{S}(\tau')$.   □

It has been convenient to define black holes in terms of the event horizon $J^-(\mathscr{I}^+, \bar{\mathscr{M}})$, because this is a null hypersurface with a number of nice properties. However this definition depends on the whole future behaviour of the solution; given the partial Cauchy surface $\mathscr{S}(\tau)$, one cannot find where the event horizon is without solving the Cauchy problem for the whole future development of the surface. It is therefore useful to define a different sort of horizon which depends only on the properties of space–time on the surface $\mathscr{S}(\tau)$.

One knows from proposition 9.2.1 that any closed trapped surface on $\mathscr{S}(\tau)$ in a regular predictable space developing from a partial Cauchy surface $\mathscr{S}$ must be in $\mathscr{B}(\tau)$. This result depends only on the fact that the outgoing null geodesics orthogonal to the two-surface are converging. It does not matter whether the ingoing null geodesics are converging or not. We shall therefore say that an orientable compact spacelike two-surface in $D^+(\mathscr{S})$ is an *outer trapped surface* if the expansion θ of the outgoing null geodesics orthogonal to it is non-positive. (We include the case $\theta = 0$ for convenience.) In order to define which is the outgoing family of null geodesics we make use of property (β) of the partial Cauchy surfaces $\mathscr{S}(\tau)$. Let $\mathbf{X}$ be the timelike vector field of proposition 9.2.3. Then any compact orientable spacelike two-surface $\mathscr{P}$ in $D^+(\mathscr{S})$ can be mapped by the integral curves of $\mathbf{X}$ into a compact orientable two-surface $\mathscr{P}'$ in $\mathscr{S}(\tau)$, for any given value of τ. Let λ be a curve in $\mathscr{S}(\tau) \cup \mathscr{Q}(\tau)$ from $\mathscr{Q}(\tau)$ to $\mathscr{P}'$ which intersects $\mathscr{P}'$ only at its endpoint. Then one can define the outgoing direction on $\mathscr{P}'$ in $\mathscr{S}(\tau)$ as the direction for which λ approaches $\mathscr{P}'$. As $\mathscr{S}(\tau)$ is simply connected, this definition is unique. The outgoing family of null geodesics orthogonal to $\mathscr{P}$ is then that family which is mapped by $\mathbf{X}$ onto curves in $\mathscr{S}(\tau)$ which are outgoing for $\mathscr{P}'$.

Knowing the solution on the surface $\mathscr{S}(\tau)$, one can find all the outer trapped surfaces $\mathscr{P}$ which lie in $\mathscr{S}(\tau)$. We shall define the *trapped region* $\mathcal{T}(\tau)$ in the surface $\mathscr{S}(\tau)$ as the set of all points $q \in \mathscr{S}(\tau)$ such that there is an outer trapped surface $\mathscr{P}$ lying in $\mathscr{S}(\tau)$, through q. As is shown by the following result, the existence of the trapped region $\mathcal{T}(\tau)$ implies the existence of a black hole $\mathscr{B}(\tau)$, and in fact $\mathcal{T}(\tau)$ lies in $\mathscr{B}(\tau)$ for each value of τ.

Proposition 9.2.8

Let $(\mathscr{M}, \mathbf{g})$ be a regular predictable space developing from a partial Cauchy surface $\mathscr{S}$, in which $R_{ab}K^aK^b \geqslant 0$ for any null vector K^a. Then an outer trapped surface $\mathscr{P}$ in $D^+(\mathscr{S})$ does not intersect $J^-(\mathscr{I}^+, \bar{\mathscr{M}})$.

The proof is similar to that of proposition 9.2.1. Suppose $\mathscr{P}$ intersects $J^-(\mathscr{I}^+, \bar{\mathscr{M}})$. Then $\dot{J}^+(\mathscr{P}, \bar{\mathscr{M}})$ would intersect $\mathscr{I}^+$. To each point of $\mathscr{I}^+ \cap \dot{J}^+(\mathscr{P}, \bar{\mathscr{M}})$ there would be a past-directed null geodesic generator of $\dot{J}^+(\mathscr{P}, \bar{\mathscr{M}})$ which had past endpoint on $\mathscr{P}$, and which contained no point conjugate to $\mathscr{P}$. By (4.35) the expansion θ of these generators would be non-positive, as it is non-positive at $\mathscr{P}$ and as $R_{ab}K^aK^b \geqslant 0$. Thus the area of a two-dimensional cross-section of the generators would always be less than or equal to the area of $\mathscr{P}$. This establishes a contradiction, as the area of $\mathscr{I}^+ \cap \dot{J}^+(\mathscr{P}, \bar{\mathscr{M}})$ is infinite, as it is at infinity. □

We shall call the outer boundary $\partial \mathcal{T}_1(\tau)$ of a connected component $\mathcal{T}_1(\tau)$ of the trapped region $\mathcal{T}(\tau)$, an *apparent horizon*. By the previous result, the existence of an apparent horizon $\partial \mathcal{T}_1(\tau)$ implies the existence of a component $\partial \mathscr{B}_1(\tau)$ of the event horizon outside it, or coinciding with it. However the converse is not necessarily true: there may not be outer trapped surfaces within an event horizon.

On the other hand, there may be more than one connected component of $\mathcal{T}(\tau)$ within one component $\partial \mathscr{B}_1(\tau)$ of the event horizon. These possibilities are illustrated in figure 59. A similar situation arises when one considers the collision and merger of two black holes. On an initial surface $\mathscr{S}(\tau_1)$, one would have two separate trapped regions $\mathcal{T}_1(\tau_1)$ and $\mathcal{T}_2(\tau_1)$ contained in black holes $\mathscr{B}_1(\tau)_1$ and $\mathscr{B}_2(\tau)_1$ respectively. As they approached each other, the two components $\partial \mathscr{B}_1(\tau)$ and $\partial \mathscr{B}_2(\tau)$ of the event horizon would amalgamate to form a single black hole $\mathscr{B}_3(\tau_2)$ on a later surface $\mathscr{S}(\tau_2)$. The apparent horizons $\partial \mathcal{T}_1(\tau)$ and $\partial \mathcal{T}_2(\tau)$ would however not join up immediately. Instead what would happen is that a third trapped region $\mathcal{T}_3(\tau)$ would develop surrounding

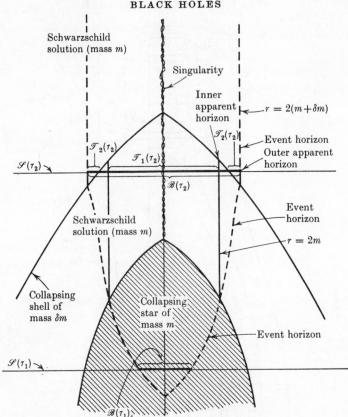

FIGURE 59. The spherical collapse of a star of mass m, followed by the spherical collapse of a shell of matter of mass δm; the exterior solution will be a Schwarzschild solution of mass m after the collapse of the star, and a Schwarzschild solution of mass $m + \delta m$ after the collapse of the shell. At time τ_1 there is an event horizon but no apparent event horizon; at time τ_2 there are two apparent horizons within the event horizon.

them both (figure 60). At some later time, $\mathcal{T}_1$, $\mathcal{T}_2$ and $\mathcal{T}_3$ might merge together.

We shall only outline the proofs of the principal properties of the apparent horizon. First of all one has:

Proposition 9.2.9

Each component of $\partial \mathcal{T}(\tau)$ is a two-surface such that the outgoing orthogonal null geodesics have zero convergence θ on $\partial \mathcal{T}(\tau)$. (We shall call such a surface, a *marginally outer trapped surface*.)

If θ were positive in a neighbourhood in $\partial \mathcal{T}(\tau)$ of a point $p \in \partial \mathcal{T}(\tau)$, then there would be a neighbourhood $\mathcal{U}$ of p such that any outer

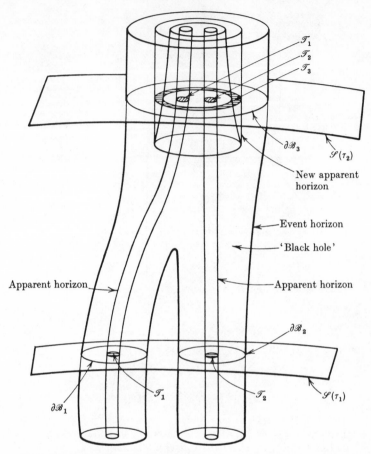

FIGURE 60. The collision and merging of two black holes. At time τ_1, there are apparent horizons $\partial\mathscr{T}_1$, $\partial\mathscr{T}_2$ inside the event horizons $\partial\mathscr{B}_1$, $\partial\mathscr{B}_2$ respectively. By time τ_2, the event horizons have merged to form a single event horizon; a third apparent horizon has now formed surrounding both the previous apparent horizons.

trapped surface in $\mathscr{S}(\tau)$ which intersected $\mathscr{U}$ would also intersect $\partial\mathscr{T}(\tau)$. Thus $\theta \leqslant 0$ on $\partial\mathscr{T}(\tau)$.

If θ were negative in a neighbourhood in $\partial\mathscr{T}(\tau)$ of a point $p \in \partial\mathscr{T}(\tau)$, one could deform $\partial\mathscr{T}(\tau)$ outwards in $\mathscr{S}(\tau)$ to obtain an outer trapped surface outside $\partial\mathscr{T}(\tau)$. $\square$

The null geodesics orthogonal to the apparent horizon $\partial\mathscr{T}(\tau)$ on a surface $\mathscr{S}(\tau)$ will therefore start out with zero convergence. However if they encounter any matter or any Weyl tensor satisfying the generality condition (§ 4.4), they will start converging, and so their intersection with a later surface $\mathscr{S}(\tau')$ will lie inside the apparent horizon $\partial\mathscr{T}(\tau')$. In other words, the apparent horizon moves outwards at least as fast as light; and moves out faster than light if any matter or radiation falls through it. As the example above shows, the apparent horizon can also jump outwards discontinuously. This makes it harder to work with than the event horizon, which always moves in a continuous manner. We shall show in the next section that the event and apparent horizons coincide when the solution is stationary. One would therefore expect them to be very close together if the solution is nearly stationary for a long time. In particular, one would expect their areas to be almost the same under such circumstances. If one has a solution which passes from an initial nearly stationary state through some non-stationary period to a final nearly stationary state, one can employ proposition 9.2.7 to relate the areas of the initial and final horizons.

9.3   The final state of black holes

In the last section, we assumed that one could predict the future far away from a collapsing star. We showed that this implied that the star passed inside an event horizon which hid the singularities from an outside observer. Matter and energy which crossed the event horizon would be lost for ever from the outside world. One would therefore expect that there would be a limited amount of energy available to be radiated to infinity in the form of gravitational waves. Once most of this energy had been emitted, one would expect the solution outside the horizon to approach a stationary state. In this section we shall therefore study black hole solutions which are exactly stationary, in the expectation that the exterior regions will closely represent the final states of solutions outside collapsed objects.

More precisely, we shall consider spaces $(\mathscr{M}, \mathbf{g})$ which satisfy the following conditions:

(1) $(\mathscr{M}, \mathbf{g})$ is a regular predictable space developing from a partial Cauchy surface $\mathscr{S}$.

(2) There exists an isometry group $\theta_t\colon \mathscr{M} \to \mathscr{M}$ whose Killing vector $\mathbf{K}$ is timelike near $\mathscr{I}^+$ and $\mathscr{I}^-$.

(3) $(\mathscr{M}, \mathbf{g})$ is empty or contains fields like the electromagnetic field or scalar field which obey well-behaved hyperbolic equations, and satisfy the dominant energy condition: $T_{ab}N^aL^b \geqslant 0$ for future-directed timelike vectors $\mathbf{N}, \mathbf{L}$.

We shall call a space satisfying these conditions, a *stationary regular predictable space*. We expect that for large values of τ, the region $J^-(\mathscr{I}^+, \overline{\mathscr{M}}) \cap J^+(\mathscr{S}(\tau))$ of a regular predictable space containing collapsing stars will be almost isometric to a similar region of a stationary regular predictable space.

The justification for condition (3) is that one would expect any non-zero rest-mass matter eventually to fall through the horizon. Only long range fields like the electromagnetic field would be left. Conditions (2) and (3) imply that $(\mathscr{M}, \mathbf{g})$ is analytic in the region near infinity where the Killing vector field **K** is timelike (Müller zum Hagen (1970)). We shall take the solution elsewhere to be the analytic continuation of this outer region. The stationary solutions we are considering here will not have asymptotically simple pasts, as they represent only the final state of the system and not the earlier dynamical stage. However we shall be concerned only with the future properties of these solutions, and not their past properties. These might not be the same, as there is no *a priori* reason why they should be time reversible, though in fact it will be a consequence of the results we shall prove that they are time reversible.

In a stationary regular predictable space, the area of a two-section of the horizon will be time independent. This gives the following fundamental result:

Proposition 9.3.1

Let $(\mathscr{M}, \mathbf{g})$ be a stationary, regular predictable space–time. Then the generators of the future event horizon $\dot{J}^-(\mathscr{I}^+, \overline{\mathscr{M}})$ have no past endpoints in $J^+(\mathscr{I}^-, \overline{\mathscr{M}})$. Let $Y_1{}^a$ be the future-directed tangent vectors to these generators; then in $J^+(\mathscr{I}^-, \overline{\mathscr{M}})$, $Y_1{}^a$ has zero shear $\hat{\sigma}$ and expansion $\hat{\theta}$, and satisfies

$$R_{ab}Y_1{}^a Y_1{}^b = 0 = Y_{1[e}C_{a]bc[d}Y_{1f]}Y_1{}^b Y_1{}^c.$$

In order not to break up the discussion we shall defer the proof of this and other results to the end of this section. This proposition shows that in a stationary space–time, the apparent horizon coincides with the event horizon.

We shall now present some results which indicate that the Kerr family of solutions (§ 5.6) are probably the only empty stationary regular predictable space–times. We shall not give the proofs of the theorems of Israel and Carter here, but shall refer to the literature. The other results will be proved at the end of this section. Because of these results, we expect that the solution outside an uncharged collapsed object will settle down to a Kerr solution. If the collapsed body had a net electric charge, we would expect the solution to approach one of the charged Kerr solutions.

Proposition 9.3.2

Each connected component in $J^+(\mathscr{I}^-, \bar{\mathscr{M}})$ of the horizon $\partial\mathscr{B}(\tau)$ in a stationary regular predictable space is homeomorphic to a two-sphere.

It is possible that there could be several connected components of $\partial\mathscr{B}(\tau)$ representing several black holes at constant distances from each other. This situation can occur in the limiting case where the black holes have charge e equal to their mass m, and are non-rotating (Hartle and Hawking (1972 a)). It seems probable that this is the only case in which one can get a sufficiently strong repulsive force to balance the gravitational attraction between the black holes. We shall therefore consider solutions where $\partial\mathscr{B}(\tau)$ has only one connected component.

Proposition 9.3.3

Let $(\mathscr{M}, \mathbf{g})$ be a stationary regular predictable space. Then the Killing vector K^a is non-zero in $J^+(\mathscr{I}^-, \bar{\mathscr{M}}) \cap \overline{J^-}(\mathscr{I}^+, \bar{\mathscr{M}})$, which is simply connected. Let τ_0 be such that $\mathscr{S}(\tau_0) \cap \overline{J^-}(\mathscr{I}^+, \bar{\mathscr{M}})$ is contained in $J^+(\mathscr{I}^-, \bar{\mathscr{M}})$. If $\partial\mathscr{B}(\tau_0)$ has only one connected component, then $J^+(\mathscr{I}^-, \bar{\mathscr{M}}) \cap \overline{J^-}(\mathscr{I}^+, \bar{\mathscr{M}}) \cap \mathscr{M}$ is homeomorphic to $[0, 1) \times S^2 \times R^1$.

The discussion now takes one of two possible courses, depending on whether or not the Killing vector K^a has zero curl, $K_{a;b}K_c\eta^{abcd}$, everywhere. If the curl is zero, the solution is said to be a *static regular predictable space–time*. Roughly speaking, one would expect the solution to be static if the black hole is not rotating in some sense.

Proposition 9.3.4

In a static regular predictable space–time, the Killing vector $\mathbf{K}$ is timelike in the exterior region $J^+(\mathscr{I}^-, \bar{\mathscr{M}}) \cap J^-(\mathscr{I}^+, \bar{\mathscr{M}})$ and is non-zero and directed along the null generators of $\dot{J}^-(\mathscr{I}^+, \bar{\mathscr{M}})$ on

$$\dot{J}^-(\mathscr{I}^+, \bar{\mathscr{M}}) \cap J^+(\mathscr{I}^-, \bar{\mathscr{M}}).$$

Since the curl of $\mathbf{K}$ vanishes, it is hypersurface orthogonal, i.e. there is a function ξ such that K_a is proportional to $\xi_{;a}$. One can then decompose the metric in the exterior region in the form $g_{ab} = f^{-1}K_aK_b + h_{ab}$ where $f \equiv K^aK_a$ and h_{ab} is the induced metric in the surfaces

$\{\xi = \text{constant}\}$ and represents the separation of the integral curves of K^a. The exterior region therefore admits an isometry which sends a point on a surface ξ to the point on the surface $-\xi$ on the same integral curve of **K**. This isometry reverses the direction of time, and a space admitting such an isometry will be said to be *time symmetric*. Thus if the analytic extension of the exterior region contains a future event horizon $\dot{J}^-(\mathscr{I}^+, \overline{\mathscr{M}})$, it will also contain a past event horizon $\dot{J}^+(\mathscr{I}^-, \overline{\mathscr{M}})$. These event horizons may or may not intersect; the Schwarzschild solution and the Reissner–Nordström solution with $e^2 < m^2$ are examples where they do intersect, and the Reissner–Nordström solution with $e^2 = m^2$ is an example where they do not. The gradient of f is zero on the horizon in the latter case, but not in the former cases. The significance of this comes from the fact that on the future horizon $\dot{J}^-(\mathscr{I}^+, \overline{\mathscr{M}}) \cap J^+(\mathscr{I}^-, \overline{\mathscr{M}})$, $K_{a;b}K^b = \frac{1}{2}f_{;a} = \beta K_a$, where $\beta \geqslant 0$ is constant along the null geodesic generators of $\dot{J}^-(\mathscr{I}^+, \overline{\mathscr{M}})$. Let v be a future-directed affine parameter along such a generator. Then $\mathbf{K} = \alpha \, \partial/\partial v$ where α is a function along the generator which obeys $d\alpha/dv = \beta$. If $\beta \neq 0$ and the generator is geodesically complete in the past direction, α and the Killing vector **K** will be zero at some point. This point cannot lie in $J^+(\mathscr{I}^-, \overline{\mathscr{M}})$, and so will be a point of intersection of the future event horizon $\dot{J}^-(\mathscr{I}^+, \overline{\mathscr{M}})$ and the past event horizon $\dot{J}^+(\mathscr{I}^-, \overline{\mathscr{M}})$ (Boyer (1969)). If $\beta = 0$, **K** will always be non-zero and there will be no such point where the horizon bifurcates.

Israel (1967) has shown that a static regular predictable space–time must be a Schwarzschild solution if:

    (a) $T_{ab} = 0$;

    (b) the magnitude $f \equiv K^a K_a$ of the Killing vector has non-zero gradient everywhere in $J^+(\mathscr{I}^-, \overline{\mathscr{M}}) \cap J^-(\mathscr{I}^+, \overline{\mathscr{M}})$;

    (c) the past event horizon $\dot{J}^+(\mathscr{I}^-, \overline{\mathscr{M}})$ intersects the future event horizon $\dot{J}^-(\mathscr{I}^+, \overline{\mathscr{M}})$ in a compact two-surface $\mathscr{F}$.

(It follows from (c) and proposition 9.3.2 that $\mathscr{F}$ is connected and has the topology of a two-sphere. Israel did not give the conditions in this precise form, but these are equivalent.) Israel (1968) has further shown that the solution must be a Reissner–Nordström solution if the empty space condition (a) is replaced by the requirement that the energy–momentum tensor is that of an electromagnetic field. Müller zum Hagen, Robinson and Seifert (1973) have removed condition (b) in the vacuum case.

From these results we expect that if the final state of the solution outside the event horizon is static, then the metric in the exterior region will be that of a Schwarzschild solution.

We shall now consider the case where the final state of the exterior solution is stationary but not static. We would expect this to be the case when the object that collapsed was rotating initially.

Proposition 9.3.5

In an empty stationary regular predictable space which is not static, the Killing vector K^a is spacelike in part of the exterior region $J^+(\mathscr{I}^-, \overline{\mathscr{M}}) \cap J^-(\mathscr{I}^+, \overline{\mathscr{M}})$.

The region of $\overline{J}^+(\mathscr{I}^-, \overline{\mathscr{M}}) \cap \overline{J}^-(\mathscr{I}^+, \overline{\mathscr{M}})$ on which K^a is spacelike, is called the *ergosphere*. From proposition 9.3.4 it follows that there is no ergosphere if the solution is static. The significance of the ergosphere is that in it, it is impossible for a particle to move on an integral curve of the Killing vector K^a, i.e. to remain at rest as viewed from infinity. Since the ergosphere is outside the horizon it is still possible for such a particle to escape to infinity. An example of a stationary non-static regular predictable space with an ergosphere is the Kerr solution for $a^2 \leqslant m^2$ (§ 5.6).

Penrose (1969), Penrose and Floyd (1971) have pointed out that one can extract a certain amount of energy from a black hole with an ergosphere, by throwing a particle from infinity into the ergosphere. Since the particle moves on a geodesic, $E_0 \equiv -p_0{}^a K_a > 0$ is constant along its trajectory

$$((p_0{}^a K_a)_{;b} p_0{}^b = (p_0{}^a{}_{;b} p_0{}^b) K_a + p_0{}^a K_{a;b} p_0{}^b = 0,$$

as $p_0{}^a$ is a geodesic vector and K^a is a Killing vector), where $p_0{}^a = m v_0{}^a$ is the momentum vector of the particle, m is its rest-mass and $\mathbf{v}_0$ is the unit tangent to the particle world-line. The particle is then supposed to split into two particles with momentum vectors $p_1{}^a$ and $p_2{}^a$, where $p_0{}^a = p_1{}^a + p_2{}^a$. Since K^a is spacelike, it is possible to choose $p_1{}^a$ to be a future pointing timelike vector such that $E_1 \equiv -p_1{}^a K_a < 0$. Then $E_2 \equiv -p_2{}^a K_a$ will be greater than E_0. This means that the second particle can escape to infinity where it will have more energy than the original particle that was thrown in. One has thus extracted a certain amount of energy from the black hole.

The particle with negative energy cannot escape to infinity, but must remain in the region where K^a is spacelike. Suppose that the ergosphere did not intersect the event horizon $\dot{J}^-(\mathscr{I}^+, \overline{\mathscr{M}})$. Then the particle would have to remain in the exterior region. By repeating the process, one could continue to extract energy from the solution. As one did this, one would expect the solution to change gradually. However the ergosphere cannot shrink to zero, as there has to be somewhere for these negative energy particles to exist. It therefore appears that either one could extract an infinite amount of energy (which seems improbable), or that the ergosphere would eventually have to intersect the horizon. We shall show that in the latter case the solution would spontaneously become either axisymmetric or static without any further extraction of energy by the Penrose process. Either the possibility of the extraction of an infinite amount of energy or the occurrence of a spontaneous change would seem to indicate that the original state of the black hole was unstable. It therefore seems reasonable to assume that in any realistic black hole situation the ergosphere intersects the horizon.

Hajicek (1973) has shown that the stationary limit surface, which is the outer boundary of the ergosphere, will contain at least two integral null geodesic curves of K^a. If the gradient of f is non-zero on these curves, and if they are geodesically complete in the past, they will contain points where K^a is zero. However there can be no such points in the exterior region (see proposition 9.3.3), so the ergosphere must intersect the horizon in this case. However although it might be reasonable to assume that the integral curves of K^a were complete in the future, it does not seem reasonable to assume that they are complete in the past, since that would be to assume something about the past region of the solution which, as we said before, is not physically significant. In the static case one could show that the solution was time symmetric, but there is no *a priori* reason why a stationary non-static solution should be time symmetric. For this reason we shall rely on the energy extraction argument above rather than on Hajicek's results, to justify our assumption that the ergosphere intersects the horizon.

One can explain the significance of the ergosphere touching the horizon as follows. Let $\mathcal{Q}_1$ be one connected component of

$$J^-(\mathscr{I}^+, \bar{\mathscr{M}}) \cap J^+(\mathscr{I}^-, \bar{\mathscr{M}})$$

and let $\mathscr{G}_1$ be the quotient of $\mathcal{Q}_1$ by its generators. By propositions 9.3.1 and 9.3.2, this will be homeomorphic to a two-sphere. By proposition 9.3.1, the spatial separation of two neighbouring generators is constant along the generators, and so can be represented by an induced metric $\mathbf{h}$ on $\mathscr{G}_1$. The isometry θ_t moves generators into generators, and so acts as an isometry group of $(\mathscr{G}_1, \mathbf{h})$. If the ergosphere intersects the horizon, K^a will be spacelike somewhere on the horizon and the action of θ_t on $(\mathscr{G}_1, \mathbf{h})$ is non-trivial. Therefore it must correspond to a rotation of the sphere $\mathscr{G}_1$ around an axis, and the orbits of the group in $\mathscr{G}_1$ will be two points, corresponding to the poles, and a family of circles. A particle moving along one of the generators of the horizon would therefore appear to be moving relative to the frame defined by K^a which is stationary at infinity. One could therefore say that the horizon was *rotating* with respect to infinity.

The next result shows that a rotating black hole must be axisymmetric.

Proposition 9.3.6

Let $(\mathscr{M}, \mathbf{g})$ be a stationary non-static regular predictable space, in which the ergosphere intersects $\dot{J}^-(\mathscr{I}^+, \bar{\mathscr{M}}) \cap J^+(\mathscr{I}^-, \bar{\mathscr{M}})$. Then there is a one-parameter cyclic isometry group θ_ϕ $(0 \leqslant \phi \leqslant 2\pi)$ of $(\mathscr{M}, \mathbf{g})$ which commutes with θ_t, and whose orbits are spacelike near $\mathscr{I}^+$ and $\mathscr{I}^-$.

The method of proof of proposition 9.3.6 is to use the analyticity of the metric $\mathbf{g}$ to show that there is an isometry θ_ϕ in a neighbourhood of the horizon. One then extends the isometry by analytic continuation. The method would therefore work even if the metric were not analytic in isolated regions away from the horizon, for example if there were a ring of matter or a frame of rods around the black hole. This leads to an apparent paradox. Consider a rotating star surrounded by a stationary square frame of rods. Suppose that the star collapsed to form a rotating black hole. If the black hole approached a stationary state, it would follow from proposition 9.3.6 that the metric $\mathbf{g}$ was axisymmetric except where it was non-analytic at the rods. However the gravitational effect of the rods would prevent the metric being axisymmetric. The resolution of the paradox seems to be that the black hole would not be in a stationary state while it was rotating. What would happen is that the gravitational effect of the rods would distort the black hole slightly. The back reaction on the frame would cause it to start rotating and so to radiate angular momentum. Eventually the rotation of both the black hole and the frame would be damped out and the solution would approach a static state. A static black hole need not be axisymmetric if the space outside it is not empty, i.e. if condition (*a*) of Israel's theorem is not satisfied.

The above discussion indicates that a realistic black hole will never be exactly stationary while it is rotating, as the universe will not be exactly axisymmetric about it. However in most circumstances, the rate of slowing down of the rotation of the black hole is extremely slow (Press (1972), Hartle and Hawking (1972b)). Thus it is a good approximation to neglect the small asymmetries produced by matter at a distance from the black hole, and to regard the rotating black hole as being in a stationary state. We shall therefore now consider the properties of a rotating axisymmetric black hole.

The following result of Papapetrou (1966), generalized by Carter (1969), shows that the Killing vectors K^a corresponding to the time translation θ_t and $\tilde{K}^a$ corresponding to the angular rotation θ_ϕ are both orthogonal to families of two-surfaces.

Proposition 9.3.7

Let $(\mathcal{M}, \mathbf{g})$ be a space–time which admits a two-parameter abelian isometry group with Killing vectors ξ_1 and ξ_2. Let $\mathscr{V}$ be a connected open set of $\mathcal{M}$, and let $w_{ab} \equiv \xi_{1[a}\xi_{2b]}$. If

(a) $w_{ab}R^b{}_c\eta^{cdef}w_{ef} = 0$ on $\mathscr{V}$,

(b) $w_{ab} = 0$ at some point of $\mathscr{V}$,

then $w_{[ab;c}w_{d]e} = 0$ on $\mathscr{V}$.

Condition (b) is satisfied in a stationary axisymmetric space–time on the axis of axisymmetry, i.e. the set of points where $\tilde{K}^a = 0$. Condition (a) is satisfied in empty space, and when the energy–momentum tensor is that of a source-free electromagnetic field (Carter (1969)). By Frobenius' theorem (Schouten (1954)), the vanishing of $w_{[ab;c}w_{d]e}$ is, when $w_{ab} \neq 0$, the condition that there should exist locally a family of two-surfaces which are orthogonal to w_{ab}, i.e. to any linear combination of ξ_1 and ξ_2. In the case of a stationary axisymmetric space–time, this means that one can locally introduce coordinates (t, ϕ, x^1, x^2) such that $\mathbf{K} = \partial/\partial t$, $\tilde{\mathbf{K}} = \partial/\partial\phi$, and $K^a x^m{}_{;a} = 0 = \tilde{K}^a x^m{}_{;a}$ for $m = 1, 2$. The metric then locally admits the isometry $(t, \phi, x^1, x^2) \to (-t, -\phi, x^1, x^2)$, which reverses the direction of time, i.e. it is time-symmetric. Thus if the analytic extension of metric near infinity of an empty stationary regular predictable space–time contains a future event horizon, it will also contain a past event horizon.

In analogy with proposition 9.3.4, one has

Proposition 9.3.8 (*cf. Carter* (1971*b*))

Let $(\mathscr{M}, \mathbf{g})$ be a stationary axisymmetric regular predictable space-time in which $w_{[ab;c}w_{d]e} = 0$, where $w_{ab} \equiv K_{[a}\tilde{K}_{b]}$. Then at any point in the exterior region $J^+(\mathscr{I}^-, \bar{\mathscr{M}}) \cap J^-(\mathscr{I}^+, \bar{\mathscr{M}})$ off the axis $\tilde{\mathbf{K}} = 0$, $h \equiv w_{ab}w^{ab}$ is negative. On the horizons $\dot{J}^-(\mathscr{I}^+, \bar{\mathscr{M}}) \cap J^+(\mathscr{I}^-, \bar{\mathscr{M}})$ and $\dot{J}^+(\mathscr{I}^-, \bar{\mathscr{M}}) \cap J^-(\mathscr{I}^+, \bar{\mathscr{M}})$, h is zero but $w_{ab} \neq 0$ except on the axis.

This shows that at each point off the axis in the exterior region, there is some linear combination of the Killing vectors K^a and $\tilde{K}^a$ which is timelike. Outside the ergosphere, K^a itself is timelike, but between the stationary limit surface and the horizon one has to add a multiple of $\tilde{K}^a$ to obtain a timelike Killing vector. On the horizon there is no linear combination which is timelike, but there is a linear combination which is null, and is directed along the null generators of the horizon. Off the axis $\tilde{\mathbf{K}} = 0$, one can locally characterize the horizon as the set of points on which $h \equiv w_{ab}w^{ab} = 0$.

We now come to the theorem of Carter (1971*b*) which indicates that the Kerr solutions are probably the only empty stationary black holes. He considered stationary regular predictable spaces which satisfy:

(*a*) $T_{ab} = 0$,

(*b*) they are axisymmetric,

(*c*) the past event horizon $\dot{J}^+(\mathscr{I}^-, \bar{\mathscr{M}})$ intersects the future event horizon $\dot{J}^-(\mathscr{I}^+, \bar{\mathscr{M}})$ in a compact connected two-surface $\mathscr{F}_1$.

(By proposition 9.3.2, this will be a two-sphere.) He showed that such solutions fall into disjoint families, each of which depends only on two parameters. The two parameters can be taken to be the mass m and angular momentum L as measured from infinity. One such family is known, namely the Kerr solutions for $m \geqslant 0$, $a^2 \leqslant m^2$, where $a = L/m$. (The Kerr solutions with $a^2 > m^2$ contain naked singularities and so are not regular predictable spaces.) It seems unlikely that there are any other disjoint families. It has been conjectured, therefore, that the solution outside an uncharged collapsed object will settle down to a Kerr solution with $a^2 \leqslant m^2$. This conjecture is supported by analyses of linear perturbations from a spherical collapse by Regge and Wheeler (1957), Doroshkevich, Zel'dovich and Novikov (1966), Vishveshwara (1970), and Price (1972).

Assuming the validity of this Carter–Israel conjecture, one would expect the area of the two-surface $\partial \mathscr{B}(\tau)$ in the event horizon to approach the area of a two-surface in the event horizon $r = r_+$ of a

Kerr solution with the same mass and angular momentum, as measured at $\mathscr{Q}(\tau)$ on $\mathscr{I}^+$. This area is $8\pi m(m + (m^2 - a^2)^{\frac{1}{2}})$, where m is the mass of the Kerr solution and ma is the angular momentum. (If the collapsing body has a net electrical charge e one would expect the solution to settle down to a charged Kerr solution. The area of a two-surface in the event horizon of such a solution is

$$4\pi(2m^2 - e^2 + 2m(m^2 - a^2 - e^2)^{\frac{1}{2}}).$$

Using this expression one can generalize our results to charged black holes.) Consider a collapse situation which by a surface $\mathscr{S}(\tau_1)$ has settled down to a Kerr solution with mass m_1 and angular momentum $m_1 a_1$. Suppose one now lets the black hole interact with particles or radiation for a finite time. The solution will eventually settle down, by a surface $\mathscr{S}(\tau_2)$, to a different Kerr solution with parameters m_2, a_2. From the discussion of § 9.2, the area of $\partial \mathscr{B}(\tau_2)$ must be greater than or equal to the area of $\partial \mathscr{B}(\tau_1)$. In fact it must be strictly greater than, since θ can be zero only if no matter or radiation crosses the horizon. This then implies that

$$m_2(m_2 + (m_2{}^2 - a_2{}^2)^{\frac{1}{2}}) > m_1(m_1 + (m_1{}^2 - a_1{}^2)^{\frac{1}{2}}). \tag{9.4}$$

If $a_1 \neq 0$, then the inequality (9.4) allows m_2 to be less than m_1. Since there is a conservation law for total energy and momentum in an asymptotically flat space–time (Penrose (1963)), this would mean that one had extracted a certain amount of energy from the black hole. One way of doing this would be to construct a square frame of rods about the black hole and employ the torque exerted by the rotating black hole on the frame to do work. Alternatively, one could use Penrose's process of throwing a particle into the ergosphere, where it divides into two particles, one of which escapes to infinity with greater energy than the original particle. The other particle will fall through the event horizon and reduce the angular momentum of the solution. One can thus regard the process as extracting rotational energy from the black hole. Christodoulou (1970) has shown that one can achieve a result arbitrarily near the limit set by the inequality (9.4). In fact the maximum energy extraction occurs when $a_2 = 0$; then the available energy $(m_1 - m_2)$ is less than

$$m_1\left\{1 - \frac{1}{\sqrt{2}}\left(1 + \left(1 - \frac{a_1{}^2}{m_1{}^2}\right)^{\frac{1}{2}}\right)^{\frac{1}{2}}\right\}.$$

Consider now a situation in which two stars a long way apart collapse to produce black holes. There is thus some τ' such that $\partial \mathscr{B}(\tau')$ consists of two separate two-spheres $\partial \mathscr{B}_1(\tau')$ and $\partial \mathscr{B}_2(\tau')$. Since these are a long way apart, one can neglect their interaction and assume that the solutions near each are close to Kerr solutions with parameters m_1, a_1 and m_2, a_2 respectively. Thus the areas of $\partial \mathscr{B}_1(\tau')$ and $\partial \mathscr{B}_2(\tau')$ will be approximately $8\pi m_1(m_1 + (m_1{}^2 - a_1{}^2)^{\frac{1}{2}})$ and $8\pi m_2(m_2 + (m_2{}^2 - a_2{}^2)^{\frac{1}{2}})$ respectively. Now suppose that these black holes fall towards each other, collide and coalesce. In such a collision a certain amount of gravitational radiation will be emitted. The system will eventually settle down by a surface $\mathscr{S}(\tau'')$ to resemble a single Kerr solution with parameters m_3, a_3. By the same argument as previously, the area of $\partial \mathscr{B}(\tau'')$ must be greater than the total area of $\partial \mathscr{B}(\tau')$, which is the sum of the areas $\partial \mathscr{B}_1(\tau')$ and $\partial \mathscr{B}_2(\tau')$. Thus

$$m_3(m_3 + (m_3{}^2 - a_3{}^2)^{\frac{1}{2}}) > m_1(m_1 + (m_1{}^2 - a_1{}^2)^{\frac{1}{2}}) + m_2(m_2 + (m_2{}^2 - a_2{}^2)^{\frac{1}{2}}).$$

By the conservation law for asymptotically flat spaces, the amount of energy carried away to infinity by gravitational radiation is

$$m_1 + m_2 - m_3.$$

This is limited by the above inequality. The efficiency

$$\epsilon \equiv (m_1 + m_2 - m_3)(m_1 + m_2)^{-1}$$

of conversion of mass to gravitational radiation is always less than $\frac{1}{2}$. If $a_1 = a_2 = 0$, then $\epsilon < 1 - 1/\sqrt{2}$. It should be stressed that these are upper limits; the actual efficiency might be much less, although the mere existence of a limit might suggest that one could attain an appreciable fraction of it.

We have shown that the fraction of mass which can be converted to gravitational radiation in the coalescence of one pair of black holes is limited. However if there were initially a large number of black holes, these could combine in pairs and then the resulting holes could combine, and so on. On dimensional grounds one would expect the efficiency to be the same at each stage. Thus one would eventually convert a very large fraction of the original mass to gravitational radiation. (This argument was suggested by C. W. Misner and M. J. Rees.) At each stage, the energy emitted in gravitational radiation would be larger. This might be able to explain Weber's recent observations of short bursts of gravitational radiation.

We now give the proofs of the propositions we have stated in this section. For convenience, we repeat the statements of the propositions.

Proposition 9.3.1

Let $(\mathcal{M}, \mathbf{g})$ be a stationary, regular predictable space–time. Then the generators of the future event horizon $\dot{J}^-(\mathcal{I}^+, \bar{\mathcal{M}})$ have no past endpoints in $J^+(\mathcal{I}^-, \bar{\mathcal{M}})$. Let $Y_1{}^a$ be the future-directed tangent vector to these generators; then in $J^+(\mathcal{I}^-, \bar{\mathcal{M}})$, $Y_1{}^a$ has zero shear $\hat{\sigma}$ and expansion $\hat{\theta}$, and satisfies

$$R_{ab}Y_1{}^aY_1{}^b = 0 = Y_{1[e}C_{a]bc[d}Y_{1f]}Y_1{}^bY_1{}^c.$$

Let $\mathscr{C}$ be a spacelike two-sphere on $\mathcal{I}^-$. Then one can cover $\mathcal{I}^-$ by a family of two-spheres $\mathscr{C}(t)$ obtained by moving $\mathscr{C}$ up and down the generators of $\mathcal{I}^-$ under the action of θ_t, i.e. $\mathscr{C}(t) = \theta_t(\mathscr{C})$. We now define the function x at the point $p \in J^+(\mathcal{I}^-, \bar{\mathcal{M}})$ to be the greatest value of t such that $p \in J^+(\mathscr{C}(t), \bar{\mathcal{M}})$. Let $\mathcal{U}$ be a neighbourhood of $\mathcal{I}^+$ and $\mathcal{I}^-$ which is isometric to a corresponding neighbourhood of an asymptotically simple space–time. Then x will be continuous and have some lower bound x' on $\mathscr{S} \cap \mathcal{U}$. From this it follows that x will be continuous in the region of $\overline{J^-}(\mathcal{I}^+, \bar{\mathcal{M}})$ where it is greater than x'. Let $p \in J^+(\mathcal{I}^-, \bar{\mathcal{M}}) \cap J^-(\mathcal{I}^+, \bar{\mathcal{M}})$. Then under the isometry θ_t, p will be moved into the region of $\overline{J^-}(\mathcal{I}^+, \bar{\mathcal{M}})$, where $x > x'$. However

$$x|_{\theta_t(p)} = x|_p + t.$$

Therefore x will be continuous at p.

Let $\tau_0 > 0$ be such that $\mathscr{S}(\tau_0) \cap \overline{J^-}(\mathcal{I}^+, \bar{\mathcal{M}})$ is contained in $J^+(\mathcal{I}^-, \bar{\mathcal{M}})$. Let λ be a generator of $\dot{J}^-(\mathcal{I}^+, \bar{\mathcal{M}})$ which intersects $\mathscr{S}(\tau_0)$. Suppose there were some finite upper bound x_0 to x on λ. Since the space is weakly asymptotically simple, $x \to \infty$ as one approaches $\mathcal{Q}(\tau_0)$ on $\mathscr{S}(\tau_0)$. Thus there will be some lower bound x_1 of x on

$$\mathscr{S}(\tau_0) \cap \overline{J^-}(\mathcal{I}^+, \bar{\mathcal{M}}).$$

Under the action of the group θ_t, λ is moved into another generator $\theta_t(\lambda)$. As the generators of $\dot{J}^-(\mathcal{I}^+, \bar{\mathcal{M}})$ have no future endpoints, the past extension of $\theta_t(\lambda)$ will still intersect $\mathscr{S}(\tau_0) \cap \overline{J^-}(\mathcal{I}^+, \bar{\mathcal{M}})$. This leads to a contradiction, since the upper bound of x on $\theta_t(\lambda)$ would be less than x_1 if $t < x_1 - x_0$.

Let x_2 be the upper bound of x on $\mathscr{S}(\tau_0) \cap \dot{J}^-(\mathcal{I}^+, \bar{\mathcal{M}})$. Then every generator λ of $\dot{J}^-(\mathcal{I}^+, \bar{\mathcal{M}})$ which intersects $\mathscr{S}(\tau_0)$ will intersect $\mathscr{F}(t) \equiv \dot{J}^+(\mathscr{C}(t), \bar{\mathcal{M}}) \cap \dot{J}^-(\mathcal{I}^+, \bar{\mathcal{M}})$ for $t \geq x_2$. Every generator of $\dot{J}^-(\mathcal{I}^+, \bar{\mathcal{M}})$ which intersects $\mathscr{F}(t')$ will intersect $\theta_t(\mathscr{S}(\tau_0))$ for $t \geq t' - x_1$. But $\theta_t(\mathscr{S}(\tau_0)) \cap \dot{J}^-(\mathcal{I}^+, \bar{\mathcal{M}}) = \theta_t(\mathscr{S}(\tau_0) \cap \dot{J}^-(\mathcal{I}^+, \bar{\mathcal{M}}))$ is compact. Thus $\mathscr{F}(t)$ is compact.

Now consider how the area of $\mathscr{F}(t)$ varies as t increases. Since $\theta \geqslant 0$ the area cannot decrease. If θ were > 0 on an open set, the area would increase. Also if the generators of the horizon had past endpoints on $\mathscr{F}(t)$ the area would increase. However as $\mathscr{F}(t)$ is moving under the isometry θ_t, the area must remain the same. Therefore $\theta = 0$, and there are no past endpoints on the region of $\dot{J}^-(\mathscr{I}^+, \mathscr{M})$ for which $x \geqslant x_2$. However since each point of $\dot{J}^-(\mathscr{I}^+, \mathscr{M}) \cap J^+(\mathscr{I}^-, \mathscr{M})$ can be moved by the isometry θ_t to where $x > x_2$, this result applies to the whole of $\dot{J}^-(\mathscr{I}^+, \mathscr{M}) \cap J^+(\mathscr{I}^-, \mathscr{M})$. From the propagation equations (4.35) and (4.36) one then finds $\hat{\sigma}_{mn} = 0$, $R_{ab}Y_1^aY_1^b = 0$ and $Y_{1[e}C_{a]bc[d}Y_{1f]}Y_1^bY_1^c = 0$, where Y_1^a is the future-directed tangent vector to the null geodesic generators of the horizon. $\square$

Proposition 9.3.2

Each connected component in $J^+(\mathscr{I}^-, \mathscr{M})$ of the horizon $\partial \mathscr{B}(\tau)$ in a stationary, regular predictable space is homeomorphic to a two-sphere.

Consider how the expansion of the outgoing null geodesics orthogonal to $\partial \mathscr{B}(\tau)$ behaves if one deforms $\partial \mathscr{B}(\tau)$ slightly outwards into $J^-(\mathscr{I}^+, \mathscr{M})$. Let Y_2^a be the other future-directed null vector orthogonal to $\partial \mathscr{B}(\tau)$, normalized so that $Y_1^aY_{2a} = -1$. This leaves the freedom $\mathbf{Y}_1 \to \mathbf{Y}_1' = e^\nu \mathbf{Y}_1$, $\mathbf{Y}_2 \to \mathbf{Y}_2' = e^{-\nu}\mathbf{Y}_2$. The induced metric on the space-like two-surface $\partial \mathscr{B}(\tau)$ is $\hat{h}_{ab} = g_{ab} + Y_{1a}Y_{2b} + Y_{2a}Y_{1b}$. Define a family of surfaces $\mathscr{F}(\tau, w)$ by moving each point of $\partial \mathscr{B}(\tau)$ a parameter distance w along the null geodesic curve with tangent vector Y_2^a. The vectors Y_1^a will be orthogonal to $\mathscr{F}(\tau, w)$ if they propagate according to

$$\hat{h}_{ab}Y_1^b{}_{;c}Y_2^c = -\hat{h}_a{}^b Y_{2c;b}Y_1^c \quad \text{and} \quad Y_1^aY_{2a} = -1.$$

Then

$$(Y_1^a{}_{;b}\hat{h}_a{}^c\hat{h}^b{}_d)_{;g}Y_2^g\hat{h}_c{}^s\hat{h}^d{}_t = \hat{h}^{sa}p_{a;b}\hat{h}^b{}_t + p^sp_t$$

$$- \hat{h}^s{}_aY_1^a{}_{;g}\hat{h}^{ge}Y_{2e;b}\hat{h}^b{}_t + R^a{}_{ceb}Y_2^eY_1^ch_a{}^sh^b{}_t, \quad (9.5)$$

where $p^a \equiv -\hat{h}^{ba}Y_{2c;b}Y_1^c$. Contracting with $\hat{h}_s{}^t$, one obtains

$$\frac{\mathrm{d}\theta}{\mathrm{d}w} = (Y_1^a{}_{;b}\hat{h}^b{}_a)_{;c}Y_2^c$$

$$= p_{b;d}\hat{h}^{bd} - R_{ac}Y_1^aY_2^c + R_{adcb}Y_1^dY_2^cY_2^aY_1^b + p_ap^a$$

$$- Y_1^a{}_{;c}\hat{h}^c{}_dY_2^d{}_{;b}\hat{h}^b{}_a.$$

On the horizon, $Y_1^a{}_{;c}\hat{h}^{cd}\hat{h}^b{}_a$ is zero, as the shear and divergence of the horizon are zero. Under a rescaling transformation $\mathbf{Y}_1' = e^\nu \mathbf{Y}_1$,

$Y_2' = e^{-y} Y_2$, the vector p^a changes to $p'^a = p^a + \hat{h}^{ab} y_{;b}$, and so $d\theta/dw|_{w=0}$ changes to

$$\left.\frac{d\theta'}{dw'}\right|_{w=0} = p_{b;a}\hat{h}^{bd} + y_{;bd}\hat{h}^{bd} - R_{ac}Y_1{}^a Y_2{}^c$$
$$+ R_{adcb}Y_1{}^d Y_2{}^c Y_2{}^a Y_1{}^b + p'^a p'_a. \quad (9.6)$$

The term $y_{;bd}\hat{h}^{bd}$ is the Laplacian of y in the two-surface $\partial\mathcal{B}(\tau)$. By a theorem of Hodge (1952), one can choose y so that the sum of the first four terms on the right of (9.6) is a constant on $\partial\mathcal{B}(\tau)$. The sign of this constant will be determined by that of the integral of

$$(-R_{ac}Y_1{}^a Y_2{}^c + R_{adcb}Y_1{}^d Y_2{}^c Y_2{}^a Y_1{}^b)$$

over $\partial\mathcal{B}(\tau)$ ($p_{b;a}\hat{h}^{bd}$, being a divergence, has zero integral). This integral can be evaluated using the Gauss–Codacci equations for the scalar curvature $\hat{R}$ of the two-surface with metric $\hat{h}$:

$$\hat{R} = R_{ijkl}\hat{h}^{ik}\hat{h}^{jl} = R - 2R_{ijkl}Y_1{}^i Y_2{}^j Y_1{}^k Y_2{}^l + 4R_{ij}Y_1{}^i Y_2{}^j,$$

since $\theta = \hat{\sigma} = 0$ on $\partial\mathcal{B}(\tau)$. By the Gauss–Bonnet theorem (Kobayashi and Nomizu (1969))

$$\int_{\partial\mathcal{B}(\tau)} \hat{R}\, d\hat{S} = 2\pi\chi,$$

where $d\hat{S}$ is the surface area element of $\partial\mathcal{B}(\tau)$ and χ is the Euler number of $\partial\mathcal{B}(\tau)$. Thus

$$\int_{\partial\mathcal{B}(\tau)} (-R_{ab}Y_1{}^a Y_2{}^b + R_{adcb}Y_1{}^d Y_2{}^c Y_2{}^a Y_1{}^b)\, d\hat{S}$$
$$= -\pi\chi + \int_{\partial\mathcal{B}(\tau)} (\tfrac{1}{2}R + R_{ab}Y_1{}^a Y_2{}^b)\, d\hat{S}. \quad (9.7)$$

By the Einstein equations,

$$\tfrac{1}{2}R + R_{ab}Y_1{}^a Y_2{}^b = 8\pi T_{ab}Y_1{}^a Y_2{}^b,$$

which is $\geqslant 0$ by the dominant energy condition. The Euler number χ is $+2$ for the sphere, zero for the torus, and negative for any other compact orientable two-surface ($\partial\mathcal{B}(\tau)$ has to be orientable as it is a boundary). Hence the right-hand side of (9.7) can be negative only if $\partial\mathcal{B}(\tau)$ is a sphere.

Suppose that the right-hand side of (9.7) was positive. Then one could choose y so that $d\theta'/dw'|_{w=0}$ was positive everywhere on $\partial\mathcal{B}(\tau)$. For small negative values of w' one would obtain a two-surface in $J^-(\mathcal{I}^+, \bar{\mathcal{M}})$ such that the outgoing null geodesics orthogonal to the surface were converging. This would contradict proposition 9.2.8.

Suppose now that χ was zero and that $T_{ab}Y_1^aY_2^b$ was zero on $\partial\mathscr{B}(\tau)$. Then one could choose y so that the sum of the first four terms on the right of (9.6) was zero on $\partial\mathscr{B}(\tau)$. Then

$$p'^a{}_{;b}\hat{h}^b{}_a + R_{abcd}Y_1^aY_2^bY_1^cY_2^d = 0$$

on $\partial\mathscr{B}(\tau)$. If $R_{abcd}Y_1^aY_2^bY_1^cY_2^d$ was non-zero somewhere on $\partial\mathscr{B}(\tau)$, then the term $p'^ap'_a$ in (9.6) would be non-zero somewhere and one could change y slightly so as to make $\mathrm{d}\hat{\theta}'|\mathrm{d}w'|_{w=0}$ positive everywhere. This would again lead to a contradiction.

Now suppose that $R_{abcd}Y_1^aY_2^bY_1^cY_2^d$ and p'^a were zero everywhere on $\partial\mathscr{B}(\tau)$. One could move the two-surface $\partial\mathscr{B}(\tau)$ back along Y_2^a, choosing the rescaling parameter y at each stage so that

$$p'^a{}_{;b}\hat{h}^b{}_a + R_{abcd}Y_1^aY_2^bY_1^cY_2^d$$
$$-\tfrac{1}{2}R - 2R_{ab}Y_1^aY_2^b = p'^a{}_{;b}\hat{h}^b{}_a - \tfrac{1}{2}\hat{R} = 0.$$

If $T_{ab}Y_1^aY_2^b$ or p'^a were non-zero for $w' < 0$ then one could adjust y to obtain a two-surface in $J^-(\mathscr{I}^+, \bar{\mathscr{M}})$ with $\theta < 0$. This would contradict proposition 9.2.8. On the other hand if $T_{ab}Y_1^aY_2^b$ and p'^a were zero everywhere for $w' < 0$, one would obtain a two-surface in $J^-(\mathscr{I}^+, \bar{\mathscr{M}})$ with $\theta = 0$ which again contradicts proposition 9.2.8.

One avoids a contradiction only if $\chi = 2$, i.e. if $\partial\mathscr{B}(\tau)$ is a two-sphere. □

Proposition 9.3.3

Let $(\mathscr{M}, \mathfrak{g})$ be a stationary regular predictable space–time. Then the Killing vector K^a is non-zero in $J^+(\mathscr{I}^-, \bar{\mathscr{M}}) \cap \overline{J^-}(\mathscr{I}^+, \bar{\mathscr{M}})$, which is simply connected. Let τ_0 be such that $\mathscr{S}(\tau_0) \cap \overline{J^-}(\mathscr{I}^+, \bar{\mathscr{M}})$ is contained in $J^+(\mathscr{I}^-, \bar{\mathscr{M}})$. If $\partial\mathscr{B}(\tau_0)$ has only one connected component, then $J^+(\mathscr{I}^-, \bar{\mathscr{M}}) \cap \overline{J^-}(\mathscr{I}^+, \bar{\mathscr{M}}) \cap \mathscr{M}$ is homeomorphic to $[0, 1) \times S^2 \times R^1$.

The function x defined in proposition 9.3.1 is continuous on $J^+(\mathscr{I}^-, \bar{\mathscr{M}}) \cap \overline{J^-}(\mathscr{I}^+, \bar{\mathscr{M}})$, and has the property that $x|_{\theta_t(p)} = x|_p + t$. This shows that **K** cannot be zero in $J^+(\mathscr{I}^-, \bar{\mathscr{M}}) \cap \overline{J^-}(\mathscr{I}^+, \bar{\mathscr{M}})$. The integral curves of **K** establish a homeomorphism between two of the surfaces

$$\dot{J}^+(\mathscr{C}(t), \bar{\mathscr{M}}) \cap \bar{J}^-(\mathscr{I}^+, \bar{\mathscr{M}}) \cap \mathscr{M} \quad (-\infty < t < \infty).$$

The region $\dot{J}^+(\mathscr{I}^-, \bar{\mathscr{M}}) \cap \overline{J^-}(\mathscr{I}^+, \bar{\mathscr{M}}) \cap \mathscr{M}$ is covered by these surfaces, and so is homeomorphic to $R^1 \times \dot{J}^+(\mathscr{C}(t'), \bar{\mathscr{M}}) \cap \overline{J^-}(\mathscr{I}^+, \bar{\mathscr{M}}) \cap \mathscr{M}$ for any t'. Choose t to be large enough that $\dot{J}^+(\mathscr{C}(t), \bar{\mathscr{M}})$ intersects

$\mathcal{S}(\tau_0)$ in the neighbourhood $\mathcal{U}$ of $\mathcal{I}^+$ which is isometric to a similar neighbourhood in an asymptotically simple space. The integral curves of $\mathbf{K}$ establish a homeomorphism between

$$\dot{J}^+(\mathscr{C}(t),\overline{\mathscr{M}})\cap \overline{J^-}(\mathscr{I}^+,\overline{\mathscr{M}})\cap \mathscr{M} \quad \text{and} \quad \mathscr{S}(\tau_0)\cap \overline{J^-}(\mathscr{I}^+,\overline{\mathscr{M}}).$$

By property (α) and proposition 9.3.2, this is simply connected. If further $\partial\mathscr{B}(\tau)$ has only one connected component, then

$$\mathscr{S}(\tau_0)\cap \overline{J^-}(\mathscr{I}^+,\overline{\mathscr{M}})$$

has the topology $[0, 1) \times S^2$. Thus $J^+(\mathscr{I}^-,\overline{\mathscr{M}})\cap \overline{J^-}(\mathscr{I}^+,\overline{\mathscr{M}})\cap \mathscr{M}$ has the topology $[0, 1) \times S^2 \times R^1$.   □

Proposition 9.3.4

In a static regular predictable space–time, the Killing vector $\mathbf{K}$ is timelike in the exterior region $J^+(\mathscr{I}^-,\overline{\mathscr{M}})\cap J^-(\mathscr{I}^+,\overline{\mathscr{M}})$ and is non-zero and directed along the null generators of $\dot{J}^-(\mathscr{I}^+,\overline{\mathscr{M}})$ on

$$\dot{J}^-(\mathscr{I}^+,\overline{\mathscr{M}})\cap J^+(\mathscr{I}^-,\overline{\mathscr{M}}).$$

The event horizon $\dot{J}^-(\mathscr{I}^+,\overline{\mathscr{M}})$ is mapped into itself by the isometry θ_t. Thus on $\dot{J}^-(\mathscr{I}^+,\overline{\mathscr{M}})\cap J^+(\mathscr{I}^-,\overline{\mathscr{M}})$, $\mathbf{K}$ must be null or spacelike. Let τ_0 be such that $\mathscr{S}(\tau_0)\cap \overline{J^-}(\mathscr{I}^+,\overline{\mathscr{M}})$ is contained in $J^+(\mathscr{I}^-,\overline{\mathscr{M}})$. Then $f \equiv K^a K_a$ must be zero on some closed set $\mathscr{N}$ in

$$J^+(\mathscr{S}(\tau_0))\cap \overline{J^-}(\mathscr{I}^+,\overline{\mathscr{M}}).$$

From the fact that K^a is a Killing vector and curl $\mathbf{K} = 0$, it follows that

$$fK_{a;b} = K_{[a}f_{;b]}. \tag{9.8}$$

By proposition 9.3.3, K^a is non-zero on the simply connected set $J^+(\mathscr{I}^-,\overline{\mathscr{M}})\cap \overline{J^-}(\mathscr{I}^+,\overline{\mathscr{M}})$. By Frobenius' theorem, it follows from the condition curl $\mathbf{K} = 0$, that there is a function ξ on this region such that $K_a = -\alpha\xi_{;a}$, where α is some positive function.

Let p be a point of $\mathscr{N}$ and let $\lambda(v)$ be a curve through p lying in the surface of constant ξ through p. Then by (9.8),

$$\tfrac{1}{2}K^a\frac{\mathrm{d}}{\mathrm{d}v}\log f = \frac{\mathrm{D}}{\partial v}K^a.$$

If $\lambda(v)$ left $\mathscr{N}$, the left-hand side of this equation would be unbounded. However the right-hand side is continuous; therefore $\lambda(v)$ must lie in $\mathscr{N}$, so $\mathscr{N}$ must contain the surface $\xi = \xi|_p$. However f cannot be zero on an open neighbourhood of p, since it would then be zero everywhere. Thus the connected component of $\mathcal{N}$ through p is the three-surface $\xi = \xi|_p$. Suppose $p \in J^+(\mathcal{I}^-, \bar{\mathcal{M}}) \cap J^-(\mathcal{I}^+, \bar{\mathcal{M}})$. Then there would be a future-directed timelike curve $\gamma(u)$ from $\mathcal{I}^-$ through p to $\mathcal{I}^+$. On $\xi = \xi|_p$, K^a would be future-directed. Thus $(\partial/\partial u)_\gamma \xi > 0$ when $\xi = \xi|_p$. This leads to a contradiction as $\xi = \xi|_p$ cannot intersect $\mathcal{I}^+$ or $\mathcal{I}^-$ since K^a is timelike near infinity. Thus near $\mathcal{I}^+$ and $\mathcal{I}^-$, either ξ is greater than $\xi|_p$ or less than $\xi|_p$.   □

Proposition 9.3.5

In an empty regular predictable space–time which is not static, the Killing vector K^a is spacelike in part of the exterior region

$$J^+(\mathcal{I}^-, \bar{\mathcal{M}}) \cap J^-(\mathcal{I}^+, \bar{\mathcal{M}}).$$

The function x introduced in proposition 9.3.1 is continuous on $J^+(\mathcal{I}^-, \bar{\mathcal{M}}) \cap \overline{J^-}(\mathcal{I}^+, \bar{\mathcal{M}})$, and is such that along each integral curve of K^a, $\partial x/\partial t = 1$. One can approximate the surface $x = 0$ in $J^+(\mathcal{I}^-, \bar{\mathcal{M}}) \cap \overline{J^-}(\mathcal{I}^+, \bar{\mathcal{M}})$ by a smooth surface $\mathcal{K}$ which is nowhere tangent to K^a. One can then define a smooth function $\bar{x}$ on $J^+(\mathcal{I}^-, \bar{\mathcal{M}}) \cap \overline{J^-}(\mathcal{I}^+, \bar{\mathcal{M}})$ by specifying that $\bar{x} = 0$ on $\mathcal{K}$ and $\bar{x}_{;a} K^a = 1$. One can express the gradient of the Killing vector as

$$f K_{a;b} = \eta_{abcd} K^c \omega^d + K_{[a} f_{;b]},$$

where $f \equiv K^a K_a$ is the magnitude of the Killing vector, and

$$\omega^a \equiv \tfrac{1}{2} \eta^{abcd} K_b K_{c,d}.$$

The second derivatives of $\mathbf{K}$ satisfy

$$2 K_{a;[bc]} = R_{dabc} K^d.$$

However $K_{a;bc} = K_{[a;b]c}$. Therefore

$$K_{a;bc} = R_{dcba} K^d$$

which implies $$K^{a;b}{}_b = -R^a{}_d K^d. \tag{9.9}$$

The vector $q_a \equiv f^{-1} K_a - \bar{x}_{;a}$ is orthogonal to K^a. Multiplying (9.9) by q_a and integrating over the region $\mathcal{L}$ of $J^-(\mathcal{I}^+, \bar{\mathcal{M}})$ bounded by the surfaces $\mathcal{N}_1$ and $\mathcal{N}_2$ defined by $\bar{x} = x_2 + 1$ and $x = x_2 + 2$, where x_2 is as in proposition 9.3.1, one finds

$$\int_{\mathcal{L}} R_{ab} K^a q^b \, dv = -\int_{\mathcal{L}} (K^{a;b} q_a)_{;b} \, dv + \int_{\mathcal{L}} K_{a;b} q^{a;b} \, dv$$

$$= -\int_{\partial \mathcal{L}} K^{a;b} q_a \, d\sigma_b - 2 \int_{\mathcal{L}} f^{-2} \omega^a \omega_a \, dv. \tag{9.10}$$

The boundary $\partial \mathscr{L}$ of $\mathscr{L}$ consists of the surfaces $\partial \mathscr{L}_1 \equiv \mathscr{N}_1 \cap \bar{J}^-(\mathscr{I}^+, \bar{\mathscr{M}})$, $\partial \mathscr{L}_2 \equiv \mathscr{N}_2 \cap \bar{J}^-(\mathscr{I}^+, \bar{\mathscr{M}})$, the portion $\partial \mathscr{L}_3$ of $\dot{J}^-(\mathscr{I}^+, \bar{\mathscr{M}})$ between $\mathscr{N}_1$ and $\mathscr{N}_2$, and the portion $\partial \mathscr{L}_4$ of $\mathscr{I}^-$ between $\mathscr{N}_1$ and $\mathscr{N}_2$. The surface integral over $\partial \mathscr{L}_1$ is minus that over $\partial \mathscr{L}_2$, since these surfaces are carried into each other by the isometry θ_1.

Near $\mathscr{I}^-$, $f = -1 + (2m/r) + O(r^{-2})$ and $\omega^a \omega_a = O(r^{-6})$, where r is some suitable radial coordinate. Thus the surface integral over $\partial \mathscr{L}_4$ at $\mathscr{I}^-$ vanishes. Suppose now that K^a were timelike everywhere in $\mathscr{L}$, becoming null on the horizon. Then ω^a, being orthogonal to $\mathbf{K}$, would be spacelike everywhere in $\mathscr{L}$. Therefore if $\boldsymbol{\omega}$ is non-zero, i.e. the solution is non-static, the last term on the right of (9.10) will be negative. This leads to a contradiction if the space is empty and if the integral over $\partial \mathscr{L}_3$ is zero.

To evaluate this integral, one has to apply a limiting procedure. Let z be a function on the surface $\mathscr{N}_1$ which is zero on the horizon but such that the gradient of z in $\mathscr{N}_1$ is not zero on the horizon. The function z can be defined on $\mathscr{L}$ by the condition $z_{;a} K^a = 0$. One can express the gradient of z as

$$z_{;a} = \bar{x}_{;b} z^{;b} (K_a + fR_a),$$

where R^a is a vector field tangent to the surfaces $\{\bar{x} = \text{constant}\}$ and normalized so that $R^a K_a = -1$. One now takes $\int K^{a;b} q_a \, \mathrm{d}\sigma_b$ over the surface $\{z = \text{constant}\}$ between $\mathscr{N}_1$ and $\mathscr{N}_2$. Then $\mathrm{d}\sigma_b = \mathrm{d}\sigma z_{;b}$, where $\mathrm{d}\sigma$ is some continuous measure. Thus

$$\int K^{a;b} q_a \, \mathrm{d}\sigma_b = \int (\tfrac{1}{2}\bar{x}_{;a}(f)^{;a} - \bar{x}_{;a} K^a_{;b} R^b f + \tfrac{1}{2} f_{;b} R^b) \bar{x}_{;b} z^{;b} \, \mathrm{d}\sigma.$$

Since the horizon was the surface $f = 0$ and since K^a was directed along the null generators of the horizon, $f_{;a}$ is proportional to K^a on the horizon. Therefore

$$\int_{\partial \mathscr{L}_3} K^{a;b} q_a \, \mathrm{d}\sigma_b = 0.$$

This gives a contradiction which shows that K^a must be spacelike somewhere in $\mathscr{L}$ if the space is empty. $\square$

Proposition 9.3.6

Let $(\mathscr{M}, \mathbf{g})$ be a stationary non-static regular predictable space–time in which the ergosphere intersects $\dot{J}^-(\mathscr{I}^+, \bar{\mathscr{M}}) \cap J^+(\mathscr{I}^-, \bar{\mathscr{M}})$. Then there is a one-parameter cyclic isometry group θ_ϕ $(0 \leqslant \phi \leqslant 2\pi)$ of $(\mathscr{M}, \mathbf{g})$ which commutes with θ_t, and whose orbits are spacelike near $\mathscr{I}^+$ and $\mathscr{I}^-$.

Let $\mathscr{Q}_1$ be one connected component of $J^-(\mathscr{I}^+, \mathscr{M}) \cap J^+(\mathscr{I}^-, \mathscr{M})$, and let $\mathscr{G}_1$ be the quotient of $\mathscr{Q}_1$ by its generators. Then the orbits of the isometry θ_t in the horizon $\mathscr{Q}_1$ will be spirals which repeatedly intersect the same generators. Let $t_1 > 0$ be such that θ_{t_1} is one rotation of $\mathscr{G}_1$. Then if $p \in \mathscr{Q}_1$, $\theta_{t_1}(p)$ will lie on the same generator of $\mathscr{Q}_1$. It will lie to the future of p, since
$$x|_{\theta_{t_1}(p)} = x|_p + t_1.$$

One can now choose the future-directed null vector $\mathbf{Y}_1$ to be directed along the generators, and scaled so that

(i) $Y_{1a;b}Y_1^{\ b} = 2\epsilon Y_{1a}$, where $\epsilon_{;a}Y_1^{\ a} = 0$,
(ii) if v is a parameter along the generators such that $\mathbf{Y}_1 = \partial/\partial v$, then
$$v|_{\theta_{t_1}(p)} = v|_p + t_1.$$

The vector field $\mathbf{Y}_1$ defined in this way is invariant under the isometry θ_t, i.e. $L_{\mathbf{K}}\mathbf{Y}_1 = 0$. One can now define a spacelike vector field $\mathbf{Y}_3$ in $\mathscr{Q}_1$ by $\mathbf{Y}_3 \equiv \mathbf{K} - \mathbf{Y}_1$; then $L_{\mathbf{K}}\mathbf{Y}_3 = 0$ and $L_{\mathbf{Y}_1}\mathbf{Y}_3 = 0$ (note that $\mathbf{Y}_3$ is not a unit vector, and in fact it will vanish on the generators γ_1 and γ_2 corresponding to the poles of $\mathscr{G}_1$). The integral curves of $\mathbf{Y}_3$ in $\mathscr{Q}_1$ will be circles which degenerate to points on γ_1 and γ_2.

Let μ be a curve in $\mathscr{Q}_1$ from γ_1 to γ_2 orthogonal to $\mathbf{Y}_1$ and $\mathbf{Y}_3$, and such that the orbits of $\mathbf{Y}_3$ which intersect μ form a smooth spacelike two-surface $\mathscr{P}$ in $\mathscr{Q}_1$. Let $\mathscr{P}(v)$ be the family of spacelike two-surfaces in $\mathscr{Q}_1$ obtained by moving each point of $\mathscr{P}$ a parameter distance v up the generators of $\mathscr{Q}_1$. $\mathscr{P}(v)$ is also equal to $\theta_v(\mathscr{P})$. Let $\mathbf{Y}_2$ be the other null vector orthogonal to $\mathscr{P}(v)$, normalized so that $Y_1^{\ a}Y_{2a} = -1$ (see figure 61); then $L_{\mathbf{K}}\mathbf{Y}_2 = 0$.

Let $\mathbf{Y}_4$ be a spacelike vector on μ, tangent to μ. Then one can define $\mathbf{Y}_4$ on $\mathscr{Q}_1$ by dragging it along by $\mathbf{K}$ and $\mathbf{Y}_1$, i.e. $L_{\mathbf{K}}\mathbf{Y}_4 = 0 = L_{\mathbf{Y}_1}\mathbf{Y}_4$. (These are compatible because $L_{\mathbf{K}}\mathbf{Y}_1 = 0$.) $\mathbf{Y}_4$ will be orthogonal to $\mathbf{Y}_1$ on $\mathscr{Q}_1$ because $L_{\mathbf{K}}(Y_4^{\ a}g_{ab}Y_1^{\ b}) = 0$, and
$$(Y_4^{\ a}Y_{1a})_{;b}Y_1^{\ b} = Y_1^{\ a}{}_{;b}Y_4^{\ b}Y_{1a} + Y_1^{\ a}{}_{;b}Y_{4a}Y_1^{\ b}.$$

The first term is zero because $\mathbf{Y}_1$ is null and the second term equals $2\epsilon Y_{1a}Y_4^{\ a}$. Thus $Y_{1a}Y_4^{\ a}$, being zero initially, remains zero. $\mathbf{Y}_4$ will be orthogonal to $\mathbf{Y}_2$ on $\mathscr{Q}_1$ because it lies in the surface $\mathscr{P}(v)$, and $\mathbf{Y}_2$ is normal to the surface. It will also be orthogonal to $\mathbf{Y}_3$ on $\mathscr{Q}_1$ because $L_{\mathbf{K}}(Y_3^{\ a}g_{ab}Y_4^{\ b}) = 0$, and
$$(Y_3^{\ a}Y_{4a})_{;b}Y_1^{\ b} = Y_1^{\ a}{}_{;b}Y_3^{\ b}Y_{4a} + Y_1^{\ a}{}_{;b}Y_4^{\ b}Y_{3a} = 0$$
since $Y_{1a;b}\hat{h}^{ac}\hat{h}^{bd} = 0$.

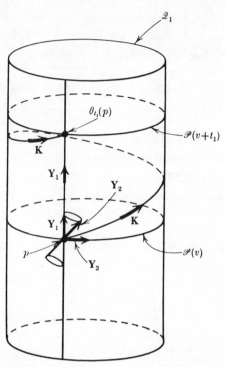

FIGURE 61. The isometry θ_{t_1} moves the point p and the surface $\mathscr{P}(v)$ into the point $\theta_{t_1}(p)$ and the surface $\mathscr{P}(v+t_1)$ in the horizon $\mathscr{Q}_1$. $\mathbf{Y}_1$ is tangent to a null geodesic generator of $\mathscr{Q}_1$, $\mathbf{Y}_2$ is a null vector orthogonal to $\mathscr{P}(v)$, and $\mathbf{Y}_3$ lies in $\mathscr{P}(v)$. $\mathbf{K}$ is the Killing vector field on $\mathscr{Q}_1$ which generates the isometry group θ_t.

In a neighbourhood of $\mathscr{Q}_1$, there will be a unique null geodesic λ orthogonal to a surface $\mathscr{P}(v)$ through a given point r. One can then define coordinates (v, w, θ, ϕ) for the point r, where w is the affine distance (as measured by $\mathbf{Y}_2$) along μ, and (v, θ, ϕ) have their values at $\mu \cap \mathscr{Q}_1$, where θ and ϕ are spherical polar coordinates for the generators of $\mathscr{Q}_1$ such that $Y_3{}^a \theta_{,a} = 0$, $Y_4{}^a \phi_{,a} = 0$. (In other words, we choose $\mathbf{Y}_3 = (2\pi/t_1)\,\partial/\partial\phi$ and $\mathbf{Y}_4 = \partial/\partial\theta$ on $\mathscr{Q}_1$.) We shall take the basis $\{\mathbf{Y}_1, \mathbf{Y}_2, \mathbf{Y}_3, \mathbf{Y}_4\}$ to be parallelly propagated along the null geodesics with tangent vector $\mathbf{Y}_2$. Then $\mathbf{Y}_2 = \partial/\partial w$. We define the vector $\hat{\mathbf{K}}$ to be $\partial/\partial v$. This means that the Lie derivative of $\hat{\mathbf{K}}$ by $\mathbf{Y}_2$ is zero. We define the vector Z^a to be

$$Z^a = \frac{1}{\sqrt{2}}\left\{\frac{Y_3{}^a}{(Y_3{}^b Y_{3b})^{\frac{1}{2}}} + \mathrm{i}\,\frac{Y_4{}^a}{(Y_4{}^b Y_{4b})^{\frac{1}{2}}}\right\}.$$

Then $\qquad Z^a Z_a = 0, \quad Z^a \bar{Z}_a = 1, \quad \bar{Z}^a \bar{Z}_a = 0,$

where $^-$ denotes the complex conjugate.

One can define on $\mathscr{Q}_1$ a family $\{\mathbf{g}_n\}$ of tensor fields, where

$$\mathbf{g}_0 = \mathbf{g} \quad \text{and} \quad \mathbf{g}_n = \underbrace{L_{\mathbf{Y}_2}(L_{\mathbf{Y}_2}(\ldots(L_{\mathbf{Y}_2}\mathbf{g})\ldots))}_{n \text{ terms}}.$$

In the coordinates given above, $g_{n\,ab} = \partial^n(g_{ab})/\partial w^n$. Since the solution is analytic, it is completely determined by the family $\mathbf{g}_n$ on $\mathscr{Q}_1$. We shall show that on $\mathscr{Q}_1$, the Lie derivatives with respect to $\hat{\mathbf{K}}$ of all the $\mathbf{g}_n$ vanish. Then the Lie derivative of the $\mathbf{g}_n$ with respect to $\tilde{\mathbf{K}} = \hat{\mathbf{K}} - \mathbf{K}$ will also vanish. This shows that the solution will admit a one-parameter group $\hat{\theta}_\phi$ generated by $\tilde{\mathbf{K}}$. For simplicity we shall consider only the empty space case, but similar arguments hold in the presence of matter fields, like the electromagnetic or scalar fields, which obey well-behaved hyperbolic equations.

By our choice of coordinates, the components of $L_{\hat{\mathbf{K}}}\mathbf{g}$ are the partial derivatives with respect to v of the coordinate components g_{ab}. These are all constant on $\mathscr{Q}_1$, so $L_{\hat{\mathbf{K}}}\mathbf{g}\big|_{\mathscr{Q}_1} = 0$. We shall show below $L_{\hat{\mathbf{K}}}\mathbf{g}_1\big|_{\mathscr{Q}_1} = 0$, and then proceed by a method of induction. Suppose that

$$L_{\hat{\mathbf{K}}}\mathbf{g}_n\big|_{\mathscr{Q}_1} = 0, \quad n \geqslant 1.$$

It then follows from the construction of the basis that $L_{\hat{\mathbf{K}}}$ of the nth covariant derivatives of all the basis vectors $\mathbf{Y}_1, \mathbf{Y}_2, \mathbf{Z}, \overline{\mathbf{Z}}$ are zero. Now

$$g_{n+1\,ab} = g_{n\,ab;c}Y_2{}^c + g_{n\,cb}Y_2{}^c{}_{;a} + g_{n\,ac}Y_2{}^c{}_{;b}.$$

The Lie derivative with respect to $\hat{\mathbf{K}}$ of the second and third terms on the right are zero. The first term involves covariant derivatives of $\mathbf{Y}_2$ of order $(n+1)$ and lower orders. The Lie derivative with respect to $\hat{\mathbf{K}}$ of all the lower order terms are zero. The terms involving $(n+1)$ covariant derivatives are

$$(Y_{2a;bef\ldots ghc} + Y_{2b;aef\ldots ghc})Y_2{}^e Y_2{}^f \ldots Y_2{}^h Y_2{}^c$$
$$= (Y_{2a;be}Y_2{}^e + Y_{2b;ae}Y_2{}^e)_{;f\ldots ghc}Y_2{}^f \ldots Y_2{}^c + \text{lower order terms}$$
$$= ((Y_{2a;e}Y_2{}^e)_{;b} + R_{pabe}Y_2{}^p Y_2{}^e + (Y_{2b;e}Y_2{}^e)_{;a} + R_{pbae}Y_2{}^p Y_2{}^e)_{;f\ldots gh}$$
$$\times Y_2{}^f \ldots Y_2{}^c + \text{lower order terms}.$$

The Lie derivatives with respect to $\hat{\mathbf{K}}$ of this expression will be zero, if the Lie derivative with respect to $\hat{\mathbf{K}}$ of the Riemann tensor and its covariant derivatives to order $(n-1)$ vanish. Then $L_{\hat{\mathbf{K}}}\mathbf{g}_{n+1}\big|_{\mathscr{Q}_1}$ will be zero.

To show that the Lie derivatives with respect to $\hat{\mathbf{K}}$ of $\mathbf{g}_1$ and of the covariant derivatives of the Riemann tensor are zero, it is convenient to use some notation introduced by Newman and Penrose (1962).

This involves using a pseudo-orthonormal basis with the two spacelike vectors $\mathbf{Y}_3$ and $\mathbf{Y}_4$ combined to give a single complex null vector $\mathbf{Z}$, giving each component of the connection and the curvature tensor a separate symbol, and writing out all the Bianchi identities and the defining equations for the curvature tensor explicitly without summation. These relations are combined in pairs to form half the number of complex equations. The symbols for the connection components are:

$$\kappa = Y_{1a;\,b}Z^aY_1^b, \qquad\qquad \pi = -Y_{2a;\,b}\bar{Z}^aY_1^b,$$

$$\rho = Y_{1a;\,b}Z^a\bar{Z}^b, \qquad\qquad \lambda = -Y_{2a;\,b}\bar{Z}^a\bar{Z}^b,$$

$$\sigma = Y_{1a;\,b}Z^aZ^b, \qquad\qquad \mu = -Y_{2a;\,b}\bar{Z}^aZ^b,$$

$$\tau = Y_{1a;\,b}Z^aY_2^b, \qquad\qquad \nu = -Y_{2a;\,b}\bar{Z}^aY_2^b,$$

$$\epsilon = \tfrac{1}{2}(Y_{1a;\,b}Y_2^aY_1^b - Z_{a;\,b}\bar{Z}^aY_1^b), \quad \alpha = \tfrac{1}{2}(Y_{1a;\,b}Y_2^a\bar{Z}^b - Z_{a;\,b}\bar{Z}^a\bar{Z}^b),$$

$$\beta = \tfrac{1}{2}(Y_{1a;\,b}Y_2^aZ^b - Z_{a;\,b}\bar{Z}^aZ^b), \quad \gamma = \tfrac{1}{2}(Y_{1a;\,b}Y_2^aY_2^b - Z_{a;\,b}\bar{Z}^aY_2^b).$$

The symbols for the Weyl tensor are:

$$\Psi_0 = -C_{abcd}Y_1^aZ^bY_2^cZ^d,$$

$$\Psi_1 = -C_{abcd}Y_1^aY_2^bY_1^cZ^d,$$

$$\Psi_2 = -\tfrac{1}{2}C_{abcd}(Y_1^aY_2^bY_1^cY_2^d - Y_1^aY_2^bZ^c\bar{Z}^d),$$

$$\Psi_3 = C_{abcd}Y_1^aY_2^bY_2^c\bar{Z}^d$$

$$\Psi_4 = -C_{abcd}Y_2^a\bar{Z}^bY_2^c\bar{Z}^d.$$

We are considering empty space, so the Ricci tensor is zero (i.e. $\Phi_{AB} = 0 = \Lambda$ in the Newman–Penrose formalism). Since the basis is parallelly propagated along $\mathbf{Y}_2$, $\nu = \gamma = \tau = 0$. As $\mathbf{Y}_2$ is the gradient of the coordinate v, $\pi = \bar{\beta} + \alpha$ and $\mu = \bar{\mu}$. Furthermore on $\mathcal{Q}_1$, $\kappa = \rho = \sigma = 0$, $\epsilon = \bar{\epsilon}$, $Y_1(\epsilon) = 0$ and $\Psi_0 = 0$.

The equations we shall need are:

$$Y_1(\alpha) - \bar{Z}(\epsilon) = (\rho + \bar{\epsilon} - 2\epsilon)\alpha + \beta\bar{\sigma} - \bar{\beta}\epsilon - \kappa\lambda + (\epsilon + \rho)\pi, \qquad (9.11\,a)$$

$$Y_1(\beta) - Z(\epsilon) = (\alpha + \pi)\sigma + (\bar{\rho} - \bar{\epsilon})\beta - \mu\kappa - (\bar{\alpha} - \bar{\pi})\epsilon + \Psi_1, \qquad (9.11\,b)$$

$$Y_1(\lambda) - \bar{Z}(\pi) = \rho\lambda + \bar{\sigma}\mu + \pi^2 + (\alpha - \bar{\beta})\pi - (3\epsilon - \bar{\epsilon})\lambda, \qquad (9.11\,c)$$

$$Y_1(\mu) - Z(\pi) = \bar{\rho}\mu + \sigma\lambda + \pi\bar{\pi} - (\epsilon + \bar{\epsilon})\mu - \pi(\bar{\alpha} - \beta) + \Psi_2, \qquad (9.11\,d)$$

$$Z(\rho) - \bar{Z}(\sigma) = \rho(\bar{\alpha} + \beta) - \sigma(3\alpha - \bar{\beta}) - \Psi_1 \qquad (9.11\,e)$$

(these are obtained from the Newman–Penrose equations (4.2)), and:

$$Y_1(\Psi'_1) - \bar{Z}(\Psi'_0) = -3\kappa\Psi'_2 + (2\epsilon + 4\rho)\Psi'_1 - (-\pi + 4\alpha)\Psi'_0, \qquad (9.12\,a)$$

$$Y_1(\Psi'_2) - \bar{Z}(\Psi'_1) = -2\kappa\Psi'_3 + 3\rho\Psi'_2 - (-2\pi + 2\alpha)\Psi'_1 - \lambda\Psi'_0, \qquad (9.12\,b)$$

$$Y_1(\Psi'_3) - \bar{Z}(\Psi'_2) = -\kappa\Psi'_4 - (2\epsilon - 2\rho)\Psi'_3 + 3\pi\Psi'_2 - 2\lambda\Psi'_1, \qquad (9.12\,c)$$

$$Y_1(\Psi'_4) - \bar{Z}(\Psi'_3) = -(4\epsilon - \rho)\Psi'_4 + (4\pi + 2\alpha)\Psi'_3 - 3\lambda\Psi'_2, \qquad (9.12\,d)$$

$$Y_2(\Psi'_0) - Z(\Psi'_1) = -\mu\Psi'_0 - 2\beta\Psi'_1 + 3\sigma\Psi'_2 \qquad (9.12\,e)$$

(these are obtained from the Newman–Penrose equations (4.5)).

From (9.11 e), $\Psi_1 = 0$ on $\mathcal{Q}_1$. Then from (9.12 b), $Y_1(\Psi'_2) = \hat{K}(\Psi'_2) = 0$ on $\mathcal{Q}_1$. Adding (9.11 a) to the complex conjugate of (9.11 b), one obtains

$$Y_1(\pi) = Y_1(\alpha + \bar{\beta}) = \bar{Z}(\epsilon) + Z(\bar{\epsilon}) + 2\pi\rho + 2\overline{\pi}\overline{\sigma} - \pi(\epsilon - \bar{\epsilon}) - \kappa\lambda - \overline{\kappa}\overline{\mu} + \bar{\Psi}_1.$$

On $\mathcal{Q}_1$, this becomes $\qquad Y_1(\pi) = \bar{Z}(\epsilon) + Z(\bar{\epsilon})$.

Therefore $Y_1(Y_1(\pi)) = Y_1(\bar{Z}(\epsilon) + Z(\bar{\epsilon}))$ on $\mathcal{Q}_1$. But on $\mathcal{Q}_1$, $L_{\mathbf{Y}_1}\mathbf{Z} = 0$ and $Y_1(\epsilon) = 0$. Thus $Y_1(Y_1(\pi)) = 0$ on $\mathcal{Q}_1$. This shows that $\pi = A + Bv$ on $\mathcal{Q}_1$, where A and B are constant along a generator of $\mathcal{Q}_1$. However $\pi|_p = \pi|_{\theta_{t_1}(p)}$; therefore π is a constant along the generators of $\mathcal{Q}_1$. Subtracting the complex conjugate of (9.11 b) from (9.11 a), one finds that $(\alpha - \bar{\beta})$ is constant along the generators.

One now applies similar arguments to (9.11 c) and (9.11 d) to show that μ and λ are constant along the generators of $\mathcal{Q}_1$. Since π, μ and λ determine the covariant derivative of $\mathbf{Y}_2$, it follows that $L_{\hat{\mathbf{K}}}Y_2{}^a{}_{;b} = 0$ on $\mathcal{Q}_1$ and hence that $L_{\hat{\mathbf{K}}}\mathbf{g}_1 = 0$ on $\mathcal{Q}_1$.

One can also apply the above kind of argument to (9.12 c) and (9.12 d) to show that $Y_1(\Psi'_3) = Y_1(\Psi'_4) = 0$ on $\mathcal{Q}_1$. Thus $L_{\hat{\mathbf{K}}} R_{abcd} = 0$ on $\mathcal{Q}_1$ and so the Lie derivative with respect to $\hat{\mathbf{K}}$ of the second derivatives of the basis vectors are zero. In particular $\mathbf{Y}_1\,\mathbf{Y}_2$ acting on any of the components of the connection gives zero.

From (9.12 e), $\hat{K}(Y_2(\Psi'_0)) = Y_1 Y_2(\Psi'_0) = 0$ on $\mathcal{Q}_1$. One now operates with $\mathbf{Y}_1\,\mathbf{Y}_2$ on (9.12 a). The commutator $\mathbf{Y}_1\,\mathbf{Y}_2 - \mathbf{Y}_2\,\mathbf{Y}_1$ involves only the first covariant derivatives of the basis vectors. Thus

$$L_{\hat{\mathbf{K}}}(\mathbf{Y}_1\,\mathbf{Y}_2 - \mathbf{Y}_2\,\mathbf{Y}_1) = 0 \quad \text{on} \quad \mathcal{Q}_1.$$

From this it follows by an argument like that given above that

$$\hat{K}(Y_2(\Psi'_1)) = Y_1(Y_2(\Psi'_1)) = 0 \quad \text{on} \quad \mathcal{Q}_1.$$

One now repeats the argument for (9.10 b), (9.10 c) and (9.10 d) to show that $\hat{K}(Y_2(\Psi'_2)) = \hat{K}(Y_2(\Psi'_3)) = \hat{K}(Y_2(\Psi'_4)) = 0$ on $\mathcal{Q}_1$. This shows that the Lie derivatives with respect to $\hat{K}$ of the first covariant derivatives of the Riemann tensor vanish. One then repeats the process, showing that $\hat{K}(Y_2(Y_2(\Psi_0))) = 0$ on $\mathcal{Q}_1$, and so on. $\qquad\square$

Proposition 9.3.7

Let $(\mathcal{M}, \mathbf{g})$ be a space–time which admits a two-parameter abelian isometry group with Killing vectors $\boldsymbol{\xi}_1$ and $\boldsymbol{\xi}_2$. Let $\mathcal{V}$ be a connected open set of $\mathcal{M}$, and let $w_{ab} = \xi_{1[a}\xi_{2b]}$. If

(a) $w_{ab}R^b{}_c\eta^{cdef}w_{ef} = 0$ on $\mathcal{V}$,

(b) $w_{ab} = 0$ at some point of $\mathcal{V}$,

then $\qquad w_{[ab;\,c}w_{d]e} = 0 \quad$ on $\quad \mathcal{V}$.

Let $_{(1)}\chi = \xi_{1a;\,b}w_{cd}\eta^{abcd}$, and $_{(2)}\chi = \xi_{2a;\,b}w_{cd}\eta^{abcd}$. Then

$$\eta^{abcd}{}_{(1)}\chi = -4!\,\xi_1{}^{[a;\,b}\xi_1{}^c\xi_2{}^{d]}$$

$$= 3!\,\xi_1{}^d\xi_2{}^{[a}\xi_1{}^{b\,;\,c]} - 3!\,\xi_2{}^d\xi_1{}^{[a}\xi_1{}^{b\,;\,c]} - 2\times 3!\,\xi_1{}^{[a}\xi_2{}^b\xi_1{}^{c]\,;\,d}.$$

Therefore

$$(3!)^{-1}\eta^{abcd}{}_{(1)}\chi_{;\,d} = \xi_1{}^d{}_{;\,d}\xi_2{}^{[a}\xi_1{}^{b\,;\,c]} + \xi_1{}^d\xi_2{}^{[a}{}_{;\,d}\xi_1{}^{b\,;\,c]}$$

$$+ \xi_1{}^d\xi_2{}^{[a}\xi_1{}^{b\,;\,c]}{}_{;\,d} - \xi_2{}^d{}_{;\,d}\xi_1{}^{[a}\xi_1{}^{b\,;\,c]} - \xi_2{}^d\xi_1{}^{[a}{}_{;\,d}\xi_1{}^{b\,;\,c]}$$

$$- \xi_2{}^d\xi_1{}^{[a}\xi_1{}^{b\,;\,c]}{}_{;\,d} - 2\xi_1{}^{[a}{}_{;\,d}\xi_2{}^b\xi_1{}^{c]\,;\,d}$$

$$- 2\xi_1{}^{[a}\xi_2{}^b{}_{;\,d}\xi_1{}^{c]\,;\,d} - 2\xi_1{}^{[a}\xi_2{}^b\xi_1{}^{c]\,;\,d}{}_{;\,d}. \qquad (9.13)$$

The first and fourth terms vanish because $\boldsymbol{\xi}_1$ and $\boldsymbol{\xi}_2$ are Killing vectors; the second and fifth terms cancel each other because $\boldsymbol{\xi}_1$ and $\boldsymbol{\xi}_2$ commute. Because $\boldsymbol{\xi}_1$ is a Killing vector, $L_{\xi_1}\xi_{1a;\,b} = 0$. This implies that the third term vanishes. Similarly $L_{\xi_2}\xi_{1a\,;\,b} = 0$ because $\boldsymbol{\xi}_2$ is a Killing vector which commutes with $\boldsymbol{\xi}_1$. This implies that the sixth and eighth terms cancel. The seventh term vanishes because $\xi_1{}^a{}_{;\,d}\xi_1{}^{c\,;\,d}$ is symmetric; and because of the relation $\xi_{a;\,bc} = R_{dcba}\xi^d$ satisfied by any Killing vector, $\xi^{a\,;\,d}{}_d = -R^a{}_b\xi^b$. Equation (9.13) is therefore

$$\eta^{abcd}{}_{(1)}\chi_{;\,d} = 2\,.\,3!\,\xi_1{}^{[a}\xi_2{}^b R^{c]}{}_d\xi_1{}^d.$$

By condition (a), the right-hand side of this equation vanishes on $\mathcal{V}$. Thus $_{(1)}\chi$ is a constant on $\mathcal{V}$; in fact it will be zero on $\mathcal{V}$ since it must vanish when w_{ab} does. Similarly $_{(2)}\chi$ will be zero on $\mathcal{V}$. However the vanishing of $_{(1)}\chi$ and $_{(2)}\chi$ is the necessary and sufficient condition that

$$w_{[ab;\,c}w_{d]e} = 0. \qquad\square$$

Proposition 9.3.8

Let $(\mathcal{M}, \mathbf{g})$ be a stationary axisymmetric regular predictable space–time in which $w_{[ab;\,c}w_{d]e} = 0$, where $w_{ab} \equiv K_{[a}\tilde{K}_{b]}$. Then at any point in the exterior region $J^+(\mathcal{I}^-, \bar{\mathcal{M}}) \cap J^-(\mathcal{I}^+, \bar{\mathcal{M}})$ off the axis $\tilde{K} = 0$, $h \equiv w_{ab}w^{ab}$ is negative. On the horizons $\dot{J}^-(\mathcal{I}^+, \bar{\mathcal{M}}) \cap J^+(\mathcal{I}^-, \bar{\mathcal{M}})$ and $\dot{J}^+(\mathcal{I}^-, \bar{\mathcal{M}}) \cap J^-(\mathcal{I}^+, \bar{\mathcal{M}})$, h is zero but $w_{ab} \neq 0$ except on the axis.

By proposition 9.3.3, K^a is non-zero in $J^+(\mathcal{I}^-, \bar{\mathcal{M}}) \cap \overline{J^-}(\mathcal{I}^+, \bar{\mathcal{M}})$. Let λ be an S^1 which is a non-zero integral curve of the vector field $\tilde{K}$ in $J^+(\mathcal{I}^-, \bar{\mathcal{M}}) \cap J^-(\mathcal{I}^+, \bar{\mathcal{M}})$. Under the isometry θ_t, λ can be moved into $D^+(\mathcal{S})$. As there are no closed non-spacelike curves in $D^+(\mathcal{S})$, λ must be a spacelike curve, and hence $\tilde{K}^a$ must be spacelike in

$$J^+(\mathcal{I}^-, \bar{\mathcal{M}}) \cap \overline{J^-}(\mathcal{I}^+, \bar{\mathcal{M}})$$

except on the axis where it is zero. Suppose there were some point p at which $\tilde{K}^a$ and K^a were both non-zero and in the same direction. As $\tilde{K}^a$ and K^a commute, the integral curves of $\tilde{K}^a$ through p would coincide with those of K^a. However the former is closed while the latter is not. Thus $\tilde{K}^a$ and K^a are linearly independent where they are non-zero. Thus w_{ab} is non-zero in $J^+(\mathcal{I}^-, \bar{\mathcal{M}}) \cap \overline{J^-}(\mathcal{I}^+, \bar{\mathcal{M}})$ except on the axis.

The axis will be a two-dimensional surface. Let $\mathcal{Y}$ be the set $J^+(\mathcal{I}^-, \bar{\mathcal{M}}) \cap \overline{J^-}(\mathcal{I}^+, \bar{\mathcal{M}}) - $ (the axis), and let $\mathcal{X}$ be the quotient of $\mathcal{Y}$ by θ_ϕ. As the integral curves of K^a are closed and spacelike in $\mathcal{Y}$, the quotient $\mathcal{X}$ will be a Hausdorff manifold. On $\mathcal{X}$, there will be a Lorentz metric $\tilde{h}_{ab} = g_{ab} - (\tilde{K}^c\tilde{K}_c)^{-1}\tilde{K}_a\tilde{K}_b$. One can project the Killing vector K^a by $\tilde{h}_{ab}$ to obtain a non-zero vector field $\tilde{h}_{ab}K^b$ in $\mathcal{X}$ which is a Killing vector field for the metric $\tilde{h}_{ab}$. The condition $w_{[ab;c}w_{d]e} = 0$ in $\mathcal{M}$ implies that in $\mathcal{X}$, $(K^b\tilde{h}_{b[c})_{|d}\tilde{h}_{e]f}K^f = 0$, where $|$ denotes the covariant derivative with respect to $\tilde{h}$. This is just the condition that there should exist a function ξ on $\mathcal{X}$ such that $K^b\tilde{h}_{ba} = -\alpha\xi_{|a}$. The argument is then similar to that in proposition 9.3.4. One shows that if $K_aK_b\tilde{h}^{ab} = 0$ at a point $p \in \mathcal{X}$, then the surface $\xi = \xi|_p$ is a null surface in $\mathcal{X}$ with respect to the metric $\tilde{h}$. The function ξ on $\mathcal{X}$ induces a function ξ on $\mathcal{Y}$, with the property: $\xi_{;a}K^a = 0$. Thus $\xi = \xi|_p$ will be a null surface in $\mathcal{M}$ with respect to the metric $\mathbf{g}$.

Suppose p corresponded to an integral curve λ of $\tilde{K}^a$ which did not lie on $\dot{J}^-(\mathcal{I}^+, \bar{\mathcal{M}})$. Let $q \in \mathcal{M}$ be a point of λ. Then there would be a future-directed timelike curve $\gamma(v)$ from $\mathcal{I}^-$ through q to $\mathcal{I}^+$. If this curve intersected the axis, it could be deformed slightly to avoid it. One would then obtain a contradiction similar to that in proposition 9.3.4. □

10

The initial singularity in the universe

The expansion of the universe is in many ways similar to the collapse of a star, except that the sense of time is reversed. We shall show in this chapter that the conditions of theorems 2 and 3 seem to be satisfied, indicating that there was a singularity at the beginning of the present expansion phase of the universe, and we discuss the implications of space–time singularities.

In §10.1 we show that past-directed closed trapped surfaces exist if the microwave background radiation in the universe has been partially thermalized by scattering, or alternatively if the Copernican assumption holds, i.e. we do not occupy a special position in the universe. In §10.2 we discuss the possible nature of the singularity and the breakdown of physical theory which occurs there.

10.1 The expansion of the universe

In §9.1 we showed that many stars would eventually collapse and produce closed trapped surfaces. If one goes to a larger scale, one can view the expansion of the universe as the time reverse of a collapse. Thus one might expect that the conditions of theorem 2 would be satisfied in the reverse direction of time on a cosmological scale, providing that the universe is in some sense sufficiently symmetrical, and contains a sufficient amount of matter to give rise to closed trapped surfaces. We shall give two arguments to show that this indeed seems to be the case. Both arguments are based on the observations of the microwave background, but the assumptions made are rather different.

Observations of radio frequencies between 20 cm and 1 mm indicate that there is a background whose spectrum (shown in figure 62 (i)) seems to be very close to that of a black body at 2.7 °K (see, for example, Field (1969)). This background appears to be isotropic to within 0.2 % (figure 62 (ii); see, for example, Sciama (1971) and references given there for further discussion). The high degree of isotropy indicates that it cannot come from within our own galaxy (we

[348]

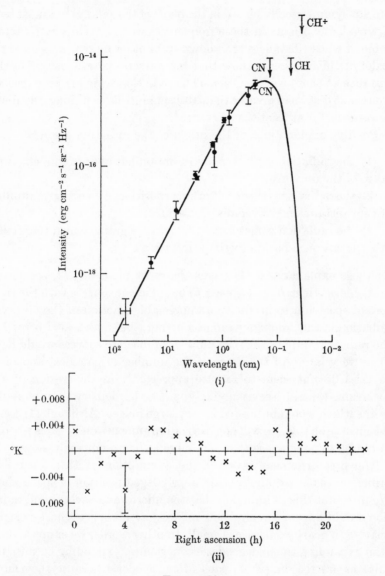

FIGURE 62

(i) The spectrum of the microwave background radiation. The plotted points show the observed values of the ' excess' background radiation. The solid line is a Planck spectrum corresponding to a temperature of 2.7 °K.

(ii) The isotropy of the microwave background radiation. The temperature distribution along the celestial equator is shown; more than two years of data have been averaged to obtain these points.

From D. W. Sciama, *Modern Cosmology*, Cambridge University Press, 1971.

are not symmetrically placed in the plane of the galaxy) but must be of extragalactic origin. At these frequencies we can see discrete sources some of whose distances are known from other evidence to be of the order of 10^{27} cm, so we know that the universe is transparent to this distance at these wavelengths. Thus radiation which is produced by sources at distances greater than 10^{27} cm must have propagated freely towards us for at least that distance.

Possible explanations of the origin of the radiation are:

(1) the radiation is black body radiation left over from a hot early stage of the universe;

(2) the radiation is the result of superposition of a very large number of very distant unresolved discrete sources;

(3) the radiation comes from intergalactic grains which thermalize other forms of radiation (perhaps infra-red).

Of these explanations, (1) seems the most plausible. (2) seems improbable, as there do not appear to be sufficient sources with the right sort of spectrum to produce an appreciable fraction of the observed radiation in this frequency range. Further, the small scale isotropy of the radiation implies that the number of discrete sources would have to be very large (of the order of the number of galaxies) and most galaxies do not seem to radiate appreciably in this region of the spectrum. (3) also seems unlikely, since the density of interstellar grains which would be needed is very large indeed. Although (1) seems the most probable, we will not base our arguments on it, since to do so would be to presuppose that the universe had a hot early stage.

The first argument involves the assumption of the Copernican principle, that we do not occupy a privileged position in space–time. We interpret this as implying that the microwave background radiation would appear equally isotropic to any observer whose velocity relative to nearby galaxies is small. In other words, we suppose there is an expanding timelike geodesic congruence (expanding because the galaxies are receding from each other, geodesic because they move under gravity alone with unit tangent vector V^a, say), representing the average motion of the galaxies, relative to which the microwave radiation appears almost isotropic. From the Copernican principle it also follows that most of the microwave background has propagated freely towards us from a very long distance ($\sim 3 \times 10^{27}$ cm). This is because the contribution to the background arising from a spherical shell of thickness dr and radius r about us will be approximately independent of r, since the amount produced in the shell will be proportional to r^2 and the reduction of intensity due to distance will be inversely proportional to r^2. This will be the case until the redshift of the sources becomes appreciable, source evolution takes place, or curvature effects become significant. These effects will however only come in at a distance of the order of the Hubble radius, $\sim 10^{28}$ cm. Thus the bulk of the radiation will have travelled freely towards us from a distance $\gtrsim 10^{27}$ cm. From the fact that it remains isotropic travelling over such a long distance, we can conclude that on a large scale the metric of the universe is close to one of the Robertson–Walker metrics (§ 5.3). This follows from a result of Ehlers, Geren and Sachs (1968), which we will now describe.

The microwave radiation can be described by a distribution function $f(u, \mathbf{p})$ ($u \in \mathcal{M}$, $\mathbf{p} \in T_u$) defined on the null vectors in $T(\mathcal{M})$, which can be regarded as the phase space of the photons. If the distribution function $f(u, \mathbf{p})$ is exactly isotropic for an observer moving with four-velocity V^a, it will have the form $f(u, E)$ where $E \equiv -V^a p_a$. Since the radiation is freely propagating, f must obey the Liouville equation in $T(\mathcal{M})$. This states that f is constant along integral curves of the horizontal vector field $\mathbf{X}$, i.e. along any curve $(u(v), \mathbf{p}(v))$ where $u(v)$ is a null geodesic in $\mathcal{M}$ and $\mathbf{p} = \partial/\partial v$.

Because $f(u, E)$ is non-negative and must tend to zero as $E \to \infty$ (since otherwise the energy density of radiation would be infinite), there must be an open interval of E for which $\partial f/\partial E$ is non-zero. In this interval, one can express E as a function of f: $E = g(u, f)$. Then Liouville's equation implies that

$$dE/dv = g_{;a} p^a \qquad (10.1)$$

on each null geodesic, where one regards g as a function on $\mathcal{M}$ with f fixed. Also, $\quad dE/dv = -d(V^a p_a)/dv = -V_{a;b} p^a p^b. \qquad (10.2)$

One can decompose p^a into a part along V^a and a part orthogonal to V^a: $p^a = E(V^a + W^a)$, where $W^a W_a = 1$, $W^a V_a = 0$. Then from (10.1) and (10.2),

$$dg/dt + \tfrac{1}{3}\theta g + (g\dot{V}_a + g_{;a}) W^a + g\sigma_{ab} W^a W^b = 0$$

holds for all unit vectors W^a orthogonal to V^a, where dg/dt is the rate of change of g along the integral curves of $\mathbf{V}$. Separating out spherical harmonics,

$$\sigma_{ab} = 0, \qquad (10.3\,a)$$

$$\dot{V}_a + (\log g)_{;a} = \alpha V_a, \qquad (10.3\,b)$$

$$\tfrac{1}{3}\theta = -d(\log g)/dt. \qquad (10.3\,c)$$

Since we assumed that $\dot{V}_a$ was zero, $(10.3b)$ shows that V_a is orthogonal to the surfaces $\{g = \text{constant}\}$, and this implies that the vorticity ω_{ab} is zero. As $\dot{V}^a = 0$, $V_{[a,\,b]} = 0$. Thus one can write V_a as the gradient of a function t: $V_a = -t_{,a}$.

The energy–momentum tensor of the radiation will have the form

$$T_{ab} = \tfrac{4}{3}\mu_r V_a V_b + \tfrac{1}{3}\mu_r g_{ab},$$

where $\mu_r = \int fE^3 \, dE$. Since the motion of the galaxies relative to the integral curves of V^a is small, their contribution to the energy–momentum tensor can be approximated by a smooth fluid with density μ_G, four-velocity V_a and negligible pressure. It now follows that the geometry of the space–time is the same as that of a Robertson–Walker model. To see this, note that

$$(V^a{}_{;b})_{;a} = \tfrac{1}{3}(\theta(\delta^a{}_b + V^a V_b))_{;a}$$
$$= (V^a{}_{;a})_{;b} + R^{ca}{}_{ba} V_c = \theta_{;b} + R_{ba} V^a.$$

Multiplying this equation by $h^b{}_c = g^b{}_c + V^b V_c$, one finds

$$h^{bc} R_{ca} V^a = -\tfrac{2}{3} h^{bc} \theta_{;c}.$$

The left-hand side vanishes by the field equations. Thus θ is constant on the surfaces of constant t (which are also the surfaces of constant g). One can define a function $S(t)$ from θ by $S^{\cdot}/S = \tfrac{1}{3}\theta$; then the Raychaudhuri equation (4.26) takes the form

$$3S^{\cdot\cdot}/S + 4\pi\mu - \Lambda = 0,$$

which implies that $\mu = \mu_G + 2\mu_R$ is also constant on the surfaces $\{t = \text{constant}\}$. From the definition of μ_R we see that the terms μ_G and μ_R are separately constant on these surfaces.

The trace-free part of (4.27) shows that $C_{abcd} V^b V^d = 0$. The Gauss–Codacci equations (§2.7) now give for the Ricci tensor of the three-spaces $\{t = \text{constant}\}$ the formula

$$R^3{}_{ab} = h_a{}^c h_b{}^d R_{cd} + R_{acbd} V^c V^d + \theta\theta_{ab} + \theta_{ac}\theta^c{}_b$$
$$= 2h_{ab}(-\tfrac{1}{3}\theta^2 + 8\pi\mu + \Lambda).$$

However for a three-dimensional manifold, the Riemann tensor is completely determined by the Ricci tensor, as

$$R^3{}_{abcd} = \eta_{ab}{}^e(-R^3{}_{ef} + \tfrac{1}{2}R^3 h_{ef})\eta^f{}_{cd}.$$

This shows that each three-space $\{t = \text{constant}\}$ is a three-space of constant curvature $K(t) = \tfrac{1}{3}(8\pi\mu + \Lambda - \tfrac{1}{3}\theta^2)$. Integrating the Raychaudhuri equation shows that

$$K(t) = \tfrac{1}{3}(8\pi\mu + \Lambda - 3S^{\cdot 2}/S^2) = k/S^2, \qquad (10.4)$$

where k is a constant. By normalizing S, one can set $k = +1$, 0 or -1. The four-dimensional space–time manifold is the orthogonal product of these three-spaces and the t-line. Thus the metric can be written in comoving coordinates as

$$ds^2 = -dt^2 + S^2(t)\,d\gamma^2,$$

where $d\gamma^2$ is the metric of a three-space of constant curvature k. But this is just the metric of a Robertson–Walker space (see §5.3).

We shall now show that in any Robertson–Walker space containing matter with positive energy density and $\Lambda = 0$ there is a closed trapped surface lying in any surface $\{t = \text{constant}\}$. To see this, we express $d\gamma^2$ in the form

$$d\gamma^2 = d\chi^2 + f^2(\chi)\,(d\theta^2 + \sin^2\theta\,d\phi^2)$$

where $f(\chi) = \sin\chi$, χ or $\sinh\chi$ if $k = +1$, 0 or -1 respectively. Consider a two-sphere $\mathcal{T}$ of radius χ_0 lying in the surface $t = t_0$. The two families of past-directed null geodesics orthogonal to $\mathcal{T}$ will intersect the surfaces $\{t = \text{constant}\}$ in two two-spheres of radius

$$\chi = \chi_0 \pm \int_{t_0}^{t} dt/S(t). \qquad (10.5)$$

The surface area of a two-sphere of radius χ is $4\pi S^2(t)f^2(\chi)$. Thus both families of null geodesics will be converging into the past if, at $t = t_0$,

$$\frac{d}{dt}(S^2(t)f^2(\chi)) > 0$$

holds for both values of χ given by (10.5). This will be the case if

$$\frac{S^{\cdot}(t_0)}{S(t_0)} > \pm\frac{f'(\chi_0)}{S(t_0)f(\chi_0)}.$$

But by (10.4), this holds if

$$(\tfrac{8}{3}\pi\mu(t_0)\,S^2(t_0) - k)^{\frac{1}{2}} > \pm f'(\chi_0)/f(\chi_0).$$

This will be the case if $S(t_0)\,\chi_0$ is taken to be greater than $\sqrt{(3/8\pi\mu_0)}$ for $k = 0$ or -1, and to be greater than $\min(\sqrt{(3/8\pi\mu_0)}, \tfrac{1}{2}\pi)$ if $k = +1$.

An intuitive way of viewing this result is that at time t_0 a sphere of coordinate radius χ_0 will contain a mass of the order of $\tfrac{4}{3}\pi\mu_0 S^3(t_0)\chi_0{}^3$, and so will be within its Schwarzschild radius if $S(t_0)\chi_0$ is less than $\tfrac{8}{3}\pi\mu_0 S(t_0)^3\chi_0{}^3$, i.e. if $S(t_0)\chi_0$ is greater than the order of $\sqrt{(3/8\pi\mu_0)}$. We shall call $\sqrt{(3/8\pi\mu_0)}$ the *Schwarzschild length* of matter density μ_0.

So far, we have assumed the microwave radiation is exactly isotropic. This is of course not the case; and this corresponds to the fact that the universe is not exactly a Robertson–Walker space. However, the large scale structure of the universe should be close to that of a Robertson–Walker model, at least back to the time when the radiation was emitted or last scattered. (One can in fact use the deviations of the microwave radiation from exact isotropy to estimate how large the departures from a Robertson–Walker universe are.) For a sufficiently large sphere, the existence of local irregularities should not significantly affect the amount of matter in the sphere, and hence should not affect the existence of a closed trapped surface round us at the present time.

The above argument did not depend on the spectrum of the microwave radiation, but it did involve the assumption of the Copernican principle. The argument we shall now give does not involve the Copernican principle, but does to a certain extent depend on the shape of the spectrum. We shall assume that the approximately black body nature of the spectrum and the high degree of small scale isotropy of the radiation indicate that it has been at least partially thermalized by repeated scattering. In other words, there must be enough matter on each past-directed null geodesic from us to cause the opacity to be high in that direction. We shall now show that this matter will be sufficient to make our past light cone reconverge.

Consider a point p representing us at the present time, and let W^a be a past-directed unit vector parallel to our four-velocity.

The affine parameter v on the past-directed null geodesics through p may be normalized by $K^a W_a = -1$, where $\mathbf{K} = \partial/\partial v$ is the tangent vector to the null geodesics. The expansion θ of these null geodesics will obey (4.35) with $\hat{\omega} = 0$. Thus, providing $R_{ab} K^a K^b \geqslant 0$, θ will be less than $2/v$. It follows that at $v = v_1 > v_0$,

$$\int_{v_0}^{v_1} R_{ab} K^a K^b \, dv - 2/v_0 > \theta,$$

so θ will become negative if there is some v_0 such that

$$\int_{v_0}^{v_1} R_{ab} K^a K^b \, dv > 2/v_0.$$

Using the field equations with $\Lambda = 0$, this becomes

$$\tfrac{1}{2} v_0 \int_{v_0}^{v_1} 8\pi T_{ab} K^a K^b \, dv > 1. \tag{10.6}$$

At centimetre wavelengths, the largest ratio of opacity to density for matter at reasonable densities is that given by Thomson scattering off free electrons in ionized hydrogen. Thus the optical depth to a distance v will be less than

$$\int_0^v \kappa\rho(K^aV_a)\,dv,$$

where κ is the Thomson scattering opacity per unit mass, ρ is the density of the matter, and V_a is the local velocity of the gas. The redshift z of the matter is given by $z = K^aV_a - 1$. Since no matter has been seen with significant blue-shifts, we shall assume K^aV_a is always greater than one on our past light cone, out to an optical depth unity. As galaxies are observed at these wavelengths with redshifts of 0.3, most of the scattering must occur at redshifts greater than this. (In fact if quasars really are cosmological, the scattering must occur at redshifts greater than two.) With a Hubble constant of $100\,\mathrm{Km/sec/}$ Mpc ($\sim 10^{10}\,\mathrm{years^{-1}}$), a redshift of 0.3 corresponds to a distance of about $3 \times 10^{27}\,\mathrm{cm}$. Taking this value for v_0, the contribution to the integral (9.9) of the matter causing the scattering is

$$3.7 \times 10^{28} \int_{v_0}^{v_1} \rho(K_a V^a)^2\,dv,$$

while the optical depth of the matter between v_0 and v_1 is less than

$$6.6 \times 10^{27} \int_{v_0}^{v_1} \rho(K^aV_a)\,dv.$$

Since $K^aV_a \geqslant 1$, it can be seen that the inequality (10.6) will be satisfied at an optical depth of less than 0.2. If the optical depth of the universe was less than 1, one would not expect either an almost black body spectrum or such a high degree of small scale isotropy, unless there was a very large number of discrete sources which covered only a small fraction of the sky and each of which had a spectrum roughly the same as a 3 °K black body but with much higher intensity. This seems rather unlikely. Thus we believe that the condition (4)(iii) of theorem 2 is satisfied, and so there should be a singularity somewhere in the universe provided the other conditions hold.

Because of its generality, theorem 2 does not tell us whether the singularity will be in our past or in the future of our past. Although it might seem obvious that the singularity should be in our past, one can construct an example in which it is in the future: consider a Robertson–Walker universe with $k = +1$ which collapses to a singularity at some time $t = t_0$, and which asymptotically approaches an Einstein static universe for $t \to -\infty$. This satisfies the energy assumption, and contains points whose past light cones start reconverging (because they meet up around the back). However the singularity is in the future. Of course this is a rather unreasonable example but it shows that one has to be careful. We shall therefore give an argument based on theorem 3 which indicates that the universe contains a singularity in our past, providing that the Copernican principle holds. Theorem 3 is similar to theorem 2, but requires that all the past-directed timelike geodesics from a point shall start to reconverge, instead of all the null geodesics. This condition is not satisfied in the example given above, though it is there satisfied by the future-directed geodesics from any point.

By an argument similar to that given above for the null geodesics, the convergence $\theta(s)$ of the past-directed timelike geodesics from a point p will be less than

$$\frac{3}{s_0} - \int_{s_0}^{s} R_{ab} V^a V^b \, ds,$$

where s is proper distance along the geodesics, $V = \partial/\partial s$ and $s > s_0$. Let W be a past-directed timelike unit vector at p, and let $c \equiv - V^a W_a|_p$ (so $c \geqslant 1$). Then θ will become less than $-c$ within a distance R_1/c along any geodesic if there is some R_0, $R_1 > R_0 > 0$, such that

$$\int_{R_0/c}^{R_1/c} R_{ab} V^a V^b \, ds > c(3/R_0 + \epsilon) \tag{10.7}$$

along that geodesic. Condition (3) of theorem 3 will then be satisfied with $b = \max(R_1, (3\epsilon)^{-1})$.

To make (10.7) appear more similar to (10.6), we shall introduce an affine parameter $v = s/c$ along the timelike geodesics; then (10.7) becomes

$$\tfrac{1}{3} R_0 \int_{R_0}^{R_1} R_{ab} K^a K^b \, dv > 1 + \tfrac{1}{3} R_0 \epsilon, \tag{10.8}$$

where $K = \partial/\partial v$ and $K^a W_a|_p = -1$. We cannot verify this condition directly by observation as in the case of (10.6) because it refers to timelike geodesics. We therefore have to appeal to the arguments given in the first part of this section to show that the universe is close to a Robertson–Walker universe model at least back to the time the microwave background radiation was last scattered.

In a Robertson–Walker model, let W be the vector $-\partial/\partial t$. Along a past-directed timelike geodesic through p,

$$\frac{d}{dv}(W_a K^a) = W_{a;b} K^a K^b$$

$$= -\frac{1}{S} \frac{dS}{dt} \{(W^a K_a)^2 - 1/c^2\}.$$

Therefore, providing that $\mathrm{d}S/\mathrm{d}t > 0$, $W_a K^a \leqslant -1$. However

$$W^a K_a = \mathrm{d}t/\mathrm{d}v;$$

thus for some $\epsilon > 0$, (10.8) will be satisfied for every geodesic provided that there are times t_2, t_3 with $t_2 < t_3 < t_p$ such that

$$\frac{t_p - t_3}{3} \int_{t_2}^{t_3} R_{ab} K^a K^b (-W_c K^c)^{-1} \, \mathrm{d}t > 1. \tag{10.9}$$

By the field equations with $\Lambda = 0$,

$$R_{ab} K^a K^b = 8\pi\{(\mu + p)(W_a K^a)^2 - \tfrac{1}{2}(\mu - p)c^{-2}\}.$$

Therefore, providing $p \geqslant 0$,

$$R_{ab} K^a K^b \geqslant 4\pi\mu(W_a K^a)^2.$$

Thus (10.9) will be satisfied if

$$\frac{t_p - t_3}{3} \int_{t_2}^{t_3} 4\pi\mu \, \mathrm{d}t > 1. \tag{10.10}$$

Assuming that the microwave radiation has a black body spectrum at $2.7\,°\mathrm{K}$, its energy density is about $10^{-34}\,\mathrm{gm\,cm^{-3}}$ at the present time. If this radiation is primaeval, its energy density will be proportional to S^{-4}. Since $S^{-1} = O(t^{-\frac{1}{2}})$ as t tends to zero, one can see that (10.10) can be satisfied by taking t_3 to be $\frac{1}{2}t_p$ and t_2 to be sufficiently small. How small t_2 has to be depends on the detailed behaviour of S, which in turn depends on the density of matter in the universe. This is somewhat uncertain, but seems to lie between $10^{-31}\,\mathrm{gm\,cm^{-3}}$ and $5 \times 10^{-29}\,\mathrm{gm\,cm^{-3}}$. In the former case, t_2 will have to be such that $S(t_p)/S(t_2) \geqslant 30$, and in the latter case, $S(t_p)/S(t_2) \geqslant 300$. Since the microwave radiation seems to be all pervasive, any past-directed timelike geodesic must pass through it. Thus an estimate based on the Robertson–Walker models should be a good approximation for its contribution to (10.10), provided that the radiation was not emitted more recently than t_2, and provided that a Robertson–Walker model is a good approximation back that far. From the arguments at the beginning of this section, the latter should be the case provided that the radiation has propagated freely towards us since t_2. However there may be ionized intergalactic gas present with a density as high as $5 \times 10^{-29}\,\mathrm{gm\,cm^{-3}}$, in which case the radiation could be last scattered at a time t such that $S(t_p)/S(t) \sim 5$. The optical depth back to a time t is

$$\int_t^{t_p} \kappa\mu_{\mathrm{gas}} \, \mathrm{d}t, \tag{10.11}$$

where κ is at most 0.5 if μ is measured in $\mathrm{gm\,cm^{-3}}$ and t in cm.

As before, there can be no significant opacity back to $t = t_p - 10^{17}$ sec, since we see objects at distances of at least 3×10^{27} cm. Taking t_3 to have this value, we see that the gas density will cause (10.11) to be satisfied for a value of t_2 corresponding to an optical depth of at most 0.5.

Thus the position is as follows. We assume the Copernican principle, and that the microwave radiation has been emitted either before a time t_2 such that $S(t_p)/S(t_2) \approx 300$, or before the time corresponding to the optical depth of the universe being unity, if this is less than t_2. In the former case, condition (2) of theorem 3 will be satisfied by the radiation density, and in the latter case by the gas density. Thus if the usual energy conditions and causality conditions hold, we can conclude that there should be a singularity in our past (i.e. there should be a past-directed non-spacelike geodesic from us which is incomplete).

Suppose one takes a spacelike surface which intersects our past light cone and takes a number of points on that surface; can one say that there is a singularity in each of their pasts? This will be the case if the universe is sufficiently homogeneous and isotropic in the past to converge all the past-directed timelike geodesics from these points. In view of the close connection between the convergence of timelike geodesics and closed trapped surfaces, we would expect this to be the case if the universe is homogeneous and isotropic at that time on the scale of the Schwarzschild length $(3/8\pi\mu)^{\frac{1}{2}}$.

We have direct evidence of the homogeneity of the universe in our past from the measurements of Penzias, Schraml and Wilson (1969), who found that the intensity of the microwave background is isotropic to within 4 % for a beam width of 1.4×10^{-3} square degrees. Assuming that the microwave radiation has not been emitted since a surface in our past corresponding to optical depth unity, the observed intensity will be proportional to $T^4/(1+z)^4$ where T is the effective temperature of the observed point on the surface and z is its redshift. Variations in the observed intensity can arise in four ways:

(1) by a Doppler shift caused by our own motion relative to the black body radiation (Sciama (1967), Stewart and Sciama (1967));

(2) by variations in the gravitational redshift caused by inhomogeneities in the distribution of matter between us and the surface (Sachs and Wolfe (1967), Rees and Sciama (1968));

(3) by Doppler shifts caused by local velocity disturbances of the matter at the surface; and

(4) by variations of the effective temperature of the surface.

(In fact the division between (1), (2) and (3) depends on the standard of reference and has heuristic value only.) Thus the observations indicate that irregularities in the temperature with an angular size of 3' of arc have relative amplitudes of less than 1 %, and that there are no local fluctuations of the velocity of the matter, on the same scale, of greater than 1 % of the velocity of light. A region on the surface which had an angular diameter 3' of arc would correspond to a region which had a diameter now of about 10^7 light years. If the surface of optical depth unity is at a redshift of about 1000 (this is the most it could be), the Schwarzschild length at that time would correspond to a region whose present diameter was about 3×10^8 light years. Thus it would seem that every point on the surface of optical depth unity should have a singularity in its past.

More indirect evidence on the degree of homogeneity of the universe in the early stages comes from the fact that observations of the helium content of a number of objects agree with calculations of helium production by Peebles (1966), and Wagoner, Fowler and Hoyle (1968), who assumed the universe was homogeneous and isotropic at least back to a temperature of about 10^9 °K. On the other hand calculations of anisotropic models have shown that in these models very different amounts of helium are produced. Thus if one accepts that there is a fairly uniform density of helium in the universe (there are some doubts about this), and that this helium was produced in the early stages of the universe, one can conclude that the universe was effectively isotropic and hence homogeneous when the temperature was 10^9 °K. One would therefore expect a singularity to occur in the past of each point at this time.

Misner (1968) has shown that if the temperature reaches 2×10^{10} °K a large viscosity arises from collisions between electrons and neutrinos. This viscosity would damp out inhomogeneities whose lengths correspond to present values of 100 light years, and reduce anisotropy to a comparatively small value. Thus if one accepts this as the explanation for the present isotropy of the universe (and it is a very attractive one), one would conclude that there should be a singularity in the past of every point when the temperature was about 10^{10} °K.

10.2   The nature and implications of singularities

One might hope to learn something about the nature of the singularities that are likely to occur by studying exact solutions with singularities. However although we have shown that the *occurrence* of a singularity is not prevented by small perturbations of the initial conditions, it is not clear that the *nature* of the singularity which occurs will be similarly stable. Although we have shown in §7.5 that the Cauchy problem is stable under small perturbations of the initial conditions, this stability applies only to compact regions of the Cauchy development, and a region containing a singularity is non-compact unless the singularity corresponds to imprisoned incompleteness. In fact we can give an example where the nature of the singularity is not stable. Consider a uniform spherically symmetric cloud of dust collapsing to a singularity. The metric inside the dust will be similar to that of part of a Robertson–Walker universe, while that outside will be the Schwarzschild metric. Both inside and outside the dust, the singularity will be spacelike (figure 63 (i)). Suppose now one adds a small electric charge density to the dust. The metric outside the dust now becomes part of the Reissner–Nordström solution for $e^2 < m^2$ (figure 63 (ii)). There will be a singularity inside the dust, as a sufficiently small charge density will not prevent the occurrence of infinite density. The nature of the singularity inside the dust will presumably depend on the charge distribution. However the important point is that once the surface of the dust has passed a point p inside $r = r_+$, whatever happens inside the dust cannot affect the portion sq of the timelike singularity.

If one now increases the charge density so that it becomes greater than the matter density, it is possible for the cloud to pass through the two horizons at $r = r_+$ and $r = r_-$ and to re-expand into another universe without any singularity occurring inside the dust, although there is a timelike singularity outside the dust (J. M. Bardeen, unpublished), as indeed there ought to be by theorem 2 (see figure 63 (iii)).

This example is very important as it shows that there can be timelike singularities, that the matter can avoid hitting the singularities, and that it can pass through a 'wormhole' into another region of space–time or into another part of the same space–time region. Of course one would not expect to have such a charge density on a collapsing star, but since the Kerr solution is so similar to the Reissner–Nordström solution one might expect that angular momentum could produce a similar wormhole. One might speculate therefore that prior to the present expansion phase of the universe there was a contraction phase in which local inhomogeneities grew large and isolated singu-

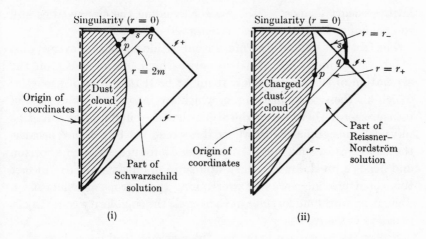

(i)   (ii)

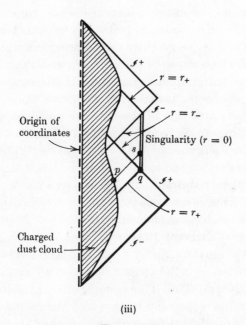

(iii)

FIGURE 63

(i) Collapse of a spherical dust cloud.

(ii) Collapse of a charged dust cloud, where the charge is too small to prevent the occurrence of a singularity in the dust.

(iii) Collapse of a charged dust cloud, where the charge is large enough to prevent the occurrence of a singularity in the dust cloud; the singularity occurs outside the dust, which bounces and re-expands into a second asymptotically flat space.

larities occurred, most of the matter avoiding the singularities and re-expanding to give the present observed universe.

The fact that singularities must occur within the past of every point at an early time when the density was high, places limits on the separation of the singularities. It might be that the set of geodesics which hit these singularities (i.e. which are incomplete) was a set of measure zero. Then one might argue that the singularities would be physically insignificant. However this would not be the case because the existence of such singularities would produce a Cauchy horizon and hence a breakdown of one's ability to predict the future. In fact this could provide a way of overcoming the entropy problem in an oscillating world model since at each cycle the singularity could inject negative entropy.

So far, we have been exploring the mathematical consequences of taking a Lorentz manifold as the model for space–time, and requiring that the Einstein field equations (with $\Lambda = 0$) hold. We have shown that according to this theory, there should be singularities in our past associated with the collapse of the universe, and singularities in the future associated with the collapse of stars. If Λ is negative, the above conclusions would be unaffected. If Λ is positive, observations of the rate of change of expansion of the universe (Sandage, (1961, 1968)) indicate that Λ cannot be greater than $3 \times 10^{-55} \, \mathrm{cm}^{-2}$. This is equivalent to a negative energy density of $3 \times 10^{-27} \, \mathrm{gm} \, \mathrm{cm}^{-3}$. Such a value of Λ could have an effect on the expansion of the whole universe, but it would be completely swamped by the positive matter density in a collapsing star. Thus it does not seem that a Λ term can enable us to avoid facing the problem of singularities.

It may be that General Relativity does not provide a correct description of the universe. So far it has only been tested in situations in which departures from flat space are very small (radii of curvature of the order of $10^{12} \, \mathrm{cm}$). Thus it is a tremendous extrapolation to apply it to situations like collapsing stars where the radius of curvature becomes less than $10^6 \, \mathrm{cm}$. On the other hand the theorems on singularities did not depend on the full Einstein equations but only on the property that $R_{ab} K^a K^b$ was non-negative for any non-spacelike vector K^a; thus they would apply also to any modification of General Relativity (such as the Brans–Dicke theory) in which gravity is always attractive.

It seems to be a good principle that the prediction of a singularity by a physical theory indicates that the theory has broken down, i.e. it no longer provides a correct description of observations. The question is: when does General Relativity break down? One would expect it to break down anyway when quantum gravitational effects become important; from dimensional arguments it seems that this should not happen until the radius of curvature becomes of the order of 10^{-33} cm. This would correspond to a density of 10^{94} gm cm^{-3}. However one might question whether a Lorentz manifold is an appropriate model for space–time on length scales of this order. So far experiments have shown that assuming a manifold structure for lengths greater than 10^{-15} cm gives predictions in agreement with observations (Foley *et al.* (1967)), but it may be that a breakdown occurs for lengths between 10^{-15} and 10^{-33} cm. A radius of 10^{-15} cm corresponds to a density of 10^{58} gm cm^{-3} which for all practical purposes could be regarded as a singularity. Thus maybe one should construct a surface by Schmidt's procedure (§8.3) around regions where the radius of curvature is less than, say, 10^{-15} cm. On our side of this surface a manifold picture of space–time would be appropriate, but on the other side an as yet unknown quantum description would be necessary. Matter crossing the surface could be thought of as entering or leaving the universe, and there would be no reason why that entering should balance that leaving.

In any case, the singularity theorems indicate that the General Theory of Relativity predicts that gravitational fields should become extremely large. That this happened in the past is supported by the existence and black body character of the microwave background radiation, since this suggests that the universe had a very hot dense early phase.

The theorems on the existence of singularities could possibly be refined somewhat, but on our view they are already adequate. However they tell us very little about the nature of the singularities. One would like to know what kind of singularities could occur in generic situations in General Relativity. A possible way of approaching this would be to refine the power series expansion technique of Lifshitz and Khalatnikov, and to clarify its validity. It may also be that there is some connection between the singularities studied in General Relativity and those studied in other branches of physics (cf. for instance, Thom's theory of elementary catastrophes (1969)). Alternatively one might try to proceed by brute force, integrating the Einstein equations numerically on a computer. However this will probably have to wait for a new generation of computers. One would like to know also whether the singularities produced by collapse from a non-singular asymptotically flat situation would be naked, i.e. visible from infinity, or whether they would be hidden behind an event horizon.

The other main problem is to formulate a quantum theory of space–time which will be applicable to strong fields. Such a theory might be based on a manifold, or might allow changes of topology. Some preliminary attempts in this line have been made by de Witt (1967), Misner (1969, 1971), Penrose (see Penrose and MacCallum (1972)), Wheeler (1968), and others. However the interpretation of a quantum theory of space–time, and its relation to singularities, are still very obscure.

Speculation and discussion on the subject of this book is not new. Laplace essentially predicted the existence of black holes: 'Other stars have suddenly appeared and then disappeared after having shone for several months with the most brilliant splendour . . . All these stars . . . do not change their place during their appearance. Therefore there exists, in the immensity of space, opaque bodies as considerable in magnitude, and perhaps equally as numerous as the stars.' (M. Le Marquis de Laplace: 'The system of the world'. Translated by Rev. H. Harte. Dublin, 1830, Vol. 2, p. 335.) As we have seen, our present understanding of the situation is remarkably similar.

The creation of the Universe out of nothing has been argued, indecisively, from early times; see for example Kant's first Antinomy of Pure Reason and comments on it (Smart (1964), pp. 117–23 and 145–59; North (1965), pp. 389–406). The results we have obtained support the idea that the universe began a finite time ago. However the actual point of creation, the singularity, is outside the scope of presently known laws of physics.

Appendix A
Translation of an essay by
Peter Simon Laplace[†]

Proof of the theorem, that the attractive force of a heavenly body could be so large, that light could not flow out of it.[‡]

(1) If v is the velocity, t the time and s space which is uniformly moving during this time, then, as is well known, $v = s/t$.

(2) If the motion is not uniform, to obtain the value of v at any instant one has to divide the elapsed space $\mathrm{d}s$ and this time interval $\mathrm{d}t$ into each other, namely $v = \mathrm{d}s/\mathrm{d}t$, since the velocity over an infinitely small interval is constant and thus the motion can be taken as uniform.

(3) A continuously working force will strive to change the velocity. This change of the velocity, namely $\mathrm{d}v$, is therefore the most natural measure of the force. But as any force will produce double the effect in double the time, so we must divide the change in velocity $\mathrm{d}v$ by the time $\mathrm{d}t$ in which it is brought about by the force $\mathbf{P}$, and one thus obtains a general expression for the force $\mathbf{P}$, namely

$$P = \frac{\mathrm{d}v}{\mathrm{d}t} = \frac{\mathrm{d}.\dfrac{\mathrm{d}s}{\mathrm{d}t}}{\mathrm{d}t}.$$

Now if $\mathrm{d}t$ is constant,

$$\mathrm{d}.\frac{\mathrm{d}s}{\mathrm{d}t} = \frac{\mathrm{d}.\mathrm{d}s}{\mathrm{d}t} = \frac{\mathrm{d}\mathrm{d}s}{\mathrm{d}t};$$

accordingly

$$P = \frac{\mathrm{d}\mathrm{d}s}{\mathrm{d}t^2}.$$

[†] *Allgemeine geographische Ephemeriden herausgegeben von F. von Zach.* IV Bd, I St., I Abhandl., Weimar 1799. We should like to thank D.W. Dewhirst for providing us with this reference. See also note at end of this Appendix.

[‡] This theorem, that a luminous body in the universe of the same density as the earth, whose diameter is 250 times larger than that of the sun, can by its attractive power prevent its light rays from reaching us, and that consequently the largest bodies in the universe could remain invisible to us, has been stated by Laplace in his *Exposition du Système du Monde*, Part II, p. 305, without proof. Here is the proof. Cf. *A.G.E.* May 1798, p. 603. v. Z.

(4) Let the attractive force of a body $= M$; a second body, for example a particle of light, finds itself at distance r; the action of the force M on this light particle will be $- M/rr$; the negative sign occurs because the action of M is opposite to the motion of the light.

(5) Now according to (3) this force also equals $\mathrm{d}\mathrm{d}r/\mathrm{d}t^2$, hence

$$-\frac{M}{rr} = \frac{\mathrm{d}\mathrm{d}r}{\mathrm{d}t^2} = - Mr^{-2}.$$

Multiplying by $\mathrm{d}r$, $\dfrac{\mathrm{d}r\,\mathrm{d}\mathrm{d}r}{\mathrm{d}t^2} = - M\,\mathrm{d}r r^{-2};$

integrating, $\dfrac{1}{2}\dfrac{\mathrm{d}r^2}{\mathrm{d}t^2} = C + Mr^{-1}$

where C is a constant quantity, or

$$\left(\frac{\mathrm{d}r}{\mathrm{d}t}\right)^2 = 2C + 2Mr^{-1}.$$

Now by (2) $\mathrm{d}r/\mathrm{d}t$ is the velocity v, accordingly

$$v^2 = 2C + 2Mr^{-1}$$

holds, where v is the velocity of the light particle at the distance r.

(6) To now determine the constant C, let R be the radius of the attracting body, and a the velocity of the light at the distance R, hence on the surface of the attracting body; then one obtains from (5) $a^2 = 2C + 2M/R$, therefore $2C = a^2 - 2M/R$. Substituting this in the previous equation gives

$$v^2 = a^2 - \frac{2M}{R} + \frac{2M}{r}.$$

(7) Let R' be the radius of another attracting body, its attractive power be iM, and the velocity of the light at a distance r be v', then according to the equation in (6)

$$v'^2 = a^2 - \frac{2iM}{R'} + \frac{2iM}{r}.$$

(8) If one makes r infinitely large, the last term in the previous equation vanishes and one obtains

$$v'^2 = a^2 - \frac{2iM}{R'}.$$

The distance of the fixed stars is so large, that this assumption is justified.

(9) Let the attractive power of the second body be so large that light cannot escape from it; this can be expressed analytically in the following way: the velocity v' of the light is equal to zero. Putting this value of v' in the equation (8) for v', gives an equation from which the mass iM for which this occurs can be derived. One has therefore

$$0 = a^2 - \frac{2iM}{r'} \quad \text{or} \quad a^2 = \frac{2iM}{R'}.$$

(10) To determine a, let the first attracting body be the sun; then a is the velocity of the sun's light on the surface of the sun. The attractive power of the sun is however so small in comparison with the velocity of light, that one can take this velocity as uniform. From the phenomena of aberration it appears that the earth travels $20''\frac{1}{4}$ in its path while the light travels from the sun to the earth, accordingly: let V be the average velocity of the earth in its orbit, then one has $a : V = $ radius (expressed in seconds) $: 20''\frac{1}{4} = 1 : \text{tang. } 20''\frac{1}{4}$.

(11) My assumption made in *Expos. du Syst. du Monde*, Part II, p. 305, is $R' = 250R$. Now the mass changes as the volume of the attracting body multiplied by its density; the volume, as the cube of the radius; accordingly the mass as the cube of the radius multiplied by the density. Let the density of the sun $= 1$; that of the second body $= \rho$; then

$$M : iM = 1R^3 : \rho R'^3 = 1R^3 : \rho 250^3 R^3$$

or

$$1 : i = 1 : \rho(250)^3$$

or

$$i = (250)^3 \rho.$$

(12) One substitutes the values of i and R' in the equation $a^2 = 2iM/R'$, and thus obtains

$$a^2 = \frac{2(250)^3 \rho M}{250R} = 2(250)^2 \rho \frac{M}{R}$$

or

$$\rho = \frac{a^2 R}{2(250)^2 M}.$$

(13) To obtain ρ, one must still determine M. The force M of the sun is equal at a distance D to M/D^2. Let D be the average distance of the

earth, V the average velocity of the earth; then this force is also equal to V^2/D (see Lande's *Astronomy*, III, §3539). Hence $M/D^2 = V^2/D$ or $M = V^2D$. Substituting this in the equation (12) for ρ gives

$$\rho = \frac{a^2R}{2(250)^2V^2D} = \frac{8}{(1000)^2}\left(\frac{a}{V}\right)^2\left(\frac{R}{D}\right),$$

$$\frac{a}{V} = \frac{\text{vel. of light}}{\text{vel. of earth}} = \frac{1}{\text{tang. } 20''\tfrac{1}{4}} \quad \text{according to (10)},$$

$$\frac{R}{D} = \frac{\text{absolute radius of } \odot}{\text{average distance of } \odot} = \text{tan average apparent radius of } \odot.$$

Hence

$$\rho = 8\frac{\text{tang. } 16'\,2''}{(1000\,\text{tang. } 20''\tfrac{1}{4})^2}$$

from which ρ is approximately 4, or as large as the density of the earth.

D. W. Dewhirst adds:

The *Allgemeine geographische Ephemeriden* was a journal founded by F. X. von Zach, of which 51 volumes were published between 1798 and 1816. The footnote (‡) is a translation of that added by von Zach to the original paper which is however not very helpful to the modern reader.

There are no less than 10 different editions of Laplace's *Exposition du Système du Monde* published between 1796 and 1835, some in one quarto volume and some in two volumes octavo. In the earlier editions the 'statement without proof' comes a few pages before the end of Book 5, Chapter 6, though Laplace removed the specific statement from later editions.

The reference by von Zach to *A.G.E.* May 1798, p. 603, seems to be a mistake on von Zach's part; he was perhaps intending to refer to *A.G.E.* Vol. I, p. 89, 1798 where there is an extensive essay review of the first edition of Laplace's *Exposition du Système du Monde*.

Appendix B
Spherically symmetric solutions and Birkhoff's theorem

We wish to consider Einstein's equations in the case of a spherically symmetric space–time. One might regard the essential feature of a spherically symmetric space–time as the existence of a world-line $\mathscr{L}$ such that the space–time is spherically symmetric about $\mathscr{L}$. Then all points on each spacelike two-sphere $\mathscr{S}_d$ centred on any point p of $\mathscr{L}$, defined by going a constant distance d along all geodesics through p orthogonal to $\mathscr{L}$, are equivalent. If one permutes directions at p by use of the orthogonal group $SO(3)$ leaving $\mathscr{L}$ invariant, the space–time is, by definition, unchanged, and the corresponding points of $\mathscr{S}_d$ are mapped into themselves; so the space–time admits the group $SO(3)$ as a group of isometries, with the orbits of the group the spheres $\mathscr{S}_d$. (There could be particular values of d such that the surface $\mathscr{S}_d$ was just a point p'; then p' would be another centre of symmetry. There can be at most two points (p' and p itself) related in this way.)

However, there might not exist a world-line like $\mathscr{L}$ in some of the space–times one would wish to regard as spherically symmetric. In the Schwarzschild and Reissner–Nordström solutions, for example, space–time is singular at the points for which $r = 0$, which might otherwise have been centres of symmetry. We shall therefore take the existence of the group $SO(3)$ of isometries acting on two-surfaces like $\mathscr{S}_d$ as the characteristic feature of a spherically symmetric space–time. Thus we shall say that space–time is *spherically symmetric* if it admits the group $SO(3)$ as a group of isometries, with the group orbits spacelike two-surfaces. These orbits are then necessarily two-surfaces of constant positive curvature.

For each point q in any orbit $\mathscr{S}(q)$, there is a one-dimensional subgroup I_q of isometries which leaves q invariant (when there is a central axis $\mathscr{L}$, this is the group of rotations about p which leaves the geodesic pq invariant). The set $\mathscr{C}(q)$ of all geodesics orthogonal to $\mathscr{S}(q)$ at q locally form a two-surface left invariant by I_q (since I_q, which permutes directions in $\mathscr{S}(q)$ about q, leaves invariant directions perpendicular to $\mathscr{S}(q)$). At any other point r of $\mathscr{C}(q)$, I_q again permutes directions orthogonal to $\mathscr{C}(q)$, as it leaves $\mathscr{C}(q)$ invariant; since I_q must operate in the group orbit $\mathscr{S}(r)$ through r, this orbit is orthogonal to $\mathscr{C}(q)$. Thus (Schmidt (1967)) the group orbits $\mathscr{S}$ are orthogonal to the surfaces $\mathscr{C}$. Further these surfaces define locally a one–one map between the group orbits, where the image $f(q)$ of q in $\mathscr{S}(r)$ is the intersection of $\mathscr{C}(q)$ and $\mathscr{S}(r)$. Since this map is invariant under I_q, vectors of equal magnitude in $\mathscr{S}(q)$ at q are mapped into vectors of equal magnitude in $\mathscr{S}(r)$ at $f(q)$; and since all the points of $\mathscr{S}(q)$ are equivalent, the same magnitude multiplication factor occurs for the maps of vectors from any point in $\mathscr{S}(q)$ to its image in $\mathscr{S}(r)$. Thus (Schmidt (1967)) the orthogonal surfaces $\mathscr{C}$ map the trajectories $\mathscr{S}$ conformally onto each other.

If one chooses coordinates $\{t, r, \theta, \phi\}$ so that the group orbits $\mathscr{S}$ are the surfaces $\{t, r = \text{constant}\}$ and the orthogonal surfaces $\mathscr{C}$ are the surfaces $\{\theta, \phi = \text{constant}\}$, it now follows that the metric takes the form $\mathrm{d}s^2 = \mathrm{d}\tau^2(t, r) + Y^2(t, r)\,\mathrm{d}\Omega^2(\theta, \phi)$, where $\mathrm{d}\tau^2$ is an indefinite two-surface and $\mathrm{d}\Omega^2$ is a surface of positive constant curvature. If one further chooses the functions t, r so that the curves $\{t = \text{constant}\}$, $\{r = \text{constant}\}$ are orthogonal in the two-surfaces $\mathscr{C}$ (cf. Bergmann, Cahen and Komar (1965)), one can write the metric in the form

$$\mathrm{d}s^2 = \frac{-\mathrm{d}t^2}{F^2(t, r)} + X^2(t, r)\,\mathrm{d}r^2 + Y^2(t, r)\,(\mathrm{d}\theta^2 + \sin^2\theta\,\mathrm{d}\phi^2). \qquad \text{(A 1)}$$

(Note that this still leaves the freedom to choose arbitrarily either r or t in these surfaces.)

Let an observer moving along the t-lines measure an energy density μ, an isotropic pressure p, an energy flux q, and no anisotropic pressures. Then the field equations for the metric (A 1) may be written in the form

$$-8\pi q = \frac{2X}{F}\left(\frac{Y''}{Y} - \frac{X'Y'}{XY} + \frac{Y'F'}{YF}\right), \qquad \text{(A 2)}$$

$$8\pi\mu = \frac{1}{Y^2} + \frac{2}{X}\left(-\frac{Y'}{XY}\right)' - 3\left(\frac{Y'}{XY}\right)^2 + 2F^2\frac{X\dot{}\,Y\dot{}}{XY} + F^2\left(\frac{Y\dot{}}{Y}\right)^2, \qquad \text{(A 3)}$$

$$-8\pi p = \frac{1}{Y^2} + 2F\left(F\frac{Y\dot{}}{Y}\right)\dot{} + 3\left(\frac{Y\dot{}}{Y}\right)^2 F^2 + \frac{2}{X^2}\frac{Y'F'}{YF} - \left(\frac{Y'}{XY}\right)^2, \qquad \text{(A 4)}$$

$$4\pi(\mu + 3p) = \frac{1}{X}\left(-\frac{F'}{FX}\right)' - F\left(F\frac{X\dot{}}{X}\right)\dot{} - 2F\left(F\frac{Y\dot{}}{Y}\right)\dot{} - F^2\left(\frac{X\dot{}}{X}\right)^2$$
$$- 2F^2\left(\frac{Y\dot{}}{Y}\right)^2 + \frac{1}{X^2}\left(\frac{F'}{F}\right)^2 - \frac{2}{X^2}\frac{Y'F'}{YF}, \qquad \text{(A 5)}$$

where $'$ denotes $\partial/\partial r$ and $\dot{}$ denotes $\partial/\partial t$.

We first consider the *empty space* field equations $R_{ab} = 0$; this means that in (A 2)–(A 5) we must set $\mu = p = q = 0$. The local solution depends on the nature of the surfaces $\{Y = \text{constant}\}$; these surfaces may be timelike, spacelike or null, or they may not be defined (if Y is constant). In the exceptional case when $Y^{;a}Y_{;a} = 0$ on some open set $\mathscr{U}$ (this includes the case when Y is constant),

$$\frac{Y'}{X} = FY^{\cdot} \qquad (A\,6)$$

holds in $\mathscr{U}$. However when (A 6) holds, the value of $Y^{\cdot\cdot}$ determined by (A 2) is inconsistent with (A 3). Thus we may consider some point p where $Y^{;a}Y_{;a} < 0$ or $Y^{;a}Y_{;a} > 0$; the same inequality must hold in some open neighbourhood $\mathscr{U}$ of p.

Consider first the situation when $Y^{;a}Y_{;a} < 0$. Then the surfaces $\{Y = \text{constant}\}$ are timelike in $\mathscr{U}$, and one can choose Y to be the coordinate r. (Then r is an *area coordinate*, as the area of the two-surfaces $\{r, t = \text{constant}\}$ is $4\pi r^2$.) Thus $Y^{\cdot} = 0$, $Y' = 1$ and (A 2) shows that $X^{\cdot} = 0$. Further (A 4) shows that $(F'/F)^{\cdot} = 0$, so one can choose a new time coordinate $t'(t)$ in such a way as to set $F = F(r)$. Then one has $F = F(r)$, $X = X(r)$, $Y = r$; the solution is *necessarily static*. Equation (A 3) now shows $\mathrm{d}(r/X^2)/\mathrm{d}r = 1$, so solutions are of the form $X^2 = (1 - 2m/r)^{-1}$ where $2m$ is a constant of integration. Equation (A 4) can be integrated, with a suitable choice of a constant of integration, to give $F^2 = X^2$, and then (A 5) is identically satisfied. With these forms of F and X the metric (A 1) becomes

$$\mathrm{d}s^2 = -\left(1 - \frac{2m}{r}\right)\mathrm{d}t^2 + \frac{\mathrm{d}r^2}{\left(1 - \dfrac{2m}{r}\right)} + r^2(\mathrm{d}\theta^2 + \sin^2\theta\,\mathrm{d}\phi^2); \qquad (A\,7)$$

this is the Schwarzschild metric for $r > 2m$.

Now suppose $Y^{;a}Y_{;a} > 0$. Then the surfaces $\{Y = \text{constant}\}$ are spacelike in $\mathscr{U}$, and one can choose Y to be the coordinate t. Then $Y^{\cdot} = 1$, $Y' = 0$ and (A 2) shows $F' = 0$. One can choose the r-coordinate so that $X = X(t)$; then $F = F(t)$, $X = X(t)$, $Y = t$ and the solution is *spatially homogeneous*. Now (A 4) and (A 5) can be integrated to find the solution

$$\mathrm{d}s^2 = -\frac{\mathrm{d}t^2}{\left(\dfrac{2m}{t} - 1\right)} + \left(\frac{2m}{t} - 1\right)\mathrm{d}r^2 + t^2(\mathrm{d}\theta^2 + \sin^2\theta\,\mathrm{d}\phi^2). \qquad (A\,8)$$

This is part of the Schwarzschild solution inside the Schwarzschild radius, for the transformation $t \to r'$, $r \to t'$ transforms this metric into the form (A 7) with $r' < 2m$. Finally, if the surfaces $\{Y = \text{constant}\}$ are spacelike in some part of an open set $\mathscr{V}$ and timelike in another part, one can obtain solutions (A 8) and (A 7) in these parts, and then join them together across the surfaces where $Y;{}^{a}Y_{;a} = 0$ as in §5.5, obtaining a part of the maximal Schwarzschild solution which lies in $\mathscr{V}$. Thus we have proved *Birkhoff's theorem*: any C^2 solution of Einstein's empty space equations which is spherically symmetric in an open set $\mathscr{V}$, is locally equivalent to part of the maximally extended Schwarzschild solution in $\mathscr{V}$. (This is true even if the space is C^0, piecewise C^1; see Bergmann, Cahen and Komar (1965).)

We now consider spherically symmetric *static perfect fluid* solutions. Then one can find coordinates $\{t, r, \theta, \phi\}$ such that the metric has the form (A 1), the fluid moves along the t-lines (so $q = 0$), and $F = F(r)$, $X = X(r)$, $Y = Y(r)$. The field equations (A 3), (A 4) now show that if $Y' = 0$, then $\mu + p = 0$; we exclude this as being unreasonable for a physical fluid, so we assume $Y' \neq 0$. One may therefore again choose Y as the coordinate r; the metric then has the form

$$\mathrm{d}s^2 = -\frac{\mathrm{d}t^2}{F^2(r)} + X^2(r)\,\mathrm{d}r^2 + r^2(\mathrm{d}\theta^2 + \sin^2\theta\,\mathrm{d}\phi^2). \tag{A 9}$$

The contracted Bianchi identities $T^{ab}{}_{;b} = 0$ now shows

$$p' - (\mu + p)\,F'/F = 0; \tag{A 10}$$

(A 5) is identically satisfied if (A 3), (A 4) and (A 10) are satisfied. Equation (A 3) can be directly integrated to show

$$X^2 = \left(1 - \frac{2\hat{M}}{r}\right)^{-1}, \tag{A 11}$$

where

$$\hat{M}(r) \equiv 4\pi \int_0^r \mu r^2\,\mathrm{d}r,$$

and the boundary condition $X(0) = 1$ has been used (i.e. the fluid sphere has a regular centre). With (A 10), (A 11), equation (A 4) takes the form

$$\frac{\mathrm{d}p}{\mathrm{d}r} = -\frac{(\mu + p)\,(\hat{M} + 4\pi p r^3)}{r(r - 2\hat{M})} \tag{A 12}$$

which determines p as a function of r, if the equation of state is known. Finally (A 10) shows that

$$F(r) = C\exp\int_{p(0)}^{p(r)} \frac{\mathrm{d}p}{\mu + p}, \tag{A 13}$$

where C is a constant. Equations (A 11)–(A 13) determine the metric inside the fluid sphere, i.e. up to the value r_0 of r representing the surface of the fluid.

References

Ames,W.L., and Thorne,K.S. (1968), 'The optical appearance of a star that
   is collapsing through its gravitational radius', *Astrophys. J.* **151**, 659–70.
Arnett,W.D. (1966), 'Gravitational collapse and weak interactions',
   Can. J. Phys. **44**, 2553–94.
Auslander,L., and Markus,L. (1958), 'Flat Lorentz manifolds', Memoir 30,
   Amer. Math. Soc.
Avez,A. (1963), 'Essais de géométrie Riemannienne hyperbolique globale.
   Applications à la Relativité Générale', *Ann. Inst. Fourier (Grenoble)*,
   132, 105–90.
Belinskii,V.A., Khalatnikov,I.M., and Lifshitz,E.M. (1970), 'Oscillatory
   approach to a singular point in relativistic cosmology', *Adv. in Phys.*
   19, 523–73.
Bergmann,P.G., Cahen,M., and Komar,A.B. (1965), 'Spherically symmetric
   gravitational fields', *J. Math. Phys.* **6**, 1–5.
Bianchi,L. (1918), *Lezioni sulla teoria dei gruppi continui finiti transformazioni*
   (Spoerri, Pisa).
Bludman,S.A., and Ruderman,M.A. (1968), 'Possibility of the speed of
   sound exceeding the speed of light in ultradense matter', *Phys. Rev.* **170**,
   1176–84.
Bludman,S.A., and Ruderman,M.A. (1970), 'Noncausality and instability
   in ultradense matter', *Phys. Rev.* D **1**, 3243–6.
Bondi,H. (1960), *Cosmology* (Cambridge University Press, London).
Bondi,H. (1964), 'Massive spheres in General Relativity', *Proc. Roy. Soc.
   Lond.* A **282**, 303–17.
Bondi,H., and Gold,T. (1948), 'The steady-state theory of the expanding
   universe', *Mon. Not. Roy. Ast. Soc.* **108**, 252–70.
Bondi,H., Pirani,F.A.E., and Robinson,I. (1959), 'Gravitational waves in
   General Relativity, III. Exact plane waves', *Proc. Roy. Soc. Lond.* A **251**,
   519–33.
Boyer,R.H. (1969), 'Geodesic Killing orbits and bifurcate Killing horizons',
   Proc. Roy. Soc. Lond. A **311**, 245–52.
Boyer,R.H., and Lindquist,R.W. (1967), 'Maximal analytic extension of the
   Kerr metric', *J. Math. Phys.* **8**, 265–81.
Boyer,R.H., and Price,T.G. (1965), 'An interpretation of the Kerr metric
   in General Relativity', *Proc. Camb. Phil. Soc.* **61**, 531–4.
Bruhat,Y. (1962), 'The Cauchy problem', in *Gravitation: an introduction to
   current research*, ed. L. Witten (Wiley, New York), 130–68.
Burkill,J.C. (1956), *The Theory of Ordinary Differential Equations* (Oliver
   and Boyd, Edinburgh).
Calabi,E., and Marcus,L. (1962), 'Relativistic space forms', *Ann. Math.* **75**,
   63–76.
Cameron,A.G.W. (1970), 'Neutron stars', in *Ann. Rev. Astronomy and
   Astrophysics*, eds. L.Goldberg, D.Layzer, J.G.Phillips (Ann. Rev. Inc.,
   Palo Alto, California), 179–208.

Carter,B. (1966), 'The complete analytic extension of the Reissner–Nordström metric in the special case $e^2 = m^2$', *Phys. Lett.* **21**, 423–4.

Carter,B. (1967), 'Stationary axisymmetric systems in General Relativity', *Ph.D.Thesis*, Cambridge University.

Carter,B. (1968a), 'Global structure of the Kerr family of gravitational fields', *Phys. Rev.* **174**, 1559–71.

Carter,B. (1968b), 'Hamilton–Jacobi and Schrödinger separable solutions of Einstein's equations', *Comm. Math. Phys.* **10**, 280–310.

Carter,B. (1969), 'Killing horizons and orthogonally transitive groups in space–time', *J. Math. Phys.* **10**, 70–81.

Carter,B. (1970), 'The commutation property of a stationary axisymmetric system', *Comm. Math. Phys.* **17**, 233–8.

Carter,B. (1971a), 'Causal structure in space–time', *J. General Relativity and Gravitation*, **1**, 349–91.

Carter,B. (1971b), 'Axisymmetric black hole has only two degrees of freedom', *Phys. Rev. Lett.* **26**, 331–2.

Choquet-Bruhat,Y. (1968), 'Espace–temps Einsteiniens gèneraux, chocs gravitationnels', *Ann. Inst. Henri Poincaré*, **8**, 327–38.

Choquet-Bruhat,Y. (1971), 'Equations aux derivées partielles-solutions C^∞ d'equations hyperboliques non-lineaires', *C. R. Acad. Sci.* (Paris).

Choquet-Bruhat,Y., and Geroch,R.P. (1969), 'Global aspects of the Cauchy problem in General Relativity', *Comm. Math. Phys.* **14**, 329–35.

Christodoulou,D. (1970), 'Reversible and irreversible transformation in black hole physics', *Phys. Rev. Lett.* **25**, 1596–7.

Clarke,C.J.S. (1971), 'On the geodesic completeness of causal space–times', *Proc. Camb. Phil. Soc.* **69**, 319–24.

Colgate,S.A. (1968), 'Mass ejection from supernovae', *Astrophys. J.* **153**, 335–9.

Colgate,S.A. and White,R.H. (1966), 'The hydrodynamic behaviour of supernovae explosions', *Astrophys. J.* **143**, 626–81.

Courant,R., and Hilbert,D. (1962), *Methods of Mathematical Physics. Volume II : Partial Differential Equations* (Interscience, New York).

Demianski,M., and Newman,E. (1966), 'A combined Kerr–NUT solution of the Einstein field equations', *Bull. Acad. Pol. Sci. (Math. Ast. Phys.)* **14**, 653–7.

De Witt,B.S. (1967), 'Quantum theory of gravity: I. The canonical theory', *Phys. Rev.* **160**, 1113–48; 'II. The manifestly covariant theory', *Phys. Rev.* **162**, 1195–1239; 'III. Applications of the covariant theory', *Phys. Rev.* **162**, 1239–56.

Dicke,R.H. (1964), *The theoretical significance of Experimental Relativity* (Blackie, New York).

Dionne,P.A. (1962), 'Sur les problèmes de Cauchy hyperboliques bien posés', *Journ. d'Analyses Mathematique*, **10**, 1–90.

Dirac,P.A.M. (1938), 'A new basis for cosmology', *Proc. Roy. Soc. Lond.* A **165**, 199–208.

Dixon,W.G. (1970), 'Dynamics of extended bodies in General Relativity: I. Momentum and angular momentum', *Proc. Roy. Soc. Lond.* A **314**, 499–527; 'II. Moments of the charge–current vector', *Proc. Roy. Soc. Lond.* A **319**, 509–47.

Doroshkevich,A.G., Zel'dovich,Ya.B., and Novikov,I.D. (1966), 'Gravitational collapse of non-symmetric and rotating masses', *Sov. Phys. J.E.T.P.* **22**, 122–30.

Ehlers,J., Geren,P., and Sachs,R.K. (1968), 'Isotropic solutions of the Einstein–Liouville equations', *J. Math. Phys.* **8**, 1344–9.

Ehlers,J., and Kundt,W. (1962), 'Exact solutions of the gravitational field equations', in *Gravitation : an Introduction to Current Research*, ed. L.Witten (Wiley, New York), 49–101.

Ehresmann,C. (1957), 'Les connexions infinitesimales dans un espace fibre differentiable', in *Colloque de Topologie (Espaces Fibres) Bruxelles 1950* (Masson, Paris), 29–50.

Ellis,G.F.R., and Sciama, D.W. (1972), 'Global and non-global problems in cosmology', in *Studies in Relativity* (Synge Festschrift), ed. L.O'Raiffeartaigh (Oxford University Press, London).

Field,G.B. (1969), 'Cosmic background radiation and its interaction with cosmic matter', *Rivista del Nuovo Cimento*, **1**, 87–109.

Foley,K.J, Jones,R.S., Lindebaum,S.J., Love,W.A., Ozaki,S., Platner,E.D., Quarles,C.A., and Willen,E.H. (1967), 'Experimental test of the pion–nucleon forward dispersion relations at high energies', *Phys. Rev. Lett.* **19**, 193–8, and 622.

Geroch,R.P. (1966), 'Singularities in closed universes', *Phys. Rev. Lett.* **17**, 445–7.

Geroch,R.P. (1967a), 'Singularities in the space–time of General Relativity', *Ph.D. Thesis* (Department of Physics, Princeton University).

Geroch,R.P. (1967b), 'Topology in General Relativity', *J. Math. Phys.* **8**, 782–6.

Geroch,R.P. (1968a), 'Local characterization of singularities in General Relativity', *J. Math. Phys.* **9**, 450–65.

Geroch,R.P. (1968b), 'What is a singularity in General Relativity?', *Ann. Phys.* (New York), **48**, 526–40.

Geroch,R.P. (1968c), 'Spinor structure of space–times in General Relativity. I', *J. Math. Phys.* **9**, 1739–44.

Geroch,R.P. (1970a), 'Spinor structure of space–times in General Relativity. II', *J. Math. Phys.* **11**, 343–8.

Geroch,R.P. (1970b), 'The domain of dependence', *J. Math. Phys.* **11**, 437–9.

Geroch,R.P. (1970c), 'Singularities', in *Relativity*, ed. S.Fickler, M. Carmeli and L. Witten (Plenum Press, New York), 259–91.

Geroch,R.P. (1971), 'Space–time structure from a global view point', in *General Relativity and Cosmology*, Proceedings of International School in Physics 'Enrico Fermi', Course XLVII, ed. R. K. Sachs (Academic Press, New York), 71–103.

Geroch,R.P., Kronheimer,E.H., and Penrose,R. (1972), 'Ideal points in space–time', *Proc. Roy. Soc. Lond.* A **327**, 545–67.

Gibbons,G., and Penrose,R. (1972), to be published.

Gödel,K. (1949), 'An example of a new type of cosmological solution of Einstein's field equations of gravitation', *Rev. Mod. Phys.* **21**, 447–50.

Gold,T. (1967), ed., *The Nature of Time* (Cornell University Press, Ithaca).

Graves,J.C., and Brill,D.R. (1960), 'Oscillatory character of Reissner–Nordström metric for an ideal charged wormhole', *Phys. Rev.* **120**, 1507–13.

Grischuk,L.P. (1967), 'Some remarks on the singularities of the cosmological solutions of the gravitational equations', *Sov. Phys. J.E.T.P.* **24**, 320–4.

Hajicek,P. (1971), 'Causality in non-Hausdorff space–times', *Comm. Math. Phys.* **21**, 75–84.

Hajicek,P. (1973), 'General theory of vacuum ergospheres', *Phys. Rev.* D**7**, 2311–16.

Harrison,B.K., Thorne,K.S., Wakano,M., and Wheeler,J.A. (1965),
Gravitation Theory and Gravitational Collapse (Chicago University Press,
Chicago).

Hartle,J.B., and Hawking,S.W. (1972a), 'Solutions of the Einstein–Maxwell
equations with many black holes', *Commun. Math. Phys.* **26**, 87–101.

Hartle,J.B., and Hawking,S.W. (1972b), 'Energy and angular momentum
flow into a black hole', *Commun. Math. Phys.* **27**, 283–90.

Hawking,S.W. (1966a), 'Perturbations of an expanding universe',
Astrophys. J. **145**, 544–54.

Hawking,S.W. (1966b), 'Singularities and the geometry of space–time',
Adams Prize Essay (unpublished).

Hawking,S.W. (1967), 'The occurrence of singularities in cosmology. III.
Causality and singularities', *Proc. Roy. Soc. Lond.* A **300**, 187–201.

Hawking,S.W., and Ellis,G.F.R. (1965), 'Singularities in homogeneous
world models', *Phys. Lett.* **17**, 246–7.

Hawking,S.W., and Penrose,R. (1970), 'The singularities of gravitational
collapse and cosmology', *Proc. Roy. Soc. Lond.* A **314**, 529–48.

Heckmann,O., and Schücking, E. (1962), 'Relativistic cosmology', in
Gravitation : an Introduction to Current Research, ed. L.Witten (Wiley,
New York), 438–69.

Hocking,J.G., and Young,G.S. (1961), *Topology* (Addison-Wesley, London).

Hodge,W.V.D. (1952), *The Theory and Application of Harmonic Integrals*
(Cambridge University Press, London).

Hogarth,J.E. (1962), 'Cosmological considerations on the absorber theory
of radiation', *Proc. Roy. Soc. Lond.* A **267**, 365–83.

Hoyle,F. (1948), 'A new model for the expanding universe', *Mon. Not. Roy.
Ast. Soc.* **108**, 372–82.

Hoyle, F., and Narlikar,J.V. (1963), 'Time-symmetric electrodynamics and
the arrow of time in cosmology', *Proc. Roy. Soc. Lond.* A **277**, 1–23.

Hoyle,F., and Narlikar,J.V. (1964), 'A new theory of gravitation', *Proc. Roy.
Soc. Lond.* A **282**, 191–207.

Israel,W. (1966), 'Singular hypersurfaces and thin shells in General
Relativity', *Nuovo Cimento*, **44**B, 1–14; erratum, *Nuovo Cimento*, **49**B,
463 (1967).

Israel,W. (1967), 'Event horizons in static vacuum space–times', *Phys. Rev.*
164, 1776–9.

Israel,W. (1968), 'Event horizons in static electrovac space–times', *Comm.
Math. Phys.* **8**, 245–60.

Jordan,P. (1955), *Schwerkraft und Weltall* (Friedrich Vieweg, Braunschweig).

Kantowski,R., and Sachs,R.K. (1967), 'Some spatially homogeneous
anisotropic relativistic cosmological models', *J. Math. Phys.* **7**, 443–6.

Kelley,J.L. (1965), *General Topology* (van Nostrand, Princeton).

Khan,K.A., and Penrose,R. (1971), 'Scattering of two impulsive gravitational
plane waves', *Nature*, **229**, 185–6.

Kinnersley,W., and Walker,M. (1970), 'Uniformly accelerating charged mass
in General Relativity', *Phys. Rev.* D **2**, 1359–70.

Kobayashi,S., and Nomizu,K. (1963), *Foundations of Differential Geometry :
Volume I* (Interscience, New York).

Kobayashi, S., and Nomizu,K. (1969), *Foundations of Differential Geometry :
Volume II* (Interscience, New York).

Kreuzer,L.B. (1968), 'Experimental measurement of the equivalence of
active and passive gravitational mass', *Phys. Rev.* **169**, 1007–12.

Kronheimer,E.H., and Penrose,R. (1967), 'On the structure of causal spaces', *Proc. Camb. Phil. Soc.* **63**, 481–501.

Kruskal,M.D. (1960), 'Maximal extension of Schwarzschild metric', *Phys. Rev.* **119**, 1743–5.

Kundt,W. (1956), 'Trägheitsbahnen in einem von Gödel angegebenen kosmologischen Modell', *Zs. f. Phys.* **145**, 611–20.

Kundt,W. (1963), 'Note on the completeness of space-times', *Zs. f. Phys.* **172**, 488–9.

Le Blanc,J.M., and Wilson,J.R. (1970), 'A numerical example of the collapse of a rotating magnetized star', *Astrophys. J.* **161**, 541–52.

Leray,J. (1952), 'Hyperbolic differential equations', duplicated notes (Princeton Institute for Advanced Studies).

Lichnerowicz,A. (1955), *Theories Relativistes de la Gravitation et de l'Electromagnétisme* (Masson, Paris).

Lifschitz,E.M., and Khalatnikov,I.M. (1963), 'Investigations in relativistic cosmology', *Adv. in Phys.* (*Phil. Mag. Suppl.*) **12**, 185–249.

Löbell,F. (1931), 'Beispele geschlossener drei-dimensionaler Clifford–Kleinsche Räume negativer Krümmung', *Ber. Verhandl. Sächs. Akad. Wiss. Leipzig, Math. Phys. Kl.* **83**, 167–74.

Milnor,J. (1963), *Morse Theory*, Annals of Mathematics Studies No. 51 (Princeton University Press, Princeton).

Misner,C.W. (1963), 'The flatter regions of Newman, Unti and Tamburino's generalized Schwarzschild space', *J. Math. Phys.* **4**, 924–37.

Misner,C.W. (1967), 'Taub–NUT space as a counterexample to almost anything', in *Relativity Theory and Astrophysics I : Relativity and Cosmology*, ed. J.Ehlers, Lectures in Applied Mathematics, Volume 8 (American Mathematical Society), 160–9.

Misner,C.W. (1968), 'The isotropy of the universe', *Astrophys. J.* **151**, 431–57.

Misner,C.W. (1969), 'Quantum cosmology. I', *Phys. Rev.* **186**, 1319–27.

Misner,C.W. (1972), 'Minisuperspace', in *Magic without Magic*, ed. J. R. Klauder (Freeman, San Francisco).

Misner,C.W., and Taub,A.H. (1969), 'A singularity-free empty universe', *Sov. Phys. J.E.T.P.* **28**, 122–33.

Müller zum Hagen,H. (1970), 'On the analyticity of stationary vacuum solutions of Einstein's equations', *Proc. Camb. Phil. Soc.* **68**, 199–201.

Müller zum Hagen,H., Robinson,D.C., and Seifert,H.J. (1973), 'Black holes in static vacuum space-times', *Gen. Rel. and Grav.* **4**, 53.

Munkres,J.R. (1954), *Elementary Differential Topology*, Annals of Mathematics Studies No. 54 (Princeton University Press, Princeton).

Newman,E.T., and Penrose,R. (1962), 'An approach to gravitational radiation by a method of spin coefficients', *J. Math. Phys.* **3**, 566–78.

Newman,E.T., and Penrose,R. (1968), 'New conservation laws for zero-rest mass fields in asymptotically flat space-time', *Proc. Roy. Soc. Lond.* A **305**, 175–204.

Newman,E.T., Tamburino,L., and Unti,T.J. (1963), 'Empty space generalization of the Schwarzschild metric', *Journ. Math. Phys.* **4**, 915–23.

Newman,E.T., and Unti,T.W.J. (1962), 'Behaviour of asymptotically flat empty spaces', *J. Math. Phys.* **3**, 891–901.

North,J.D. (1965), *The Measure of the Universe* (Oxford University Press, London).

Ozsváth,I., and Schücking,E. (1962), 'An anti-Mach metric', in *Recent Developments in General Relativity* (Pergamon Press – PWN), 339–50.

Papapetrou,A. (1966), 'Champs gravitationnels stationnares à symmétrie axiale', *Ann. Inst. Henri Poincaré*, A iv, 83–105.

Papapetrou,A., and Hamoui,A. (1967), 'Surfaces caustiques dégénérées dans la solution de Tolman. La Singularité physique en Relativité Générale', *Ann. Inst. Henri Poincaré*, vi, 343–64.

Peebles,P.J.E. (1966), 'Primordial helium abundance and the primordial fireball. ii', *Astrophys. J.* 146, 542–52.

Penrose,R. (1963), 'Asymptotic properties of fields and space–times', *Phys. Rev. Lett.* 10, 66–8.

Penrose,R. (1964), 'Conformal treatment of infinity', in *Relativity, Groups and Topology*, ed. C.M.de Witt and B.de Witt, Les Houches Summer School, 1963 (Gordon and Breach, New York).

Penrose,R. (1965a), 'A remarkable property of plane waves in General Relativity', *Rev. Mod. Phys.* 37, 215–20.

Penrose,R. (1965b), 'Zero rest-mass fields including gravitation: asymptotic behaviour', *Proc. Roy. Soc. Lond.* A 284, 159–203.

Penrose,R. (1965c), 'Gravitational collapse and space–time singularities', *Phys. Rev. Lett.* 14, 57–9.

Penrose,R. (1966), 'General Relativity energy flux and elementary optics', in *Perspectives in Geometry and Relativity* (Hlavaty Festschrift), ed. B.Hoffmann (Indiana University Press, Bloomington), 259–74.

Penrose,R. (1968), 'Structure of space–time', in *Battelle Rencontres*, ed. C.M. de Witt and J.A. Wheeler (Benjamin, New York), 121–235.

Penrose,R. (1969), 'Gravitational collapse: the role of General Relativity', *Rivista del Nuovo Cimento*, 1, 252–76.

Penrose,R. (1972a), 'The geometry of impulsive gravitational waves', in *Studies in Relativity* (Synge Festschrift), ed. L.O'Raiffeartaigh (Oxford University Press, London).

Penrose,R. (1972b), 'Techniques of differential topology in relativity' (Lectures at Pittsburgh, 1970), A.M.S. Colloquium Publications.

Penrose,R., and MacCallum,M.A.H. (1972), 'A twistor approach to space–time quantization', *Physics Reports (Phys. Lett. Section C)*, 6, 241–316.

Penrose,R., and Floyd,R.M. (1971), 'Extraction of rotational energy from a black hole', *Nature*, 229, 177–9.

Penzias,A.A., Schraml,J., and Wilson,R.W. (1969), 'Observational constraints on a discrete source model to explain the microwave background', *Astrophys. J.* 157, L49–L51.

Pirani,F.A.E. (1955), 'On the energy–momentum tensor and the creation of matter in relativistic cosmology', *Proc. Roy. Soc.* A 228, 455–62.

Press,W.H. (1972), 'Time evolution of a rotating black hole immersed in a static scalar field', *Astrophys. Journ.* 175, 245–52.

Price,R.H. (1972), 'Nonspherical perturbations of relativistic gravitational collapse. i: Scalar and gravitational perturbations. ii: Integer spin, zero rest-mass fields', *Phys. Rev.* 5, 2419–54.

Rees,M.J., and Sciama,D.W. (1968), 'Large-scale density inhomogeneities in the universe', *Nature*, 217, 511–16.

Regge,T., and Wheeler,J.A. (1957), 'Stability of a Schwarzschild singularity', *Phys. Rev.* 108, 1063–9.

Riesz,F., and Sz-Nagy,B. (1955), *Functional Analysis* (Blackie and Sons, London).

Robertson,H.P. (1933), 'Relativistic cosmology', *Rev. Mod. Phys.* 5, 62–90.

Rosenfeld,L. (1940), 'Sur le tenseur d'impulsion–energie', *Mem. Roy. Acad. Belg. Cl. Sci.* **18**, No. 6.

Ruse,H.S. (1937), 'On the geometry of Dirac's equations and their expression in tensor form', *Proc. Roy. Soc. Edin.* **57**, 97–127.

Sachs,R.K., and Wolfe,A.M. (1967), 'Perturbations of a cosmological model and angular variations of the microwave background', *Astrophys. J.* **147**, 73–90.

Sandage,A. (1961), 'The ability of the 200-inch telescope to discriminate between selected world models', *Astrophys. J.* **133**, 355–92.

Sandage,A. (1968), 'Observational cosmology', *Observatory*, **88**, 91–106.

Schmidt,B.G. (1967), 'Isometry groups with surface-orthogonal trajectories', *Zs. f. Naturfor.* **22**a, 1351–5.

Schmidt,B.G. (1971), 'A new definition of singular points in General Relativity', *J. Gen. Rel. and Gravitation*, **1**, 269–80.

Schmidt,B.G. (1972), 'Local completeness of the *b*-boundary', *Commun. Math. Phys.* **29**, 49–54.

Schmidt,H. (1966), 'Model of an oscillating cosmos which rejuvenates during contraction', *J. Math. Phys.* **7**, 494–509.

Schouten,J.A. (1954), *Ricci Calculus* (Springer, Berlin).

Schrödinger,E. (1956), *Expanding Universes* (Cambridge University Press, London).

Sciama,D.W. (1953), 'On the origin of inertia', *Mon. Not. Roy. Ast. Soc.* **113**, 34–42.

Sciama,D.W. (1967), 'Peculiar velocity of the sun and the cosmic microwave background', *Phys. Rev. Lett.* **18**, 1065–7.

Sciama,D.W. (1971), 'Astrophysical cosmology', in *General Relativity and Cosmology*, ed. R.K. Sachs, Proceedings of the International School of Physics 'Enrico Fermi', Course XLVII (Academic Press, New York), 183–236.

Seifert,H.J. (1967), 'Global connectivity by timelike geodesics', *Zs. f. Naturfor.* **22**a, 1356–60.

Seifert,H.J. (1968), 'Kausal Lorentzräume', *Doctoral Thesis*, Hamburg University.

Smart,J.J.C. (1964), *Problems of Space and Time*, Problems of Philosophy Series, ed. P. Edwards (Collier–Macmillan, London; Macmillan, New York).

Sobolev,S.L. (1963), *Applications of Functional Analysis to Physics*, Vol. 7, Translations of Mathematical Monographs (Am. Math. Soc., Providence).

Spanier,E.H. (1966), *Algebraic Topology* (McGraw Hill, New York).

Spivak,M. (1965), *Calculus on Manifolds* (Benjamin, New York).

Steenrod,N.E. (1951), *The Topology of Fibre Bundles* (Princeton University Press, Princeton).

Stewart,J.M.S., and Sciama,D.W. (1967), 'Peculiar velocity of the sun and its relation to the cosmic microwave background', *Nature*, **216**, 748–53.

Streater,R.F., and Wightman,A.S. (1964), *P.C.T., Spin, Statistics, and All That* (Benjamin, New York).

Thom,R. (1969), *Stabilité Structurelle et Morphogenése* (Benjamin, New York).

Thorne,K.S. (1966), 'The General Relativistic theory of stellar structure and dynamics', in *High Energy Astrophysics*, ed. L. Gratton, Proceedings of the International School in Physics 'Enrico Fermi', Course XXXV (Academic Press, New York), 166–280.

Tsuruta,S. (1971), 'The effects of nuclear forces on the maximum mass of neutron stars', in *The Crab Nebula*, ed. R. D. Davies and F. G. Smith (Reidel, Dordrecht).

Vishveshwara,C.V. (1968), 'Generalization of the "Schwarzschild Surface" to arbitrary static and stationary Metrics', *J. Math. Phys.* **9**, 1319–22.

Vishveshwara,C.V. (1970), 'Stability of the Schwarzschild metric', *Phys. Rev.* D **1**, 2870–9.

Wagoner,R.V., Fowler,W.A., and Hoyle,F. (1968), 'On the synthesis of elements at very high temperatures', *Astrophys. J.* **148**, 3–49.

Walker,A.G. (1944), 'Completely symmetric spaces', *J. Lond. Math. Soc.* **19**, 219–26.

Weymann,R.A. (1963), 'Mass loss from stars', in *Ann. Rev. Ast. and Astrophys.* Vol. 1 (Ann. Rev. Inc., Palo Alto), 97–141.

Wheeler,J.A. (1968), 'Superspace and the nature of quantum geometro-dynamics', in *Batelle Rencontres*, ed. C. M. de Witt and J. A. Wheeler (Benjamin, New York), 242–307.

Whitney,H. (1936), 'Differentiable manifolds', *Annals of Maths.* **37**, 645.

Yano,K. and Bochner,S. (1953), 'Curvature and Betti numbers', *Annals of Maths. Studies* No. 32 (Princeton University Press, Princeton).

Zel'dovich,Ya.B., and Novikov,I.D. (1971), *Relativistic Astrophysics. Volume I : Stars and Relativity*, ed. K. S.Thorne and W. D. Arnett (University of Chicago Press, Chicago).

Notation

Numbers refer to pages where definitions are given

≡	definition	⇒	implies	
∃	there exists	Σ	summation sign	
	□	end of a proof		

Sets

∪ $A \cup B$, union of A and B

∩ $A \cap B$, intersection of A and B

⊃ $A \subset B$, $B \supset A$, A is contained in B

− $A - B$, B subtracted from A

∈ $x \in A$, is a member of A

∅   the empty set

Maps

$\phi : \mathcal{U} \to \mathcal{V}$, ϕ maps $p \in \mathcal{U}$ to $\phi(p) \in \mathcal{V}$

$\phi(\mathcal{U})$ image of $\mathcal{U}$ under ϕ

ϕ^{-1} inverse map to ϕ

$f \circ g$ composition, g followed by f

ϕ_*, ϕ^* mappings of tensors induced by map ϕ, 22–4

Topology

$\bar{A}$ closure of A

$A^{\cdot}$ boundary of A, 183

$\operatorname{int} A$ interior of A, 209

Differentiability

$C^0, C^r, C^{r-}, C^\infty$ differentiability conditions, 11

Manifolds

$\mathcal{M}$ n-dimensional manifold, 11

$(\mathcal{U}_\alpha, \phi_\alpha)$ local chart determining local coordinates x^a, 12

$\partial \mathscr{M}$　boundary of $\mathscr{M}$, 12
R^n　Euclidean n-dimensional space, 11
$\frac{1}{2}R^n$　lower half $x^1 \leqslant 0$ of R^n, 11
S^n　n-sphere, 13
$\times$　Cartesian product, 15

Tensors

$(\partial/\partial t)_\lambda, \mathbf{X}$　vectors, 15
$\boldsymbol{\omega}, \mathrm{d}f$　one-forms, 16, 17
$\langle \boldsymbol{\omega}, \mathbf{X} \rangle$　scalar product of vector and one-form, 16
$\{\mathbf{E}_a\}, \{\mathbf{E}^a\}$　dual bases of vectors and one-forms, 16, 17
$T^{a_1 \ldots a_r}{}_{b_1 \ldots b_s}$,　components of tensor $\mathbf{T}$ of type (r, s), 17–19
$\otimes$　tensor product, 18
$\wedge$　skew product, 21
$()$　symmetrization (e.g. $T_{(ab)}$), 20
$[\,]$　skew symmetrization (e.g. $T_{[ab]}$), 20
$\delta^a{}_b$　Kronecker delta ($+1$ if $a = b$, 0 if $a \neq b$)
$T_p, T^*{}_p$　tangent space at p and dual space at p, 16
$T^r_s(p)$　space of tensors of type (r, s) at p, 18
$T^r_s(\mathscr{M})$　bundle of tensors of type (r, s) on $\mathscr{M}$, 51
$T(\mathscr{M})$　tangent bundle to $\mathscr{M}$, 51
$L(\mathscr{M})$　bundle of linear frames on $\mathscr{M}$, 51

Derivatives and connection

$\partial/\partial x^i$　partial derivatives with respect to coordinate x^i
$(\partial/\partial t)_\lambda$　derivative along curve $\lambda(t)$, 15
d　exterior derivative, 17, 25
$L_{\mathbf{X}}\mathbf{Y}$, $[\mathbf{X}, \mathbf{Y}]$　Lie derivative of $\mathbf{Y}$ with respect to $\mathbf{X}$, 27–8
∇, $\nabla_{\mathbf{X}}$, $T_{ab;c}$　covariant derivative, 30–2
$\mathrm{D}/\partial t$　covariant derivative along curve, 32
$\Gamma^i{}_{jk}$　connection components, 31
exp　exponential map, 33

Riemannian spaces

$(\mathscr{M}, \mathbf{g})$　manifold $\mathscr{M}$ with metric $\mathbf{g}$ and Christoffel connection
$\boldsymbol{\eta}$　volume element, 48
R_{abcd}　Riemann tensor, 35
R_{ab}　Ricci tensor, 36

R curvature scalar, 41
C_{abcd} Weyl tensor, 41
$O(p, q)$ orthogonal group leaving metric G_{ab} invariant, 52
G_{ab} diagonal metric diag $(\underbrace{+1, +1, ..., +1,}_{p \text{ terms}} \underbrace{-1, ..., -1}_{q \text{ terms}})$
$O(\mathcal{M})$ bundle of orthonormal frames, 52

Space–time

Space–time is a 4-dimensional Riemannian space $(\mathcal{M}, \mathbf{g})$ with metric normal form diag $(+1, +1, +1, -1)$. Local coordinates are chosen to be (x^1, x^2, x^3, x^4).

T_{ab} energy momentum tensor of matter, 61
$\Psi_{(i)}{}^{a...b}{}_{c...d}$ matter fields, 60
L Lagrangian, 64
Einstein's field equations take the form
$$R_{ab} - \tfrac{1}{2}Rg_{ab} + \Lambda g_{ab} = 8\pi T_{ab},$$
where Λ is the cosmological constant.
$(\mathcal{S}, \boldsymbol{\omega})$ is an initial data set, 233

Timelike curves

$\perp$ perpendicular projection, 79
$\mathrm{D}_F/\partial s$ Fermi derivative, 80–1
θ expansion, 83
$\omega^a, \omega_{ab}, \omega$ vorticity, 82–4
σ_{ab}, σ shear, 83–4

Null geodesics

$\hat{\theta}$ expansion, 88
$\hat{\omega}_{ab}, \hat{\omega}$ vorticity, 88
$\hat{\sigma}_{ab}, \hat{\sigma}$ shear, 88

Causal structure

I^+, I^- chronological future, past, 182
J^+, J^- causal future, past, 183
E^+, E^- future, past horismos, 184
D^+, D^- future, past Cauchy developments, 201
H^+, H^- future, past Cauchy horizons, 202

Boundary of space–time

$\mathcal{M}^* = \mathcal{M} \cup \Delta$ where Δ is the c-boundary, 220

$\mathcal{I}^+, \mathcal{I}^-, i^+, i^-$ c-boundary of asymptotically simple and empty spaces, 122, 225

$\overline{\mathcal{M}} = \mathcal{M} \cup \partial \mathcal{M}$ when $\mathcal{M}$ is weakly asymptotically simple; the boundary $\partial \mathcal{M}$ of $\mathcal{M}$ consists of $\mathcal{I}^+$ and $\mathcal{I}^-$, 221, 225

$\mathcal{M}^+ = \mathcal{M} \cup \partial$ where ∂ is the b-boundary, 283

Index